MONARCHS IN A CHANGING WORLD

MONARCHS IN A CHANGING WORLD

BIOLOGY AND CONSERVATION OF AN ICONIC BUTTERFLY

Edited by

Karen S. Oberhauser
Kelly R. Nail
Sonia Altizer

COMSTOCK PUBLISHING ASSOCIATES, *a division of*

CORNELL UNIVERSITY PRESS

Ithaca and London

Copyright © 2015 by Cornell University except for portions of Chapters 2 and 13, which were written by federal employees and cannot be copyrighted.

All rights reserved. Except for brief quotations in a review, this book, or parts thereof, must not be reproduced in any form without permission in writing from the publisher. For information, address Cornell University Press, Sage House, 512 East State Street, Ithaca, New York 14850.

First published 2015 by Cornell University Press

Printed in the United States of America

Library of Congress Cataloging-in-Publication Data

Monarchs in a changing world : biology and conservation of an iconic
  butterfly / edited by Karen S. Oberhauser, Kelly R. Nail, Sonia Altizer.
    pages cm
  Includes bibliographical references and index.
  ISBN 978-0-8014-5315-1 (cloth : alk. paper)
  1. Monarch butterfly. 2. Monarch butterfly—Conservation—North America. I. Oberhauser, Karen Suzanne, editor. II. Nail, Kelly R., editor. III. Altizer, Sonia M., editor. IV. Oberhauser, Karen Suzanne. Monarchs and people. Container of (work):
  QL561.D3M6635   2015
  595.78'9—dc23   2014034844

Cornell University Press strives to use environmentally responsible suppliers and materials to the fullest extent possible in the publishing of its books. Such materials include vegetable-based, low-VOC inks and acid-free papers that are recycled, totally chlorine-free, or partly composed of nonwood fibers. For further information, visit our website at www.cornellpress.cornell.edu.

Cloth printing   10 9 8 7 6 5 4 3 2 1

Contents

Preface   ix

PART I   *Model Programs for Citizen Science, Education, and Conservation: An Overview*
Karen S. Oberhauser   1

1. Environmental Education and Monarchs: Reaching across Disciplines, Generations, and Nations
   Elisabeth Young-Isebrand, Karen S. Oberhauser, Kim Bailey, Sonya Charest, Brian Hayes, Elizabeth Howard, Jim Lovett, Susan Meyers, Erik Mollenhauer, Eneida B. Montesiños-Patino, Ann Ryan, Orley R. Taylor, and Rocío Treviño   5

2. Contributions to Monarch Biology and Conservation through Citizen Science: Seventy Years and Counting
   Karen S. Oberhauser, Leslie Ries, Sonia Altizer, Rebecca V. Batalden, Janet Kudell-Ekstrum, Mark Garland, Elizabeth Howard, Sarina Jepsen, Jim Lovett, Mía Monroe, Gail Morris, Eduardo Rendón-Salinas, Richard G. RuBino, Ann Ryan, Orley R. Taylor, Rocío Treviño, Francis X. Villablanca, and Dick Walton   13

3. Monarch Habitat Conservation across North America: Past Progress and Future Needs
   Priya C. Shahani, Guadalupe del Río Pesado, Phil Schappert, and Eligio García Serrano   31

PART II   *Monarchs as Herbivores, Prey, and Hosts: An Overview*
Jacobus C. de Roode   43

4. Macroevolutionary Trends in the Defense of Milkweeds against Monarchs: Latex, Cardenolides, and Tolerance of Herbivory
   Anurag A. Agrawal, Jared G. Ali, Sergio Rasmann, and Mark Fishbein   47

5. Invertebrate Natural Enemies and Stage-Specific Mortality Rates of Monarch Eggs and Larvae
   Alma De Anda and Karen S. Oberhauser   60

6. Lacewings, Wasps, and Flies—Oh My: Insect Enemies Take a Bite out of Monarchs
   Karen S. Oberhauser, Michael Anderson, Sophia Anderson, Wendy Caldwell, Alma De Anda, Mark Hunter, Matthew C. Kaiser, and Michelle J. Solensky   71

7. Monarchs and Their Debilitating Parasites: Immunity, Migration, and Medicinal Plant Use
   Sonia Altizer and Jacobus C. de Roode   83

PART III   *Monarchs in a Changing Climate: An Overview*
Kelly R. Nail and Karen S. Oberhauser   95

8. What's Too Hot and What's Too Cold? Lethal and Sublethal Effects of Extreme Temperatures on Developing Monarchs
   Kelly R. Nail, Rebecca V. Batalden, and Karen S. Oberhauser   99

9. Microclimatic Protection of Overwintering Monarchs Provided by Mexico's High-Elevation Oyamel Fir Forests: A Review
   Ernest H. Williams and Lincoln P. Brower   109

10. Effect of the 2010–2011 Drought on the Lipid Content of Monarchs Migrating through Texas to Overwintering Sites in Mexico
    Lincoln P. Brower, Linda S. Fink, Ridlon J. Kiphart, Victoria Pocius, Raúl R. Zubieta, and M. Isabel Ramírez   117

11. Estimating the Climate Signal in Monarch Population Decline: No Direct Evidence for an Impact of Climate Change?
    Myron P. Zalucki, Lincoln P. Brower, Stephen B. Malcolm, and Benjamin H. Slager   130

PART IV *Conserving North American Monarch Butterflies: An Overview*
    Lincoln P. Brower and Linda S. Fink   143

12. Understanding and Conserving the Western North American Monarch Population
    Sarina Jepsen and Scott Hoffman Black   147

13. Threats to the Availability of Overwintering Habitat in the Monarch Butterfly Biosphere Reserve: Land Use and Climate Change
    M. Isabel Ramírez, Cuauhtémoc Sáenz-Romero, Gerald Rehfeldt, and Lidia Salas-Canela   157

14. Monarch Butterflies and Agriculture
    John M. Pleasants   169

15. Fires and Fire Management in the Monarch Butterfly Biosphere Reserve
    Héctor Martínez-Torres, Mariana Cantú-Fernández, M. Isabel Ramírez, and Diego R. Pérez-Salicrup   179

16. Project Milkweed: A Strategy for Monarch Habitat Conservation
    Brianna Borders and Eric Lee-Mäder   190

17. Grassland and Roadside Management Practices Affect Milkweed Abundance and Opportunities for Monarch Recruitment
    Kristen A. Baum and Elisha K. Mueller   197

PART V *New Perspectives on Monarch Migration, Evolution, and Population Biology: An Overview*
    Andrew K. Davis and Sonia Altizer   203

18. Tracking the Fall Migration of Eastern Monarchs with Journey North Roost Sightings: New Findings about the Pace of Fall Migration
    Elizabeth Howard and Andrew K. Davis   207

19. Potential Changes in Eastern North American Monarch Migration in Response to an Introduced Milkweed, *Asclepias curassavica*
    Rebecca V. Batalden and Karen S. Oberhauser   215

20. Migration and Host Plant Use by the Southern Monarch, *Danaus erippus*
    Stephen B. Malcolm and Benjamin H. Slager   225

21. Monarchs in the Mist: New Perspectives on Monarch Distribution in the Pacific Northwest
    Robert Michael Pyle   236

22. Monarchs across the Atlantic Ocean: What's Happening on the Other Shore?
    Juan Fernández-Haeger, Diego Jordano, and Myron P. Zalucki   247

23. Unraveling the Mysteries of Monarch Migration and Global Dispersal through Molecular Genetic Techniques
    Amanda A. Pierce, Sonia Altizer, Nicola L. Chamberlain, Marcus R. Kronforst, and Jacobus C. de Roode   257

24. Connecting Eastern Monarch Population Dynamics across Their Migratory Cycle
    Leslie Ries, Douglas J. Taron, Eduardo Rendón-Salinas, and Karen S. Oberhauser   268

   References   *283*
   Contributors   *313*
   Index   *317*

   Color plates follow page 132.

Preface

Whether you are a monarch citizen scientist, an entomologist, a population biologist, a conservation policy maker, a teacher, or just interested in monarchs' amazing biology and their impressive annual migratory cycle, you are reading this book because monarchs fascinate you. With contributions from dozens of individuals across the globe, *Monarchs in a Changing World: Biology and Conservation of an Iconic Butterfly* highlights the unique and remarkable natural history of monarchs and their complex and multifaceted interactions with people.

Five international conferences have been held on monarch biology and conservation, four of which triggered the creation of edited volumes: the Symposium on the Biology and Conservation of the Monarch Butterfly (Morelos, Mexico, 1981), the Second International Conference on the Monarch Butterfly (Los Angeles, California, 1986; Malcolm and Zalucki 1993), the North American Conference on the Monarch Butterfly (Morelia, Michoacán, 1997; Hoth et al. 1999), the Monarch Population Dynamics Conference (Lawrence, Kansas, 2001; Oberhauser and Solensky 2004), and the Monarch Biology and Conservation Meeting (Minneapolis, Minnesota, 2012; this volume).

With three existing compendia of monarch biology and conservation, why do we need another one? There has been a veritable explosion of knowledge about monarchs in the last decade. Monarchs continue to provide a window into some of the most fascinating questions facing biologists and the public, and we've compiled recent findings that utilize cutting-edge genetic tools and analytical techniques, as well as tried-and-true methods in laboratory and field biology. Since publication of the last volume in 2004, the North American Monarch Conservation Plan was published (CEC 2008) and the Monarch Joint Venture was formed (Monarch Joint Venture 2013). These efforts have brought even more attention to monarch conservation, with a concomitant increase in local, regional, national, and international conservation efforts. The past decade has also brought an explosion of interest by citizens in collecting scientific data; these citizen scientists invest tens of thousands of volunteer hours in monitoring monarchs every year. At the same time, monarch numbers are decreasing in response to environmental changes brought on by habitat loss and other stressors; like many other organisms, monarchs require our attention and care to survive these changes. Now is a crucial time for monarchs, and this book is not only a celebration of their amazing biology and our love of this charismatic insect, but also a call to arms.

The cutting-edge scientific developments described in the following chapters build on a long tradition of research. Monarchs have helped answer fundamental biological questions about how organisms migrate, find and digest food, and cope with a world in which natural enemies attack them as eggs, larvae, pupae, and adults. They've also been the focus of studies that have helped elucidate how genes are translated into molecules that allow organisms to function appropriately as they age and face a diversity of environmental conditions. New aspects of monarch biology have come to light in the past few years, including understanding of large-scale trends in monarch population sizes and their response to environmental and human perturbations, how monarchs might respond to future global climate change, and patterns of genetic variation and evolutionary divergence among wild monarch populations. New findings on monarch biology are accumulating at a fast rate, and this book includes both summaries of recent published work and findings that are included here for the first time.

Despite all the work represented by this book and the papers and books that preceded it, there is still a great deal we don't understand about monarchs. Most chapters include a preview of the next steps in our understanding of monarch biology and conservation challenges, offering a glimpse into where monarchs might lead us next.

How should you approach *Monarchs in a Changing World*? First, note that it does not include a basic introduction to monarch biology. If your goal is to learn about the monarch life cycle or the basics of their migration, we invite you to read the overviews in the previous book (Oberhauser and Solensky 2004), or visit one of the many excellent websites on monarchs. Next, we don't recommend reading it like a novel. Rather, skim the brief summaries that will orient you to the content in each chapter, find the chapters that interest you most, let that information sink in, and then move on to new subjects. The structure of chapters varies deliberately; most chapters that present new findings are set up like scientific papers (in a traditional introduction, methods, results, conclusions format), while those that synthesize knowledge or summarize many projects are each structured uniquely in order to best present their specific information.

Monarchs in a Changing World is divided into five sections, each with an overview that introduces the chapters and puts them in the context of what we already know about the topic. Part 1, "Monarchs and People: Model Programs for Citizen Science, Education, and Conservation," focuses on interactions between people and monarchs, illustrating the long fascination people have had with monarchs, and introducing themes that continue through the rest of the book. Part 2, "Monarchs as Herbivores, Prey, and Hosts," focuses on how monarchs interact with a complex food web that includes their milkweed host plants and an amazing array of natural enemies that range from microscopic organisms to much larger insects and even birds. Part 3, "Monarchs in a Changing Climate," highlights how monarchs respond to temperature and precipitation extremes throughout their life cycle. These chapters address climate from a variety of perspectives, using analyses of citizen science data, experiments in the lab and field, and simple mathematical models. Part 4, "Conserving North American Monarch Butterflies," builds on the overview of conservation efforts in the first section by highlighting examples of on-the-ground conservation programs, and threats imposed by specific anthropogenic changes. Part 5, "New Perspectives on Monarch Migration, Evolution, and Population Biology," addresses new findings that are pushing scientific boundaries in the areas of genetics, migration, herbivore use of non-native plants, and population dynamics.

Chapter authors include monarch biologists and conservation practitioners who work and live in the United States, Canada, Mexico, Spain, Portugal, Bolivia, Argentina, Morocco, New Zealand, and Australia. These authors work for universities, government agencies, nonprofit organizations, and citizen science programs and include the top experts and practitioners in the fields of monarch biology, outreach, and conservation.

While individuals and organizations are acknowledged at the end of each chapter, the editors would like to acknowledge the Monarch Joint Venture for supporting the 2012 meeting that brought so many of the authors and citizen scientists together; Patrick Guerra, Jessica Hellman, and Steve Reppert for their constructive comments that led to important improvements in the book; and all the people who study and support monarchs. We are especially indebted to the monarch and butterfly citizen scientists whose countless hours of observing have contributed to our knowledge of monarchs and the best ways to conserve them.

MONARCHS IN A CHANGING WORLD

PART I

Model Programs for Citizen Science, Education, and Conservation

An Overview

KAREN S. OBERHAUSER

Monarchs are arguably the most popular insect species in the world, and the fact that they are an icon for conservation and environmental education programs has both contributed to and resulted from this popularity. People's affection for monarchs has many origins. First, monarchs are widespread and familiar; caterpillars and adults are easily recognizable as visitors to gardens in North America and beyond, and many people welcome the return of these beautiful insects year after year. Second, as evidenced by the chapters in this book, monarchs are exceedingly interesting. They have fascinating relationships with natural enemies and milkweed plants, use an amazing variety of strategies to live in very different habitats throughout the world, and have an unusual (for butterflies) mating system (Brower et al. 2007). Third, there is still a great deal of mystery surrounding the ways in which monarchs navigate long distances, find appropriate breeding and wintering locations, and survive in so many different habitats. Finally, monarchs are admired for their tenacity in enduring an incredible migration from breeding locations in Canada and the U.S. to overwintering sites in Mexico, and surviving, at least for now, all of the changes that humans have imposed on their breeding, migrating, and wintering habitats. The combination of these features makes monarchs a focus of education, conservation, and scientific research programs, and many of the research programs engage members of the public through citizen science. Thus, the lives of monarchs and thousands of people in North America are intertwined in many ways through science education, biological conservation, and scientific research.

People's fascination with butterflies and monarchs is not new. Egyptian hieroglyphs contain pictures of butterflies, and the pre-Hispanic cultures that lived in Mexico, especially the Teotihuacan, Mixtec, Cholultec, and Aztec cultures, observed and studied butterflies. In many cultures, butterflies symbolize rebirth; they seemingly die when the caterpillar turns into a pupa, and re-emerge in a new form. In fact, the ancient Greek word for butterfly is *psyche*, which can be translated as "soul."

The aboriginal groups that lived in the monarch overwintering area in central Mexico included the Otomí, Mazahua, Matlazinca, and Purepecha. A mural found in Zitácuaro, a city close to the overwintering sites, is dedicated to the god of creation, and depicts the belief that "from the mouth of the god of Creation each day the sun appeared and during the winter its rays became butterflies" (Beutelspacher 1988). These butterflies were believed to clothe the earth, fertilize the soil, pollinate flowers, and decorate both life and the air. It is very likely that the butterflies to which this myth refers are monarchs, which arrived as winter began and departed as it ended. The Purepecha Indians thought that the monarchs were the souls of the dead, and their arrival near November 2nd, the Day of the Dead, announced visits by departed loved ones. The Mazahua and Otomí Indians associated monarchs' arrival with agricultural production cycles, calling them reapers because the crop was ready when the butterflies arrived and when they left it was time to prepare for planting. People in the area continue to use monarchs in embroidery, knitting, poetry, music, and other artistic venues.

The three chapters in this section illustrate the long fascination people have had with monarchs, and introduce themes that will continue through the rest of the book. The chapters describe educational, scientific, and conservation programs that focus on monarchs from many perspectives.

First, Young-Isebrand and colleagues summarize school- and non-school based education programs for children and adults throughout North America. In 1992, several educational programs that continue today started in Canada (Monarchs without Borders at the Insectarium of Montréal), the United States (Monarchs in the Classroom at the University of Minnesota and Monarch Watch at the University of Kansas), and Mexico (Correo Real through the non-profit organization Profauna). Thanks to these programs and many others that began later, schoolchildren, visitors to nature centers, and many other people have developed strong personal connections to both monarchs and the natural world that supports them. These connections can benefit monarchs, as generation after generation develops awareness and appreciation of this amazing insect.

Next, Oberhauser and colleagues introduce citizen science efforts that address all phases of the monarchs' annual cycle—breeding, migrating and overwintering—and summarize how these programs have contributed to scientific knowledge of monarch biology, thus previewing key data sources used in the rest of the book. While many scientists have dedicated their professional lives to increasing our understanding of monarch biology, important advances stem directly from the work of citizen scientists, people who are not professional scientists. Thanks to this 'research army,' we understand where North American monarchs go in the fall, winter, and spring; how monarch numbers vary from year to year; and details about their interactions with other species. Brief descriptions of these projects and their history are included in this section, and their scientific findings feature heavily in other chapters (Altizer and De Roode, Nail et al., Jepsen and Black, Pleasants, Howard and Davis, Batalden and Oberhauser, and Ries et al., this volume, Chapters 7, 8, 12, 14, 18, 19, and 24). One could argue that the input of citizen scientists to our scientific understanding of monarchs is greater than for any other single organism; certainly no other species has as many different citizen science programs dedicated to it. In Chapter 2 we describe 11 different programs dedicated solely to monarchs, and nine projects that collect data on other species, but for which monarchs are key focal subjects.

Finally, Shahani and colleagues summarize government and NGO-based efforts to conserve monarchs throughout North America. Because the continuing migratory phenomenon of monarchs flying to both the coast of California and central Mexico depends on the continuing availability of habitat for egg laying and caterpillar development, migration, and overwintering, all three North American countries must cooperate to preserve this endangered phenomenon (Wells et al. 1983). Monarch habitat loss (summarized by Jepsen and Black, Ramirez et al., and Pleasants, this volume, Chapters 12, 13, and 14) has led to declines in the both the eastern (Brower et al. 2011) and western (Jepsen and Black, this volume, Chapter 12) monarch populations. It is clear that now is a crucial time for monarchs, and the other, less well-documented species with which they share habitats. Shahani et al. summarize actions by government agencies and non-profit organizations, although organizational action alone will not be enough to preserve monarchs. Individual home owners, farmers, and businesses have control over land that is potential monarch habitat, and there is much that we can do collectively to ensure that this land continues to be available to future generations of monarchs, and people. People living in and near habitats used by monarchs are united by the common goal of monarch conservation. Like the great-grandchildren of the monarchs that return to Mexico year after year, our great-grandchildren deserve to experience these amazing creatures.

1

Environmental Education and Monarchs

Reaching across Disciplines, Generations, and Nations

Elisabeth Young-Isebrand, Karen S. Oberhauser, Kim Bailey, Sonya Charest, Brian Hayes, Elizabeth Howard, Jim Lovett, Susan Meyers, Erik Mollenhauer, Eneida B. Montesiños-Patino, Ann Ryan, Orley R. Taylor, and Rocío Treviño

Educational programs across North America utilize monarchs to engage thousands of children, educators, and people of all ages in environmental learning. Although the audiences and methods of the programs vary, they are unified in their efforts to conserve monarchs through education and action. Here we summarize the goals and achievements of eight successful programs. Key features of these programs include connections to formal K-12 education through curriculum and teacher professional development, emphasis on conservation through programs to restore and improve monarch and pollinator habitats, connections with citizen science programs, cross-cultural connections, and the use of living monarchs. We discuss risks and benefits of mass rearing of monarchs for educational purposes, and conclude that, in some cases, the biological knowledge and environmental literacy that result from monarch rearing are valuable enough to outweigh these risks, with careful attention to release practices. We also highlight the value of activities that do not rely on captive-reared monarchs, as these can promote greater awareness of the monarch's natural behavior, population biology, and habitat needs.

INTRODUCTION

Environmental educators across North America use monarchs to make connections between science education, conservation, and research. These educators, working in formal K-12 classrooms and non-formal settings at parks and nature centers, engage youth and adults in understanding elements of basic organismal biology, environmental conservation, and connections between humans and nature. For example, preschoolers might learn about the developmental stages of monarchs, elementary students can investigate or enhance monarch habitat, middle and high school students might conduct original research on monarch ecology, and children and adults can contribute to scientific research by participating in monarch citizen science projects (Plate 1). Many teachers bring live monarchs into their classrooms to inspire students' natural curiosity for science. Monarchs are captivating subjects owing to their easily observed life cycle, fascinating migration, and aesthetic beauty, all of which contribute to the success of projects described in this chapter.

The goals of environmental education programs that use monarchs vary with the audience and setting. Many educators use monarchs to foster both a personal connection with the natural world and increased student achievement. Often, monarch study takes place outside, and learning outdoors in a natural setting boosts children's confidence and academic performance, especially in science (Lieberman and Hoody 1998; Carrier 2009). Interacting with nature can reduce stress and increase cognitive functioning (Berman et al. 2008) and nurture self-discipline (Taylor et al. 2001). Environmental educators also utilize monarchs to promote multi-disciplinary learning in natural settings. For example, a student studying monarch migration can learn concepts and skills from math, geography, meteorology, social studies, languages, reading, and writing in the context of a real-world project that begins in a school, garden, or local nature center.

A key benefit of environmental education efforts is that they enhance awareness of conservation challenges and motivate people to contribute to solutions for a healthy planet. Encouraging children and adults to spend time outdoors can also counter recent trends of reduced physical activity and disconnection from nature (Pergams and Zaradic 2007). For example, a 2010 survey (Kaiser Family Foundation 2010) revealed that 8 to 18 year-olds in the United States spend about 7.5 hours a day plugged into electronic media. The resulting lack of direct experience with the natural world could impact the way future generations make decisions about the environment, since direct experience with nature as a child can lead to a life of action on behalf of the natural world (Chalwa 1998; Louv 2008). We argue that monarchs provide a powerful antidote to apathy toward the natural world; the programs described below were developed on the premise that increasing awareness and appreciation of monarchs will help foster concern for other organisms (plants, animals, and microbes) that share monarchs' habitat.

Here, we review eight monarch-based programs that offer people across North America opportunities to engage in environmental outreach (Table 1.1), increase environmental literacy among diverse audiences, and give both children and adults the awareness and understanding necessary to meet the complex challenges of biodiversity conservation. Collectively, these programs cover diverse regions and emphasize that the environmental challenges facing monarchs and humans alike are best resolved when people work together across borders, languages, and cultures. To provide a historical perspective on the development of these programs, we describe them in chronological order, beginning with the earliest founding dates.

MONARCH WATCH

Monarch Watch, based at the University of Kansas and directed by Orley "Chip" Taylor, is a cooperative network of students, teachers, volunteers, and researchers dedicated to the study of monarchs. The program was launched in 1992 as a monarch tagging project; citizen science aspects involving tag-recovery data are described by Oberhauser et al. (this volume, Chapter 2). Monarch Watch also provides information about monarch biology and their spectacular migration to the general public, with the goal of promoting monarch habitat restoration and protection throughout North America. The program is supported through the sale of promotional and educational materials (tagging kits, rearing kits, Monarch Waystation signage, etc.), private donations, and grants from the Monarch Joint Venture.

In 2005, Monarch Watch initiated the Monarch Waystation program, designed to encourage citizens to plant milkweeds and suitable nectar plants in gardens. This program reaches homeowners and gardeners, business owners, park personnel, zoos, and nature centers through online resources that include an international habitat registry and milkweed profiles. By fall 2013, more than 7300 Monarch Waystations had been certified. In 2010 the "Bring Back the Monarchs" program was launched to promote restoration of 20 region-specific milkweed species throughout the United States. This program has led to collaborations with the National Resources Conservation Service (NRCS), U.S. Forest Service, U.S. Environmental Protection Agency, native plant societies, and nurseries. Monarch Watch now coordinates seed collection and germination as well as the distribution and planting of milkweeds in regionally appropriate areas.

Monarch Watch sells Monarch Rearing Kits to educational institutions and individuals in locations east of the Rocky Mountains; no sales are made to locations west of the Rocky Mountains to prevent eastern and western monarch populations from mixing. These kits include monarch larvae and rearing instructions and reach thousands of children in schools across the eastern monarch flyway each year. The goal is to provide hands-on monarch rearing experiences and allow children to follow the developmental life cycle. Some of the adult monarchs reared from these kits are tagged and released during the fall migration in conjunction with the Monarch Watch tagging program.

The Monarch Watch website provides free monarch biology and conservation resources; it also gives visitors access to communication forums via social media sites, a blog, and an e-mail listserv dedicated to monarchs. Local public engagement occurs through events each year (such as a fall open house) that

Table 1.1. Monarch environmental education programs highlighted in this chapter.

Project (year founded)	Focus	Web address
Monarch Watch (1992)	Citizen science, distribution of live monarchs for educational purposes, Monarch Waystation program, sale of monarchs, Bring Back the Monarchs campaign.	monarchwatch.org
Correo Real (1992)	Professional development for educators, citizen science, curriculum and online resources.	Profauna.org.mx/monarca/
Monarchs in the Classroom (1992)	Professional development for educators, educational materials, youth research fair, citizen science, garden grants.	monarchlab.org
Monarchs Without Borders (1992)	Distribution of live monarchs for classroom rearing, professional development for educators, general insect education for the public.	ville.montreal.qc.ca/insectarium
Journey North (1994)	Online educational resources about monarchs and other migratory organisms and seasonal phenomena, citizen science.	learner.org/jnorth/
Monarchs Across Georgia (2000)	Professional development for educators, pollinator habitat education, literacy promotion in Mexican villages of the overwintering grounds, and educational travel to Mexico.	monarchsacrossgeorgia.org
Monarch Teacher Network (2001)	Professional development for educators in the US and Canada, trilingual curriculum and educational materials. Educational tours of Mexico and California overwintering sites.	monarchteachernetwork.org
MBBR Workshops (2006)	Professional development for adults in Monarch Butterfly Biosphere Reserve (Mexico), curriculum and other resources, guide training.	monarchbutterflyfund.org

attract thousands of visitors. Staff also give tours of their facility at the University of Kansas to school or gardening groups, and they visit local schools, zoos, and nature centers. Most recently, a team of monarch and habitat advocates called Monarch Conservation Specialists was formed to promote monarch conservation in their regions.

CORREO REAL

Correo Real, started in 1992, is run by Project Director Rocio Treviño and is part of the Mexican nonprofit organization Protección de la Fauna Mexicana (Profauna). While Correo Real is primarily a citizen science program (see Oberhauser et al., this volume, Chapter 2), the extensive training it provides to teachers and children through manuals, online curriculum, and other materials has made monarchs an important focus of environmental education programs in parks, museums, and schools. Based in Saltillo, Coahuila, the program has fostered school festivals and media reports focused on monarchs, especially in the states of Coahuila and Nuevo León (in Northern Mexico). In Coahuila, the Secretariat of Public Education designated October as "monarch butterfly conservation month"; during this month, when monarchs begin entering Mexico in fall migration, children study monarchs and create posters to inform their communities about monarch migration and the importance of its conservation. Correo Real's participation in the Journey North symbolic migration (see below) has enabled children and teachers in Coahuila and Nuevo León to exchange messages of good will with children in the United States and Canada.

MONARCHS IN THE CLASSROOM

Monarchs in the Classroom (MITC) was started in 1992 by Karen Oberhauser and Elizabeth Goehring at the University of Minnesota. The program focuses on improving K–12 student achievement in science, connections to nature, and awareness of monarch biology and conservation. MITC also includes programs for informal educators and is closely linked to a citizen science project (the Monarch Larva Monitoring Project, MLMP; see Oberhauser et al., this volume, Chapter 2). MITC has been supported by the National Science Foundation, the Minnesota Office of Higher Education, the Medtronic Foundation, and the U.S. Forest Service. Materials to support student learning include curriculum guides, field guides, and digital media for presentations.

MITC staff (including scientists and K–12 teachers) have conducted teacher workshops every summer since 1997. In 2002, the focus of these intensive 10-day workshops was expanded beyond monarchs to include other classroom-friendly and locally abundant invertebrates. These workshops help teachers prepare to engage their students in natural history learning and research, with a strong focus on activities that can be conducted in schoolyards. More than 800 teachers from more than 400 schools, mostly in Minnesota, have completed the courses. A partnership with the U.S. Forest Service International Programs has engaged teachers from throughout the United States in a related program called the North American Monarch Institutes (NAMI). These three-day workshops bring teachers, nonformal science educators, and scientists together to study monarchs and their habitat and to build lasting collaborations. More than 350 participants attended NAMI workshops from 2010–2013. Participants from El Valor, a community-based organization in Chicago for preschool children, illustrate the impacts of NAMI workshops. El Valor has sent 21 educators to NAMI workshops who have returned to lead professional development for all staff, including home-based teachers and parents.

To support outdoor instruction, MITC developed the Schoolyard Garden Grant program in 2006 to support easily accessible habitats for outdoor learning. From 2006 to 2013, more than 80 schools received grants to plant or improve schoolyard habitats (native plant or vegetable gardens, or natural areas). MITC also sponsors an annual student research fair, at which students (in grades 3–12) present research projects and receive feedback from scientists and peers. From 1996 to 2013, more than 3300 students participated in the Research Fair.

Finally, the Driven to Discover (D2D) project, supported by a National Science Foundation grant, provides 10–14-year-olds the opportunity for authentic science inquiry by extending work in established citizen science projects. D2D builds on the structure of existing citizen science projects by training adult mentors—teachers, Master Gardeners, Scout leaders, 4H leaders, Master Naturalists, and parents—to lead youth research clubs focused on monarchs (through the MLMP) or birds (through the Cornell Lab of Ornithology). From 2009 to 2013, more than 300 youth learned to collect data, draw conclusions based on their data, and report the results of their investigations to peers at an annual Research Summit. The eventual goal is to adapt the curriculum for a wide variety of citizen science projects.

MONARCHS WITHOUT BORDERS

The Monarchs Without Borders program was developed in 1992 by the Insectarium of Montréal. The goal of the program is to facilitate the rearing of live monarchs in classrooms. For the first two years, the Insectarium's involvement was modest; about 30 monarch-rearing kits were distributed in schools. Interest in the program grew quickly and kit sales increased dramatically over the years. Today, the Insectarium sells about 1000 kits every fall.

Participants pick up their kits in person, receiving one milkweed plant, four caterpillars, two pupae, and Monarch Watch tags. The monarchs are reared in captivity, with new parent stock (originating in Quebec) each summer. On distribution day, an information kiosk answers participants' questions about monarchs and the rearing process. Insectarium staff also provide group presentations and encourage participants to take part in workshops on topics that include tagging, rearing monarchs, and monarch ecology. Participants have exclusive access to a blog containing steps to successful monarch rearing and teaching; the blog addresses topics such as monarch metamorphosis and migration, includes regular posts by experts, and allows participants to ask questions and share experiences.

Most participants (70%) in Monarchs without Borders are elementary school teachers and their students, but the number of other participants is growing. Various social institutions, including hospitals, prisons, and organizations in low-income neighborhoods, have purchased monarch rearing kits. In one hospital, a group of girls being treated for anorexia were allowed to take care of the caterpillars when their eating habits stabilized. The girls observed the caterpillars consume milkweed leaves, grow in size and complete their metamorphosis. The Insectarium is working to develop connections with people along the migratory path, and in Mexico. In 2013, it launched a Monarch Gardens certification program for Quebec.

JOURNEY NORTH

Journey North is a nonprofit organization whose mission is to engage citizen scientists in a global study of migration and seasonal change, focusing on a range of species including monarchs, hummingbirds, whales, and flowering plants. It was established in 1994 by Elizabeth Howard, with a grant from the National Fish and Wildlife Foundation. Ongoing support is provided by the Annenberg Foundation. Journey North reaches 980,000 students at 45,000 sites through free, web-based resources. Citizen science components of the program are described by Oberhauser et al. (this volume, Chapter 2); here we focus on the program's educational aspects.

Journey North summarizes the status of the monarch migration through weekly online newsletters during the spring and fall migrations, and it provides news from the overwintering region in Mexico. Every week, citizen scientists tell the monarch's story through images and observations shared from across North America. The website is rich with educational resources, images, video clips, maps, activities, and lesson plans built on citizen science observations. While other programs described in this chapter have branched out to include other plants and animals living in monarch habitat, Journey North is distinct in its focus on phenology and the inclusion of organisms that are far removed from monarchs.

Journey North also coordinates an annual cultural exchange between children in Mexico, the United States, and Canada. More than 60,000 students in the United States and Canada create symbolic paper butterflies and send them to Mexico for the winter. Schoolchildren in Mexico protect the butterflies and return them to the north in the spring. Through the Symbolic Migration, children across North America are united by monarchs and a continental celebration of their spectacular migration. The symbolic migration is tied to authentic lessons of ambassadorship, conservation, and international cooperation. Estela Romero, who coordinates the Symbolic Migration in the overwintering region in Mexico, describes the strong impact of this program on students in this region: "Students are proud to learn that our part of the world is unique and important. Now they see the forest that surrounds us with new eyes, and they see themselves as part of an international community."

MONARCHS ACROSS GEORGIA

The mission of Monarchs Across Georgia (MAG) is to inspire future caretakers of the natural environment through education about monarchs and other pollinators. MAG was founded in 2000 by a group of educators led by Susan Meyers and Kim Bailey; this group became MAG's steering committee and, later, a working committee of the Environmental Education Alliance (EEA) of Georgia. Today, the MAG steering committee includes environmental education volunteers and professionals. An advisory committee includes horticulturists, educators, and research scientists from universities in Georgia, Kansas, and Minnesota. Funding for MAG comes from membership dues and event fees paid to the EEA, as well as from donations, grants, program fees, and plant and merchandise sales.

MAG's target audience includes teachers, students, families, businesses, gardeners, nature enthusiasts, and others interested in studying monarchs and restoring pollinator habitat. MAG uses many outreach strategies, including community presentations, a website, and strong social media presence, and a biannual newsletter called The Chrysalis. Other activities directly benefit monarch habitat, including selling native nectar plants and milkweed and sponsoring a pollinator habitat certification program that has resulted in more than 115 certified habitats as of 2013. Other MAG activities aimed at K-12 educators in the United States and Mexico include field trips to the monarch wintering colonies in Mexico, donations of books and supplies to Mexican schools, and professional development workshops for Georgia educators.

From 2000 to 2013, MAG held more than 30 educator workshops in 15 counties throughout Georgia, reaching nearly 450 teachers who have carried the message of monarchs to more than 20,000 students across grades K-12. The workshops, approved by the Georgia Department of Education for teaching license renewal credits, include lessons from the MITC curriculum and promote engagement in monarch citizen science projects.

One of MAG's greatest successes has been its Mexico Book Project. Since 2004, MAG has donated more than 1000 Spanish language books and hundreds of dollars in supplies to schools near the monarch sanctuaries in Mexico. In appreciation of MAG's efforts to advance literacy in these rural areas, the book supplier, Scholastic Mexico, matched the number of books purchased in 2008, doubling the number of books donated that year.

MONARCH TEACHER NETWORK

The Monarch Teacher Network (MTN), started in 2001, includes teachers, educators, and others who bring the monarchs' story to classrooms and communities across North America. It is sponsored by the Educational Information Resource Center (EIRC), a New Jersey public agency, and was created by EIRC employee Erik Mollenhauer. MTN work is supported by grants from private foundations and includes collaborations with a variety of public and private schools, universities, environmental groups, and other organizations. The MTN of Canada was formed in 2003, sponsored by the Toronto Region Conservation Authority, and MTN–Western Canada began shortly thereafter. Both Canadian networks work closely with their U.S. counterpart.

Each summer MTN conducts a series of two-day workshops across the United States and Canada. These workshops provide materials and knowledge that support the use of monarchs as an interdisciplinary learning tool for people of all ages and abilities. The monarchs' story is used to explore the past, present, and future of North America, its people, and the land that sustains us all. By August 2013, about 4900 people from 30 U.S. states, 8 Canadian provinces, and 5 other countries had attended MTN workshops. Currently about 500 people are trained at MTN workshops annually. Workshop participants experience two days of hands-on instruction and receive rearing cages, a curriculum, a trilingual film called *Journeys and Transformations*, and a trilingual monarch life cycle poster. After the workshops, MTN provides ongoing support and networking opportunities through a variety of social media, and it organizes tours of monarch winter colonies in Mexico or California.

MONARCH BUTTERFLY BIOSPHERE RESERVE WORKSHOPS

In a program sponsored by the Monarch Butterfly Fund (MBF 2013), free monarch workshops have been designed, coordinated, and conducted by Eneida Montesiños within and near the Monarch Butterfly Biosphere Reserve (MBBR), which houses the key monarch overwintering sites in central Mexico. Beginning in 2006, the three-day workshops have offered training on monarch biology and forest conservation for people who live near the monarch overwintering areas. Local tour guides and other community members participate in the workshops; because community members make decisions that often affect the surrounding forests, it is important that they understand monarchs and their habitat needs. Additionally, guides are better prepared to provide accurate biology and conservation information to tourists who visit the monarch wintering sites. The curriculum also includes monitoring protocols with lessons and field activities to train people to help with research in the reserve.

From 2006 to 2013, 898 people from 7 communities were trained in 30 workshops. Many participants now engage in conservation activities to protect monarch habitat and work with local schools and visitors on environmental education issues. Events in the MBBR attract a great deal of attention in the Mexican media, and the potential of these workshops to influence public knowledge in Mexico was illustrated by an interview that aired on a Mexican television channel. The reporter interviewed a community member as an expert on monarch migration and biology. This workshop participant could not read or write, but she was able to communicate the importance of the wintering sites in the monarchs' annual migratory cycle. In fall 2012, Montesiños joined MBBR and World Wildlife Fund–Mexico personnel to plan and conduct a series of presentations for local schoolchildren in towns near the MBBR. Although the children are very familiar with the overwintering phenomenon, they have no experience with the breeding phase of the monarch annual cycle, and the workshops clarified the big picture of the monarch migratory phenomenon. These presentations, also sponsored by the Monarch Butterfly Fund, engaged more than 900 children from six schools.

OUTLOOK AND CHALLENGES FOR MONARCH OUTREACH AND EDUCATION

Environmental education using monarchs has the potential to develop concerned, knowledgeable advocates who can effectively address present and future challenges of monarch survival. As these programs illustrate, monarchs capture the attention and curiosity of people of all ages and walks of life. The educational programs highlighted here have provided thousands of participants with personal connections to monarchs and their habitat, and they have prepared these participants to support monarch conservation in a variety of ways.

While these programs have different goals and audiences, they have many features in common that have led to their popularity and both educational and conservation value. Most have ties to formal education through curricula that are available in hard copy or on the web, and many also conduct workshops for K-12 teachers. The curricula and workshops have helped spread these programs into classrooms across the continent, resulting in dissemination of knowledge to many additional individuals. Most programs also promote monarch habitat restoration and protection, and many programs combine education, research, and conservation by including citizen science as an integral part. These data collection efforts have the educational benefit of engaging people in authentic research, thus fostering knowledge of the practice of science, as well as providing data to inform conservation efforts. Finally, many programs have strong cross-cultural components that emphasize monarchs' trinational movement, and thus the shared nature of monarch populations and the need for cooperative conservation efforts.

Many programs promote monarch rearing for educational and nature engagement purposes. Of the programs reviewed here, Monarch Watch and Monarchs Without Borders distribute live monarchs for educational use (Monarchs in the Classroom stopped distributing monarchs in 2012). Many butterfly breeders (IBBA 2012) also sell monarchs for educational purposes and for releases at events not related to education. While information on the number of monarchs released in these programs is not available, Pyle et al. (2010) cite a *New York Times* editorial from 2006 suggesting that 11 million human-reared butterflies, mostly monarchs and painted ladies, were released annually in North America. This practice is not without critics. A recent review highlighted concerns about the butterfly house industry (suppliers to vivaria in which "curious people and beautiful insects share the same space," Boppré and Vane-Wright 2012, p. 286). Many of these concerns apply to the rearing and release of purchased monarchs.

Perhaps the most substantial concerns of captive rearing and releases noted by scientists involve the proliferation of disease and effects on the genetic composition of wild populations. Collection of monarchs from one area, subsequent mass breeding, and translocation to other areas could distort estimates of genetic diversity and gene flow in wild monarchs (Brower et al. 1995, 1996). While selling monarchs across state lines in the United States requires permits from the USDA, and in some cases a state permit as well, a large number of growers have obtained such permits, and a repository for records on the numbers of monarchs shipped and sold for release (and the locations of transfer) is lacking. Moreover, releases that occur within the monarchs' native range but outside their normal timing, or during times of the year when natural monarch abundance is low (such as in early spring), can confuse monitoring efforts or have unusually large influences on monarch ecology and genetics. The study of monarch biogeography, especially in western North America, is still in its infancy, and many important questions critical to monarch conservation are still largely unanswered. Releasing monarchs into the landscape without monitoring interferes with our ability to answer those questions.

Brower et al. (1995, 1996) raised concerns about the translocation of monarchs between the western and eastern United States because of potential genetically distinct populations. Such movement is illegal, as the USDA will not issue permits for monarch transfers across the Rocky Mountains continental divide. While recent studies suggest that large genetic differences between monarchs from these two regions are lacking (Pierce et al., this volume, Chapter 23), the movement could result in the long-distance spread of novel pathogens. Indeed, eastern and western monarchs also harbor genetically distinct strains of the debilitating protozoan *Ophryocystis elektroscirrha* (reviewed by Altizer and de Roode, this volume, Chapter 7).

More generally, captive rearing of monarchs can create conditions for the proliferation of disease, including the protozoan parasite *Ophryocystis elektroscirrha* (*OE*), and anecdotal reports indicate the potential for unexplained monarch die-offs, presumably associated with disease, under local captive rearing operations. Although *OE* can be readily monitored by scientists and knowledgeable breeders, not all pathogens affecting monarchs are as well known. Teachers and other members of the public who obtain reared monarchs might be less aware of problems caused by pathogens, and breeders have no requirement to follow specific disease-preventing protocols, nor are there resources in place for routinely testing captive stock for most diseases.

We acknowledge the serious risks posed by mass breeding and translocation and do not condone the release of monarchs where recreation or amusement are the primary goals (Boppré and Vane-Wright 2012); however, like Boppré and Vane-Wright, we think that the biological knowledge, conservation awareness, and environmental literacy resulting from monarch rearing programs are valuable. Careful attention to education, breeding practices, and the source and destination of monarchs will minimize many risks. Whenever possible, we recommend that educators raise monarch eggs and caterpillars that they themselves collect in local milkweed patches, instead of purchasing mass-reared individuals. As long as the monarchs are reared with care and released locally, this practice can avoid many of the risks described above. Since the risks associated with sales of mass-reared monarchs cannot be completely eliminated, it is important that monarchs are sold only when clear educational benefits result.

Clearly, monarchs capture the attention of many people. Indeed, a recent survey of U.S. households suggests that Americans are willing to support monarch butterfly conservation at high levels, up to about $6.5 billion if extrapolated to all U.S. households (Diffendorfer et al. 2013). If even a small percentage of people acted on this willingness, the cumulative effort would translate into a large, untapped potential for conservation of this iconic butterfly. It is likely that the monarch education programs described in this chapter have played a role in people's willingness to support monarch conservation at such high levels; for this reason, it behooves us to do all we can to ensure that monarch education programs champion effective conservation practices and tap into people's desire to preserve this amazing insect. It will be especially important to continue to promote public engagement in monarch conservation in the face of declining populations (Brower et al. 2012a; Rendón-Salinas and Tavera-Alonso 2013; Jepsen and Black, this volume, Chapter 12).

ACKNOWLEDGMENTS

We thank all the individuals who take part in monarch environmental education programs as funders, participants, leaders, and coordinators. Sarina Jepsen, Michelle Solensky, Sonia Altizer, and Patrick Guerra provided comments on earlier drafts of the manuscript.

2

Contributions to Monarch Biology and Conservation through Citizen Science

Seventy Years and Counting

Karen S. Oberhauser, Leslie Ries, Sonia Altizer, Rebecca V. Batalden, Janet Kudell-Ekstrum, Mark Garland, Elizabeth Howard, Sarina Jepsen, Jim Lovett, Mía Monroe, Gail Morris, Eduardo Rendón-Salinas, Richard G. RuBino, Ann Ryan, Orley R. Taylor, Rocío Treviño, Francis X. Villablanca, and Dick Walton

The public's fascination with monarchs has inspired and sustained a rich array of monarch citizen science programs, beginning with Dr. Fred Urquhart's tagging program in the 1950s–1990s that led to the discovery of monarch wintering grounds in central Mexico. No other single species has garnered such a wide following of personally involved educators, conservation advocates, and citizen scientist contributors. The tens of thousands of hours per year invested by these volunteers have allowed scientists to answer basic questions about how and when monarchs use available habitat, how their numbers change within and among years, how environmental perturbations affect these changes, and how monarch populations are responding to contemporary global change and conservation efforts. Here, we review past and current programs, focusing on characteristics of successful programs and their wide-reaching scientific, environmental, and educational outcomes. We also present a data gap analysis to ask what locations and times of year have limited data on monarch biology, to inform the targeted recruitment of monarch citizen scientists into current and future programs.

INTRODUCTION

For decades, people have monitored monarchs in many locations using diverse methods. Citizen scientists—armed with data sheets and pencils, apps, hand lenses, butterfly nets, and binoculars (Plate 2)—have been and continue to be key players in monarch monitoring programs. Many programs assess monarch numbers, ranging from local densities of different life stages, to the numbers and locations of butterflies at migratory stopover sites, to areas occupied by monarchs at overwintering sites. Other programs track the timing and location of the fall and spring migrations, measure attack rates by natural enemies, and document milkweed emergence. These programs have allowed scientists to answer crucial questions at large spatial and temporal scales, including how and when monarchs use available habitat, how their numbers change over time, how environmental perturbations affect these changes, and how monarch populations are responding to environmental change and conservation efforts (CEC 2009).

Unlike most current scientific research, citizen science involves amateurs instead of professional scientists (although most scientific research was conducted by amateurs prior to the late nineteenth century; Vetter 2011). Citizen monitoring has a long history, with records of locust outbreaks in China dating back at least 3500 years (Tian et al. 2011). Many other programs originated more than a hundred years ago (Miller-Rushing et al. 2012). In North America, the oldest large-scale biodiversity monitoring project driven by citizen observations is the Christmas Bird Count, started in 1900 (Audubon 2012). A recent surge of interest in the value of citizen science is evidenced by the devotion of an entire issue (Aug. 2012) of *Frontiers in Ecology and the Environment* and a book (*Citizen Science: Public Participation in Environmental Research*, Dickinson and Bonney 2012) to the topic of citizen science, as well as by multiple conferences, symposia, and workshops on the subject.

Butterfly monitoring by citizen scientists also has a long history. The field notes, reports, and specimens from Victorian collectors contributed important knowledge of butterfly ranges, behaviors, and abundance. One of the earliest coordinated citizen science projects involving butterflies focused on monarchs. Beginning in the 1950s, Dr. Fred Urquhart's tagging program engaged hundreds of volunteers in a hunt for the winter destination of eastern North American migratory monarchs, a goal ultimately achieved in early 1975 (Urquhart 1976). Monarch followers in North America are less likely to be familiar with Dr. Courtenay Smithers, who engaged volunteers in Australia in a mark-release program from 1966 to 1970. Volunteers in about 200 locations made nearly 6300 observations of monarch presence and absence from Adelaide to Brisbane, helping Smithers (1977) document monarch distributions. Their observations showed that monarchs in Australia cover a wide area during the summer, and contract into three regions during the winter: a coastal strip from northeastern New South Wales to the Cape York Peninsula, the Sydney basin, and Adelaide. Smithers concluded that monarch movement allows them to use both seasonally and permanently suitable habitats in Australia (Smithers 1977).

Why is this single insect species the subject of such intense scientific study and public interest? We suggest that monarchs enjoy an almost iconic status, inspiring people to contribute considerable time to understanding their biology. This iconic status may stem from the ease with which monarchs are recognized and the accessibility of their habitats, as well as from their beauty and unique biology. The long history of engagement by both citizen scientists and professional scientists who have developed monarch citizen science research programs sets a precedent for success and a tradition that involves thousands of volunteers each year. Public participation in monarch citizen science projects has grown rapidly since 1990 and now spans several programs (Table 2.1). Participation has also become more intense, with volunteers taking a stronger role in designing their own studies and collecting more intensive data. As a case in point, at a meeting of monarch researchers in 2012 (Monarch Lab 2012), more than half of 165 attendees were citizen scientists, many of whom presented their own research that had grown out of engagement in large-scale citizen science projects. We believe that this level of participation is the next frontier for citizen scientist research and engagement.

Below, we describe current North American citizen science programs that focus on various aspects of monarch biology. We also include programs that monitor all butterflies and can therefore provide data on monarch abundance and distribution. These programs collectively involve well over 14,000 volunteers (as of 2011; Table 2.1) and span thousands of sites across the eastern and western United States, Canada, and Mexico (Figures 2.1–2.3). In 2009, the leaders of several programs formed a network of monarch monitoring programs, called MonarchNet, with the goal of coordinating data management efforts. The goals and accomplishments of this effort, along with a list of peer-reviewed publications based on monarch citizen science data, can be found at the group's website (www.monarchnet.org).

Technology has played a key role in fueling the growth of butterfly and monarch citizen science programs, which rely on relatively recent advances in information technology (e.g., web-based reporting and data management, and communication forums) that would not have been possible just 20 years ago. Social media are likely to play a growing role in how people obtain and communicate information relevant to their engagement in citizen science, including data collection and transmission. A very active e-mail discussion list (Monarch Watch Dplex-L, Monarch Watch 2013a) is a good example of citizen engagement via technology; while the list is not centered on a specific citizen science project, participants provide reports of monarchs in their areas in addition to sharing in a lively and interactive community.

TRACKING A MOVING TARGET: PROGRAMS FOCUSED ON MIGRATION

Citizen scientists study monarch migration through tagging, individual observations, and sur-

Table 2.1. Summary of current monarch and butterfly citizen science projects.

Project and web address	Start year	Type of data	No. volunteers, 2011[a]	No. observations, 2011[a]
Projects that monitor migration				
Journey North www.learner.org/jnorth	1994	Monarch sightings, spring and fall migrations throughout range	>4000	6737
Correo Real www.profauna.org.mx/monarca	1992	Monarch sightings, spring and fall migrations in Mexico	128	400
Monarch Watch www.monarchwatch.org	1992	Tagging, fall migration throughout range	>10,000	~83,000
Southwest Monarch Study swmonarchs.org	2003	Fall tagging, habitat reports in Arizona	Tagging: 52 Monitoring: 8	1850 20
Monarch Monitoring Project www.monarchmonitoringproject.com	1990	Fall censuses, tagging demonstrations in Cape May NJ	Censuses: 16 Tagging demos: 4	170 23
St. Marks National Wildlife Refuge	1988	Fall tagging in northern FL	Not available	1588
Peninsula Point Monarch Research Project	1996	Fall censuses, egg and larva monitoring, tagging in Michigan Upper Peninsula	Censuses: 1 Monitoring: 4	91 24
Projects that monitor overwintering sites				
Western Monarch Thanksgiving Count www.xerces.org/	1997	Counts at western overwintering sites	85	119
Monarch Alert monarchalert.calpoly.edu	2001	Fall and winter censuses and tagging at western overwintering sites	9	355
MBBR Surveys www.wwf.org.mx	1992	Biweekly measurements of area occupied by monarchs during winter in Mexico	NA	
Projects that monitor during all phases				
Monarch Larva Monitoring Project www.mlmp.org	1996	Weekly surveys of eggs and larvae, habitat, and parasitism rates throughout range	Density: 129 Parasitism: 27 Sightings: 23	1710 1719 69
MonarchHealth www.monarchparasites.org	2006	Sampling adult monarchs for OE spores throughout range	130	2839
Butterfly or Butterfly and Moth projects				
North American Butterfly Association www.naba.org	1975	Annual butterfly censuses within 15-mile-diameter circles, individual observations throughout range		
Butterfly Monitoring Networks (IL, OH, FL, IA, MI) Various websites	1987 and later	Multiple counts within years along fixed transects at locally appropriate times		
Butterflies and Moths of North America Sightings Program butterfliesandmoths.org	2005	Photos submitted to project website and verified by experts throughout range		
eButterfly www.ebutterfly.ca	2012	Photos submitted to project website and verified by experts throughout range		

[a] Volunteer and observation numbers shown for monarch-only projects, using 2011 numbers to illustrate annual participation.

veys in specific locations. Tagging has been used by multiple programs to study patterns and timing of monarch movement. While details of the programs vary, in each of them volunteers apply small tags to monarch wings. A unique code and program contact information are printed on each tag; taggers record the date and location when they tag a monarch and send this information to program coordinators. Individuals who find tagged butterflies send the identifying codes, recovery date, and location to the program. The first tagging program was the Insect Migration Association, established in 1952 by Fred Urquhart to determine where monarchs from the eastern population go in the winter. This program lasted until 1994 and involved schoolchildren, naturalists, and others in observing, capturing, and tagging monarchs (Urquhart 1960, 1987; Urquhart and Urquhart 1977). In 1975, volunteers Kenneth Brugger and his wife, Cathy Aguado, helping Urquhart in Mexico, found the monarch wintering grounds in central Mexico. Although the sites had been known by local citizens, until then no one understood that the monarchs that blanketed these mountaintops had flown from as far away as the northern United States and southern Canada.

Several programs monitor the size and timing of the autumn migration at specific locations. Most of these programs take place on peninsulas, where monarchs often cluster during the migration. The programs use a variety of methods to count monarchs along predetermined transects or at specific stops. In addition to the projects highlighted below, others take place on Long Point on the north shore of Lake Erie, Ontario (begun in 1995, ongoing; Crewe et al. 2007); Chincoteague National Wildlife Refuge on Assateague Island, Virginia (1997–2006); and the Coastal Virginia Wildlife Observatory and the Eastern Shore of Virginia National Wildlife Refuge (1998–2000).

Journey North

Arguably the largest and best-known active monarch monitoring program involves volunteers who report individual sightings of spring and fall migrating monarchs. Journey North, founded in 1994, is supported by the Annenberg Foundation and directed by Elizabeth Howard. Citizen scientists report their first sightings of multiple organisms, thus tracking the moving front of migrations and other seasonal phenomena in real time. Journey North's goal is to help scientists and the general public understand how migratory species respond to climate and changing seasons.

First spring sightings of adult monarchs, as reported by volunteers, are shown on a live migration map on the project website. In the fall, volunteer reports of overnight roosts are also visualized on a real-time map. Journey North collects and archives other data pertaining to monarchs, including the first eggs, larvae, and milkweed that volunteers observe each year. Data can be reported via the Internet or a mobile smartphone application. All sightings are reviewed by experts and clarifications sought when necessary to ensure data quality. Staff spend more than 1000 hours a year clarifying instructions, reviewing data, and confirming accuracy with volunteers. From 1997 to 2014, Journey North observers contributed data on 72,299 monarch sightings (see Figures 2.1 and 2.2 for locations of sightings in 2011).

Journey North materials reach an audience far beyond the people who contribute data; for example, scientists and teachers use animated versions of the real-time migration maps to illustrate monarch migration patterns to many audiences. Other widely distributed products are the weekly migration updates which are distributed electronically to 45,000 subscribers. Almost 600,000 students are enrolled in registered Journey North classrooms, and it is likely that many others are exposed to the findings.

Since 2000, scientific analyses based on Journey North data have clarified the order in which U.S. states are recolonized by monarchs during the spring (Howard and Davis 2004), the speed of the spring migration over several years (Howard and Davis 2011), characteristics of fall stopover sites (Davis et al. 2012a), fall migration flyways (Howard and Davis 2009), the pace of the fall migration (Howard and Davis, this volume, Chapter 18), mortality from a storm event (Howard and Davis 2012), and monarch overwintering in the southern United States (Howard et al. 2010).

Correo Real

Correo Real, started in 1992, is the Mexican counterpart to Journey North. Project founder and director Rocio Treviño, from the nonprofit organization Protección de la Fauna Mexicana (Profauna), manages a network that currently includes more than 200 volunteers who collect data on the fall monarch migration through northern Mexico. Most Correo Real observations come from schools in the states of Coahuila and Nuevo Leon (see Figure 2.1c), although contributions from other groups (e.g., biologists, engineers, and agronomists) have increased since 2003. When the program began, all communication took place by mail, thus the name Correo Real (Spanish for "Royal Mail"). The goal of Correo Real is to understand and conserve habitat along the monarchs' migration route through Mexico. Treviño provides directions for reporting monarch observations during workshops for teachers and other educators.

Participants record the number of butterflies they observe; the time, location, and date of their observations; monarch behavior (such as flying, feeding, or resting); plants on which the butterflies feed or rest; and weather conditions. Most observations are made in urban areas, but an increasing number come from rural areas as the project is now integrated into the technical program of Natural Protected Areas (NPAs) in northern Mexico. Correo Real provides information on monarch movement

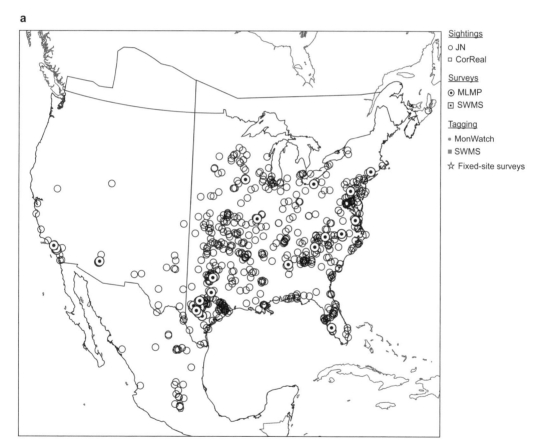

Figure 2.1. Maps of monarch citizen science data collection sites in 2011 based on location data provided by projects. The vertical line through the middle of the maps roughly divides the eastern and western populations (see Pyle this volume, Chapter 21, for a discussion of the "fuzziness" of this line). Northern boundary shows monarch summer range, determined using data from butterfly sightings programs. Note that dates do not correspond exactly to calendar seasons. (a) Spring migration and breeding (8 March–16 May; Journey North, Monarch Larva Monitoring Project). (b) Summer expansion and breeding (17 May–15 August; Journey North, Monarch Larva Monitoring Project). (c) Autumn migration (16 August–31 October; Journey North, Monarch Larva Monitoring Project, Correo Real, Southwest Monarch Study, Monarch Watch, Cape May, Peninsula Point, St. Marks). (d) Winter (1 November–7 March; Journey North, Monarch Larva Monitoring Project, Western Thanksgiving Monarch Count [inset], Monarch Butterfly Biosphere Reserve).

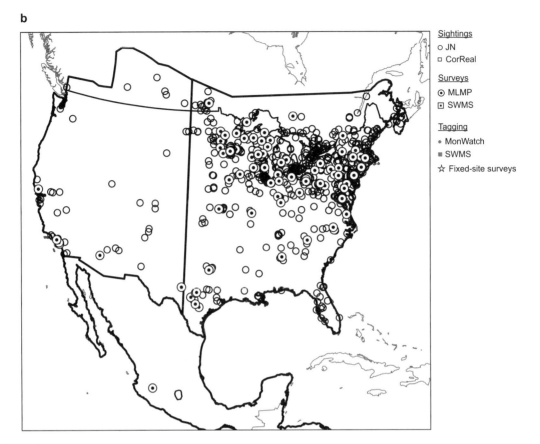

Figure 2.1b.

through Mexico, a passage of utmost importance that has received less attention from both citizen scientists and researchers than other portions of the migration. Findings are communicated to government officials, educators, and the media.

Monarch Watch

Monarch Watch, under the direction of Orley "Chip" Taylor, established a citizen-science tagging program in 1992 to address the dynamics of monarch migration, orientation, and navigation. Monarch Watch engages students, teachers, volunteers, and researchers in the cooperative study of the monarch's fall migration, with additional goals of promoting science education and monarch conservation. In 1997, Monarch Watch introduced an improved tagging system using circular, lightweight, and weatherproof tags with a strong adhesive; this system is now used by other tagging programs to provide localized data on migratory patterns. Monarch Watch is funded by the sale of tagging kits, educational materials, and promotional items, as well as donations and support from the Monarch Joint Venture.

Every fall, uniquely coded wing tags are issued to thousands of participants who tag 30,000–105,000 monarchs throughout the range of the eastern migratory monarch population (see Figure 2.1c for 2011 tagging locations). The recovery rate for tagged butterflies in Mexico ranges from 0.6 to 4.5% per year, depending on the size of the overwintering population and winter mortality. Information from the recovered tags can be used to estimate the origins, timing, sex ratio, and pattern of the migration, and to understand influences of weather events. Data from Monarch Watch's tagging program, along with new analyses of Urquhart's tagging data, have addressed influences of weather and other environmental factors that vary from year to year (Rogg et al. 1999), and regional migratory patterns (McCord and

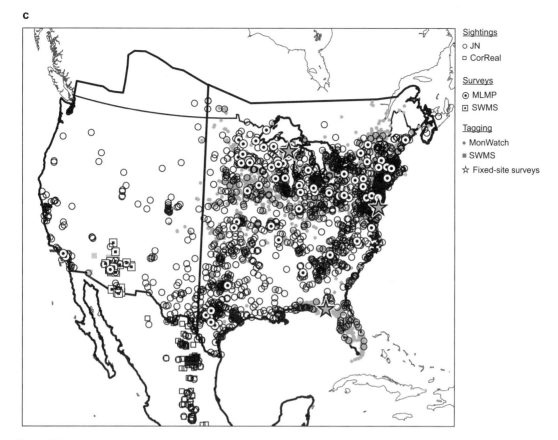

Figure 2.1c.

Davis 2010). General observations have been used in several papers (Brower et al. 1996, 2006, 2012b). Monarch Watch has produced a dataset with records of more than 1 million tagged butterflies and 16,000 recoveries (Monarch Watch 2013a). Nine years of recoveries were published in season summaries (1993–2001, Monarch Watch 2001) and the entire recovery dataset is available via a searchable database on the Monarch Watch website.

Southwest Monarch Study

The Southwest Monarch Study (SWMS), based in Arizona, was founded in 2003 by Chris Kline and is currently led by Gail Morris. Goals are to identify and describe migration and breeding patterns of monarchs in the southwestern United States and to encourage monarch conservation. While it is mainly a tagging program (with tagging sites shown in Figure 2.1c), SWMS volunteers also monitor milkweeds, nectar plants, and adult and immature monarchs in about 20 sites in Arizona. One focus of the SWMS is to determine the migratory destination of monarchs in the desert Southwest. To achieve this goal, both wild and "farmed" (commercially raised in California and purchased by the Desert Botanical Garden in Phoenix) monarchs are tagged and released. The project website distinguishes between wild and farmed monarchs to avoid confusing data interpretation (see Pyle, this volume, Chapter 21).

SWMS findings are posted on their website and have been shared in several presentations. A key finding of the SWMS is that monarchs in Arizona migrate to both Mexico and the coast of California during their fall migration (Morris 2012). Additionally, a small number of monarchs winter along the Salt River in Phoenix and the Colorado River in Yuma. Monarchs at the Rio Salado Habitat Restoration Area (in Phoenix) appear to be in diapause during the winter and can survive a hard freeze.

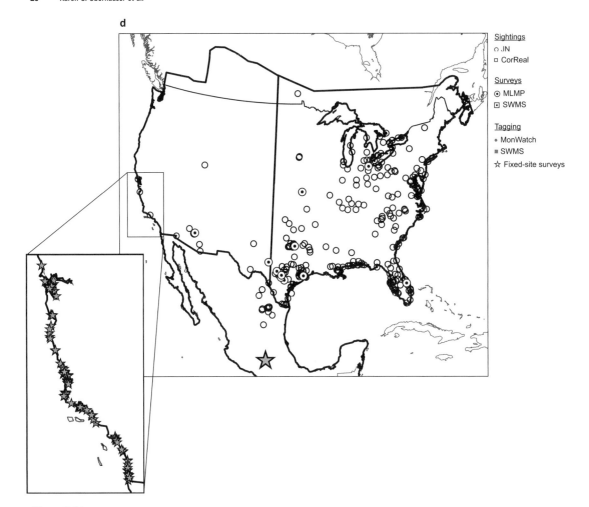

Figure 2.1d.

A small number of breeding monarchs are sometimes observed in Phoenix and Yuma during warm winters. Monarchs use three evergreen milkweeds (*Asclepias subulata*, *A. angustifolia*, and *A. linaria*, which grow throughout southern Arizona and in urban gardens) as fall host plants, and breeding and oviposition are common from September through mid-October. Despite the frequency of fall sightings, observations of monarchs in the spring are limited.

Monarch Monitoring Project

The Monarch Monitoring Project (MMP), affiliated with the Cape May Bird Observatory and New Jersey Audubon Society, was established in 1990 to focus on monarch fall migration along the Atlantic coast. MMP volunteers census monarchs moving through West Cape May and Cape May Point, New Jersey, a peninsula bordered by the Atlantic Ocean and Delaware Bay (see Figure 2.1c). They tag monarchs; conduct driving censuses along a standard route that traverses beach, dune, light residential, garden, grassland, marsh, and forest habitat throughout September and October; and conduct informational programs on monarch biology and tagging that are open to the public. Initiated by Dick Walton and Lincoln Brower, the MMP is managed by a field coordinator, communications director, and intern.

Scientists have used MMP annual count data to document fluctuating monarch numbers at Cape May (Walton and Brower 1996; Walton et al. 2005), and to compare trends at different fall monitoring sites (Gibbs et al. 2006). Data are available on the MMP and MonarchNet websites (Monarch Net

2012). Cape May is a destination for many people interested in nature viewing, and the MMP engages thousands of visitors with formal and informal presentations about monarch biology, migration, and conservation.

Fall monarch tagging at St. Marks National Wildlife Refuge

St. Marks National Wildlife Refuge (SMNWR) covers about 68,000 acres across parts of three counties in the Florida panhandle: Wakulla, Jefferson, and Taylor (Figure 2.1c). The SMNWR monarch tagging program was started in 1988 by Tonya Van Hook, who managed it from 1988 to 1993, and again in 1998 (there was no monitoring from 1993 to 1997). Richard RuBino oversaw the program from 1999 to 2003, and David Cook from 2004 to the present. Project goals are to monitor fall migrating monarchs, provide management recommendations to refuge staff, and provide experience-oriented environmental education opportunities (Van Hook 1990). While records of volunteer participation are not kept, somewhere between 100 and 200 volunteer days (one volunteer working on one day) are tallied each year. From 1988 to 2014, about 50,000 monarchs were counted, and about 30,000 tagged and released.

An important finding of the project is that very few monarchs migrating through this area successfully reach Mexico. Only 0.03% of the monarchs tagged at the Refuge from 1988 to 2011 (R. RuBino unpublished) were found in Mexico, an order of magnitude below the overall Monarch Watch return rate reported above. This has led to speculation that these monarchs may be flying to other locations, such as the Caribbean or Central American countries.

Peninsula Point Monarch Research Project

The Peninsula Point Monarch Research Project, started in 1994, is located on the northern shore of Lake Michigan at a migratory stopping point (see Figure 2.1c) in Hiawatha National Forest. It was founded by C. J. Meitner and Anne Okonek and has been led by C. J. Meitner, Gina Badgett, Pat Landry, Therese Fix, and Sue Jamison. Volunteers census roosting monarchs three times a day, traveling along a walking transect in early morning, late morning, and early afternoon from mid-August through mid-September. They also participate in monarch tagging (through Monarch Watch) and the Monarch Larva Monitoring Project (MLMP, described below), and conduct a variety of habitat improvement projects. The project has received funding from Wildlife Unlimited of Delta County, The Nature Conservancy, and the Superior Watershed Partnership.

Since the project began, 1646 censuses have been conducted. Peninsula Point monarch numbers are correlated with wind direction, temperature, and cloud cover but do not demonstrate a consistent timing of peak migration as was observed at Cape May (Meitner et al. 2004). Moreover, total annual migration count numbers at Peninsula Point do not correlate well with numbers seen at east coast flyway sites, indicating that the size of breeding monarch populations in the north central region might not match well to monarchs in the northeastern United States.

ALL TOGETHER NOW: OVERWINTERING MONARCH COUNTS

Western Monarch Thanksgiving Count

The Western Monarch Thanksgiving Count (WMTC) estimates total abundance at overwintering sites in California, documenting changes in monarch occupancy of individual sites and the overall status of monarchs in the western United States. Volunteers monitor forested groves in California (inset, Figure 2.1d) that have hosted or currently host seasonal monarch aggregations. Most of the sites contain non-native *Eucalyptus* trees, but many also contain native trees, such as Monterey cypress, Monterey pine, and western sycamore. Monitoring locations extend throughout much of Southern and Central California, ranging from San Francisco to Santa Monica, and include several high-profile sites that are frequently visited by the public as well as many smaller, less-visited sites on public and private land.

Citizen scientists have been monitoring western overwintering sites since the 1980s, but the WMTC did not formally begin until 1997. The project was proposed by Walt Sakai and other monarch scientists, initiated by the Monarch Program, and adopted by the Xerces Society in 2000. Mía Monroe, acting as a Xerces volunteer, coordinates volunteers and Dennis Frey manages the data, which are maintained by the Xerces Society (Xerces 2012b). Volunteers use a standardized protocol for estimating monarch abundance and receive training from Monarch Alert (see below) or the Xerces Society. A conservative estimate of the total number of people involved since the project began is 500; these individuals made 2011 site visits from 1997 to 2014.

WMTC data provide the only overall picture of the status of the western monarch population to date and have been used in scientific publications (Frey and Schaffner 2004; Stevens and Frey 2010), unpublished reports (Frey et al. 2003a; Stevens and Frey 2004) and legal documents (IELP 2012). Findings are also used by the media, in docent talks at overwintering sites, and to prioritize outreach to landowners who manage habitat with monarch aggregations. WMTC data include information on how monarchs historically and currently use overwintering sites and can thus guide conservation decision-making when projects with potential to impact overwintering sites are proposed.

Monarch Alert

Monarch Alert is also focused on monitoring overwintering monarchs in California. It was started in 2001 by Sarah Hamilton and Sarah Stock of the Ventana Wildlife Society, and Dennis Frey of California Polytechnic State University. The size of its volunteer program, extent of geographic focus, and research objectives have varied over the years. In its smallest incarnations the focus has been on population census and monitoring in Monterey County. At its largest it has addressed postwinter dispersal patterns through tagging studies, censuses, and monitoring of overwintering populations in San Luis Obispo and Monterey counties; microclimate attributes of trees and wintering groves; and population dynamics and connectivity between individual overwintering sites (e.g., Frey et al. 2003a; Frey and Schaffner 2004; Griffiths 2006).

Volunteer opportunities are currently directed at museum docents and students, who tag and sight monarchs to monitor within and between colony movement, and enter data. While professional researchers have conducted most of the censuses, citizen scientists contribute counts of overwintering populations for some locations and some years. These counts, and replication by Monarch Alert researchers, provide for calibrated census data. Such calibration and training for citizen scientists has allowed Monarch Alert to maintain a well-prepared volunteer force that makes a substantial contribution to the WMTC.

Monarch Butterfly Biosphere Reserve Monitoring

Since the early 1990s, CONANP (Comisión Nacional de Áreas Naturales Protegidas) personnel in the Monarch Butterfly Biosphere Reserve (MBBR, see Figure 2.1d) and staff of the World Wildlife Fund–Mexico (WWF-Mexico) have measured the areas occupied by monarchs at multiple overwintering sites in Mexico throughout the wintering season (Garcia-Serrano et al. 2004; Rendón-Salinas and Tavera-Alonso 2013). Beginning in 2004 and currently coordinated by Eduardo Rendón of WWF-Mexico, monitoring activities include biweekly measurements from November to March. Goals of MBBR monitoring include assessing the overall size of the eastern migratory population, and determining rates and causes of overwintering mortality. Because this is the only time that most of the eastern migratory population can be assessed together, the findings are used widely to describe monarch population trends (e.g., Brower et al. 2012a, 2012b; Journey North 2012a, Monarch Watch 2012; Pleasants and Oberhauser 2012; Rendón-Salinas and Tavera-Alonso 2013; Ries et al., this volume, Chapter 24).

This monitoring project is conducted within a protected preserve, thus limiting the ability of citizen scientists to participate. We include the effort here because it provides key data that allow tracking of population numbers from year to year. Further, while most MBBR monitoring is conducted by paid personnel, the efforts are aided by local residents. To carry out research within the Reserve, a formal request, establishing the protocol and goals for the project, must be submitted and approved. Local residents can take part in specific activities if the permits allow this participation, and a current program supported by the Monarch Butterfly Fund (MBF 2013) includes residents in monitoring activities that support those of WWF-Mexico.

AN EXPANDING POPULATION: MONARCH REPRODUCTION AND DISEASE

Traditionally, butterfly citizen science projects involve counting or observing adults and, less often, tracking movement. Recently, a new focus has emerged for citizen science programs to engage "super" volunteers in the collection of more process-oriented data (Dickinson et al. 2012). Monarch researchers have been pioneers in such efforts, with volunteers who collect data that delve deeply into the biology of the organisms they study. Here, we describe two programs that exemplify the development of super volunteers, the Monarch Larva Monitoring Project and Monarch Health. These projects collect data on larval development, parasitism, and disease. Other smaller projects involve local groups in studies of monarch development. For example, Richard RuBino and Ron Nelson led the Eden Monarch Fields Spring Migration Augmentation Project in Tallahassee, Florida. Volunteers raise monarchs with the primary goal of augmenting the migrating population and a secondary goal of studying the spring migration in the Tallahassee area. In the 12 years of this project, more than 1700 monarchs have been raised to maturity and tagged as part of the Monarch Watch Tagging Program.

Monarch Larva Monitoring Project

The Monarch Larva Monitoring Project (MLMP) was begun by Michelle Prysby and Karen Oberhauser at the University of Minnesota in 1996. Goals of the MLMP are to understand the factors that affect monarch reproduction and development during the breeding season, and to enhance volunteers' appreciation and concern for monarchs and their habitat. MLMP volunteers conduct up to six data collection activities. Every year, they provide (1) a site description and (2) an estimate of milkweed density. In addition to these annual activities, they (3) conduct weekly surveys of monarchs and milkweeds. Additional, optional activities include (4) comparing plants occupied by monarchs to random plants, (5) measuring rates of parasitism by parasitoids by collecting and rearing larvae, and (6) collecting rainfall data. Beginning in 2004, volunteers have also been able to report sightings of monarchs at locations other than regularly monitored sites. Volunteers choose their own monitoring sites, which include backyard gardens, abandoned fields, pastures, and restored prairies located throughout the monarch's breeding range.

Grants from the National Science Foundation have supported project dissemination, and engagement of students, teachers, and nonformal youth groups in the MLMP. Project staff compile newsletters to share MLMP findings and volunteer contributions; from 2000 to 2011 these newsletters were published annually in hard copy format and archived on the project website, and beginning in late 2011, monthly newsletters are e-mailed to participants and then archived.

From 1996 through fall 2014, over 1000 MLMP volunteers monitored over 900 sites, raised 18,158 monarchs for the parasitism study, and reported 960 other sightings (for site locations in 2011, see Figures 2.1 and 2.2).

The MLMP has produced a field guide (Rea et al. 2011) and peer-reviewed papers on spatial and temporal patterns in monarch densities (Prysby and Oberhauser 1999, 2004), predators and parasitoids (Prysby 2004; Oberhauser et al. 2007; Oberhauser 2012), potential impacts of climate change (Batalden et al. 2007) and genetically modified crops (Oberhauser et al. 2001; Pleasants and Oberhauser 2012), winter breeding by monarchs in the United States (Batalden and Oberhauser, this volume, Chapter 19), and conservation (Oberhauser and Prysby 2008) and educational (Kountoupes and Oberhauser 2008; Oberhauser and LeBuhn 2012) impacts of the MLMP. Lindsey et al. (2009) and Bartel et al. (2011) synthesized MLMP and MonarchHealth data to detect relationships between monarch density and parasite incidence. The program's website (MLMP 2013) includes real-time summaries of monarch densities and parasitoid incidence, which can be downloaded. Other data are made available on a case-by-case basis.

Monarch Health

Project MonarchHealth, started in 2006 by Sonia Altizer at the University of Georgia, engages citizen scientists in measuring the prevalence of a protozoan parasite of monarchs, *Ophryocystis elektroscirrha* (OE), throughout North America. Project goals are to track spatial and temporal variation in parasite infections, and to enhance awareness of the biology and impacts of monarch parasites. Volunteer citizen scientists sample monarchs (either adults captured from the wild or wild caterpillars they rear to adulthood) by swabbing their abdomens to collect parasite spores that are present on infected butterflies. They return swabs to the University of

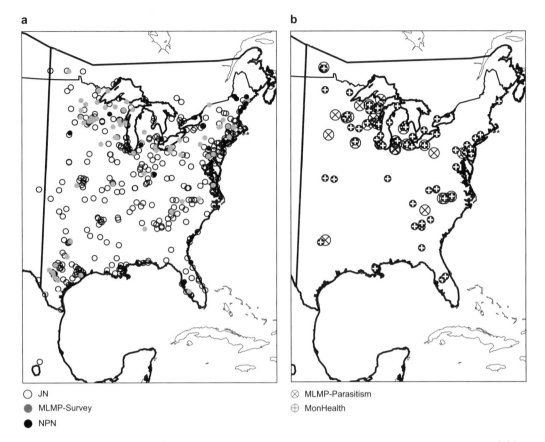

Figure 2.2. Multitrophic interactions (shown only in the eastern United States since few data from the West are available). (a) Host plant (emerging milkweed). (b) Natural enemies.

Georgia, where MonarchHealth scientists analyze the samples. Participants include families, adults, classrooms, and nature centers. They register electronically (by e-mailing monarch@uga.edu), and are mailed supplies needed to conduct sampling. Individual volunteers receive the results of their sampling efforts by mail, and the data are also posted on the project website. From 2006 to 2011, volunteers submitted 12,553 samples (for locations of sites in 2011, see Figure 2.2).

MonarchHealth data have shown that about 12% of monarchs collected by volunteers are infected with OE (Monarch Parasites 2012), but infection levels increase from early to late in the breeding season and then decline over the course of the migration (Bartel et al. 2011). Because breeding habitats are more heavily infected, migration may allow monarchs to flee areas with large numbers of parasites. Additionally, monarchs' long-distance movements might weed out infected individuals (Bartel et al. 2011), who may not be vigorous enough to travel the long distance to overwintering sites. More recently, initiatives of MonarchHealth have been launched in the southern United States (starting in 2011, to track infection levels in winter breeding monarchs at sites with tropical milkweed plants) and in the western United States (starting in 2013).

MONARCHS, AMONG OTHERS: GENERAL BUTTERFLY PROGRAMS

Many programs collect butterfly survey and sighting data. The fact that these programs are not focused on monarchs results in more reliable absence data, because participation is not driven by volunteer interest in monarchs. Absence data (information on when and where monarchs are not present) are important to understanding population dynamics. For example, communication with volunteers in the MLMP suggests that they are less likely to monitor when they feel they will not observe monarchs (K. Oberhauser, pers. observ.), and this may be true for other monarch-specific programs as well. Three main types of programs collect data on butterfly populations: counts, transects, and opportunistic sightings. Below, we describe groups that organize each of these program types (see Table 2.1 and Figure 2.3), and highlight results pertaining to monarchs. These programs have received little attention from the scientific community until recently, and have thus resulted in few published papers to date.

The North American Butterfly Association's Count Program

The longest-running butterfly monitoring program is run by the North American Butterfly Association (NABA), directed by Jeffrey Glassberg. This program, modeled after the National Audubon Society's Christmas Bird Counts, was started by the Xerces Society in 1975 (Swengel 1990) and adopted by NABA in 1992. It provides information about butterfly distributions and population sizes, promotes interactions among butterfly enthusiasts, and encourages interest in butterflies and their conservation. Each volunteer count coordinator establishes a fixed, 15-mile-diameter search area (called

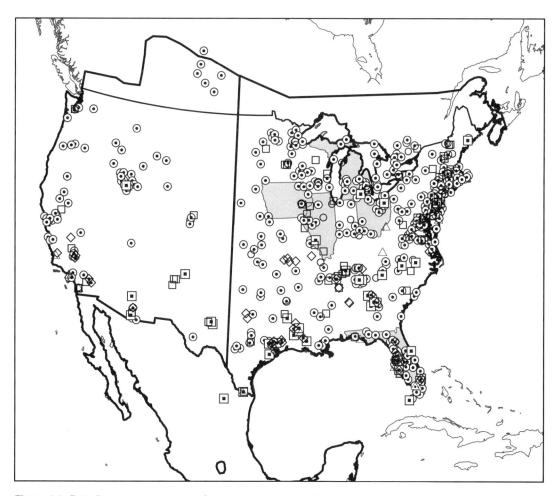

Figure 2.3. Butterfly monitoring programs. Observational monarch sightings are shown by an open symbol and come from BAMONA and both NABA sighting programs; NABA Count Surveys have a center point. Season is indicated by shape: triangle (winter), diamond (spring), circle (summer), or square (fall). Regional state-based butterfly monitoring programs (shaded) run from 10 to ~200 transects, depending on state. Massachusetts has a field trip program (shading not visible) that reports surveys of hundreds of nonfixed sites each year.

a count circle) and recruits volunteers to census all butterflies sighted within the circle on a single day. These counts are held throughout the summer depending on the optimal timing for each location. In 2008, NABA began encouraging count coordinators to conduct three counts per year, one each in the spring, fall, and summer (although most counts are still conducted only once per year).

Currently, about 450–475 counts are conducted annually (see Figure 2.3 for locations). Multiple locations are visited within a count circle on a given day, but even the largest groups cannot survey an entire circle. Groups report all adult butterflies observed and the number of party-hours (an hour spent by a group at a given site, regardless of the group size) and party-miles (a mile traveled on foot by each party) spent on the survey. NABA data were used for one of the first published accounts of year to year fluctuations in monarch numbers (Swengel 1995) and by Ries et al. (this volume, Chapter 24) to document monarch population dynamics.

Butterfly Monitoring Networks (Transect Programs)

Butterfly Monitoring Networks (BMNs) assess the abundance of butterflies in specific survey locations (Figure 2.3). These programs involve repeated measures each year along fixed transects, and during set observation periods. Compared with count programs, they use stricter protocols, adapted from those proposed by Pollard (1977), and thus provide more rigorous data on butterfly habitat use and population dynamics within and across years. All volunteers are trained in survey protocols and identification.

The first U.S. BMN began in 1987 in Illinois with support from The Nature Conservancy. The Illinois network currently monitors more than 100 sites. The program is run from the Peggy Notebaert Nature Museum in Chicago by Doug Taron, who also provides support and assistance to other BMN programs throughout the United States. The Ohio Lepidopterists' Society began conducting a similar program in 1995; Ohio BMN data were used to show the effects of climate on yearly monarch population growth in Ohio (Zipkin et al. 2012). Other regional programs are based in Florida (started in 2003, coordinated by Jaret Daniels, University of Florida), Iowa (started in the 2007, coordinated by Nathan Brockman, Iowa State University), Michigan (started in 2010, coordinated by Monika Egerer, Kalamazoo Nature Center), and Cascades National Park (started in 2011, coordinated by Regina Rochefort, National Park Service). Several other states are currently organizing their own BMNs, including Tennessee, Colorado, and southern California.

Opportunistic data collection

Three programs currently collect butterfly observations that are not part of a fixed protocol. These programs run the gamut from websites that allow people to log random observations, to organized field trips where all butterflies observed are recorded (so absences can be assumed). In addition to the programs described below, general interest nature observation programs, like iNaturalist, have also started to accumulate monarch occurrence data. While none of these programs currently constitute a major data source on monarch distributions, if the growth of bird citizen-science programs can be used as an indicator, their potential importance should not be downplayed. Currently, the main entry port for bird observation data, eBird, logs up to 3 million occurrence records per month (Kelling et al. 2013). NABA runs two national programs to collect observation data. The larger of these is a program called "Butterflies I've Seen." This program allows users to record single observations or all sightings from field trips and track their life lists. It was launched in 2001 and by 2014 had recorded data from more than 18,000 separate field trips. Another NABA program is strictly for individual sightings, with a focus on alerting members to the presence or emergence of species of interest. The "Recent Sightings" program was started in 2000 and had logged nearly 162,000 records by 2014. A similar program is run by another NABA chapter, the Southern Wisconsin Butterfly Association. This program, launched in 2005, allows users to record multiple species observed at one site along with notes and photos. In 2011, there were 1277 count postings (Wisconsin Butterflies 2012).

One of the most active opportunistic programs is run by the Massachusetts Butterfly Club, a NABA chapter that manages its program and data locally. The club organizes field trips that are similar to NABA counts, except the locations are not fixed and they are not necessarily repeated from year to year. Since 1992 approximately 9000 trips have been organized, so the available data are quite dense. A recent analysis used these data to identify species showing declines or increases near their northern and southern range limits (Breed et al. 2012). Monarchs, while trending down, showed no significant pattern.

The Butterfly and Moth Information Network, directed by Kelly Lotts and Thomas Naberhaus, hosts another growing program for sightings data called Butterflies and Moths of North America (BAMONA 2012). This web-based project originally presented presence records based on historical and current records submitted by regional coordinators. Along with distribution records, the website offers information and photos of each species and is one of the most popular butterfly sites, based on traffic and Google search results. Starting in 2005, a separate sightings database allows citizen scientists to submit photographs of butterflies and moths to the website (Figure 2.3). Quality control is provided by volunteer lepidopterists who serve as regional coordinators. Standardized data are accessible through the website via checklists, species profiles, maps displaying point occurrence data, and individual record details. Quality control is high, with the trade-off that the protocol requiring photo confirmation of each record may limit the number of observations ultimately flowing into this site.

The newest large-scale project for observations is eButterfly (eButterfly 2013), run from the Montréal Insectarium and the University of Ottawa. Contributors report sightings and upload photos of butterflies (both current and historical records) in North America using a checklist system. Despite its short history (it began in 2012), the project had already logged nearly 110,000 new observations and will have about 300,000 historical observation records in 2015. The website also offers a portal for entering new sightings. As with BAMONA, experts review all submitted records. The program provided strong documentation of the unusual spring migrations of monarchs, painted ladies, and red admirals in spring 2012. The focus on migrations could provide important information regarding monarch distributions under changing climate regimes.

DATA GAPS AND RECOMMENDATIONS FOR VOLUNTEER RECRUITMENT

A tremendous amount of data stream in from the programs detailed above, but some areas still lack data for certain stages. To identify data gaps, we focused on biological dynamics during the four phases of monarchs' annual cycle, which roughly correspond to the four seasons: overwintering, spring migration and breeding, summer expansion and breeding, and fall migration. We have used occurrence data from the programs described above to define monarch ranges, combining these ranges with our knowledge of monarch phenology to determine gaps that should be filled in each season to provide a full picture of monarchs' complex spatiotemporal population dynamics, as well as conservation threats and challenges. To determine where and when monarch populations are or are not well covered by existing programs, we overlaid monitoring locations on these maps, presented separately for each season in Figure 2.1. Our goals are to illustrate both the remarkable amount of data available and gaps that could be filled by establishment of new programs or increased recruitment for existing programs. In many cases, these data gaps probably reflect less dense concentrations of humans or differing cultural and educational priorities among regions of North America, as well as lower monarch population densities, making recruitment difficult.

An obvious conclusion is that monarch monitoring is much more dense and active in the eastern than the western United States during all seasons (Figure 2.1), likely caused by a combination of sparser human populations in the West and lower overall monarch densities, making it harder to engage volunteers. However, conservation of the western population would benefit from more data regarding monarch habitat use and numbers. We have little knowledge of pathways taken as the western population recolonizes its breeding ground, whether movement occurs from Mexico into the western United States, or even where key breeding areas are located. As a start, we recommend targeted recruitment of individuals reporting spring sightings (via Journey North and Correo Real) in northwestern Mexico and in all regions in the western United States. More western MLMP observers would provide valuable information on use of breeding habitat in this region, and help to inform conservation efforts focused on breeding habitat.

Journey North data have good spring coverage in the eastern United States (Figure 2.1a), although there are few reports from potentially important regions in North and South Dakota, the far northeastern United States, and northwestern Minnesota. Additional spring observers in northeastern Mexico (recruited through Correo Real) would help us to better understand key spring migration habitats as monarchs move northward. Broader distribution of MLMP observers in the spring would provide valuable information on the start of oviposition and development, monarch use of milkweed resources, and pressures from parasitoids in the south. Recent research suggests that the success of the spring generation produced by migrants from Mexico is critical to summer population growth (Zipkin et al. 2012), so more spring data from the South would allow us to understand the factors that drive successful breeding during this critical stage. Ideally, recruitment efforts would include northern Mexico, for which we have little understanding of the importance of spring breeding.

Summer and fall coverage is strong in the Upper Midwest and Northeast (Figures 2.1b and c) through the MLMP, Monarch Watch, and Journey North. More MLMP volunteers in the midcentral and southern regions would help to document whether monarchs actually do vacate these regions in the heat of the summer, as suggested by Prysby and Oberhauser (2004), and the dynamics of reproduction as monarchs are moving north in the early summer and south in the fall (Batalden and Oberhauser, this volume, Chapter 19).

In the winter, sampling adult presence and reproduction in the southern United States and northern Mexico will help us determine the degree to which monarchs use locations other than their traditional wintering grounds. There is fairly good eastern coverage from Journey North in the winter (Figure 2.1d). Additional MLMP volunteers could provide more detailed information about the degree to which winter reproduction is increasing or associated with environmental factors such as the presence of the non-native *A. curassavica* (Batalden and Oberhauser, this volume, Chapter 19). Given the potential importance of monarch movement through the desert Southwest (e.g., Pyle 2001; Brower and Pyle 2004; Pyle, this volume, Chapter 21), additional SWMS, Journey North, and MLMP volunteers in this region would be beneficial. The WMTC has fairly good coverage of western wintering sites, although the sites at the north and south ends of the wintering range tend to have less regular monitoring, probably because it is difficult to motivate volunteers to visit sites with fewer monarchs (Frey and Schaffner 2004). Additionally, while WWF-Mexico monitors throughout the winter in Mexico, there is less repeated monitoring of California sites. Because movement between sites appears to be an important part of western winter dynamics (Villablanca 2012a), programs that monitor throughout the season would be valuable.

Both Journey North and MLMP volunteers report milkweed emergence dates, and these reports, combined with some data from the National Phenology Network (NPN), provide fairly extensive coverage of the eastern United States (but not the West, Figure 2.2). Continuing to encourage volunteers to report milkweed phenology will be important in the face of global climate change and potential monarch range changes (Batalden et al. 2007). Data on monarch parasitoids (through the MLMP) and OE (through MonarchHealth) are clumped around the Great Lakes and the East Coast, and it would be useful to recruit more broadly for these programs.

General butterfly projects like BAMONA and NABA sample broadly and can help us locate data gaps (Figure 2.3) and potentially recruit new volunteers. For example, BAMONA sightings in southwestern Canada informed our range map (Figure 2.1); we know that there are monarchs in southern Alberta, but little monarch-specific monitoring. We encourage monarch researchers to take advantage of these valuable data and volunteer sources.

CONSERVATION IMPACTS

Many organizations that coordinate monarch citizen science projects have explicit conservation goals. For example, Monarch Watch's Monarch Waystation and Bring Back the Monarchs programs engage volunteers in activities that promote monarch habitat conservation; Xerces and Monarch Watch are working to make milkweed available for habitat restoration projects; and Xerces, Monarch Watch, and Monarchs in the Classroom (which runs the MLMP) are part of the Monarch Joint Venture (Shahani et al., this volume, Chapter 3). Here, we focus on the conservation impacts of the citizen science programs themselves; these impacts led Oberhauser and Prysby (2008) to call monarch citizen scientists a "research army for conservation."

An obvious conservation value of all programs stems from their ability to collect thousands of observations throughout monarchs' annual life cycle; these data can be used in formal analyses to advance conservation science on monarchs and inform specific actions. Data collected by monarch citizen scientists have identified main migration pathways and critical times of passage (Howard and Davis 2004), weather events or human activities that affect the population (Stevens and Frey 2010; Howard and Davis 2012; Pleasants and Oberhauser 2012), potential impacts of climate change on monarch migration and range (Batalden et al. 2007), and habitat types that are important to monarchs during the fall migration (Davis et al. 2012a).

Citizen science data have also been used to influence specific policies or actions. On a small scale, many volunteers engage in stewardship activities on the land they monitor, basing management decisions on their observations of what constitutes quality monarch habitat. For example, MLMP volunteers often plant milkweed or nectar plants at their monitoring sites, and they manage their land in more environmentally friendly ways, such as changing mowing regimes to avoid monarch presence and promote milkweed growth. On larger scales, WMTC data are used to inform management that promotes long-term use of wintering sites by monarchs and to try to stop harmful development (e.g., Villablanca 2012b). Correo Real data have illustrated the importance of the state of Coahuila on monarchs' migratory route, and resulted in the publication of a State Ministry of the Environment document that recommends suspending mosquito spraying during the migration, and alerting drivers to the importance of slowing down to avoid hitting monarchs.

Data from the fall monitoring project on the Chincoteague National Wildlife Refuge informed a plan that included habitat enhancement and management to benefit monarchs. Tragically, the primary dune that provided nectar plants and roosting shrubs on Assateague Island was wiped out by Hurricane Sandy in fall 2012. Denise Gibbs provided this account of the destruction to the Monarch Watch e-mail discussion list: "In the 50 years I have been coming here, that dune stood firm through every hurricane. It is now just wide expanses of flat sand right up to deadened edge of the maritime forest. No interdune area, no transition zones; just a lot of dead and broken loblolly pine trees or their remaining stumps." Without the data provided by Gibbs and her colleagues (2006), we would have no understanding of the impact of the hurricane on this important monarch habitat.

In addition to providing data with conservation value, monarch monitoring engages citizens in actions that affect their understanding of the importance of conservation and, we argue, fosters connections between participants and local, national, and international natural communities. Many monarch citizen scientists engage in education activities with children, and many programs are school-based (Journey North, MonarchHealth, Correo Real) or include programming for adults working with children (MLMP, Monarch Watch). Monarch citizen scientists also share their findings and knowledge about monarchs in a variety of settings, such as local presentations and with news media and elected officials. The expertise developed through volunteers' long-term involvement builds confidence and thus encourages outreach activities. Finally, many volunteers engage in environmental advocacy, often as a result of a specific experience, such as losing a monitoring site to development. Such experiences increase volunteers' awareness of habitat loss, and encourage advocacy to prevent further loss.

Citizen science programs can support volunteers' conservation actions in many ways. We can explicitly support outreach and education activities by providing materials or other support, such as tips on working with news media. We can support volunteers in their training efforts by providing materials and advertising. Finally, many volunteers ask for our help in their communications with individuals and organizations that have control over monarch habitat. For example, they ask for information about potential impacts of wind turbines or broad-scale pesticide applications to control insect pests. The degree to which we can engage citizens and provide them with tools to foster monarch conservation would greatly expand the benefits of these programs beyond simply helping us to collect data.

CONCLUSION

Every year, tens of thousands of volunteer hours are invested in monitoring monarchs, and the value of this investment is incalculable. Thanks to monarch citizen scientists, we are better equipped to face the challenge of ensuring that monarchs and their incredible migratory phenomenon are around to fascinate future generations of children, scientists, and citizens. As references noted above illustrate, the data generated in these projects provide ways to answer questions that could never be addressed through traditional academic research programs.

The assortment of monarch citizen science projects and their widespread use in scientific publications illustrate the value of both very simple reporting procedures, such as reporting a sighting of a single monarch to Journey North, and more involved ones, such as weekly site monitoring that requires reporting the stages and numbers of immature monarchs, blooming plants, and weather conditions to MLMP. This impressive range of programs provides opportunities to engage citizens at all levels of involvement.

ACKNOWLEDGMENTS

We thank the thousands of volunteer citizen scientists throughout North America who have contributed so much to our understanding of monarch biology. As leaders of these programs, we feel incredible gratitude and awe toward these individuals. We thank the past and current funding sources, all noted in the text, for supporting these programs and promoting their conservation, education, and basic research outcomes; and Priya Shahani and Scott Black for their helpful comments on a previous version of the manuscript.

3

Monarch Habitat Conservation across North America

Past Progress and Future Needs

Priya C. Shahani, Guadalupe del Río Pesado, Phil Schappert, and Eligio García Serrano

Preserving monarchs and their remarkable migration requires protecting habitats crucial for monarch breeding, migrating, and overwintering across North America. In Mexico, the Monarch Butterfly Biosphere Reserve (MBBR) was created to protect the overwintering sites, and the Monarch Fund (Fondo Monarca) compensates landowners for conservation measures taken on these lands. Twenty-five different NGOs work on conservation and community issues in the MBBR, and most of these collaborate through the Monarch Network. In the United States, 14 organizations collaborate through the Monarch Joint Venture to plant milkweed and nectar plants for monarchs, restore breeding habitats, and support citizen science projects to better understand monarch conservation needs. In Canada, legal protections and on-the-ground habitat improvement projects to conserve breeding habitat are achieved through collaboration between agencies and engaged volunteers. Future work should focus on identifying the links between monarch population size and habitat changes, and predicting shifts in monarch habitat needs in response to environmental change.

INTRODUCTION

The remarkable migration of North American monarchs relies on a wide range of habitats in Mexico, the United States, and Canada. Conserving monarch migration requires that each phase of their annual cycle of breeding, migrating, and overwintering receives conservation attention. Historically, the relatively small size of the monarchs' wintering sites in Mexico and California and human encroachment on these sites has made their protection of paramount concern (CEC 2008). In Mexico, monarch wintering sites are threatened by commercial and subsistence-scale timber harvesting (Ramírez et al., this volume, Chapter 13); in California, real estate development with lack of integrated protection and habitat management threatens the persistence of monarch wintering colonies along the coast (Jepsen and Black, this volume, Chapter 12). More recently, the proliferation of herbicide-tolerant crops and the suburbanization of agricultural land in the United States have caused the loss of milkweed and nectar plants required for monarch reproduction and migration (CEC 2008; Pleasants and Oberhauser 2012; Pleasants, this volume, Chapter 14). Prior work using stable isotopes to infer monarch natal origins showed that the U.S. Upper Midwest was the major source of the monarchs overwintering in Mexico (Wassenaar and Hobson 1998), and between-year variation in the numbers of overwintering monarchs in Mexico is predicted by the abundance of monarch caterpillars in the Midwest (Pleasants and Oberhauser 2012). These findings, combined with recent declines in overall monarch population sizes (Brower et al. 2011), underscore that habitat protection in monarch wintering sites alone is inadequate to protect the spectacular North American migration from intensifying human pressures.

Other threats monarchs face in their breeding range and along migratory routes include non-target impacts of insecticide use; frequent mowing regimes in roadsides, parks, and other lands; and contemporary climate change that can intensify severe weather events (floods, droughts) and might shift the location of suitable breeding habitats (Oberhauser and Peterson 2003; Batalden et al. 2007; Stevens and Frey 2010; Brower et al., this volume, Chapter 10; Ramírez et al., this volume, Chapter 13). The planting and proliferation of invasive species is an issue as well. As one example, monarchs will lay eggs on two non-native plant species, *Cynanchum louiseae* and *C. rossicum*, on which their larvae cannot feed and develop (Casagrande and Dacey 2007). These plants serve as sinks for monarchs, and their spread is of concern. In Canada, milkweeds are listed as noxious weeds, preventing their planting (and causing their eradication) in some places (Schappert 1996). In the United States, milkweed species are not included on the primary noxious weed list of any state (National Plant Board 2013), but some states include milkweed species on secondary lists that allow their listing at the county level. Conserving the monarch migration will require efforts in Mexico, the United States, and Canada to ensure that (1) sufficient suitable habitat is available on the monarchs' wintering grounds in California (U.S.) and Mexico; and (2) sufficient breeding and migration habitat is available in all three countries to maintain, and ideally enhance, North American monarch recruitment and survival. In this chapter, we first detail the legal status and conservation recognition afforded to monarchs internationally. We then describe legal protections and habitat conservation efforts under way in each North American country. In reviewing these efforts, we note the very different approaches taken by each country to address their unique habitat challenges. Multiple chapters in this volume, including those by Brower et al., Brower and Fink, and Ramírez et al. provide more detailed accounts of specific threats to monarchs, and yet others, including those by Borders and Lee-Mäder, Baum and Mueller, and Jepsen and Black, highlight specific effects underway aimed at monarch habitat conservation. Our aim here is to provide a broad summary of the variety of efforts undertaken to protect and conserve monarchs and their habitats across North America.

MONARCH PROTECTION EFFORTS SUPPORTED BY INTERNATIONAL ORGANIZATIONS AND COLLABORATION

In 1983, shortly after the scientific discovery of monarch wintering sites in Mexico (Urquhart 1976), concern over the loss of monarch overwintering habitat motivated the International Union for Conservation of Nature and Natural Resources (IUCN) to designate protected areas for winter roost sites in Mexico and California (Wells et al. 1983). This action was taken to protect the monarchs' migration rather than to save the species per se, and thus was the first designation of protected areas for a biological phenomenon in the history of international conservation. In other words, the action recognized that the migratory phenomenon was imperiled, even if the species as a whole was not in imminent danger of extinction.

In 1995, Canada and Mexico signed a declaration to create an International Network of Monarch Butterfly Reserves and pledged to jointly expand this network. In Mexico, the five sites included in the Monarch Butterfly Special Biosphere Reserve were also included in this network. Canada designated three migratory stopover sites in Southern Ontario, where large numbers of monarchs aggregate each fall prior to crossing the Great Lakes (Long Point National Wildlife Area, Prince Edward Point National Wildlife Area, and Point Pelee National Park).

In March 2006, the Trilateral Committee of Wildlife and Ecosystem Conservation and Management (headed by the directors of the Canadian Wildlife Service [CWS], the U.S. Fish and Wildlife Service [USFWS], and the Ministry of Environment and Natural Resources of Mexico) established a network of monarch conservation areas in each North American country, where monarch habitat conservation, research, monitoring, and environmental education would be prioritized. This "Sister Protected Area" network (see map in CEC 2008) includes 13 protected areas administered by the USFWS, US National Park Service (USNPS), CWS, Parks Canada Agency (PCA), and Mexico's National Commission of Natural Protected Areas (CONANP). While no new protections for monarchs were conferred by these designations, specific actions on each site range from outreach activities to habitat restoration and protection, and the creation of the network demonstrated an international awareness of and commitment to preservation of the monarch, its habitats and its migration.

Finally, in 2008, the Commission for Environmental Cooperation (CEC), an organization formed to support cooperation among the North American Free Trade Agreement (NAFTA) partners to address environmental issues of continental concern, commissioned Karen Oberhauser to lead an effort by monarch scientists, agency staff, and other individuals to author the North American Monarch Conservation Plan (NAMCP; CEC 2008). This document summarizes monarch biology and conservation efforts, and, based on the input of 29 experts, prioritizes recommended actions to promote monarch conservation. It has served as a stimulus and guide for monarch habitat conservation since its publication.

MONARCH CONSERVATION IN MEXICO: FROM FORESTS TO THE SURROUNDING COMMUNITY

Legal status and protections in Mexico

The forested sites where monarchs winter in Mexico were first protected by the Mexican federal government in 1980 (DOF 1980). This decree designated all monarch overwintering areas as wildlife refuges, protected indefinitely from all uses, but because no boundaries were defined for protected areas (SEMARNAT 2001), the action was difficult to understand and enforce. After requests from federal agencies, scientists, and environmental organizations for more specific protections (Brower et al. 2002), President Miguel de la Madrid issued a second decree in 1986 establishing the Monarch Butterfly Special Biosphere Reserve (DOF 1986). This reserve included five discrete conservation areas totaling 16,110 hectares, with the goal of protecting all known overwintering sites (Calvert and Brower 1986). In each site, a core area in which no logging would be allowed and a buffer zone in which controlled logging could occur were established (Missrie 2004). Design of the 1986 reserve was based on ecological information available at the time regarding the spatial extent of monarch overwintering colonies, their movement patterns through the season, and the importance of lower watershed areas (Missrie 2004). The decree also limited use of the forests by their legal owners (*ejidos*, indigenous communities, and private landowners), who criticized the resource limitations placed on use of their lands (without compensation for loss), and the top-down process by which the restrictions were imposed (Honey-Rosés et al. 2009).

In 1997, the National Ecology Institute (INE), tasked with reviewing the social and ecological efficacy of the 1986 decree, asked the World Wildlife Fund Mexico (WWF) to propose new reserve boundaries based on ecological data (Brower et al. 2002; Honey-Roses et al. 2009). The Ministry of the Environment used the WWF proposal to negotiate new core and buffer zones with the property owners (Missrie 2004; Honey-Rosés et al. 2009). In exchange for the additional restrictions on use of their land, property owners were promised compensation for meeting conservation goals (Honey-Rosés et al. 2009), and the Monarch Fund (Fondo Monarca) was established to provide this compensation (see next section).

In 2000, based on these negotiations, a third presidential decree was issued to create the current Monarch Butterfly Biosphere Reserve (MBBR), with the protected area increased to 56,259 ha, with 13,551 ha in the core and 42,707 ha in the buffer zone (DOF 2000). This larger reserve boundary protects a contiguous area of forest rather than five separate patches, with the same stipulations prohibiting resource extraction in the core, and limiting extraction in the buffer zone (Missrie 2004).

A number of smaller colonies outside the MBBR have varying degrees of protected status. These areas are administered by CONANP, but with few protective actions specifically directed at monarchs. The Forestry Commissions of the states of Michoacán and Mexico support conservation programs and actions in these locations by providing technical assistance and subsidies for conservation actions (CEC 2008), especially the annual program of reforestation and forest fire prevention.

In response to growing interest among local communities in supporting tourism at the monarch wintering sites, CONANP developed the National Strategy for Sustainable Tourism in Protected Areas in 2007. In the MBBR, the strategy focuses on controlling the harmful impacts of tourism through planning, monitoring, and regulatory activities; promoting sustainable tourism by supporting infrastructure, such as more appropriate foot paths;

and improving the knowledge base of individuals involved with tourism. Additionally, CONANP promotes year-round tourist activities that focus on the ecology and landscapes of the MBBR. The World Wildlife Fund-Telcel Alliance is working with CONANP and the *ejido* of El Rosario to develop land use and tourism business plans, and to support more sustainable tourism (CEC 2008).

Monarchs are also listed "under special protection" in the Species at Risk Norm (NOM-059-SEMARNAT-2001), and are thus considered a species that could be threatened and whose recovery and conservation should be promoted wherever it is found.

Forest conservation and restoration in Mexico

In 2000, an endowment fund of US$6.5 million was established to ensure conservation of the core area of the MBBR. This endowment led to the creation of the Monarch Fund, an initiative of WWF and Fondo Mexicano para la Conservación de la Naturaleza, A.C. (FMCN). Support for the Monarch Fund was also provided by the Packard Foundation, the Mexican Ministry of Environment and Natural Resources (SEMARNAP), and the governments of the states of México and Michoacán. In 2012, the endowment fund reached close to US$7.3 million, following contributions from the state of México. Interest earnings from investment of the endowment are provided to landowners who have signed an agreement to protect the forest in exchange for financial incentives. Two types of economic incentives are offered: first, *ejidos* (communities that collectively own land parcels), indigenous communities, and private property owners whose forestry use permits in the core area were modified by the decree are eligible to receive US$18 per year for each cubic meter of wood not harvested. Second, landowners without usage permits are eligible to receive payments of US$12 per hectare of conservation-related services. These payments are made in exchange for a commitment to conserve the core area and work in conjunction with the managing entity of the MBBR to ensure its protection. During a first phase of support (2000–2009), the Monarch Fund supported forest owners with payments totaling approximately US$2.1 million. More recently, the National Forestry Commission of Mexico (Comisión Nacional Forestal or CONAFOR) is providing matching funds to facilitate this work (García-Serrano 2012), and has committed resources to the initiative's second phase (2009–2018). Together, the Monarch Fund and CONAFOR have earmarked about US$5 million for direct disbursement to 38 landowners. For every dollar that CONAFOR contributes per hectare of preserved forest, the Monarch Fund contributes US$1.21.

The Monarch Fund has successfully secured the commitment of national and foreign civil organizations, federal and state agencies and *ejido*, community, and private landowners. While not all landowners eligible to participate in the forfeiture of logging rights and receipt of conservation payments have chosen to do so, most have, and the 2000 decree (in conjunction with the Monarch Fund) successfully reduced the intensity of logging in the MBBR and engaged community members in conservation activities. A recent assessment of the success of this effort found that disturbance to the forest canopy in the MBBR has been reduced by 11.6% since the 2000 decree (Honey-Rosés et al. 2011), and in 2011–2012, for the first time since the sites were protected, no forest loss to illegal logging was detected (WWF 2012a).

Several Mexican nongovernment organizations (NGOs) also support monarch conservation through habitat monitoring and restoration. For example, WWF has been involved in colony monitoring, forest management, community restoration, eco-tourism, and environmental education programs. La Cruz Habitat Protection Project supports the planting of pine and oyamel fir trees in the area of monarch overwintering habitat. Alternare A.C. supports local communities in and near the MBBR by promoting a variety of sustainable practices, including farming, building construction and reforestation. Similar activities are conducted in the state of Mexico by Fundación Nacional para la Conservación del Hábitat Boscoso de la Mariposa Monarca. Biocenosis' Monarca program focuses on promoting conservation of threatened species and habitats, general ecosystem conservation and management, and social monitoring. Hombre y Alas de Conservación and Gestión Ambiental y Proyectos para el Desarrollo Sustentable Monarca, NGOs based in Zitacuaro, support local communities in the MBBR through projects that include land use plans, forest management programs, sustainable development, and environmental restoration (CEC 2008).

Improving the lives and environment of people who live near monarchs

To illustrate the social and conservation activities focused on helping communities near the monarch wintering sites in Mexico, many of which are led by NGOs, we highlight the work of Alternare, as a case study, and describe a collaborative effort through "Red Monarca" (Monarch Network). Alternare was established in 1998 to promote a comprehensive community development strategy through which ecosystems within the MBBR could be restored while improving the quality of life of its human inhabitants. Alternare aims to change local attitudes toward natural resource use and management, working with community members who sign-up for their programs voluntarily. This model is based on a local instructor program, in which community members are recruited to train fellow *campesinos*, thus replicating and perpetuating the model. These hands-on trainings teach sustainable practices such as nursery plant production, vegetable gardening, and construction of chicken coops, cisterns, fuel-efficient stoves, and adobe structures. Adoption of these sustainable practices reduces the intensity of forest resource extraction, thus alleviating pressures on monarch overwintering habitat. After instructors complete the training program, their skills are assessed and certified by the Ministry of Agriculture, Livestock, Rural Development, Fisheries and Food (SAGARPA).

Alternare has established a community training center in the municipality of Aporo, Michoacán, very close to the MBBR. The center is a key element in Alternare's long-term efforts to assist *campesinos* in achieving self-sufficiency and sustainability, and in establishing permanence of these efforts in the MBBR. It can host 58 individuals, and includes classrooms, a library, computers with Internet access, a kitchen, and land for fieldwork and hands-on training. At this center, Alternare carries out its model of recovering traditional knowledge, sharing modern practices, and then drawing from both of these to design sustainable practices by which communities can make use of their natural resources.

Alternare has grown steadily over the years, from 29 families and 5 working groups in 1998 to more than 600 families and 33 groups by 2012. Through a focus on social issues and education, Alternare has achieved impressive conservation successes, including erosion prevention in land plots and rainwater retention through construction of ditches and terraces; reforestation through tree production in collective forest nurseries; decreased use of wood for cooking by 50% through promotion of fuel-efficient stoves; and decreased logging for local home building due to the use of adobe as an alternative. At the same time, participants have seen a twofold increase in basic crops (corn and beans); production of organic vegetables for family consumption, with surpluses for sale in some cases; enhanced fruit growing and production of feed for domestic animals; savings in purchasing of fertilizers, vegetables, animals, eggs, and grains; and the creation of rural enterprises and productive projects. Thus, Alternare promotes a higher standard of living for communities while simultaneously conserving monarch overwintering forests.

Alternare co-initiated Red Monarca (Monarch Network), an organization comprising 18 organizations that work in the MBBR to address human and conservation needs. These organizations have joined efforts to support communities through regional sustainable development and natural resource conservation activities, increasing participation, efficiency, and collaboration. They established a strategic framework that defined their mission and vision, as well as communication schemes, a code of ethics, and general organizational principles. The Network has facilitated information sharing through development of a project matrix which helped to identify areas of overlap and gaps. A spatially explicit database using Geographic Information System (GIS) tools helps identify and keep track of all the different projects.

The NGOs that are part of the Monarch Network presented their achievements and the GIS database to the United Nations Educational, Scientific and Cultural Organization (UNESCO) mission in Mexico in January 2011. Organized jointly with the IUCN and the MBBR's Secretariat for Natural and World Heritage Sites, this visit was carried out to evaluate the Mexican government's work in this National Protected Area. Collectively, the Monarch Network represents an opportunity to build a long-lasting relationship between NGOs and the MBBR. For example, NGOs in the Network participated in the process to update the MBBR's Management Plan.

International support for conservation in Mexico

Because the overwintering sites are so crucial to the survival of monarch migration in North America, there is strong international support for their conservation. The U.S. Fish and Wildlife Service (USFWS) Wildlife Without Borders program has partnered with Mexican authorities and organizations to protect and restore monarch wintering habitat in Mexico. Between 1995 and 2012, the USFWS awarded almost US$800,000 in grants for monarch projects. About 94% of the funds supported efforts to develop the capacity of local communities in the MBBR to sustainably manage their natural resources. USFWS partnered with Mexican authorities to support Alternare and funded programs to provide training in reforestation techniques for farmers living in the MBBR (CEC 2008).

Since 1993, the U.S. Forest Service (USFS)-International Programs has worked with MBBR managers and other partners in the region to build management capacity, provide guidance to communities for resource management, and conserve natural resources in the MBBR core zone. Staff from the Willamette National Forest and other units have provided training and consultations on forest inventory, GPS/GIS utilization, and design and maintenance of trails. A U.S. nonprofit organization, the Monarch Butterfly Fund, raises funds to support a variety of conservation efforts in the Mexican overwintering sites. These efforts include reforestation, forest monitoring, environmental education for tour guides in the Reserve and K-12 students living in nearby communities, and support for Mexican organizations such as Alternare.

MONARCH CONSERVATION IN THE UNITED STATES: DIVERSE STRATEGIES TO COUNTER HABITAT LOSS

Legal status and protections

Neither monarchs nor their habitats hold special legal status at the federal level in the United States. Monarchs are included in the State Wildlife Action Plans of California, Kansas, Rhode Island, Washington, and the District of Columbia. While this does not provide legal protection, it does open greater potential for habitat improvement projects by state land management agencies. In California, State Assembly Bill 1671 (passed in 1987) provided state support for the conservation of monarch overwintering habitat (Bell et al. 1993). In 1988, voters approved a bond initiative providing $2 million to acquire lands containing monarch aggregation sites. Monarch wintering habitats are afforded some legal protections in California via a patchwork of city ordinances, coastal zone management plans, and state law (IELP 2012; Jepsen and Black, this volume, Chapter 12).

Conservation of breeding and stopover habitats in the United States: A joint effort

Monarch habitat conservation is a significant and growing priority across the United States, involving numerous federal and state agencies, NGOs, and private citizens, visible both through a coordinated, collaborative partnership-based effort known as the Monarch Joint Venture, and through efforts of many individual programs. The North American Monarch Conservation Plan (NAMCP; CEC 2008) outlined steps needed to conserve monarchs in North America and highlighted the critical need for a trinational conservation effort. This plan provided a solid scientific foundation upon which monarch conservation efforts could be built in the United States, and it thus caught the attention of USFS personnel working on migratory species conservation. The "joint venture" model, used by the USFWS for bird habitat conservation efforts, was identified as an ideal approach for conserving migratory species. Through this model, agencies, tribes, NGOs, and corporations are invited to work together to conserve migratory species that depend on multiple habitats across large landscapes and management jurisdictions. This approach makes it possible to apply the technical expertise and financial and personnel resources of a variety of organizations to large-scale habitat management challenges. The Monarch Joint Venture (MJV) was formed in December 2008 and now has 22 partner organizations whose work is described below.

Framing the MJV as a U.S. counterpart to Mexico's Monarch Network would fail to acknowledge the Monarch Network's crucial role in enabling navigation of the complex human relationships involved in monarch conservation work in the MBBR. Still, there are significant parallels between these two associations. Each has developed bylaws and strategic plans to guide their work, and each serves as a forum for organizations working on monarch conservation to share information and develop collaborations. The MJV is taking a three-pronged approach to monarch conservation, focusing on habitat conservation, education and outreach, and research and monitoring. This chapter focuses only on habitat conservation efforts, but a list of all projects can be found online (Monarch Joint Venture 2013).

To address monarch habitat needs in the United States, the MJV is pursuing four strategies: (1) monarch breeding and migration habitat restoration and enhancement; (2) milkweed resource development for habitat enhancement on public and private lands; (3) overwintering site inventory, assessment, and habitat enhancement in California; and (4) provision of tools and guidelines to inform U.S. monarch conservation efforts. As is true of most biological conservation problems, habitat loss and degradation pose serious threats to monarch breeding (Schappert 2004; CEC 2008; Brower et al. 2011; Pleasants and Oberhauser 2012). While degraded habitat can be restored or improved, habitat transformation for urban, suburban, and commercial development is often more permanent and extreme. Fortunately, monarchs can utilize a diversity of breeding habitat types, provided that milkweed is present. To support their migration, they need nectar plants and roost trees to use as stopover points. Here we highlight current monarch habitat conservation projects in the United States, as well as promising areas for future growth.

The U.S. Forest Service had already implemented habitat projects that increase milkweed and nectar plant availability on more than 340,000 acres in 2010 (USFS 2011), and another 277,000 acres in 2011 (USFS 2012). The largest of the projects, a 340,000 acre restoration of longleaf pine-bluestem habitat in the Ouachita National Forest (Arkansas and Oklahoma), was designed in consultation with monarch experts to ensure the planting of regionally appropriate milkweed and nectar plants (Orley Taylor, pers. comm.). Other habitat improvement activities have included planting pollinator gardens at district offices and schools; planting nectar plants and milkweed along trails; implementing mowing, burning, thinning, and harvesting regimes designed to restore ecosystem structure and species composition on National Forest lands (with resultant increases in abundance of milkweed and key nectar plants); propagating milkweed for seed increase work; including milkweed in habitat restoration plantings; and seeding utility right-of-ways with native plants, including milkweed (USFS 2011, 2012). The inclusion of milkweed and nectar plants for monarchs is promoted as an easy addition to habitat restoration and pollinator conservation projects, resulting in routine adoption of these practices (Holtrop 2010; Larry Stritch, pers. comm.).

Six National Wildlife Refuges (NWRs), managed by USFWS, are included in the trinational Sister Protected Area (SPA) Network. USFWS personnel at each NWR site set priorities for monarch conservation activities, and projects thus vary from site to site. Two good examples of habitat conservation work within the SPA network come from the Flint Hills and Marais des Cygnes NWRs in Kansas. Between these two refuges, more than 6,000 acres are either maintained or restored to retain and create breeding and migration habitat for monarchs (Tim Menard, pers. comm.). Old fields are maintained in open condition through prescribed burns and mechanical removal of shrubs, preserving plant communities that support monarchs. Tallgrass prairie restorations have been underway since 1990, including planting of five milkweed species and a number of monarch nectar plants. Hydrological work at wetlands in these NWRs maintains 2500 acres of nectar plants that bloom during the fall monarch migration.

Species-focused conservation efforts on state lands generally do not include insects in their purview. With current funding limitations, habitat projects focusing specifically on monarchs on state-managed lands are rare. The MJV promotes the inclusion of milkweeds and nectar plants in habitat restoration projects as a cost-effective approach to creating vast acreage of quality monarch breeding and migration habitat. An excellent example of such work comes from the state of Iowa, where the Department of Natural Resources (Iowa DNR) reconstructs from 1500 to 2700 acres of tallgrass prairie annually on state land. The Iowa DNR partnered with the MJV to purchase milkweed seeds that were included in more than 2000 acres of tallgrass prairie plantings in 2012, and more than 2200 acres in 2013. Over time, this acreage adds up; since 2000, the Iowa DNR has planted over 24,600 acres of prairie habitat, with an estimated 63% of these acres including plants that could benefit monarchs (William Johnson, pers. comm.).

U.S. Farm Bill conservation programs provide a number of opportunities to improve habitat for pollinators on agricultural lands. Milkweeds are nectar-rich plants that benefit many bee species and other butterflies as well as monarchs. The Xerces Society and the USDA Natural Resources Conservation Service (NRCS) are working with agricultural landowners to install large-scale pollinator habitat plantings across the United States and promoting milkweed planting in their trainings and technical land management guidance.

Many acres of land in the United States are managed by utility companies as right-of-way (ROW) for power lines and pipelines, with 300,000 linear kilometers of electric utility ROWs spanning the United States alone (Wojcik and Buchmann 2012). These lands represent a key opportunity for monarch conservation, both because of the acreage included and because of their potential as corridors for wildlife movement (including monarchs). Utility ROWs are managed to maintain open conditions, thus providing habitat for shade-intolerant wildflowers. The Pollinator Partnership is leading efforts to include monarch habitat conservation in utility corridors; initial work is underway on land managed by an electrical company in North Carolina (Pollinator Partnership 2012). The Pollinator Partnership is also developing a handbook on managing ROWs for monarchs and partnering with utility companies to increase adoption of monarch-friendly habitat management practices (Laurie Davies Adams, pers. comm.).

Significant habitat potential also exists at sites managed by corporations across the United States. Many corporations are members of the Wildlife Habitat Council (WHC), a nonprofit organization that encourages construction of wildlife habitat on corporate lands. The WHC and Pollinator Partnership are creating a monarch habitat program through which corporations will be encouraged to install monarch-friendly plants and to use butterfly-friendly land management practices. Many WHC corporate partners are already creating monarch breeding habitat on the lands that they manage (Corrine Lackner Stephens, pers. comm.).

The Monarch Waystation program, launched by Monarch Watch in 2005, encourages individuals and organizations to create breeding habitat for monarchs in whatever size possible. Large or small, field or garden, all these sites provide habitat for monarchs. Monarch Watch provides guidelines regarding species to plant, and how to manage the habitat to best support monarchs. As of October 2014, more than 9000 registered Monarch Waystations have been established across the United States, at nature centers, parks, butterfly houses, and private residences (Monarch Watch 2013b).

Any discussion of monarch habitat conservation in the United States would be incomplete without touching on the tremendous work of private citizens. Many are members of monarch-focused listservs, so they can learn more and connect with other monarch enthusiasts. On these lists, one can find countless stories of the habitat conservation and outreach efforts of these individuals.

A lack of access to native milkweed plant materials is often cited on communication forums as a limitation to monarch habitat planting efforts. Commercial sources of affordable native milkweed seeds and plants are needed both for large-scale monarch habitat improvement projects and for gardens of those who value locally sourced, native plants. Monarch Watch has built a grassroots campaign to gather milkweed seeds from across the United States to bolster the availability of locally sourced milkweed seeds and plants for gardens and restoration projects. Volunteers ship seeds of prioritized milkweed species to Monarch Watch, and these seeds are coded to track their origin. To preserve the genetic integrity of native plant populations, Monarch Watch pairs these seeds, or plants grown from them, with habitat restoration projects and native plant sales in the Bailey's Level 1 Ecoregion from which seeds were sourced (Bailey 1995). The Xerces Society, NRCS, and native plant producers are increasing the commercial availability of milkweed seeds in several high-priority regions of the United States, especially the southern tier of states where the first wave of monarchs leaving Mexico breed (Borders and Lee-Mäder, this volume, Chapter 16).

Protecting monarch wintering sites in California

Conservation of monarch overwintering habitat in coastal California saw its beginnings in the 1960s and 1970s, when citizens lobbied for protection of properties when such locations were listed for sale. This grassroots advocacy led to a number of land acquisitions by state, county, and city parks programs, and by conservation organizations that protected a number of monarch mass aggregation sites from development (Mía Monroe, pers. comm.). The 1988 California bond measure (Proposition 70, Parks and Wildlife Bond) allocated funds to purchase monarch overwintering sites when public agency funds were not adequate to do so (see above), and a number of additional overwintering sites were subsequently purchased and protected from development. However, the bond measure did not designate funds for conservation planning or management, resulting in limited management to maintain habitat condition (Monroe, pers. comm.). Some sites changed ownership, and new landowners were not always aware of monarch use of these sites, resulting in some damage to and loss of overwintering sites.

The acquisition and management of overwintering sites in California were decentralized and remain so. Recent efforts by The Xerces Society to inventory and assess the condition of monarch aggregation sites, develop targeted management recommendations, and compile relevant laws and policies will be instrumental in guiding a strategic program to conserve monarch overwintering habitat in the western United States (Jepsen and Black, this volume, Chapter 12). Monarch overwintering habitat management projects are also under way at several privately and publicly owned sites, usually in consultation with one of several monarch habitat specialists and often involving work by private citizens and dedicated volunteers.

In addition to the efforts described above, MJV partner organizations are developing a number of resources to facilitate monarch habitat conservation efforts. These tools and materials include regional milkweed species recommendations, many web resources, regionally focused booklets that provide profiles of native milkweed species and address concerns about milkweed toxicity and spread, nectar plant recommendations, and habitat management guidelines.

In sum, the MJV has sought to build a strategic program to move monarch conservation forward in the United States. The MJV partnership is a long-term effort, and its goals are reevaluated on an annual basis to capture newly accessible and emerging opportunities. Over time, the MJV hopes to achieve the habitat improvement needed within the United States to support abundant monarchs into perpetuity, simultaneously increasing habitat available for a wide variety of additional plant, pollinator, and wildlife species.

MONARCH CONSERVATION IN CANADA: PROTECTING MILKWEED AND HABITATS FOR FUTURE GENERATIONS

Legal status and protections in Canada

The monarch was designated a species of Special Concern in Canada in April 1997. This designation, which means that monarchs are considered potentially threatened or vulnerable to extirpation or extinction, was reconfirmed in November 2001 and in April 2010 (COSEWIC 2010). Of the 47 species assessed by the Committee on the Status of Endangered Wildlife in Canada (COSEWIC) as of November 2012, a total of 15 species have been given this federal designation. The Species at Risk Act (SARA), enacted in 2003, protects wildlife (including monarchs) and their critical habitat on Crown Lands (federally owned and managed) across Canada. Working together with provinces and territories under the Accord for the Protection of Species at Risk, the Canadian Endangered Species Conservation Council provides national leadership for the protection of species at risk as recommended by COSEWIC.

A federal-level management plan for monarchs has been written and submitted to the Canadian Wildlife Service (Environment Canada) and is undergoing internal review as of early 2014. Active regional monarch conservation programs vary with provincial legislative and regulatory status, public interest, and the availability of NGO partner organizations. Alberta, Manitoba, Ontario, and New Brunswick currently have legislation to protect listed Lepidoptera species, while other provinces have listed butterfly species with no legislative protection. Monarchs are specifically protected only in Ontario (see below), although both Saskatchewan and British Columbia give them S3B status (special concern, vulnerable to extirpation or extinction) based on the federal rank of Special Concern.

Monarchs are among 14 butterfly species listed as "specially protected invertebrates" under Ontario's 1997 Fish and Wildlife Conservation Act (OFWCA). The OFWCA provides protection for all wildlife on Crown Lands and prohibits collecting, propagating, keeping in captivity, and buying, selling, or exchanging specimens of listed species. The Ontario

Endangered Species Act (ESA), enacted in 2007 and amended in 2009, provides complete protection, including prohibiting habitat damage, for wildlife species that are listed as extirpated, endangered, or threatened. While the ESA includes three butterfly species (all considered extirpated within the province), it lists monarchs only as a species of Special Concern. This status affords recognition, but not protection by the ESA.

Habitat conservation efforts in Canada

Monarch habitat conservation efforts in Canada function through a combination of agency, NGO, and volunteer efforts. Most provinces are, at a minimum, tracking the status of monarch populations (e.g., the Natural Heritage Information Centre in Ontario and the Atlantic Canada Conservation Data Centre in the Maritime provinces), under the umbrella of SARA if no specific provincial initiative yet exists. Of particular note was the federal designation of Monarch Protected Areas in Ontario (Point Pelee National Park, already protected under the Canada National Parks Act, and the Long Point and Prince Edward Point National Wildlife Areas) in a 1995 international agreement with Mexico. As noted above, two of these reserves, Point Pelee and Long Point, were included as member reserves of the Trilateral Monarch Butterfly Sister Protected Area Network under the trinational Trilateral Committee for Wildlife and Ecosystem Conservation and Management; however, there is little documentation of monarch habitat conservation projects at these sites.

Nongovernmental conservation efforts are carried out by national and international organizations (e.g., World Wildlife Fund Canada, Nature Canada, eButterfly, NatureServe Canada) through education, monitoring, and citizen science programs. Most of the grassroots work is accomplished by regional and local nature clubs, stewardship groups, and cooperative ventures responsible for planting and maintaining butterfly gardens, promoting organic farming, reducing pesticide use, and promoting general pollinator conservation. Extensive work is also conducted by knowledgeable volunteers and volunteer organizations that tag monarchs to track their migration, monitor and report on the occurrence and breeding success of monarch butterflies, propagate and plant milkweeds in gardens and wildlife reserves, pressure governments to enact endangered species legislation, and lobby for changes in laws and regulations that lead to reduced or eliminated herbicide and pesticide use.

One outstanding example of these grassroots efforts is the work of the Southwest Nova Biosphere Reserve (designated by UNESCO in 2001) Community Outreach Team that—along with the participation, funding, and logistical support of the Canadian Wildlife Federation, Friends of Keji (Kejimkujic National Park), the Mersey Tobeatic Research Institute, and Parks Canada—has been successfully producing milkweeds and planting monarch butterfly gardens with native milkweeds throughout southwestern Nova Scotia.

Of ongoing concern, however, is the fact that three provinces (Manitoba, Quebec, and Nova Scotia) list at least some milkweed species as "noxious weeds," subject to removal upon discovery (COSEWIC 2010). These weed control legislative instruments are often dismissed as "of no consequence" and "rarely, if ever, enforced," but until they are amended or changed to remove the monarch host plant, they could pose impediments to monarch conservation efforts. In May 2014, Ontario removed *Asclepias* from the Schedule of Noxious Weeds.

TAKING STOCK OF PROGRESS AND LOOKING TO THE FUTURE

The difficulty in accurately measuring monarch populations throughout their complicated migratory life cycle, and the challenges of understanding and predicting year-to-year variations in monarch density (Ries et al., this volume, Chapter 24) make it difficult to link monarch numbers to habitats and conditions at any given location or time of year. Still, the cumulative long-term and strategic efforts underway to preserve and create habitat for monarchs throughout North America are certain to benefit both the eastern and western monarch populations. It is our hope that these efforts are like milkweed seeds themselves; that they will continue to disperse, growing new conservation efforts for monarchs and other species as they reach new fields.

Conservation of a migratory species that relies on habitat across numerous political and landownership boundaries is by no means straightforward. It requires the effort and financial commitment of many different agencies and organizations, and the number of organizations and passionate individuals involved in monarch conservation efforts across North America is notable. Given the conservation challenges facing monarchs and the clear evidence that both the eastern and western migratory populations are declining (see Part 4 of this volume), it is vitally important that we mobilize as many people as possible, and that our efforts are carefully planned to maximize their impacts. We need to understand the links between monarch population size and habitat changes, and potential shifts in monarch habitat needs in response to environmental change. It is also crucial to engage in legislative efforts at state and province as well as federal levels; the attention to environmental services provided by pollinators is one on which we can capitalize, since creating habitat for monarchs will benefit a large suite of other pollinators (and vice versa). Through our collective efforts we hope that monarchs will indeed grow in abundance, so that their migrations can be appreciated by many generations yet to come.

ACKNOWLEDGMENTS

We thank the indigenous communities and *ejidos*, and Mónica Missrie, for their support of Alternare, as well as several foundations and government institutions. We also thank the U.S. Forest Service–International Programs for support of the Monarch Joint Venture; MJV partners for their ardent efforts to protect the monarch migration and build the MJV; and the many scientists and land managers who shared their knowledge regarding habitat management regimes under various land management jurisdictions in the United States. Finally, we thank Mace Vaughan, Karen Oberhauser, and Sonia Altizer for comments on earlier versions of this chapter.

PART II

Monarchs as Herbivores, Prey, and Hosts

An Overview

JACOBUS C. DE ROODE

Monarchs are perhaps best known for their spectacular fall migration, during which millions of individuals migrate from North America to overwinter in Mexico each year (Urquhart and Urquhart 1978). Often overlooked is the fact that only a select few monarchs that set out on their journey south successfully reach Mexico. These are the monarchs that found enough food and avoided predation, avoided or survived parasitism, and survived the toxins ingested from their milkweed host plants during larval development. Like all other organisms, monarchs do not live in isolation, but they are part of a complex network of species with which they can form close relationships. These species include the milkweeds (*Asclepias* spp.) used as larval food, the predators that attack or consume them, and the parasites that infect them. As the following four chapters show, these interacting species have major consequences for monarchs.

MONARCHS AS HERBIVORES

As specialist herbivores, monarchs form a tight association with their milkweed host plants (Ackery and Vane-Wright 1984). There are more than 100 species of milkweeds in North America alone (Woodson 1954), and monarchs are known to use more than 30 of these in the wild. Milkweeds are named after the milky latex exuded from the leaves and stems of many species following herbivore or mechanical damage. The production of latex forms a potent defense against generalist herbivores; the sticky consistency can glue up herbivore mouth parts or mire small animals, while the high concentrations of cardiac glycosides and cysteine proteases are toxic to many herbivores (Agrawal and Konno 2009b). Cardiac glycosides, also known as cardenolides, also occur in the foliage, roots, and nectar of milkweeds, albeit at lower concentrations (Agrawal et al. 2012). These chemicals interfere with the sodium-potassium channels in the cell membranes of many animals and thereby disrupt cellular functions (Malcolm 1991).

Like other milkweed specialists, monarchs have evolved a range of mechanisms to circumvent milkweed defenses. For example, larvae bite trenches in milkweed leaves and nick leaf stems to disrupt the flow of latex (Zalucki and Malcolm 1999), and they possess sodium-potassium channels that are resistant to the action of cardenolides (Dobler et al. 2012). Monarchs can sequester milkweed cardenolides, making them unpalatable to vertebrate predators (Brower et al. 1968; Malcolm 1991). The use of high-cardenolide milkweeds also protects monarchs against their protozoan parasites (Lefèvre et al. 2010; Altizer and de Roode, this volume, Chapter 7). Although monarchs utilize milkweeds as their larval food, they are not perfectly immune: first instar larvae are often found mired in latex, and they may die or experience slowed growth owing to high levels of cardenolides (Zalucki and Malcolm 1999; Agrawal 2005).

Milkweeds have clearly left their mark on monarch ecology and evolution. But as Agrawal and colleagues explain in their chapter in this section, the reverse may be true as well. A common pattern across plant taxonomic groups is that as plant species diversify, they evolve greater defenses against herbivores, but this is not the case with milkweeds. In fact, as milkweeds diversified over evolutionary time, they became less defended, having evolved lower amounts of latex exudation and lower concentrations and diversity of cardenolides (Agrawal and Fishbein 2008; Agrawal et al. 2009c). Agrawal et al. (this volume, Chapter 4) hypothesize that because milkweeds are used mostly by specialist herbivores, the maintenance of defenses that can be hijacked by these herbivores could backfire, giving protection to the herbivores against their own natural enemies. Using 53 species of milkweeds, Agrawal et al. also confirm their previous findings that more derived (recently evolved) species have evolved lower defenses. Moreover, when rearing monarch larvae for 5 days on these 53 milkweed species, the authors found that monarchs gain greater biomass from more derived species. This greater performance was due largely to declines in latex exudation. Agrawal et al. also found that more derived milkweed species have greater regrowth ability than less derived species (see also Agrawal and Fishbein 2008). Thus, it is possible that milkweeds that are less chemically defended compensate for this with greater tolerance to herbivore damage. It appears that monarchs, together with other milkweed specialists, may have driven the evolution of lower resistance and greater tolerance in their milkweed host plants.

MONARCHS AS PREY

Monarchs are able to sequester high levels of cardenolides, even when their milkweed host plants have low to moderate concentrations (Malcolm and Brower 1989). Despite this sequestration, however, monarchs still suffer high mortality from predation; it has been previously estimated that fewer than 10% of monarch eggs ultimately develop to adults (Prysby 2004), and two chapters in this section suggest that predation may be even more severe than that.

De Anda and Oberhauser (this volume, Chapter 5) present data from an observational study of monarch eggs and larvae on *A. syriaca* plants in Minnesota. Carefully keeping track of individual eggs and larvae, the authors calculated that the daily survival rates of eggs, first instars, and second instars were 63, 60, and 53%, respectively. Extrapolating from these numbers suggests that survival from egg to third instar larva is less than 2%. Although the authors did not observe a high number of predation events directly, a considerable proportion of eggs were observed to have been sucked (15%) or chewed (8.4%), indicating predation. Moreover, monarch survival was significantly reduced on plants that contained spiders, high levels of herbivory, and high numbers of aphids. Spiders were observed eating monarchs, high levels of herbivory may attract predators due to plant volatile releases (Oberhauser et al., this volume, Chapter 6), and high aphid densities are associated with aphid-tending ants, which also prey on monarchs (Prysby 2004).

Monarch predation is not confined to the early instars. In their chapter, Oberhauser and colleagues show that paper wasps, parasitoid wasps, and tachinid flies cause considerable mortality of pre-pupae and pupae. For example, the tachinid fly *Lespesia archippivora* is well known to infect large proportions of monarchs, up to 90% in some milkweed stands, and up to 30% prevalence at regional scales (Oberhauser 2012). Since the vast majority of monarchs infected by tachinids die, these data suggest that parasitoids can kill a considerable fraction of monarchs at regional scales.

Overall, the high mortality that monarchs experience from predators and parasitoids likely plays an important role in regulating monarch population densities. Similarly, monarchs can influence the population dynamics of their predators and parasitoids. For example, even though it is often assumed that the tachinid fly *L. archippivora* is a generalist parasitoid, Oberhauser (2012) found that tachinid fly prevalence in the upper Midwest correlates strongly with the density of monarchs the previous year.

MONARCHS AS HOSTS

Beyond their contribution to population regulation, natural enemies can also affect other aspects of monarch biology. In their chapter, Altizer and De Roode show that the protozoan parasite *Ophryocystis elektroscirrha* (OE) interacts with monarch migration in important ways. In particular, heavily infected monarchs suffer reduced flight ability (Bradley and Altizer 2005), and are less likely to successfully reach the Mexican overwintering sites (Bartel et al. 2011). The relationship between migration success and infection can lead to two major outcomes. On the one hand, because parasite infection carries greater costs for migratory than nonmigratory monarchs, migratory monarchs may have evolved greater levels of resistance or tolerance to their parasites (Altizer et al. 2011). On the other hand, monarch migration could be an effective mechanism to keep parasite prevalence low, and it is an interesting possibility that parasites have played a role in the selection and maintenance of monarch migration (Altizer and de Roode, this volume, Chapter 7).

The monarch-protozoan interaction is also mediated by host plant chemistry, and monarchs show evidence of fascinating self-medication behavior in response to infection. In particular, when infected monarchs are reared on milkweed species with high concentrations and diversity of cardenolides, they suffer lower parasite infection, growth, and virulence (de Roode et al. 2008a; Lefèvre et al. 2010; Sternberg et al. 2012). Moreover, when given a choice between *A. curassavica* and *A. incarnata*, infected monarchs preferentially lay their eggs on the more toxic *A. curassavica*, which reduces parasite infection and growth in their offspring (Lefèvre et al. 2010, 2012). Thus, parasites affect monarch oviposition behavior, and it is likely they have influenced the evolution of monarch host plant use (Altizer and de Roode, this volume, Chapter 7).

OUTLOOK

The interwoven ecology of milkweeds, monarchs, and their natural enemies has likely been a driving force in the evolution of each species. As the next four chapters suggest, monarch oviposition and migration, parasitoid population dynamics, and evolutionary declines in milkweed defenses can all be better understood by studying these species' interactions.

Despite constant new discoveries in monarch biology, many crucial questions remain. Although it is clear that predators cause monarch mortality, we do not yet know the relative importance of predators (top-down), milkweeds (bottom-up), and their interaction in regulating monarch population dynamics. Thus, it remains to be seen whether mortality rates from predators are similarly high on milkweed species that have greater concentrations of cardenolides. Similarly, although it is clear that milkweeds have evolved lower chemical defenses over evolutionary time, the exact role of monarch herbivory in this process is still poorly understood (Agrawal et al., this volume, Chapter 4).

Perhaps the most challenging future questions will center on how interactions in the entire food web affect monarch ecology and evolution. As community ecologists have long realized, species can affect each other directly (e.g., through predation) or indirectly, through interactions with other third species (Werner and Peacor 2003). Indirect effects can be major drivers of ecological communities and are also likely to play an important role in monarch biology. For example, monarchs are more likely to die on milkweeds with high densities of aphids (De Anda and Oberhauser, this volume, Chapter 5), not because aphids eat monarchs, but because high aphid densities attract aphid-tending ants, which subsequently consume monarch larvae (Prysby 2004). Similarly, the presence of aphids on *A. curassavica* increases the growth and virulence of OE in infected monarchs; this effect occurs because aphid presence lowers the concentrations of cardenolides in milkweed plants, leading to increased parasite growth (de Roode et al. 2011b). Finally, associations between milkweed roots and mycorrhizal fungi can change the cardenolide concentrations of milkweed foliage shoots, which may have major consequences for the performance of monarchs and their herbivores (Vannette and Hunter 2011b).

As these few examples already demonstrate, the interactions between monarchs and their natural enemies are affected not just by the milkweeds on which they feed, but also by other species that interact with these same milkweeds; thus, studies of indirect effects remain an important future frontier for understanding monarch ecology and evolution.

4

Macroevolutionary Trends in the Defense of Milkweeds against Monarchs

Latex, Cardenolides, and Tolerance of Herbivory

Anurag A. Agrawal, Jared G. Ali, Sergio Rasmann, and Mark Fishbein

Theory predicts that plants will increase their defenses against herbivores over time; however, several milkweed defenses have declined as the genus *Asclepias* has diversified. We review the evolutionary history of milkweed defense and provide new findings from an experiment on 53 *Asclepias* species. As predicted, monarch performance increased on more recently evolved species, and latex was primarily responsible for this effect. We next conducted focused analyses of eight *Asclepias* species, spanning early diverging and recently evolved species. We hypothesized that a decline in latex and cardenolides may be favored if (1) chemical defenses lose effectiveness due to sequestration, and (2) milkweeds evolve tolerance to herbivory in place of resistance traits. Monarch sequestration did not decline on derived *Asclepias* because they concentrated cardenolides from mid- to low cardenolide plants; nonetheless, derived milkweed species were more tolerant of herbivory because of enhanced investment in roots and the potential for clonal growth. The costs of producing latex and cardenolides, costs of sequestration for monarchs, and overall herbivore attack rates likely contributed to these macroevolutionary patterns.

INTRODUCTION

The availability of molecular phylogenies is transforming our ability to address classic questions in chemical ecology and coevolution (Wink 2003; Futuyma and Agrawal 2009; Kursar et al. 2009). In particular, some of our most long-standing coevolutionary hypotheses relate to the determinants of insect herbivore host shifts (Dethier 1941; Becerra 1997; Murphy and Feeny 2006), relationships between plant defense traits and diversification in plants and insects (Ehrlich and Raven 1964; Farrell et al. 1991; Agrawal et al. 2009a), and tradeoffs and synergies among plant defenses (Feeny 1976; Rudgers et al. 2004; Agrawal and Fishbein 2006). Many of the original hypotheses were based in comparative biology, and yet only recently has a phylogenetically informed comparative approach been advocated and implemented (Agrawal 2007). Because milkweeds and monarchs have been so important in the development of the research fields of chemical ecology and coevolution (Malcolm 1995), and a recent molecular phylogeny of *Asclepias* has been produced (Fishbein et al. 2011), the time is ripe to summarize what we know about the evolutionary history of milkweeds (Figure 4.1) as it relates to monarch biology.

PHYLOGENETIC PATTERNS

Our laboratories have been studying the defenses and evolutionary history of New World milkweeds (*Asclepias* spp.) for the past two decades. Several interrelated phylogenetic trends can be summarized briefly as follows: as American *Asclepias* has diversified, there have been consistent increases or declines in particular trait values (Figure 4.2), with a decelerating rate toward the present (Agrawal et al. 2009a). These directional trends have been particularly apparent in the defensive traits of milkweeds,

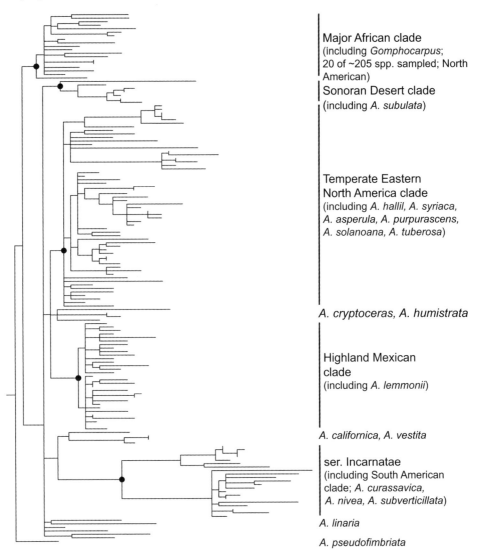

Figure 4.1. Maximum likelihood phylogeny of *Asclepias* (after Fishbein et al. 2011), in which branch lengths are proportional to expected substitutions per site in noncoding plastid DNA (introns and intergenic spacers at 3 loci). Major clades are indicated by large dots on the phylogeny and with names on the right, along with representative species discussed in the text. Other species names have been omitted for clarity. The earliest dichotomy represents the split between African *A. pseudofimbriata* and all other African and American species. The second dichotomy separates all other sampled African species (which have been classified as *Asclepias* and as many as 20 other genera) and all American *Asclepias*, which form a single clade. Relationships among the major American clades are not well supported by these data, although a single origin of all South American species from within series Incarnatae is strongly supported.

but not in other vegetative and physiological traits unrelated to defense (Agrawal and Fishbein 2008; Agrawal et al. 2009a).

In other words, milkweed defenses have tended to either increase or decrease as new species evolved. Such trends are a remarkable feature of evolutionary biology (Farrell et al. 1991; Vermeij 1994; Jablonski 2008; Futuyma and Agrawal 2009), although their causes are difficult to disentangle (Agrawal et al. 2009c). One interpretation of this pattern is that directional selection for changes in plant defense are realized only at the time of speciation (Futuyma 1987; Vermeij 1994; Pagel 1999), while another interpretation is that particular defensive phenotypes have promoted speciation (Paradis 2005; Maddison et al. 2007; Freckleton et al. 2008). These two

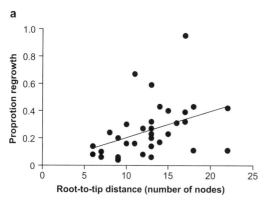 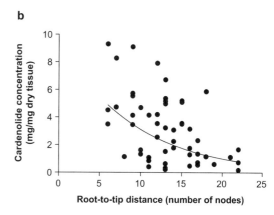

Figure 4.2. Directional trends in the macroevolution of the defense traits in *Asclepias*; root-to-tip distance indicates phylogenetic distance measured as the number of speciation events or nodes on a phylogeny. Each point is the mean of an *Asclepias* species. (a) A positive and linear relationship between trait values and the extent of phylogenetic divergence, shown for regrowth of plants after defoliation (a measure of tolerance); data based on Agrawal and Fishbein 2008. (b) A decelerating decline in cardenolide values (also known for cardenolide diversity and latex exudation); based on Agrawal et al. 2009a. These relationships demonstrate an association between speciation events and trait evolution. When a directional trend is accelerating or decelerating (as in b), this is considered a signature of adaptive radiation because the greatest change occurred during the initial stages of the lineage diversification.

alternatives suggest that defensive traits are either a consequence of evolution or the cause. Despite the difficulty in resolving the direction of causality, such patterns allow us to infer that particular traits in the lineage change directionally, and thus are important for inferring drivers of evolution. Given that monarchs and milkweeds share a long evolutionary history (A.V.Z. Brower and Jeansonne 2004), such temporal trends provide insight into the importance of monarchs (and milkweed herbivores more generally, see Plate 3) in plant evolution.

EVOLUTIONARY HISTORY OF *ASCLEPIAS*

Although milkweeds are common, an understanding of *Asclepias* diversification has begun to emerge only recently. A basic question that has long vexed taxonomists concerns the relationships among the three geographic centers of milkweed diversity: Africa (ca. 250 spp.), North America (ca. 130 spp.), and South America (ca. 9 spp.). At the turn of the twenty-first century, the prevailing view held that, while African and North American milkweeds are closely related, these continents possess independent evolutionary lineages, and many African milkweeds have been placed in as many as 20 other genera (Bullock 1952; Goyder 2001a). No unique traits characterize either African or North American milkweeds as a whole, however, and progress in classifying African species in genera other than *Asclepias* has stalled (Goyder 2009). The accommodation in *Asclepias* of the few species endemic to South America, however, has been uncontroversial. For convenience, we refer to all African milkweeds as *Asclepias*, including those formerly placed in *Gomphocarpus* and other segregate genera.

Comprehensive phylogenetic study of milkweeds based on noncoding chloroplast DNA sequences (Fishbein et al. 2011) has largely supported the conjecture that African and American milkweeds represent distinct, but closely related evolutionary lineages. We (Fishbein et al. 2011) sampled all but a handful of the American species and found that all American species form a single clade to the exclusion of all sampled African species (Figure 4.1); moreover, all South American species belong to one lineage that represents a single dispersal event from North American ancestors. American and African milkweed lineages appear to consist of sister clades stemming from a single common ancestor; however, one African species, *A. pseudofimbriata*, was placed outside of these two major milkweed clades as an early diverging lineage independent of the major African diversification (Figure 4.1). This result is not explained by any unusual morphological feature of the species (Goyder 2001b), and the unexpected placement of *A. pseudofimbriata* has been supported by subsequent phylogenetic study of African milkweeds (D. Chuba and M. Fishbein, unpublished data). Deeper phylogenetic relationships suggest that the ancestor of African and American milkweeds originated in Africa, and that the genus expanded its range to the Americas via spread across Atlantic or Beringian land bridges, or through long-distance dispersal (Fishbein et al. 2011). Because there is no fossil record for *Asclepias* (or any other milkweed genus), the timing of origin, dispersal to the Americas, and colonization of South America remain shrouded in mystery. Based on relative molecular branch lengths of the *Asclepias* phylogeny, it seems clear that the origin of the South American species has been quite recent, perhaps explaining the low diversity of this clade.

Within the American lineage, our results provide insights into the delimitation of major milkweed clades and patterns of morphological evolution, although their apparently rapid radiation leaves the details of the diversification yet to be fully resolved (Figure 4.1). Woodson (1954) proposed a classification of North American milkweeds into nine subgenera (the largest of which was further subdivided) based on flower characteristics. The implicit assumption was that the evolution of these complex flowers was conservative. Phylogenetic analysis showed this assumption to be unfounded, in that just one of Woodson's (1954) species groups is monophyletic, and that growth form and leaf morphology are perhaps more indicative of relationships (Fishbein et al. 2011). The phylogeny of *Asclepias* also exhibits a strong geographic component, in that major clades are largely distributed regionally, such as the Mexican highlands, the eastern temperate forests and plains, and the Sonoran Desert (see Figure 4.1) (Fishbein et al. 2011). Although we found strong support for these major clades based on chloroplast sequences, relationships among and within them were not supported. It should be noted that phylogenies estimated from chloroplast data alone may not accurately represent species-level relationships if incomplete lineage sorting or introgressive hybridization have occurred in milkweeds (Soltis and Kuzoff 1995; Maddison and Knowles 2006). Current efforts at resolving the *Asclepias* phylogeny are employing whole chloroplast genome sequences and multiple, independent nuclear markers (Straub et al. 2011, 2012). Results are promising, as whole genome sequences provide almost complete resolution of the relationships among major milkweed clades (S. Straub, A. Liston, M. Fishbein, R. Cronn, unpublished data). However, data from whole chloroplast genomes, complete nuclear ribosomal DNA cistrons, and mitochondrial DNA fragments show strong intergenomic conflict in the Sonoran Desert milkweeds, and the genus as a whole is probably characterized by a complex history of diversification that includes both hybridization and incomplete lineage sorting (Straub et al. 2012).

DEFENSIVE TRAITS OF MILKWEEDS

Milkweeds take their name from the characteristic latex that exudes following tissue damage (Figure 4.3). An important feature of this trait is that latex has no known function in a plant's primary metabolism (resource acquisition and allocation) and has been strongly implicated as a defense against chewing herbivores such as monarchs (Agrawal and Konno 2009). Although the defensive function of latex has historically been ascribed to the physical action of coating and gumming up the insect's mouthparts, there is also evidence for potent chemical defenses in latex. For example, many milkweeds have tremendously high concentrations of cardenolides (steroids that disrupt cellular ATPase function) in their latex (Nelson et al. 1981; Zalucki et al. 2001a, Agrawal et al. 2012b). Cardenolides are also produced throughout the plant, even in nectar and roots, which are plant parts that lack latex (Rasmann et al. 2009; Agrawal et al. 2012b; Manson et al. 2012). Like latex, cardenolides have no known primary function and have been strongly implicated in the defense against insect herbivores (Agrawal et al. 2012b). Because the chemical mode of action and ecological impacts of cardenolides have been well-reviewed, we do not elaborate on these subjects here (Seiber et al. 1983; Malcolm 1991; Agrawal et al. 2012b).

In addition, the latex of many plants, including milkweeds, contain cysteine proteases (Arribere et al. 1998; Trejo et al. 2001; Liggieri et al. 2004; Stepek et al. 2005; Agrawal et al. 2008), which have recently been implicated as toxins that degrade an essential part of the insect's gut, the peritrophic membrane (Pechan et al. 2002; Konno et al. 2004). Beyond these particular chemical defenses, milkweeds have diverse traits that may also contribute to defense. For example, saponins, pregnanes, phenolics, and

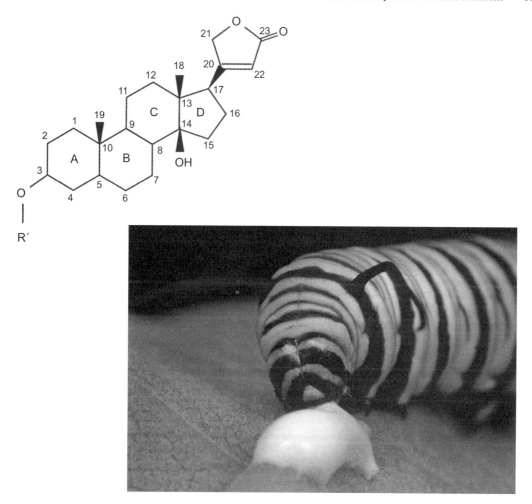

Figure 4.3. Cardenolides and latex are the two most well-studied defensive traits of milkweeds. (a) The skeleton structure of a cardenolide, composed of a core steroid (four fused rings), an oxygenated lactone group, and a glycoside group typically attached at position 3 (R'). R represents one or more sugars in glycosides or H in genins. Cardenolides function by binding to the critical animal sodium-potassium pump, a ubiquitous cellular enzyme. (b) A nearly mature monarch larva beginning to notch the midrib of an *Asclepias syriaca* leaf. In this case, the larva was met by a large droplet of pressurized latex. It proceeded to move the latex out of the way by wiping it on either side of the midrib. Occasionally, the larva will even imbibe such a large droplet of latex, presumably to remove it, and continue cutting the latex-delivering canals. Photo by Ellen Woods.

alkaloids have been reported from *Asclepias* species (Malcolm 1991; Agrawal et al. 2012b). Leaf hairs (trichomes), low nutritional quality, and volatile organic compounds that are induced by herbivory are also produced by *Asclepias*, and may impact insect herbivores (Agrawal and Fishbein 2006; Agrawal et al. 2009b; Wason et al. 2013). Given this diversity of plant defensive strategies, it is no wonder that the bulk of insect species that consume milkweed are specialists that have evolved to cope with these defenses (Agrawal and Konno 2009; Agrawal et al. 2012b; Dobler et al. 2012).

In the rest of this chapter we focus on latex and cardenolides because these have been the best studied defense traits of milkweed, and many of their impacts and interactions with monarchs have been elucidated over the past decades (e.g., Zalucki et al. 2001a).

A BRIEF REVIEW OF PHYLOGENETIC TRENDS IN *ASCLEPIAS*

Previously, we reported that the production of total cardenolides and latex has declined during milkweed diversification (Agrawal and Fishbein 2008; Agrawal et al. 2009a), while regrowth ability and production of phenolics has been enhanced (Agrawal and Fishbein 2008; Agrawal et al. 2009c). Because total cardenolide concentration and diversity of distinct cardenolide compounds show a strong evolutionary correlation (Rasmann and Agrawal 2011), it is clear that as cardenolide amounts declined, so too did the diversity of compounds. Nonetheless, we still do not know enough about the structure-function relationships of most cardenolides, and thus it may be too early to conclude that declines result in reduced toxicity. Additionally, the means by which latex is delivered is not fully understood. Latex is produced and delivered in elongated cells called laticifers. While most *Asclepias* have nonarticulated laticifers that consist of long multinucleate cells, some species such as *A. curassavica* have articulated laticifers comprising multiple cells that branch and rejoin. These two delivery methods may provide different means of defense (S. Malcolm, pers. comm.).

A further piece of evidence that the evolution of defenses promoted milkweed diversification comes from an analysis of the tempo and mode of trait evolution (Agrawal et al. 2009a). In particular, species-rich lineages of *Asclepias* underwent a proportionately greater decline in latex and cardenolides relative to species-poor lineages, and the rate of trait declines was most rapid early in milkweed diversification. Our interpretation of these results is that reduced investment in defensive traits accelerated diversification, and disproportionately so in the early adaptive radiation of milkweeds. While the declines in some traits, especially in latex and cardenolides, were surprising to us (and in opposition to theory), the statistical significance of the pattern has led us to speculate on the causes of this pattern.

First, we hypothesized that because most *Asclepias* herbivores are specialists, with adaptations to cope with latex and cardenolides (Dussourd and Eisner 1987; Holzinger and Wink 1996), the plants may be tending toward tolerance of herbivory (van der Meijden et al. 1988). In other words, given that several milkweed herbivores sequester cardenolides (Agrawal et al. 2012b), high levels of plant-produced cardenolides could backfire, as specialist herbivores may enjoy protection from natural enemies (e.g., Brower et al. 1967; Malcolm 1995; Sternberg et al. 2012). Second, if a plant cannot prevent herbivory because the herbivores are specialists, a strategy of tolerance could be favored (Strauss and Agrawal 1999).

In the rest of this chapter, we describe three experiments that help us understand whether the evolutionary trends in defensive traits of milkweeds actually impact one of the major herbivores of *Asclepias*, the monarch caterpillar. The experiments address five questions: (1) Does caterpillar performance improve on progressively derived *Asclepias* spp., as was predicted from our previous phylogenetic analyses of plant defensive traits? (2) Which defensive traits are associated with impacts on monarch performance? (3) Do monarch caterpillars have enhanced sequestration when feeding on early diverging *Asclepias* compared with that on more derived species? (4) Are early diverging *Asclepias* less tolerant of monarch herbivory than more derived species? And, (5) is tolerance of herbivory explained by patterns of allocation to roots or shoots, and correlated with a particular life-history trait, such as the extent of clonal reproduction?

Experiment 1, monarch larval performance on 53 milkweed species

Here we report previously unpublished data on the performance of monarch larvae fed 53 different *Asclepias* spp. for which we had both phylogenetic and defense strategy data (see Figure 4.4, experimental details provided in Agrawal et al. 2009a; Rasmann and Agrawal 2011). Note that monarchs appear able to complete development on all *Asclepias* species, encounter many if not most species in the wild, and show little local adaptation because of their panmictic populations (Pierce et al., this volume, Chapter 23). We grew plants from seed in growth chambers, and when plants had grown for 30 days, we introduced a single freshly hatched monarch larva to each of 3–6 plants per species. Plants were fully randomized within the growth chambers and caterpillars were allowed to feed for 5 days, after which caterpillar mass was recorded. This time period resulted in some caterpillars being in the first instar while others were in the second, potentially contributing to variation in the data; nonetheless, we maintained a single harvest point in order to follow the mass gain for each caterpillar over a fixed amount of time, thereby integrating performance as growth per unit time.

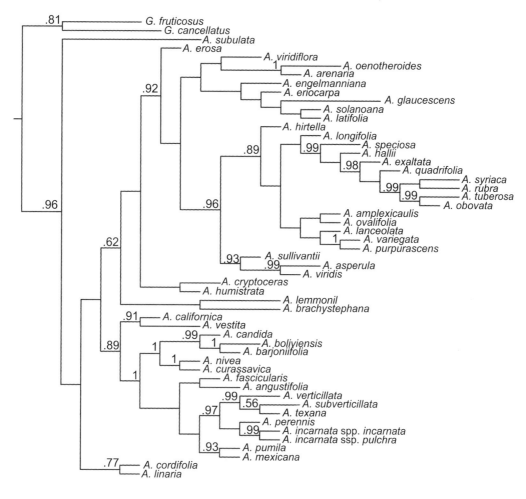

Figure 4.4. A pruned phylogeny of milkweed (53 species) based on the all-compatible consensus of trees sampled in a Bayesian analysis of the complete data set of 155 samples. Branch lengths are drawn in proportion to the number of speciation events between nodes; sister lineages may not be of equivalent length because multiple speciation events in the complete phylogeny may occur on a single branch of the pruned phylogeny. Modified from Agrawal et al. 2009.

We evaluated monarch performance (mean mass of the larvae on each milkweed species) as a function of phylogenetic root-to-tip distance, measured by the number of intervening nodes (Figure 4.2, see Figure 4.4 for the pruned phylogeny with branch lengths scaled to this measure) using BayesTraits (Pagel 1999). Phylogenetic signal (Pagel's λ) was estimated using maximum likelihood and tested against fixed models of $\lambda = 0$ and $\lambda = 1$. A λ value of 1 indicates phylogenetic conservatism consistent with a random walk model of larval performance as milkweeds diversified (i.e., similarity in performance is directly proportional to the extent of shared milkweed evolutionary history). A λ value of 0 indicates no influence of shared ancestry on performance (i.e., phylogenetic independence). Further details of our analyses can be found in published studies (Agrawal and Fishbein 2008; Agrawal et al. 2009a).

Short-term monarch growth varied more than 10-fold across the milkweed species, and caterpillars feeding on species that are more derived showed proportionally greater mass (Figure 4.5, likelihood ratio test, maximum likelihood estimate of $\lambda = 0$, LR = 4.112, $P = 0.043$). We next used a phylogenetically informed multiple regression to predict monarch mass using three predictors: cardenolides, latex

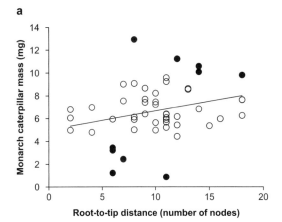

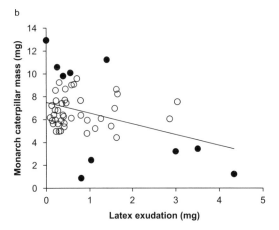

Figure 4.5. Impacts of evolutionary history and latex on short-term monarch growth. The five best- and five worst-quality milkweed species used for qualitative comparisons (Table 4.1) are shown in filled symbols. (a) More derived *Asclepias* species (those with a greater root-to-tip distance, see Figure 4.1) supported greater monarch larval growth. (b) In a phylogenetically informed analysis, latex was the only significant predictor of monarch growth. These data are consistent with earlier results suggesting phylogenetic declines in milkweed defenses.

exudation, and trichome density measured on the same plants (methods are given in Agrawal et al. 2009a). Briefly, foliar cardenolides were measured by high performance liquid chromatography, latex was measured as wet mass exuded from a 2 mm cut leaf tip on filter paper, and trichomes were measured as the number of leaf hairs along a transect under a dissection microscope. Our full model explained a modest 16% of the variation in monarch mass, and only latex was a significant predictor (Figure 4.5b, $\lambda = 0$, LR = 7.201, $P = 0.007$). Thus, phylogenetic declines in latex improve monarch performance, while declines in cardenolides or trichome density were not implicated in promoting monarch growth in this experiment.

To characterize the biology of the best and worst food plants for monarchs, we next examined the top five and bottom five species for monarch growth based on our experiment that included 53 species. On average, the top five species supported monarchs that grew nearly five times the mass of those fed the bottom five species (Figure 4.5, Table 4.1). The phylogenetic position of our best food species for monarchs was decidedly derived (mean ± SE nodes from root-to-tip, 13.2 ± 1.3 compared with our worst species: 7.2 ± 1.3). Again, our best predictor of monarch mass was latex (mean ± SE latex exudation in mg, best species, 0.52 ± 0.52 compared with our worst species: 2.53 ± 0.52, Table 4.1). Despite the fact that our poorest quality plant species had, on average, nearly five times the trichome density and nearly 80% higher cardenolides than our best quality *Asclepias*, these two variables were not significantly different among the two extreme plant categories (Table 4.1). Nonetheless, we note two interesting points relating to deployment of plant defenses in milkweeds as a suite of synergic traits. First, latex and trichomes show positive correlated evolution (Agrawal and Fishbein 2006; Agrawal et al. 2009b), suggesting that, on average, species with high latex also have dense trichomes. Second, latex delivers a concentrated dose of cardenolides, often more than 10-fold that of leaves per se. Thus, these traits are are not necesarrily acting independently, and we still have much to learn about the combined effects of latex, cardenolides, and trichomes (see also Zalucki et al. 2001a, 2012).

A further assessment of the phylogenetic, geographic, and climatic attributes of these 10 "best" and "worst" species reveals some interesting patterns. First, the worst species are not strongly phylogenetically clustered (although *A. californica* and *A. vestita* are sister taxa), despite the fact that all were relatively early diverging species (Table 4.1, Figure 4.4); nonetheless, they are native to arid regions of the western United States and northern Mexico (although *A. lemmonii* is most abundant in wet pine-oak forest in the Sierra Madre Occidental). For *A. asperula*, it appears that a slightly different path has been taken, given that it is the most derived species in the well-defended group, and has maintained high cardenolides, despite having low latex.

Table 4.1. Characteristics of the five best and five worst *Asclepias* food plants from Experiment 1

	Monarch mass(mg dry)	Root-to-tip distance	Trichome index	Latex exudation (mg)	Cardenolide (mg/g dry mass)
Poorest host plants					
A. asperula	0.874	11	0.384	0.802	6.612
A. lemmonii	1.231	6	3.248	4.329	3.224
A. cryptoceras	2.446	7	0.489	1.030	1.029
A. vestita	3.215	6	117.82	2.995	3.098
A. californica	3.444	6	42.807	3.498	4.853
mean ± SE	2.24 ± 0.52	7.2 ± 0.97	32.95 ± 22.69	2.53 ± 0.69	3.76 ± 094
Best host plants					
A. syriaca	9.815	18	7.980	0.388	1.533
A. hallii	10.088	14	0	0.553	1.560
A. purpurascens	10.586	14	5.734	0.251	0.349
A. solanoana	11.230	12	19.195	1.406	2.720
A. nivea	12.930	8	1.6002	0	4.385
mean ± SE	10.93 ± 0.56	13.2 ± 1.62	6.90 ± 3.39	0.52 ± 0.24	2.11 ± 0.68
ANOVA	**131.4****	**10.1***	**1.3^{ns}**	**7.5***	**2.0^{ns}**

Source: Data from Agrawal et al. 2009a.
Notes: Values to the right of each species name indicate the species means, usually based on 5 replicates (except the root-to-tip distance, which is simply based on the phylogenetic position of the species). The row in gray, directly below each group, represents the means ± standard error of that group; the final row, marked ANOVA, provides an analysis of variance result with an F value (df = 1,8 for all tests).
* < 0.05, ** < 0.001, ^{ns} = not significant

The best *Asclepias* species for monarchs are dominated by the eastern North American clade (except *A. solanoana*), and are highly variable in terms of geography and habitat. *Asclepias solanoana* occurs only on dry and barren serpentine soils in northern California. It is possible that it typically accumulates toxic metals from the soil that were not available in our common potting mix, or that this edaphic specialist typically escapes herbivores because of its geographic location or life in harsh environments. The other derived species are scattered from the eastern United States to the Rocky Mountains (and also include the Caribbean island endemic, *A. nivea*) and are more typical of open field, mesic habitats than the early diverging, poor food plant species; nonetheless, *A. purpurascens* and *A. nivea* can be abundant in open forest understories and edges, and *A. hallii* grows in seasonally wet but open riparian habitats.

Asclepias syriaca is in the "best" food group. In addition to having relatively modest levels of trichomes, latex, and cardenolides, it is by far the most common host plant of monarchs across North America. Indeed, it has been estimated that more than 90% of butterflies that migrate to the Mexican overwintering grounds developed on *A. syriaca* (Malcolm et al. 1993). Thus, if monarchs are adapted to feeding on any milkweed, it is likely the common milkweed, *A. syriaca*.

In summary, there does not appear to be a strong phylogenetic constraint on the evolution of plant defenses against monarchs. Given the geographic and climatic characteristics of the best-defended species, it seems that defense trait convergence has occurred because of similar environments or the historical impact of herbivores in those regions. Some overlap exists between inhabiting relatively harsh habitats and early divergence in the milkweed phylogeny (with more derived forms having reduced resistance traits). As for the least defended species, they appear to be more diffuse in terms of the geographic or climatic conditions associated with relaxed expression of cardenolides and latex.

Experiment 2, monarch sequestration on eight milkweed species

In this experiment, we tested the hypothesis that phylogenetic trends can be observed in monarch sequestration. One possible explanation for the macroevolutionary declines in total cardenolides could be that more derived species would support reduced sequestration, and thereby enhanced control by predators and pathogens. To address this, we grew eight species of *Asclepias* under controlled conditions as described previously (Agrawal and Fishbein 2008; Agrawal et al. 2009a). *Asclepias curassavica*, *A. humistrata*, *A. linaria*, and *A. subulata* were selected as early diverging species that span the root of the American milkweed clade while belonging to low-diversity subclades (Figures 4.2, 4.4), with root-to-tip distances of 9, 9, 6, and 9 nodes, respectively. *Asclepias purpurascens*, *A. solanoana*, *A. subverticillata*, and *A. syriaca* are more derived species representing species-rich clades, with root-to-tip distances of 17, 15, 14, and 22, respectively. Caterpillars were allowed to grow for 10 days after which we starved them to empty their guts for 24 hours, and then harvested both the insects and the plants (n = 10 plants per species). Cardenolides were assessed by HPLC following the methods of Rasmann and Agrawal (2011).

We first assessed the relationship between plant cardenolides and sequestration. Our eight species varied from barely detectable cardenolides (*A. subverticillata*) to the highly toxic *A. linaria*, with more than 7 mg/g dry mass cardenolide content (Figure 4.6). As previously summarized by Malcolm (1995, focusing on adults) we found a nonlinear pattern of cardenolide sequestration in larvae. Monarchs tend to follow one of three patterns of sequestration: (1) little to no cardenolides in monarchs fed plants with little to no cardenolides, (2) concentrated cardenolides, often double the concentration in monarchs fed plants of intermediate cardenolide concentrations (plants ranging ≈0.75–2.5 mg/g), or (3) saturated cardenolides, where monarch bodies contain no higher levels (or perhaps even reduced levels) compared with the very high foliar concentrations in some plant species. Indeed, in our study, *A. linaria* exhibited typically high cardenolide concentrations, but larvae contained intermediate levels and could barely grow or sequester on *A. linaria*; they achieved only 3 mg mass (and remained in the second instar) compared with larvae on most other species, which were more than 19 mg and in the third to fourth instar.

Although it was initially thought that monarchs were capable of accumulating the highest level of cardenolides from plants with the highest concentrations (Brower et al. 1972), we now recognize that monarchs seem particularly good at accumulating and concentrating cardenolides to levels much higher than those found in their host when their host plants contain moderate to low concentrations. Plants with very high cardenolides represent a saturation point for the monarchs, or may even decrease their ability to sequester cardenolides. Our data confirm this pattern. Why monarchs have difficulty feeding on some species (e.g., *A. linaria*, *A. perennis*, *A. asperula*) appears to be related to the extremely high cardenolide content, although further studies on the toxicity and sequestration of high levels of cardenolides are needed (see also Zalucki et al. 2001a; Agrawal et al. 2012b). We further speculate that the linear, tough leaves of *A. linaria* may be nearly impossible for monarchs to notch to cut off latex flow, which may make this species even more difficult to consume.

Monarch sequestration is a dynamic process. At least five factors may explain the amount of cardenolides a monarch will sequester: sex, quantity of cardenolides in a plant (including induced compounds), biomass consumed, polarity of cardenolides, and structural attributes independent of polarity. First, females sequester higher concentrations of cardenolides than do males when reared on plant species with similar cardenolide quantities (Nelson 1993).

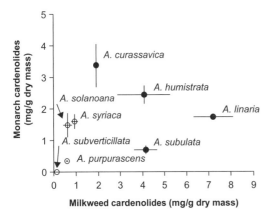

Figure 4.6. The relationship between foliar cardenolides and sequestered cardenolides in monarch larvae. Phylogenetically basal species are coded as closed circles, derived species as open circles. Note that basal plant species showed substantially higher cardenolides than derived species; however, this pattern did not translate into significant differences in larval sequestration. Shown are means ± SE.

This phenomenon and its ecological consequences are not well studied. Because the fat bodies of adult female butterflies contain nutritional reserves carried through pupation from larval feeding, and these are a major part of the resources available for somatic maintenance and gamete production, high levels of sequestration may be critical. Second, both quantitative and qualitative factors play a large role in cardenolide storage. Cardenolide quantity was discussed above. It may be less costly to store and sequester relatively polar compared to nonpolar cardenolides (Malcolm 1991). This hypothesis is supported by the observation that nonpolar cardenolides are converted into more polar forms during sequestration (Seiber et al. 1980; Brower et al. 1984; Martin et al. 1992), while polar cardenolides and those of intermediate polarity are directly stored (Frick and Wink 1995). Finally, structural attributes of cardenolides, especially the sugar groups (Figure 4.3), have a strong impact on their toxic potential. For example, genins (cardenolides lacking a sugar group) are less toxic than identical compounds with sugars (Malcolm 1991; G. Petschenka et al., unpublished).

In summary, we hypothesized that the most derived *Asclepias* species may benefit from having reduced cardenolide concentrations because of reduced sequestration by their herbivores. Indeed, cardenolide sequestration was remarkably low on the four most derived species tested. Other very low cardenolide species not tested here (i.e., the highly derived *A. tuberosa*), may provide no opportunity for monarch sequestration. Thus, although more work is needed to understand which cardenolides are most easily sequestered and are most effective as a defense against predators of monarchs, it is clear that the macroevolutionary trend toward reduced *Asclepias* cardenolides impacts sequestration and is likely to impact interactions with enemies.

Experiment 3, tolerance of herbivory in basal and derived milkweeds

We used the same set of eight species to test for differences in tolerance of herbivory between early diverging and more derived milkweeds. A previous study suggested that as milkweeds diversified and showed reduced investment in latex and cardenolides, their ability to regrow following damage was enhanced (Agrawal and Fishbein 2008). This led us to hypothesize that tolerance had evolved as an alternative strategy to resistance, and that such a strategy may be particularly useful when plants are faced with highly specialized herbivores (Agrawal and Fishbein 2008). We further demonstrated that the ability to regrow following damage was dependent on relatively high investment in roots (i.e., a high root-to-shoot ratio) (Agrawal and Fishbein 2008); however, our previous work had two weaknesses. First, our measure of tolerance was limited to the regrowth capacity of plants in aboveground tissues relative to their previous growth and not to an undamaged control. Second, we imposed catastrophic damage by cutting the plants down to the soil, but we did not use real herbivores. Real vs. mechanical damage causes distinct responses in milkweed (Mooney et al. 2008). Here, we report the results of a new study designed to test tolerance of herbivory in eight milkweeds using both root and shoot measures.

Plants were grown from seed as described above. After they had grown for 21 days following germination, we measured total leaf area by spreading each leaf over square millimeter graph paper, tracing its outline, and calculating the corresponding area. Half the plants were then treated with a single freshly hatched monarch larva and the other half were left as undamaged controls (n = 10 plants per species per treatment). After 10 days, larvae were removed from plants, and all uneaten leaves were clipped off with scissors from plants of the treatment group. Thus, real herbivores attacked all the plants for an equal amount of time, followed by 100% leaf tissue removal for plants in the damaged treatment. This combination allowed for insect-specific signals to induce plants and also for equal (catastrophic) impacts on plants. Total leaf area was remeasured on all plants after 21 days of (re)growth. Finally, above- and belowground tissues were harvested and separated (roots were washed free of their soil), dried at 45 °C, and weighed. We report three measures of tolerance for each species: (1) the proportional reduction in root mass of damaged compared with undamaged plants (the difference in root mass of control plants and damaged plants divided by the root mass of control plants), (2) the proportional reduction in shoot mass (calculated as for roots), and (3) proportional leaf area recovery. This final measure was based only on defoliated plants (relative to their starting area); we divided the amount of new leaf area produced during the 21 day (re)growth period by the initial leaf area recorded just before damage was imposed.

In the undamaged state, early diverging and derived milkweeds produced, on average, similar absolute amounts of root and shoot mass as well as leaf area (all P-values > 0.2). Thus, overall growth rates appear quite similar; nonetheless, the relative investment in roots (i.e., root-to-shoot ratio) per species was more than four times higher in the more derived compared with the early diverging species (mean ± SE early diverging species, 0.304 ± 0.276 compared with more derived: 1.383 ± 0.276, $F_{1,8} = 7.636$, $P = 0.033$). Reduction in root mass due to herbivory was strongly predicted by phylogenetic position, with more derived species showing progressively reduced proportional impacts of leaf defoliation (Figure 4.7a); however, there was no difference for shoot mass (Figure 4.7b) or leaf area growth ($F_{1,8} = 1.106$, $P = 0.334$). Thus, the early relative allocation to roots is higher in more derived milkweeds, and this seems to allow greater tolerance (i.e., reduced negative impacts of defoliation) in root tissues.

Interestingly, early diverging and more derived species recovered quite equally aboveground, and this apparently comes at a greater cost to early diverging species, since they apparently lose proportionally more root tissue. Although the specific importance of root reserves is unclear, we expect that the low tolerance of roots could lead to compromised fitness over several years of defoliation (all species of *Asclepias* are perennial). We caution that we examined only the short-term growth of young plants; thus, the degree to which the extensive "root" diversity among milkweed species (true roots, tubers, rhizomes, etc.; M. Fishbein, unpublished data) could contribute to differences in tolerance in mature plants is unknown and may not reflect patterns found during the early establishment phase.

As a final approach to understanding resistance and tolerance of milkweeds in the context of overall strategies to cope with both their biotic and abiotic environments, we consider the extent of clonality in milkweeds, which could be a proxy for tolerance by means of sprouting new ramets. As a measure of clonal potential, we examined the caudices (the underground bases of the stem) of all of the individuals from our previous study of 51 milkweed species (Agrawal et al. 2009a), and recorded the number of belowground buds, which are dormant meristems that can produce new ramets. Species known to grow as well-defined genets emerging from single root crowns (e.g., *A. humistrata*, *A. incarnata*, *A. linaria*, and *A. tuberosa*) showed very few, if any, underground stem buds, while highly clumped species known to grow in clonal patches (e.g., *A. pumila*, *A. subverticillata*, *A. speciosa*, and *A. syriaca*) produced high numbers of such underground buds. A phylogenetic analysis of directional trends revealed that progressively more derived species have enhanced clonal growth potential (Figure 4.8, LR = 12.4, $P < 0.001$). In this analysis, λ was estimated by maximum likelihood to be zero, indicating no phylogenetic signal for the clonal potential of *Asclepias*.

We emphasize that we have not attempted to disentangle the various drivers of root system evolution. Our primary conclusion is that more derived species are indeed more tolerant of herbivory, and that this tolerance is achieved by enhanced investment in the initial root system. It is also associated with the potential for clonal growth. Although it is tempting to ascribe these directional phylogenetic trends as causally correlated with the declines observed in

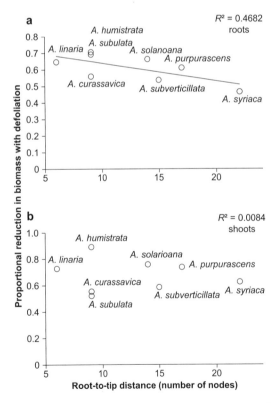

Figure 4.7. The relationship between phylogenetic position of milkweeds and tolerance to herbivory in (a) root and (b) shoot tissues. The identity of basal and derived species is given in the text under Experiment 2.

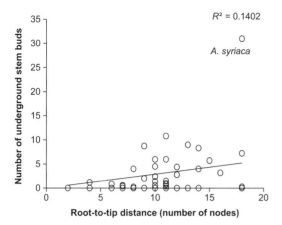

Figure 4.8. The relationship between phylogenetic position and production of stem buds produced on roots, an indicator of potential clonal growth for 51 milkweeds. Note that this relationship is significant with or without the single outlier (*Asclepias syriaca*) shown in the top right corner.

resistance traits (latex and cardenolides), we have yet to make this a strong link.

CONCLUSION AND OUTLOOK

We have been working under the assumption that monarchs and other specialist herbivores are major selective agents in milkweed evolution. Indeed, phylogenetic patterns suggest that milkweeds have evolved greater levels of tolerance of herbivory and reduced levels of latex (which is abundantly implicated in resistance to monarchs). The details of cardenolide evolution are less clear cut. Although we have observed phylogenetic declines in cardenolides, the interpretation of this pattern is complicated by the fact that monarchs sequester cardenolides, which may protect them against predators and parasites. Nonetheless, we know remarkably little about the costs of producing the various milkweed defenses, the costs of sequestration for monarchs, and the rates of herbivore attack on the majority of *Asclepias* species in the field. All these factors have likely contributed to the macroevolutionary patterns we observe.

ACKNOWLEDGMENTS

We thank Karen Oberhauser for encouraging us to contribute to this book and for comments on the manuscript. Steve Malcolm and Jaap de Roode also provided thought-provoking suggestions, and Amy Hastings provided research support. Our work on milkweeds and monarchs has been advanced by several past students and postdocs in the lab, supported by the National Science Foundation, most recently by DEB-1118783, and the John Templeton Foundation.

5

Invertebrate Natural Enemies and Stage-Specific Mortality Rates of Monarch Eggs and Larvae

Alma De Anda and Karen S. Oberhauser

We measured stage-specific immature monarch mortality rates over 24-hour intervals in the wild. We documented similar mortality rates during two summer breeding seasons, and 24-hour survival rates were slightly but significantly lower for both first- and second-instar larvae (~60%) than eggs (~63%). Monarchs had lower survival on plants with spiders or higher numbers of aphids present and on plants with more evidence of herbivory, and survival decreased over the course of the summer. Additionally, monarchs found on the bottom and apex of milkweed leaves were more likely to survive. Position on plants varied across monarch ages, with eggs more likely to be found on the bottoms of leaves, first and second instars more likely to be hidden in the leaves at the apex of the plant, and older larvae more likely to be on the tops of leaves. We argue that position on the plant may affect the likelihood of predation differently at different monarch ages. Potential predators were very common on milkweed plants, and several lines of evidence suggest that predation was responsible for most of the mortality we observed.

INTRODUCTION

Understanding the extent to which predation affects species' population dynamics has been a focus of many ecological studies, beginning with early work by Lotka (1924), Voltera (1926), and Gause (Gilbert et al. 1952). For example, predators drive population cycles in larch budmoths (*Zeiraphera diniana*) (Turchin et al. 2003) and the southern pine beetle (*Dendroctonus frontalis*) (Turchin 2003), and alter larval foraging behavior in many species (Montllor and Bernays 1994; Stamp and Bowers 2000). Monarchs incur high (~90–95%) mortality rates from egg to the fifth instar (Borkin 1982; Zalucki and Kitching 1982b; Oberhauser et al. 2001, this volume, Chapter 6; Prysby 2004). For example, fire ants (*Solenopsis invicta*) kill up to 100% of monarch larvae in some locations in Texas (Calvert 2004a). All published studies of monarch mortality during development have assessed only eggs and larvae, but parasitoids and other pupal enemies cause additional mortality (Oberhauser et al. 2007, this volume, Chapter 6; Oberhauser 2012). Thus, the studies cited above underestimate egg to adult mortality. While immature monarchs may be somewhat protected from predators by aposematism and toxicity derived from their milkweed host plants (Reichstein et al. 1968), the extent to which these defenses prevent predation by invertebrates is poorly understood (Oberhauser et al., this volume, Chapter 6).

We conducted a two-year observational study of monarch eggs and larvae in the wild. Detailed tracking of individuals allowed us to measure mortality rates of eggs and early instar larvae, and in many cases, to infer causes of mortality. We tracked several biotic and abiotic factors that we hypothesized could affect mortality (see Table 5.1). Several of these factors were chosen because they could indicate the likelihood of predation: the presence of predators on the plant, density of aphids, and the amount of the plant that had been consumed by herbivores. We predicted that all these factors would lead to decreased survival. While aphids and other herbivores do not kill monarchs directly, their presence could attract predators. For example, aphids are tended by ants, which defend the aphids against predators and are voracious predators themselves that are known to consume monarch eggs and larvae (reviewed in Oberhauser et al., this volume, Chapter 6). Additionally, herbivore damage often results in the release of volatile compounds that recruit natural enemies (Turlings et al. 1995; Takabayashi and Dicke 1996; Thaler 1999). The presence of monarchs (or other herbivores) could attract predators by stimulating the release of plant volatiles, or it could indicate a plant of high quality that had been chosen by more than one female; thus we did not have a clear basis for prediction of whether or how monarch density on the plant would affect survival.

Several plant characteristics could affect a monarch's susceptibility to predation, including its position on the plant, whether the plant has flowers or other reproductive structures, and milkweed density in the area immediately surrounding each plant. Some locations on plants are seldom searched by natural enemies, inaccessible to enemies, or impede enemy searching movement (reviewed by Price et al. 1980). For example, the predatory bug *Anthocoris confusus* searches mainly along leaf margins and midribs (Evans 1976). Plant parts can also provide structural refuges if herbivores are out of reach, for

Table 5.1. Independent variables in the logistic regression of immature monarch daily survival rates

Variable	Explanation	Predicted effect on survival	Observed effect on survival
Predator presence	Presence/absence of potential predators on plant: ants, lacewing larvae (Neuroptera), Asian lady beetle (*Harmonia axyridis*) adults and larvae, syrphid fly larvae (Syrphidae, recorded only in 2007), spiders (Salticidae, Thomisidae), various Hemiptera. Each order included as a separate categorical variable.	−	− (only spiders)
Aphid density	Ordered variable with 5 levels: 1 = 0 aphids on plant, 2 = 1–10 aphids, 3 = 11–100 aphids, 4 = 101–1000 aphids, 5 > 1000 aphids	−	−
Percent herbivory	Ordered variable with 4 levels estimating % plant damage from herbivory: Level 1 = 0%, 2 = <5%, 3 = 5–25%, 4 > 25%	−	−
Monarch density	Number of other monarch individuals on a milkweed plant (log-transformed)	?	0
Position on plant	Categorical variables for each individual's position on a milkweed plant: top of leaf, bottom of leaf, plant apex, plant stem, in flowers or buds. Because so few monarchs were on the stem or in flowers, these were not included in the model.	variable	variable
Ramet reproductive status	Categorical variables for reproductive state of milkweed plant: not flowering, flowering, finished flowering for the season with no seedpods, finished flowering with seedpods	?	No flowers > others
Milkweed density	Number of other milkweed plants within the m^2 surrounding focal plant	−	0
Start stage	Categorical variables for monarch stage at time of observation: egg, first instar, second instar. The analysis included only eggs, firsts, and seconds because of lack of certainty about survival of older individuals.	Eggs > larvae	Eggs > larvae
Age of individual	Time an individual has been in the current age class (days)	+	0
Date	The day of an observation (day of year)	?	0
Year	Categorical variable with two levels (2006 or 2007)	?	0
Average air temperature	Average of 1-h temperature readings over 24-h interval leading up to the observation	Variable	0
Average dew point	Average of 1-h dew point readings over 24-h interval leading up to the observation	?	0
Average precipitation	Average of 1-h precipitation readings over 24-h interval leading up to the observation	−	0

example in leaf folds or flower buds. Even plant trichomes can hinder some small natural enemies (e.g., van de Merendonk and van Lenteren 1978), such as predatory mites.

We hypothesized that flowers might attract predators either seeking nectar themselves or because the flowers might attract nectivores that could be potential prey. Alternatively, buds or flowers might provide refuges for monarchs. Leaf nutritive value can also affect larval performance (Zalucki et al. 2002), and can vary with the age of the plant (indicated in our study by its reproductive status) or the position of the leaves (Wilson et al. 1988), with younger plants and newer leaves being more nutritious.

Predators might be attracted to denser stands of milkweed; in fact, Zalucki and Kitching (1982b) found lower immature monarch survival rates in larger milkweed patches. Thus, we predicted that monarchs in denser milkweed stands would suffer lower survival.

Monarch developmental age is also likely to affect mortality. In a review of several studies of mortality, Hawkins et al. (1997) found that enemy-induced mortality is higher in later developmental stages of herbivores, although it was similar for eggs and early instar larvae, the stages that we measured. Zalucki et al. (2002) reviewed 141 studies of 105 Lepidoptera species, and found that, generally, first-instar mortality was higher than egg mortality, although this was not true for leaf miners or species that laid their eggs in batches. We predicted that early instar monarchs would have lower survival rates than eggs. We also predicted that an individual monarch that had successfully survived in a given stage on a given plant for more time might be more likely to survive, for the simple reason that something about its location might make it less susceptible to mortality.

Finally, weather could affect survival (reviewed in Zalucki et al. 2002) through the effects of drowning or being washed off the plant during rainstorms or being blown off during windstorms. Additionally, extremely high or low temperatures, or dry conditions can cause mortality (Nail et al., this volume, Chapter 8). We predicted that rainfall or extreme temperatures would lead to lower survival.

METHODS

We conducted this study in a 5.9 hectare city park in Falcon Heights, Minnesota (45.0°N, 93.2°W) that contains a restored prairie with naturally occurring milkweed. We monitored individual common milkweed (*Asclepias syriaca*) ramets, tracking immature monarchs throughout the summers of 2006 and 2007. From one monitoring event to the next, we assigned each monarch a fate: *survived* = still present, *gone* = completely missing, *sucked* = only chorion or exoskeleton remaining, *chewed* = pieces of monarch missing, *mired* = dead larva mired in milkweed latex, or *unknown* = dead for unknown reasons. We also recorded any predation events we observed. We assumed that eggs that were gone had been killed by predators that consumed the entire egg. We assumed that first- and second-instar larvae that were gone had been consumed by predators, for reasons addressed in the discussion. We assumed that sucked and chewed eggs and larvae were the result of predation. We were confident we were tracking individuals because there is rarely more than one individual on a plant; when there were multiple individuals we could keep track of them by their age or, if they were eggs, by their position on the plant. To increase confidence that individuals that were gone had died, we tracked data from all 5343 individual observations; if subsequent observations showed that a monarch had been missed, the data were corrected. We recorded the independent variables that we had hypothesized could influence immature monarch survival, obtaining climate data from a nearby climate station (air temperature, dew point, and precipitation) (Table 5.1).

In 2006, we visited the site five mornings a week from 31 May through 28 July. We began monitoring at approximately the same time each day, and used aluminum tags to identify milkweed ramets on which we observed monarchs. We made 2526 observations of eggs and larvae on a total of 237 milkweed ramets. Because we observed only on weekdays, we separated data by the time interval since the last observational period: either 24 hours (Tuesday–Friday observations) or 3 days (Monday observations). In 2007, we visited the park from 29 May through 7 August, monitoring 10 days in a row with 4 days between 10-day observation periods to increase the number of 24-hour intervals. To allow easier tracking of individual monarchs, we monitored 90 ramets spread throughout the park. The rest of the protocol was the same as in 2006. We made 2817 observations of eggs and larvae in 2007.

All analyses of survival focused on the observations made at 24-hour intervals. While we recorded all monarchs that we saw, survival analyses focused on eggs and first- and second-instar larvae, since frequent movement off the plant by older larvae (pers. observ.) made assigning fates difficult. We used a logistic regression analysis (stepwise method, forward selection with new variables added when deviance was significantly improved based on a likelihood ratio test) to assess the effects of the biotic and abiotic predictors that we had hypothesized could affect monarch survival. Because this was not an experimental study, but rather an exploration of the factors that might affect monarch survival, the stepwise regression is appropriate for our analysis. We conducted individual analyses of individual predictors to verify the results of the stepwise regression.

We conducted a short experiment in summer 2008 to test whether precipitation could wash eggs off host plant leaves. We used 3-day-old monarch eggs ($N = 802$), laid naturally on 6 potted tropical milkweed plants (*A. curassavica*) by laboratory females. Eggs were laid on both the top and bottom surfaces of leaves, but we did not record numbers in different positions. The eggs were exposed to "rain" through a circular-spraying sprinkler system for 30 minutes, and we used a rain gauge to measure how much water fell over the eggs. We did not manipulate or measure the strength of the "rain"; since the force of rainfall affects the likelihood that it will dislodge eggs, this experiment is preliminary.

RESULTS

Daily survival

The survival of monarch eggs and first- and second-instar larvae over a period of 24 hours was similar in both years, although second instars appeared to suffer higher mortality in 2006 (Figure 5.1). We estimated survival rates between age classes by extrapolating the higher daily survival estimates for each age class (63%, 60%, and 56% for eggs, first instars, and second instars, respectively) over the approximate time that monarchs spend in each age class (average time in each class was determined using our observations of individual monarchs, so were specific to our study; because development time varies with temperature, the values are approximate daily survival rates). We assumed constant daily mortality within each age class because the time that an individual had been in its current age class did not affect survival (see below). Extrapolating 63% daily survival over the approximately four days that a monarch is in the egg stage suggests that survival during this stage is ~16%. Similarly, applying 60% daily first-instar survival to the remaining 16%, over approximately two days for this age class, egg to second-instar survival is ~5.6%; and 56% daily second-instar survival over about two days suggests that egg to third-instar survival is ~1.7% (Figure 5.2).

In the multivariate model to assess factors that affect monarch survival, six predictors were identified with significant explanatory power: monarch age, herbivory, position on the plant, spider presence, aphid infestation level, and the presence of flowers on the plant (Table 5.2). Eggs had higher daily survival rates than first- and second-instar larvae (Figure 5.1). The negative coefficients for herbivory, spiders present, and aphid infestation level indicate that monarch survival decreases with the amount of

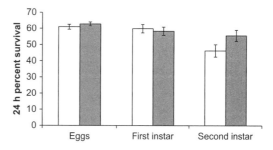

Figure 5.1. Percent survival of immature monarchs over 24-hour intervals during 2006 (open bars) and 2007 (gray bars). Error bars represent 95% binomial confidence intervals. Sample sizes for eggs, first instars, and second instars, respectively, in 2006: 1284, 239, 89; in 2007: 1229, 558, 302.

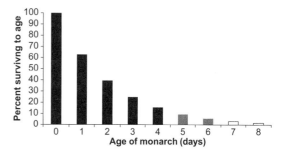

Figure 5.2. Extrapolation of daily immature monarch survival estimates for eggs (black bars) and first (gray bars) and second (open bars) instars. See text for detailed explanation.

Table 5.2. Stepwise logistic regression analyses, showing significant predictors of immature monarch survival

Parameter	Coefficient	Std error	P-value
a. (no interactions included in model)[a]			
Egg	0.306	0.079	0.0001
Herbivory	−0.435	0.069	<0.0001
Position on plant (in buds)	−0.790	0.115	<0.0001
Position on plant (top of leaf)	−0.406	0.119	0.0006
No flowers on plant	0.303	0.097	0.0018
Spiders present	−0.394	0.135	0.0035
Aphid infestation level	−0.077	0.036	0.033
b. Full model (stage*position interactions included)[b]			
Herbivory	−0.444	0.068	<0.0001
Egg*bottom of leaf	0.471	0.082	<0.0001
No flowers on plant	0.401	0.093	<0.0001
Egg*buds	−0.593	0.135	<0.0001
First*apex	0.585	0.205	0.004
Spiders present	−0.405	0.135	0.003
Second*bottom of leaf	−0.396	0.172	0.022
Second*top of leaf	−0.816	0.373	0.029

[a] All single factor predictors were included in the model.
[b] Post-hoc analysis that included stage*position interactions, conducted after observing stage differences in the effects of position (Figure 5.3d). See text for more explanation.

herbivory we measured on the plant (Figure 5.3a), in the presence of spiders (Figure 5.3b), and with higher levels of aphid infestation (Figure 5.3c). Overall, a monarch on the top of a leaf or in the flower buds (vs. on the bottom of a leaf or in the plant apex) was less likely to survive, but position on the plant affected survival differently for the different monarch ages (Figure 5.3d; note that monarchs on the stem or in flowers are not included in this analysis because of low sample sizes in these positions, see Figure 5.4). Finally, monarchs on plants without flowers (vs. those that were flowering, had finished flowering, or had seed pods) were more likely to survive (Figure 5.3e).

A post-hoc logistic regression that included, in addition to the significant variables from the full model, interactions between monarch stage and position showed that eggs on the bottom of leaves and first instars on the plant apex were more likely to survive, while eggs in buds and second instars on leaves (either the bottoms or tops) were less likely to survive (Table 5.2b, Figure 5.3d). Other stage-position interactions were not significant. The following predictors were insignificant in both models: presence of other predators (besides spiders), monarch density, date, milkweed density, time in current stage, year, air temperature, dew point, and precipitation. Individual logistic regressions of each predictor confirmed the findings of the multivariate models (data not shown).

As we monitored, we noted that monarchs at different life stages tended to be found at different positions on the plant (Figure 5.4). We tested the significance of this observation using a chi-square test. For this analysis, we counted each egg only once, deleting subsequent observations of the same egg to avoid multiple counts of a single choice of egg-laying location. We counted each observation of a larva separately, since larvae can move and thus can make multiple choices. Because we are not representing survival, the analysis includes all monarch observations, including those made after intervals greater than 24 hours, and third, fourth, and fifth instars. Most monarch eggs, first instars, and fifth instars were found on the bottom of leaves. First and second instars were more likely found on the plant apex than other stages, and monarchs appeared more likely to move to leaf tops as they got older. One could argue that the higher survival of monarchs in some positions (Table 5.2b) could account for the higher frequencies of observations in these positions. This would not be the case for eggs, since we counted eggs only the first time we saw them for this analysis. While it could affect first and second instars on the

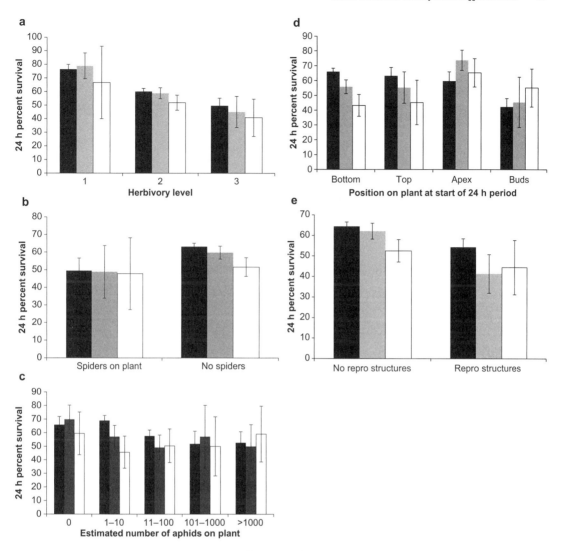

Figure 5.3. Relationship between monarch survival and (a) Herbivory (see Table 5.1 for explanation of levels of herbivory) (b) Spider presence (c) Aphid infestation level (d) Position on the plant or leaves (e) Reproductive status of plant. In all graphs, eggs = black bars, first instars = gray bars, second instars = white bars. Error bars represent 95% binomial confidence intervals.

plant apex, it appears that larvae often move to the apex after hatching, since they were observed there more commonly than eggs. Another possibility is that we were less able to detect eggs in the apex, but we rarely found first and second instars that we had not also observed as eggs, suggesting that this is not the case.

Categories of mortality

Most monarchs that did not survive from one day to the next were simply gone (Figure 5.5). More eggs than larvae were found dead without evidence of predation (our unknown category: 6.5% of eggs; 1.4 and 2.7% of first and second instars, respectively). It is possible that the dead eggs had been parasitized (Barron et al. 2003), but we did not rear them to determine whether this was the case. Many eggs had also been sucked (15%) or chewed (8.4%). We only saw one larva each in the sucked and chewed categories, and only two first-instar larvae that were mired in the milkweed latex. We observed a total of 14 predation events of eggs and larvae over the two years of the study; these predators included lacewing larvae,

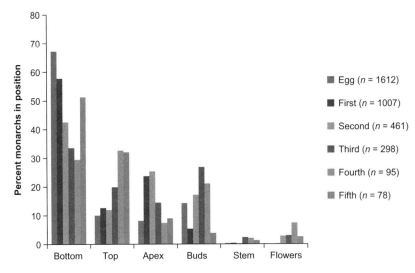

Figure 5.4. Locations of immature monarchs on milkweed plants. There is a significant association between life stage and position ($\chi^2 = 410$, df = 15, $P = 0.0001$; monarchs on stems and flowers not included in analysis because of small sample sizes). Large (> 20) contributions to χ^2: more than expected eggs on bottom, fourth and fifth instars on top, first and second instars on apex, third instars in buds; fewer than expected third instars on bottom, eggs on apex, first instars on buds.

stink bug nymphs, and spiders. We also observed ants attacking, but not killing, a third-instar larva.

While we observed few predation events, we did see many predators, with increasing prevalence of ants over the course of each summer, and many lacewing larvae (adults were not counted), with higher lacewing levels in 2007 (Figure 5.6). Aphids were present on many plants during both years; as with ant presence, aphid infestation levels increased over the course of each summer (Figure 5.7). There were more aphids in 2007.

The fact that we did not see a relationship between precipitation and mortality suggests that rain does not wash monarchs off plants. Additionally, out of the 802 eggs laid by captive females on six potted milkweed plants, all but one remained on the plant after 30 minutes of watering by a sprinkler. During this period, 25.4 cm of water fell from the sprinkler into the rain gauge, a rate higher than all but the very heaviest of rain storms.

DISCUSSION

Mortality rates

As found in earlier studies (Borkin 1982; Zalucki and Kitching 1982b; Oberhauser et al. 2001, this volume, Chapter 6; Prysby 2004), few monarchs survive their first several days of life. The daily survival rates we observed suggest that approximately 20% of eggs become larvae, fewer than 10% survive to become second instars, and fewer than 2% to become third instars. Eggs had significantly higher daily survival rates than first or second instars, as is common in Lepidoptera (Zalucki et al. 2002). Most of the eggs and larvae that did not survive from one day to the next were simply gone from the plant, as was true in other studies of lepidopteran mortality (Zalucki et al. 2002).

Most monarch eggs were found on the bottom of milkweed leaves (Figure 5.4), where they were more likely to survive (Table 5.2). Residing on the bottom of a milkweed leaf is likely to offer protection from flying predators; that these predators are important was demonstrated in exclusion studies by Prysby (2004). Alternatively or additionally, residing on the bottom of a leaf could protect monarchs from direct sunlight, which might lead to desiccation or overheating (see Nail et al., this volume, Chapter 8). This leaf position might also prevent exposure to ultraviolet radiation and reduce the chance of being washed off by rain. Precipitation, however, did not affect the chance of eggs disappearing in our observational study, and almost all eggs remained on the milkweed plant during our experimental rain treatment. Nonetheless, it would be valuable to study the effects of precipitation on larvae and on longer-term survival of eggs in more detail; we measured only

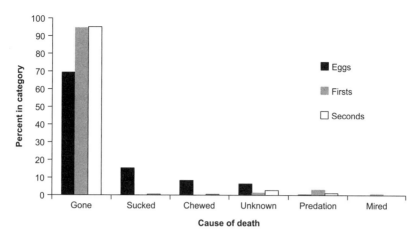

Figure 5.5. Immature monarch mortality. The percent of individuals that were dead or assumed dead in each of the categories measured.

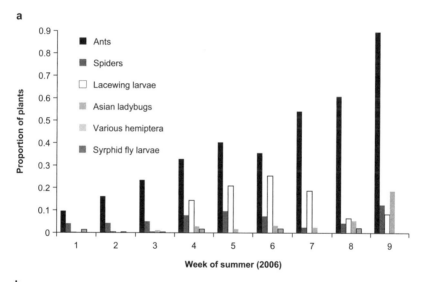

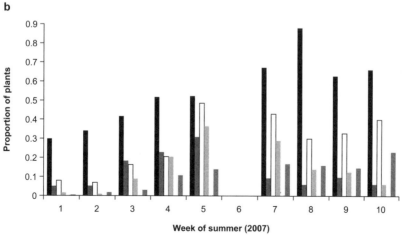

Figure 5.6. Proportion of plants with different potential predators present in (a) 2006 and (b) 2007, with Week 1 starting on 31 May in 2006 and 28 May in 2007. Years shown separately to illustrate year-to-year variation. No data shown for Week 6 in 2007 because of low sample sizes. Note that syrphid fly larvae were recorded only in 2007.

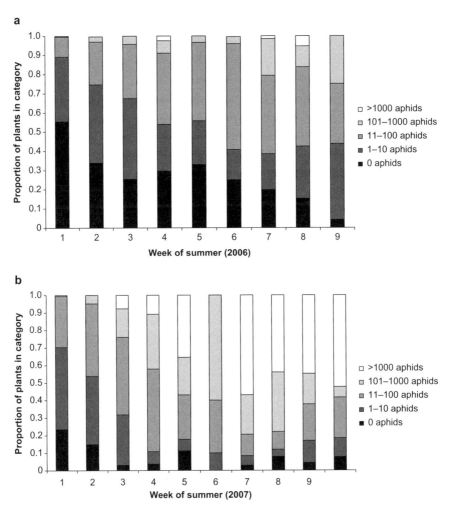

Figure 5.7. Proportion of plants with different levels of aphid infestation in (a) 2006 and (b) 2007, with Week 1 starting on 31 May in 2006 and 28 May in 2007. Years shown separately to illustrate year-to-year variation.

whether the eggs were knocked off the plant, not whether they survived, and we did not assess larvae.

While 58% and 43% of first and second instars, respectively, were also found on the bottom of leaves, many of them had moved to the apex of the milkweed plant, which often contains leaves that are tightly bound and may serve as a hiding place from predators; in fact, first instars in the plant apex were more likely to survive. It is probably difficult for ovipositing females to reach the space inside the vertical leaves on the plant apex, but small larvae can crawl into this space. Newer leaves also tend to contain more nitrogen (e.g., Wilson et al. 1988), so there may be multiple benefits to moving higher on the plant.

Third and fourth instars showed the most positional variation, but fifth instars appeared to move back to the bottom of leaves, possibly because they are so apparent to potential aerial predators. The differences in position between monarch ages, and the different effects of position on survival, at least for eggs through second instars, suggest that further research on the effects of position on the likelihood of predation would be valuable.

Importance of predators

Zalucki et al. (2002) questioned the assumption by previous researchers that immature Lepidoptera that were gone from one observation period to the next had been eaten by predators. Because most of the monarchs that we assumed had died in this study were simply gone, this point made by Zalucki et al. (2002) is relevant to our findings. We can be quite certain that monarchs that were chewed or sucked or that we saw being consumed by predators died from predation. We are less certain about those that disappeared, but several lines of evidence suggest that predators were important sources of mortality. First, the level of milkweed damage by herbivores was negatively correlated with monarch survival. Generalist invertebrate predators might home in on volatile organic compounds (Tumlinson et al. 1993) emitted by damaged plants, or on visual cues associated with herbivory. An alternative hypothesis for this correlation is that plants that had suffered high levels of herbivory were of poorer quality, but low food quality is more likely to affect growth and long-term survival than daily survival, and it would obviously not affect eggs.

Second, the presence of flowers had a negative effect on survival. Flowers are likely to attract predators, such as many beetles, that might come to the plant to obtain nectar for themselves, or to prey on nectivores.

Third, we are quite confident that most missing eggs had been consumed by predators. This assumption is problematic for eggs that hatched between observations, since larvae often fail to establish a feeding site after hatching (Cornell and Hawkins 1995; Zalucki et al. 2002). However, there was no effect of egg age on survival; in fact, survival from day 4 to day 5, when most eggs hatched, was 65% (slightly higher than overall daily egg survival). If there had been a large effect due to failure to establish a feeding site, we would have expected lower survival during this time interval.

First-instar monarchs can suffer reduced survival as a result of miring in the sticky latex exuded by damaged milkweed leaves, or from catalepsis, a symptom of cardenolide toxin poisoning (Zalucki and Brower 1992; Zalucki et al. 2001a, 2001b, 2012); indeed, monarch larvae of all ages have several strategies for avoiding latex consumption (Plate 5). Larvae could have suffered catalepsis and fallen off the plant (and been in our gone category) or died (and been in our dead for unknown causes category). Our fourth reason for arguing that predation explained most of the mortality in this study is that *A. syriaca* has relatively low levels of cardenolides and high levels of latex (Malcolm 1991; Zalucki and Malcolm 1999). We would thus expect higher levels of miring than catalepsis, but we saw only two mired caterpillars over the entire study. To study the effects of plant defenses on larval mortality, Zalucki and Malcolm (1999) severed *A. syriaca* leaves to prevent latex flow, which also decreased cardenolide consumption, since the latex contains higher levels of cardenolides than leaf tissue. Severed *A. syriaca* leaves did not increase larval survival in their study, supporting our contention that monarchs feeding on this host suffer low rates of plant-induced mortality.

Finally, we observed many predators on the plants, although monarch survival was significantly correlated only with spider presence. Spiders feed almost exclusively on insects, and rarely show prey specificity (Reichert and Lockley 1984). Jumping spiders (Salticidae), which we observed often and were the only spiders we actually saw consuming monarchs, search for their prey, a strategy that is likely to be more effective for eggs and larvae than sit-and-wait or web-building strategies. Because most spiders are generalists, the milkweed community, with its large arthropod assemblage, could result in either benefits or costs to immature monarchs. The abundance of prey could attract predatory spiders, resulting in higher predation on monarchs, or the presence of alternative prey could reduce spiders' use of monarchs.

Although ants were present throughout the summer breeding season over both years, their presence on plants did not correlate with monarch survival; however, we observed ants attacking a third-instar larva. It is possible that the act of monitoring had a negative effect on immediate ant presence, although we tried not to disturb invertebrates on the plants as we monitored. Also, we recorded only ant presence, not abundance; a more detailed index of ant numbers might have led to different results. There was an effect of aphid infestation level on monarch survival. Aphids may indirectly affect monarch survival by attracting tending ants as well as other generalist predators.

We observed many other known monarch predators (reviewed by Oberhauser et al., this volume, Chapter 6), including stink bug nymphs, lacewing larvae, syrphid fly larvae, Asian lady beetles, and several other Hemiptera (including assassin bugs), but none of these correlated with monarch survival. Our use of presence/absence data for predators for analysis was probably less likely to detect an impact than indices of abundance, and it is also likely that many predators came and left during the time we were not observing individual plants. We suggest that future research involve experiments on individual predators, like those summarized by Oberhauser et al. (this volume, Chapter 6).

Predators that suck out the contents of their prey (Hemiptera and spiders) without consuming the whole carcass may have been more common for eggs than larvae (Figure 5.5), but it is also possible that carcasses of larvae fell off the plant. The fact that most eggs and larvae were gone suggests that most of their predators consume or remove the entire individual.

It is also important to note that we did not quantify the mortality of later instars, because of the difficulty of tracking them. Larger prey are generally consumed by larger predators, whereas small prey are consumed by both small and large predators (Cohen et al. 1993). Thus, being large offers protection from many predators; however, mortality from tachinid flies, which attack monarchs during all larval instars (Oberhauser 2012), does not occur until the late-larva or early-pupa stage.

Predation, which occurs at the individual level, provides a link to population level dynamics. While this study was specific to one location, one host plant species, and a short time period, it highlights the fact that invertebrate predators can be very efficient at reducing immature monarch numbers over short time intervals.

ACKNOWLEDGMENTS

Leah Alstad, Rebecca V. Batalden, Grant Bowers, Matt Kaiser, and Wendy Macziewski provided help during various stages of this work, and Leslie Ries, Kelly Nail, Patrick Guerra, and Elizabeth Howard made helpful comments on an earlier version of the manuscript. Funding was provided by the National Science Foundation (DGE-0440517 to KSO), the Dayton and Wilkie Funds for Natural History (University of Minnesota), and the Monarchs in the Classroom Program. This research was completed in partial fulfillment of the requirements for a master's degree by Alma De Anda in the University of Minnesota's Department of Ecology, Evolution and Behavior.

6

Lacewings, Wasps, and Flies—Oh My

Insect Enemies Take a Bite out of Monarchs

Karen S. Oberhauser, Michael Anderson, Sophia Anderson, Wendy Caldwell, Alma De Anda, Mark Hunter, Matthew C. Kaiser, and Michelle J. Solensky

We review published accounts of monarch natural enemies and present new findings on three generalist enemies, a parasitoid wasp *Pteromalus* sp. (probably *P. cassotis* Walker), larvae of the green lacewing *Chrysoperla rufilabris*, and tachinid flies *Lespesia archippivora*. We discuss our findings in light of risks to nontarget species posed by generalist biocontrol agents, and protections gained by monarchs from their milkweed hosts.

INTRODUCTION: DESPITE BEING TOXIC, MONARCHS ARE CONSUMED BY MANY PREDATORS

Monarchs have a well-studied and biologically interesting interaction with their milkweed host plants, and this interaction in turn affects their interactions with natural enemies. They sequester toxins from milkweed, thus becoming toxic themselves (Brower 1984; Malcolm and Brower 1989), and their aposematic coloration presumably warns predators of this toxicity (Reichstein et al. 1968). Despite this protection, however, monarchs have many predators and parasitoids that tolerate or avoid the toxins they contain (Plate 14). In this chapter, we summarize previous research on monarch natural enemies, and present findings from three new studies.

Predation and parasitism of monarch eggs, larvae, and adults (but not pupae) have been well-studied (Table 6.1). Monarch eggs and early larval instars suffer high mortality rates (up to 90% or higher: Borkin 1982; Oberhauser et al. 2001; Prysby 2004; De Anda and Oberhauser, this volume, Chapter 5). While host plant defenses cause some larval mortality, especially of first instars (Zalucki and Brower 1992; Zalucki and Malcolm 1999; Zalucki et al. 2001a, 2001b, 2012), predators probably have larger impacts on monarch larval survival (Prysby 2004; De Anda and Oberhauser, this volume, Chapter 5).

Some studies of predation on monarch larvae have specifically addressed their toxicity. Rayor (2004) found that the paper wasp *Polistes dominulus* discriminated against monarch larvae that had consumed more toxic milkweed (*Asclepias curassavica*) compared with those fed the less toxic *A. incarnata* or *A syriaca*, and also preferred less toxic caterpillar species to monarchs. Rafter et al. (2013) found that Chinese mantids *Tenodera sinensis* gutted monarch larvae before consuming them, but not caterpillar species that were less toxic, possibly to avoid the toxic gut contents of the monarchs. In Mexico, black-backed orioles and black-headed grosbeaks consume large numbers of monarch adults (Calvert et al. 1979; Fink et al. 1983; Brower and Calvert 1985); the orioles avoid the toxins in the cuticle by slitting open the abdomen and eating the fat body, while the grosbeaks appear to be able to tolerate at least a certain amount of the toxins and consume the whole abdomen.

Early work by Brower (1984) and subsequent studies (e.g., Malcolm 1991; Martin et al. 1992; Malcolm et al. 1989; Agrawal et al. 2012b) addressed the variation in cardenolide levels within and between milkweed species, and between monarchs that consume plants with different cardenolide levels. A direct correlation is not always found between host and herbivore toxicity, but monarch cardenolide

Table 6.1. Predators and parasitoids known to attack monarchs

Stage of attack (study location)	Predator	Reference(s)	Notes
Eggs (MN, lab study)	Lacewing larvae	This paper	Up to 40 eggs consumed in 24 hours
Eggs (WI)	Ants (*Formica montana*)	Prysby 2004	*F. montana* ate introduced eggs within 90 minutes, egg survival lower on plants with ants
Eggs and larvae (TX)	Fire ants	Calvert 2004a	Exclusion study: up to 100% mortality when fire ants were not excluded
Eggs and early larvae (MN lab and field trials)	*Harmonia axyridis*	Koch et al. 2003, 2005	Beetles consumed monarchs in lab and field trials (up to 25 eggs or 15 larvae per day), fewer monarchs consumed in presence of alternate prey
Eggs and larvae (Australia)	Various insects and spiders	Smithers 1973; Zalucki and Kitching 1982b	Anecdotal observations: lady beetles, cockroach, ants (eggs); various spiders, vespid (*Rhopolidia revolutionalis* and *Polistes variabilis*) and sphecid (*Chlorion* sp.) wasps, pentatomid bug (larvae)
Larvae (US, including HI; Australia)	Tachinid flies	Arnaud 1978; Borkin 1982; Etchegaray and Nishida 1975a; Lynch and Martin 1993; Oberhauser 2012; Oberhauser et al 2007; Prysby 2004; Smithers 1973; Urquhart 1960; Zalucki 1981a	Parasitism rates from field-collected monarchs: rates range from 1% to 100% (but monarch stadium at collection not always given). Australia: *Winthemia diversa, Sturmia* sp. U.S. (incl Hawaii): *Lespesia archippivora*
Larvae (NY, lab study)	*Polistes dominulus*	Rayor 2004	Wasps captured and consumed monarch larvae in lab: preference for medium-sized larvae, and for more palatable species if given a choice
Larvae	Chinese mantids	Rafter et al. 2013	Introduced (and commercial biocontrol species) *Tenodera sinensis* gutted and consumed larvae in field and lab
Larvae (WI, MN)	Pentatomid nymphs, jumping spiders, ants	Borkin 1982; De Anda and Oberhauser, this volume	Field observations
Larvae (TX, LA)	Crab spiders, ants	Lynch and Martin 1993	Field observations
Prepupae and pupae (MN)	*Polistes dominulus*	This chapter	Wasps found and killed prepupae and pupae in a field experiment
Pupae (MN)	*Pteromalus* sp. (probably *P. cassotis*)	This chapter	Generalist pupal parasitoid attacked and killed newly pupated monarchs; rates may reach 20% in wild.
Adults (Mexico wintering sites)	Birds	Brower and Calvert 1985; Calvert et al. 1979; Fink and Brower 1981; Fink et al. 1983	Scott's oriole, black-backed oriole, black-headed grosbeak main species, others observed rarely: 60–84% of dead butterflies predated, overall rates ~9%, with more predation in smaller colonies
Adults (Mexico wintering sites)	Mice	Brower et al. 1985; Glendinning and Brower 1990	*Peromyscus melanotis* was the most important mouse predator, consuming mainly dead or moribund monarchs on the ground
Adults (CA wintering sites)	Birds	Sakai 1994	Rufous-sided towhees: ~7% annual mortality (preferred tagged monarchs)
Adults (CA wintering sites)	Wasps	Leong et al. 1990	Field observations of *Vespula vulgaris*
Adults (Australia)	Birds, mammals	Smithers 1973	Rufous whistler, currawong, pallid cuckoo, black-faced cuckoo-shrike, fantailed cuckoo observed; unknown small nocturnal mammal (speculated)

Stage of attack (study location)	Predator	Reference(s)	Notes
Adults (eastern North America)	Dragonflies	White and Sexton 1989	Observed predation events along lakeshore in Michigan
Adults (Australia and eastern North America)	Spiders	Smithers 1973	Adults trapped in webs
Adults (Hawaii)	Birds (bulbuls)	Stinson and Berman 1990	Selective predation of orange monarchs by introduced bulbuls (*Pycnonotus jacosus* and *Pycnonotus cafer*) may be driving increase in white morph
Adults (Australia)	Mantids	Smithers 1973	Anecdotal field observation

Notes: In addition to the published reports cited, we have received photographic evidence of predation on monarch adults by a leopard frog (*Rana* sp.) (from Ridland Kiphart in Texas), and praying mantises (possibly *Stagmomantis carolina*, but species not identified with certainty; Charlie Gatchell and Carole Jordan in Louisiana and West Virginia, respectively). KSO observed an adult monarch caught in an orb web and being consumed by the spider in northern WI.

levels tend to increase up to a certain point relative to the concentration in the plants they consume; once that point has been reached, monarch toxicity does not increase further with host plant toxicity (Agrawal et al., this volume, Chapter 4).

Monarch parasitoids

Parasitoids develop by feeding in or on a host organism, causing its eventual death; they are probably the most important source of mortality for herbivorous insects (Godfray 1994; Hawkins et al. 1997). About 78% of parasitoid species are Hymenoptera (wasps) and most of the remaining species (20%) are Diptera (flies) (Feener and Brown 1997). Both fly and wasp parasitoids are used as biological control agents, but a poor understanding of the host-parasitoid system can lead to negative effects of biocontrol on nontarget organisms. For example, the generalist tachinid fly *Compsilura concinnata*, introduced to control gypsy moths in 1906, has been implicated in native silk moth declines in the northeastern United States (Boettner et al. 2000).

Monarch parasitoids are reported to include 12 species of tachinid flies and at least one brachonid wasp (Arnaud 1978). The best-studied monarch parasitoid is the tachinid fly *Lespesia archippivora* (Table 6.1), which attacks larvae, resulting in the death of late-instar larvae or pupae. Many researchers have studied rates of parasitism by *L. archippivora* by collecting and rearing wild larvae (Urquhart 1960; Smithers 1973; Borkin 1982; Prysby 2004; Oberhauser et al. 2007; Oberhauser 2012). These studies found parasitism rates of up to 90% in some collection sites, with average rates for monarchs collected as fourth- or fifth-instar larvae usually between 10 and 20%. Oberhauser (2012) found that high-density monarch years tend to be followed by years with high rates of tachinid fly parasitism.

Successful parasitoids must locate and penetrate their host and circumvent the host immune response. They must then acquire adequate nutrition from the host internal environment, making is likely that host diet affects parasitoid fitness. For example, larvae of the fly parasitoid *Thelairia bryanti* were more likely to survive on woolly bear caterpillars *Platyprepia virginalis* fed lupine (less toxic) than hemlock (more toxic) (English-Loeb et al. 1993). Cardenolide sequestration by monarchs is likely to affect parasitoids, but little is known about parasitoid survival rates within monarchs (but see Hunter et al. 1996; Sternberg et al. 2011), the ability of monarchs to defend themselves from parasitoids that have penetrated their bodies, or differential attractiveness of monarchs to parasitoids based on the toxicity of their host plant.

Effect of biocontrol agents on monarchs

Many biocontrol agents introduced to control agricultural and garden pests are also likely to pose a risk to monarchs. *L. archippivora* was introduced to Hawaii in the late nineteenth century to control armyworms *Spodoptera frugiperda* (Swezey 1923, 1927, cited in Etchegaray and Nishida 1975a). The Asian lady beetle *Harmonia axyridis* was introduced to the United States multiple times beginning in 1916 (Gordon 1985) to control agricultural pests, mainly aphids. *H. axyridis* can also consume monarchs (Koch et al. 2003, 2005, 2006). Larvae of green lacewings, a diverse and widespread family (Chrysopidae) in the order Neuroptera, are frequently used for biological control in agriculture (Daane et. al. 1996, 1997; Goolsby et. al. 2000; Knutson and Tedders 2002; Fondren et. al. 2004; Day et. al. 2006) and gardens (Chang et. al. 1995; Silvers et. al. 2002). Their generalist feeding behavior suggests that they could affect nontarget species, and they are common on milkweed plants (pers. observ., all authors).

Here, we report on three new studies that address gaps in our knowledge of monarch natural enemies. These studies lead to further understanding of mortality agents during the pupal stage, the outcome of parasitism events by the tachinid fly *L. archippivora*, and the potential for deleterious impacts of green lacewings.

WHAT EATS MONARCH PUPAE?

Methods: Pupal enemies

We exposed monarch prepupae (larvae that have stopped eating) and pupae to natural predators and parasitoids by putting them outside in a series of experiments described below. In each experiment we transferred monarchs to individual 1 liter plastic containers in the laboratory after 7–9 days of outdoor exposure, and recorded their fates. If no parasitoids emerged from dead pupae, we dissected them 21 days after the last parasitoids had emerged from other monarchs in their cohort to determine whether parasitoids were present. For all experiments, we kept control pupae in the lab to measure background mortality rates; because there was almost no mortality in control groups, analyses do not include them. In a pilot study in July 2008, we documented parasitism by a wasp parasitoid *Pteromalus* sp. (family Pteromalidae, superfamily Chalcidoidea; first identified by M. W. Gates, Department of Entomology, Smithsonian Institution, as *Pteromalus puparum*. Later experimental work in our lab suggests that this wasp is actually *Pteromalus cassotis*, for which the first and only reports in monarchs were made in the late 1880s [Gillette 1888; Howard 1889]). Our experiments were conducted by A. De Anda, W. Caldwell, and K. Oberhauser (experiments 1 and 2), and by S. Anderson and M. Anderson, middle school students at the time (experiment 3). All used F1 or F2 offspring of wild-caught individuals from Minnesota.

Experiment 1: We exposed monarch pupae to natural enemies in early September 2008. Monarchs were reared in individual petri dishes until the fourth instar, then moved into 10 wood-frame screen cages (60 × 30 × 30 cm with 1 mm aperture mesh; 30–50 monarchs per cage) with fresh bouquets of *A. syriaca*. We placed cages in three garden sites in St. Paul, Minnesota, 2–3 days before the monarchs pupated, thus exposing local parasitoids to frass and plant volatile organic compounds (VOCs), potential host-finding cues (Tumlinson et al. 1993, De Moraes et al. 1998). We opened cages when almost all larvae had finished eating, and left frass and milkweed stalks in the cages.

Experiment 2: In August 2010, we assessed the effects of host plant species on *Pteromalus* parasitism. We reared experimental monarchs as before, randomly assigning half to feed on *A. syriaca* and half to *A. curassavica*, milkweed species with different levels of cardiac glycosides that result in different levels of cardenolide storage within the larvae (Malcolm 1991). We fed larvae cuttings from potted (*A. curassavica*) or wild-collected (*A. syriaca*) plants throughout their development, moving them to the wooden frame cages as fourth instars (26–27 larvae per cage). As monarchs finished feeding, cages were randomly assigned to 4 sites (all on or within 1.5 km of the University of Minnesota [UM] St. Paul campus) that contained naturally growing *A. syriaca*. Cages, all containing milkweed and frass, were placed next to each other in 5 pairs; each pair included one cage with larvae fed *A. curassavica* and one cage with larvae fed *A. syriaca*. Three pairs were placed in gardens (two on either side of a home [labeled Roseville east and south in Figure 6.1] and one on the UM campus), one in a strip of prairie plants between a UM roadside and a field with a variety of vegetable crops, and one in a small milkweed patch within the same field. Unlike the 2008 treatment, cages were kept closed throughout the experiment, after *Pteromalus* females were observed accessing monarch pupae by crawling through the screen mesh. This modification prevented predation by *Polistes* wasps (see below).

Experiment 3: We exposed uncaged, individual pupae in 2010 ($N = 50$) and 2011 ($N = 60$). These individuals had been fed cuttings of *A. syriaca* and allowed to pupate in 1 liter plastic rearing containers. Five to 30 hours after pupation, they were carefully removed from the rearing containers and placed in a restored prairie at the UM Arboretum in Chanhassen, Minnesota. In August 2010, we used a thread tied to the silk and cremaster (the black "stem" at the top of the pupa) to attach 10 monarchs separately in each of the following locations, chosen to mimic places in which we have observed monarch pupae: on milkweed, on milkweed with frass, on goldenrod (*Solidago* sp.), hidden on a human-built structure (such as a bench or shelter), or on a tree. In 2011 we used glue to attach the cremaster and silk to a wooden clothespin and exposed 15 monarchs in each of the following conditions: on milkweed with or without frass (early September), or on other plants with or without frass (early August).

Data interpretation and calculation of parasitism rates. Experiment 1 allowed some monarch escape because cages were opened while a few monarchs were still crawling, and some monarchs were not refound in Experiment 3. In Experiment 1, almost all monarchs at one site disappeared. Crawling larvae might have escaped, but larvae that were hanging in the "J-shape" or pupae must have been eaten by predators. Predators left the pupal cremasters remaining; we also recovered cremasters in Experiment 3. In a separate study in 2009 we saw paper wasps (non-native *Polistes dominula*) remove crawling larvae, prepupae, and pupae in pieces from cages. The remaining cremasters and speed of removal were similar to what we observed in Experiment 1, and we assumed that *P. dominula* were responsible for other predation.

We defined 5 experimental outcomes: *survived* = monarch adults emerged, *taken* = monarchs consumed by *P. dominula*; *died* = monarchs that did not emerge as adults and contained no visible parasitoids when dissected; *Pteromalus* = monarchs from which *Pteromalus* adults emerged or that contained *Pteromalus* larvae, pupae, or adults when they were dissected; and *missing* = fate could not be assigned with certainty (crawling larvae that either escaped from cages or were consumed by wasps in 2008, or pupae that were not relocated in Experiment 3).

The marginal parasitism rate (m_p) was calculated as

$$m_p = 1 - (1-q)^{\frac{q_p}{q}}$$

where q_p = the apparent mortality from parasitism (Pteromalus/total N in treatment) and q = the mortality from all combined causes (taken + died + Pteromalus). The marginal parasitism rate is used when there are multiple agents of mortality and no knowledge of the results of competition between these agents; it can be thought of as the level of mortality that would have occurred if the parasitoid were acting alone (Bellows et al. 1992; Elkinton et al. 1992; Barron et al. 2003). This is appropriate because *P. dominula* could have consumed parasitized monarchs, or parasitized monarchs could have died of other causes before the parasitoids were detectable.

In Experiment 1, we calculated predation rates by *P. dominula* in two ways: using only monarchs for which the fate was known (taken/[total N − missing]), or assuming that missing monarchs were killed by the wasps ([taken + missing]/[total N]). The true value lies somewhere between the two. Because it is unlikely that predators removed all evidence of monarchs in Experiment 3, we assumed that missing pupae were simply not refound and did not include them in analyses.

Results: Pupal enemies

Predation rates by *P. dominula* were variable across treatments and locations in Experiment 1 (Table 6.2), ranging from 0 to 76% (or up to 97% if we assume that all missing monarch larvae were attacked by paper wasps). All three monarch stages (larvae, prepupae, and pupae) were attacked. In the 2010 round of Experiment 3, wasps consumed 17% of the 36 pupae with known fate. We found no

Table 6.2. Summary of results in pupa mortality study

Experiment	Treatment	Date	N[a]	Missing[b]	Polistes predation w/o (w/) missing[c]	Marginal Pteromalus w/o (w/) missing[d]
1[e]	Open cages	Sept 2008	120	7	0.00 (0.058)	0.59 (0.58)
			92	41	0.039 (0.47)	0.35 (0.26)
			192	167	0.76 (0.97)	0.00 (0.00)
2[f]	A. syr: Closed	Aug 2010	134	0	NA	0.87
	A. cur: Closed		133	0	NA	0.60
3[g]	Various	Aug 2010	50	14	0.17	0.072
	Non-MW, frass	Aug 2011	15	4	0.00	0.091
	MW, frass	Sept 2011	15	2	0.00	0.62
	Non-MW	Aug 2011	15	3	0.00	0.00
	MW	Sept 2011	15	3	0.00	0.087

[a] Number of monarchs exposed
[b] Monarchs that were not found and for which there was no clear evidence of Polistes predation (remaining cremaster or a monarch that had been exposed as a hanging J that could not have crawled away)
[c] Assumes that no (or all) missing monarchs were consumed
[d] Does not (or does) include missing monarchs in the calculation of parasitism rates
[e] Experiment 1: 3 sites reported separately; monarchs exposed in open cages containing ≤50 monarchs
[f] Experiment 2: Host plant treatments but not sites reported separately; monarchs exposed in closed cages containing ≤50 monarchs
[g] Experiment 3: 2010, all treatments combined; 2011, treatments reported separately; monarchs exposed individually in both years. Because missing monarchs in Experiment 3 were unlikely to have been consumed by Polistes wasps, we did not include missing monarchs in calculations of Polistes predation for this experiment. For treatment details, see text.

remaining cremasters that indicated wasp predators in 2011.

Pteromalus parasitism rates are shown in Table 6.2. In Experiment 1, only six monarchs were recovered from one site (bottom row in Experiment 1); all the rest were either missing or eaten by paper wasps. None of the recovered monarchs from this site were parasitized, but *Pteromalus* wasps had attacked monarchs in both other sites. In Experiment 2, a significantly higher proportion of monarchs that consumed *A. syriaca* (vs. *A. curassavica*) were parasitized, and monarchs in the sites near agriculture fields on the UM campus (the prairie and ag field sites) suffered significantly more parasitism (Table 6.2 and Figure 6.1).

In 2010, two of the monarchs in Experiment 3 (out of 36 with known fates) were parasitized by *Pteromalus* wasps (Table 6.2), one on a milkweed plant with frass on the leaves, and one on a goldenrod plant. In 2011, 10 monarchs were parasitized (out of 48 with known fates), eight from milkweed plants with frass, one from another plant with frass on it, and one from a milkweed plant without frass. Marginal mortality rates suggest 7% mortality due to *Pteromalus* wasps in 2010 and 0–62% in 2011, depending on treatment (Table 6.2). In 2011, sig-

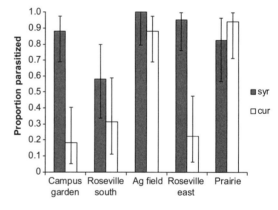

Figure 6.1. Proportion of monarchs parasitized in each site × treatment combination in Experiment 2. Errors bars represent 95% binomial CI. syr = *A. syriaca*, cur = *A. curassavica*. Unweighted logistic regression: coefficients for syr = 2.68, $P < 0.0001$; ag field = 3.8, $P = 0.0001$; prairie = 4.59, $P = 0.0001$; prairie*syr interaction = −4.01, $P = 0.002$. Overall deviance = 161.7, df = 188, $P = 0.918$.

nificantly more monarchs in the milkweed/frass treatment were parasitized (logistic regression: coefficient for frass*milkweed interaction on likelihood of being parasitized = 3.20 + 0.88 SE, $P = 0.0003$).

HOW DO HOST PLANT AND PARASITOID NUMBER AFFECT THE OUTCOME OF MONARCH/PARASITOID INTERACTIONS?

Methods: Host plant species and tachinid fly parasitoids

In fall 2007, we randomly assigned monarch larvae to plants of one of five host plant species (Table 6.3) upon hatching, and reared them on the assigned species until they died or pupated. We exposed them individually to a single mated female *L. archippivora* as second instars and inspected each larva under a dissecting microscope at 10–20× magnification to count the number of collapsed fly eggs (the number of parasitoid larvae that hatched and entered the host [Solensky unpublished data; Etchegaray and Nishida 1975a]). We moved each parasitized larva onto an individual potted plant of its assigned host plant species. This experiment was conducted by M. Solensky at the College of Wooster.

To measure host plant chemistry, we removed 12 leaf disks (5 mm diameter each) from pairs of leaves on the top, middle, and bottom of each plant. We stored the disks in 1.5 ml of 100% methanol to begin cardenolide extraction, then dried and weighed them. We then ground, vortexed, sonicated at 60 °C for an hour, centrifuged, decanted, dried, and resuspended the samples in 1 ml methanol with 0.15 mg/ml digitoxin as an internal standard. These samples were vortexed, centrifuged, and filtered through a syringe with a 2 μm pore filter, and then run through a high-performance liquid chromatography machine and mass spectrophotometer to quantify the number of distinct cardenolides and their concentrations. The plant chemistry work was done at the University of Michigan by M. Hunter.

We enclosed each plant in a cage consisting of two 60 cm bamboo stakes covered with a mesh bag, then closed by a rubber band around the pot. Host larvae were moved to new plants when they had consumed all leaf material, then into individual 700 ml plastic containers as fourth instars, the earliest that *L. archippivora* emerges from monarchs (Solensky and Oberhauser unpublished; Etchegaray and Nishida 1975a; Stapel et al. 1997), and continued to feed them their assigned milkweed species. We recorded either host or parasitoid survival (they never both survive). When parasitoid larvae emerged from their hosts, we moved them to a small petri dish (1 dish per host larva) and monitored them daily for fly emergence. We recorded the number of parasitoids that emerged from each host, and puparium mass seven days after emergence.

Results: Host plant species and tachinid fly parasitoids

There was significant variation in milkweed cardenolide concentration and number of distinct cardenolides between the different hosts (Table 6.3). This finding is consistent with past research (Malcolm 1991; Agrawal et al. 2008).

Female *L. archippivora* laid eggs in what appeared to be one swift oviposition event, and we counted one to five fly eggs on each monarch. We used a logistic regression to explore the effect of host plant and the number of fly eggs laid on the monarch on monarch survival. Monarchs on which more fly eggs were laid were less likely to survive (Figure 6.2a), and there was a marginal effect of host plant ($P = 0.079$), with monarchs fed *A. syriaca* less likely to survive (Figure 6.2b).

Even if monarchs are more likely to die when more parasitoids eggs are laid on them, from the female fly's perspective, it might be better to lay fewer eggs per host if individual eggs have a better chance of surviving. One way to address this is to assess the payoff to a female from a single oviposition event; Lack (1947) proposed that birds should produce clutch sizes that lead to the maximum offspring number per clutch, a concept that is relevant to this study. The number of emerging flies increased with the number of eggs laid (Figure 6.3a), suggesting that it is better to lay more eggs, at least within the range of egg numbers in this experiment.

Table 6.3. Plants used in tachinid study with cardenolide levels

Plant species	Mean cardenolide concentration μg/0.1 g dry tissue ± SE	Mean number cardenolides ± SE	Number plants sampled
A. verticillata	0 ± 0 [a]	0 ± 0 [a]	10
A. incarnata	0.448 ± 0.448 [a]	0.033 ± 0.033 [a]	30
A. syriaca	14.6 ± 4.41 [a]	0.50 ± 0.13 [a]	30
A. speciosa	11.2 ± 3.90 [a]	0.53 ± 0.17 [a]	15
A. curassavica	724 ± 41.2 [b]	5.45 ± 0.23 [b]	29

Note: Superscripts within each column denote values that are significantly different from each other (ANOVA $P < 0.01$).

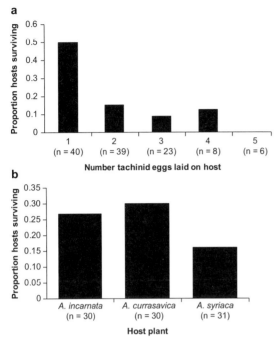

Figure 6.2. Monarch survival in tachinid fly experiments. (a) Effect of the number of *L. archippivora* eggs laid on monarch larvae on larval survival. (b) Effect of host plant species on monarch survival. Unweighted logistic regression: coefficients for number of eggs = −1.38, $P = 0.0001$; syriaca = −1.28, $P = 0.079$. Overall deviance = 105.1, df = 111, $P = 0.639$.

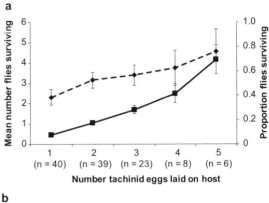

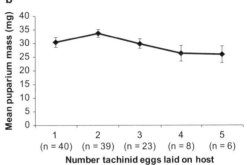

Figure 6.3. Tachinid survival and mass. (a) Mean number (solid line) and proportion (dashed line) of flies that successfully emerged from hosts on which 1–5 *L. archippivora* eggs were counted. Number of eggs laid has a significant positive effect on the proportion of eggs surviving when correcting for the effect of host plant: Arcsin transformed proportion = 0.36 + 0.11 * (number of eggs) + 0.36 * (syriaca) (P for egg number = 0.05, for syriaca = 0.008). (b) Fly puparium mass in relation to number of eggs laid. Spearman's rank correlation = −0.27, $P = 0.017$. Error bars represent SE.

Another way to address the best female strategy is to assess the payoff per egg laid; if female flies can easily find hosts, it would be in their best interest to maximize individual egg survival rather than the number of offspring emerging from each host. However, the proportion of fly eggs that successfully emerged from the monarch actually increased with the number of eggs laid (Figure 6.3a), again suggesting that the best strategy is to lay more eggs per host, at least within the range of egg numbers in this study. However, mean puparium mass from each monarch was negatively correlated with the number of fly eggs laid on a host (Figure 6.3b); if fly fitness is correlated with size, individual flies should fare better when fewer eggs are laid.

DO LACEWING LARVAE EAT MONARCH EGGS?

Methods: Lacewings and monarch eggs

We obtained *Chrysoperla rufilabris* eggs from Beneficial Insectary (Redding, CA) in summer 2007, and placed approximately 10 eggs with a moistened piece of filter paper in petri dishes. We kept the dishes in a controlled environment of 26 °C, 80% relative humidity, and 14:10 hours light:dark. We thinned newly hatched larvae to one larva per dish, and provided each with 1 mg frozen eggs of Mediterranean flour moth *Ephestia kuehniella* (from Beneficial Insectary), a fresh piece of filter paper, and a few drops of water every two days to prevent desiccation. Any remaining food was removed prior to each feeding. We tracked larval development by visual observations of molted integuments.

When lacewing larvae reached the third instar, we removed all food from their dishes for 24 hours before experimental trials. We exposed individual larvae to densities of 5, 10, 15, 20, 30, 40, or 50 prey

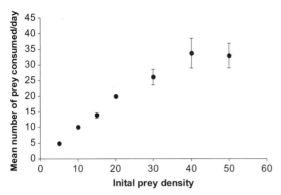

Figure 6.4. Number of monarch eggs attacked by third-instar lacewing larvae as a function of the number of prey available. Error bars represent SE.

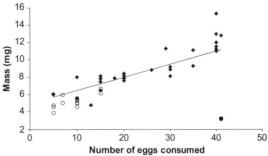

Figure 6.5. Lacewing final larval mass as a function of number of eggs consumed during third instar. Open circles represent lacewings that died before reaching the adult stage; closed circle, an outlier (that lived), not included in the regression. Larva mass = 0.177 * (eggs eaten) + 4.11 mg, $R^2 = 0.65$, df = 1,44, $P < 0.0001$.

(one- or two-day-old monarch eggs, 6–8 lacewing larvae per prey density, total $N = 49$). After 24 hours, we counted the numbers of eggs eaten or damaged and weighed each lacewing larva. For the rest of the experiment, no more food was provided, based on observations that *C. rufilabris* larvae pupate after a day of feeding in their third and final instar. We tracked the date and success of pupation, which usually occurred within one or two days after the feeding trial. This experiment was conducted by M. Kaiser at the University of Minnesota.

Results: Lacewings and monarch eggs

Lacewing larvae consumed up to 41 monarch eggs in 24 hours (Figure 6.4), with mean prey consumption rates of just under 35 eggs during their day-long feeding period in the two highest prey densities (40 and 50 monarch eggs). There was a significant correlation between final larval mass and the number of eggs consumed (Figure 6.5). Of the 15 lacewings that did not survive to the adult stage, all had consumed 15 or fewer monarch eggs and only three were in treatments that received more than 15 eggs. This finding suggests that for most of the lacewings that died, a lack of food and not the toxicity of their food killed them.

DISCUSSION

Pupal enemies

We documented two natural enemies of monarch prepupae and pupae for the first time: the paper wasp *P. dominula*, and the parasitoid wasp *Pteromalus* sp. *Polistes dominula* consumes many prey species (Cervo et al. 2000), including monarch larvae (Rayor 2004), but we are not aware of other reports of consumption of pupae of any species by this predator. Social wasps are opportunistic, generalist foragers that often return to successful hunting sites (Ravert Richter 2000). They provision their larvae with chewed arthropods; in her review of social wasp prey, Ravert Richter (2000) lists caterpillars, flies, alate ants, termites, spiders, bees, and other wasps as common prey items. While pupae are not listed, the behavior that we observed, hunting for live prey and returning to remains of killed prey and successful hunting sites, is typical for this species.

Our *Pteromalus* findings represent a new contribution to our understanding of monarch natural enemies. The wasps that we found were first identified as *P. puparum*, which had been reported as an unsuccessful parasitoid in New Zealand monarchs (Ramsay 1964). However, further studies in our lab have suggested that *P. puparum* is not the species that we found in monarchs (Carl Stenoien and Shaun McCoshun, unpublished research). The most likely candidate for the monarch parasitoid is *Pteromalus cassotis*, which was reported in monarchs by Gillette (1888) and Howard (1889); these early reports were difficult to find because they used an older name for monarchs (*Danaus archippus*). In addition to the findings reported here, we have received photo and specimen verification of successful parasitism by what is probably the same species from M. Monson (from a dead monarch pupa collected May

2006, Walker, MN), C. Johnson (from a monarch pupa observed in August 2010, Lebanon, PA), A. H. Williams (from three pupae collected in Grant and Wood Counties, WI, September 2007, 2008, and 2009), D. Satterfield (from 15 of 39 monarch pupae collected in Savannah, GA, January 2012), and S. Oberhauser (from a monarch pupa collected in Caroline, WI, June 2012).

When we exposed individual pupae, most *Pteromalus* parasitism occurred on milkweed plants with frass on them in 2011 (Table 6.2). Other researchers have identified plant VOCs and host cues (mostly frass) as signals that parasitoids use to locate hosts (Udayagiri and Jones 1992; Tumlinson et al. 1993; De Moraes et al. 1998). Plant VOCs can be detected over long distances, and common milkweed emits VOCs in response to monarch feeding (Wason et al. 2013). Furthermore, both wasps and ants are more likely to attack clay models of monarchs placed on plants on which real monarchs have been feeding (Wason and Hunter 2013). Because monarchs rarely pupate on their host plants, actual parasitism rates in the wild may be lower than those we observed, but we demonstrated that *Pteromalus* sp. can locate isolated monarch pupae even if they are not on milkweed plants with frass. Additionally, while monarch movement away from host plants before pupation could decrease detection, *Pteromalus* females sometimes ride on pre-pupal hosts in the field (Takagi 1986), possibly allowing them to take advantage of cues emitted prior to pupation. This behavior may also ensure that they do not miss the opportunity to lay eggs, since they cannot oviposit into hardened pupae (Askew 1971). In any case, caterpillar movement away from their plants is clearly not a foolproof strategy for avoiding detection by pupal parasitoids responding to cues from frass and host plants; this would be an interesting avenue for further study.

While very little is known about *P. cassotis*, its congener *P. puparum* has a wide host range (Barron et al. 2003; Barron 2007) and is thought to have been accidentally introduced into the United States from Europe in the late 1800s (Oatman 1966). It is an important biocontrol agent (Harvey et al. 2007), with parasitism rates ranging from 4 to 90% on cabbage white *Pieris rapae* and 17 to 69% on two other lepidopteran cabbage pests (Lasota and Kok 1986; Lee and Heimpel 2005). It was introduced to New Zealand in 1932 to control *P. rapae* (Muggeridge 1933), and is still common there. We do not know what caused the differences in parasitism rates in the different sites in Experiment 2 (Figure 1), but patchy distribution of parasitoids is common (Hassell 1978) and could explain differences between sites relatively close to each other; however, the potential for alternative hosts to influence parasitism rates in monarchs would be an interesting avenue for future research.

Tachinid flies

Host defenses against parasitoids include avoidance behaviors before and during egg laying, and immune responses within the host, especially encapsulation by hemocytes (blood cells). Encapsulation kills the parasitoid both through the asphyxiating effects of the dead blood cells and with toxic effects of phenoloxidase or free radicals produced as the dead cells undergo melanization (Hartzer et al. 2005; Strand 2008). Most tachinid fly larvae attach their posterior spiracles to host tracheae, ensuring a continuous air supply and possibly protecting them from encapsulation (Etchegaray and Nishida 1975a; Feener and Brown 1997). In general, Lepidoptera with strong immune responses have lower levels of parasitism (Smilanich et al. 2009b).

We demonstrated that penetration of the host by tachinid fly larvae does not always kill monarchs; the battle for survival between the host and the parasitoid continues after the penetration of the host integument. Recent work has shown that monarchs infected by the protozoan parasite *Ophryocystis elektroscirrha* are less likely to die from *L. archippivora* attacks (Sternberg et al. 2011), demonstrating that features of the monarch internal environment affect the outcome of this battle. The marginally significant effect of host plant species on monarch survival in our study suggests that host plant chemistry could also play a role in determining which player wins the battle, although this effect is complicated (see discussion below).

Both the number of flies emerging from a monarch and the chances of survival of individual offspring were positively correlated with the number of eggs laid on the monarch, suggesting that the ability of the female parasitoid to overcome host behavioral defenses (such as dropping off the plant, or making jerking motions) during oviposition is an important determinant of the outcome of this host-parasitoid interaction. Once penetration occurs, monarchs appear better able to overcome one than multiple parasitoids. It is possible that multiple parasitoid larvae are better able to overcome host immune responses.

Lacewing predators

Prey consumption by lacewing larvae saturated near 35 eggs in 24 hours. Green lacewing eggs and larvae are common on milkweed (De Anda and Oberhauser, this volume, Chapter 5), and the results of our lab study suggest that it would be worthwhile to investigate whether and how lacewings affect monarch populations in the field, where monarch densities are lower, alternative prey are available, and the substrate structure (plant shape) more complex. Clark and Messina (1998) found that substrate structure can affect the ease with which lacewing larvae find prey. Milkweed plants vary morphologically within and between species, and it is possible that host plant structure may affect susceptibility of monarchs to predation by lacewings and other predators.

Monarchs and generalist natural enemies

Together, these studies and the work cited in Table 6.1 suggest that monarch eggs, larvae, and pupae have several generalist naturalist enemies; *P. dominula*, *L. archippivora*, and *C. rufilabris* are all reported to be generalists, although Oberhauser (2012) questions whether this is actually true of *L. archippivora*. Our lacewing research was directly motivated by an interest in the potential nontarget effects of chrysopid larvae used as biological control agents; however, *L. archippivora* has also been released as a biocontrol agent, and our research demonstrates that even nontarget species that consume toxic host plants can be harmed by generalist biocontrol species. To fully understand the impacts of lacewing larvae and other biocontrol agents on monarchs, it will be important to determine preferences for monarchs, target prey species, and other prey in the environment into which the predators or parasitoids are released.

Toxic host plants can have both sublethal and lethal effects on parasitoids and other internal parasites (e.g., English-Loeb et al. 1993; Stireman and Singer 2003a; Ode 2006). Milkweed toxicity affects the outcome of interactions between monarchs and their protozoan parasite *Ophryocystis elektroscirrha*;

a more toxic host plant results in reduced parasite load and increased host life span (Lefèvre et al. 2010). Similarly, the number of tachinid flies emerging from monarch caterpillars declines as the cardenolide content of host plants increases, even though the caterpillars still die from the parasitism event (Hunter et al. 1996). However, Dyer et al. (2004) found that *Eois* sp. (Geometridae) caterpillars collected in the field and then randomly selected for diets with enhanced concentrations of secondary metabolites (amides) were more likely to produce parasitoids. They hypothesized that the increased amide levels disrupted parasitoid encapsulation. This mechanism was supported by Smilanich et al. (2009a) in their study of the common buckeye *Junonia coenia*, which sequesters toxic iridoid glycosides from host plants. Immune response was compromised by increased iridoid glycosides in the caterpillars' diet.

Our work along with that of Rayor (2004) suggests that at least some predators and parasitoids can distinguish between monarchs with different levels of toxicity. In our *Pteromalus* study, monarchs that had consumed the less toxic *A. syriaca* as larvae were more likely to be parasitized (Figure 6.1). Herbivores that consume toxic host plants tend to be attacked by specialist parasitoids (Stireman and Singer 2003b), and further studies in our lab are addressing the degree to which the *Pteromalus* species we have found is a specialist.

In our tachinid study, monarchs fed *A. syriaca* were marginally more likely to die when parasitized by *L. archippivora* (Figure 6.2b), suggesting that the parasitoids performed better in monarchs fed this low toxicity host plant; however, monarchs that consumed *A. incarnata*, also with low toxicity, had survival rates similar to those fed *A. curassavica*, a very toxic milkweed (Figure 6.2b, Table 6.3). *A. syriaca* is the most commonly used host plant species by northern monarchs (Malcolm et al. 1993), and the parasitoids used in this study were collected in the north. Recent research (Oberhauser 2012) suggests that *L. archippivora* may be more of a monarch specialist than previously thought; these findings suggest a tightly coevolved relationship between *A. syriaca*, monarchs, and *L. archippivora*.

Our findings suggest many avenues for further study of the relationships between monarchs, their natural enemies, and their host plants. For example, the array of predators and parasites throughout monarchs' life cycle are likely to interact in interesting ways. Tachinid flies attack monarchs throughout their larval stages, when they are vulnerable to many other predators (Table 6.1). Clearly, attack by parasitoids is wasted when the host is subsequently consumed, and studying the payoff to tachinid flies in different environments and at different host ages will be valuable. Pupal parasitoids attack their hosts after egg and larval mortality has occurred, likely resulting in higher parasitoid survival. In addition, the degree of protection derived from milkweed host plants by immature monarchs, and the variation in this protection with specialist and generalist enemies, provide exciting avenues for more research.

ACKNOWLEDGMENTS

Cindy Petersen, Lis Young-Iscbrand, Quentin Knutson, Leah Alstad, Kara Lillehaug, Kristina Tester, Grant Bowers, Amy Eichorn, and Joe Shekleton provided help during various stages of the *Pteromalus* work. Margot Monson, Cindy Johnson, Andrew Williams, Suzanne Oberhauser, and Dara Satterfield shared their unpublished reports, specimens, and photographs of monarchs parasitized by *Pteromalus* sp. Carl Stenoien's and Shaun McCoshun's ongoing work on *Pteromalus* and monarchs probably helped us avoid publishing the incorrect name for parasitoid wasp that we studied. Rebecca V. Batalden, Grant Bowers, Emily Nimmer, Don Alstad, and George Heimpel provided input for the lacewing study. Bethany Caldwell, Beth DeLong, David Wisseman, Emilie Rosset, Johanna Nader, and J. T. Starc helped with the tachinid study. Anurag Agrawal, Kelly Nail, and Andrew Davis provided helpful comments on an earlier draft of the manuscript. Research was supported by the Science Education Partnership of Greater Minnesota, the Monarchs in the Classroom program at the University of Minnesota, National Science Foundation DGE-0440517 and ISE 0917450 (to KSO), the Dayton and Wilkie Funds (University of Minnesota), and the Environmental Action and Analysis Program and Sophomore Research Assistant Program at the College of Wooster in Ohio.

7

Monarchs and Their Debilitating Parasites

Immunity, Migration, and Medicinal Plant Use

Sonia Altizer and Jacobus C. de Roode

Throughout their geographic range, monarchs are affected by the debilitating protozoan *Ophryocystis elektroscirrha (OE)*. This highly transmissible parasite harms monarchs by causing reduced longevity, smaller body size, wing deformities, reduced mating success, and lower flight performance. Given the substantial costs of infection, we might expect monarchs to be highly resistant to OE, yet parasite infections remain common and widespread. In this chapter, we review recent findings on the key mechanisms used by monarchs to defend themselves against parasite infection, including (1) the genetic basis for variation in resistance to OE, (2) determinants of monarch innate immunity and relevance to parasite infection, (3) plant-derived defenses and evidence for self-medication, and (4) long-distance migration as a potential behavioral defense against infection. We conclude by considering the implications of monarch defenses for host-parasite evolution and outline several challenges for future work, including understanding how future environmental changes will affect the monarch-parasite interaction.

INTRODUCTION AND OVERVIEW

A growing body of evidence indicates that monarchs, like other animals, experience significant ecological and evolutionary pressures from parasites and infectious diseases. Indeed, parasites affecting other invertebrate populations can have profound evolutionary effects, ranging from increasing host genetic diversity to selecting for behavioral and immune defenses to counter infections (Lively et al. 2004; Zuk et al. 2006; Riddell et al. 2009). The most widespread and best-studied parasite of monarchs is the protozoan *Ophryocystis elektroscirrha (OE;* Figure 7.1, inset), first reported to infect monarchs and queens (*Danaus gilippus*) in Florida in the late 1960s (McLaughlin and Myers 1970). Recent work shows that this parasite occurs in nearly every monarch population examined to date, from the New World tropics to the Pacific Islands (Figure 7.1), with the proportion of infected monarchs in some locations reaching close to 100% (Leong et al. 1997a, Altizer et al. 2000). The ubiquitous nature of this parasite and its presence in ancestral monarch populations suggest that it has a long evolutionary association with monarchs and thus may have shaped key aspects of monarch behavior and immunity, including preferences for different host plant species and the propensity for long-distance migration (Altizer et al. 2011, Lefèvre et al. 2012).

As a neogregarine parasite, OE falls within the gregarine group of the phylum Sporozoa (syn. Apicomplexa), a group that also includes human parasites such as *Cryptosporidium*, *Plasmodium*, and *Toxoplasma* (Carreno et al. 1999). Like many insect pathogens, OE infects orally, and its life cycle very closely follows that of its host (Plate 4a). Infected females scatter dormant parasite spores onto egg and milkweed surfaces during oviposition (De Roode et al. 2009). Parasite spores can be transferred vertically, from infected adults to their progeny, and horizontally, when butterflies scatter spores that are ingested by unrelated larvae (Altizer et al. 2004, De Roode et al. 2009). Larva-to-larva transmission does not occur (Leong et al. 1997b); spores that are eaten by larvae lyse in the gut, where infective sporozoites migrate to the hypoderm and undergo two phases of vegetative reproduction (McLaughlin and Myers 1970). After monarchs pupate, spore production is initiated, and about three days before adults eclose, developing spores can be seen forming through the pupal integument. Infected adults emerge covered with dormant spores on the outside of their bodies, with the highest spore densities on their abdomens (Leong et al. 1992). Importantly, parasites do not continue to replicate on adults. Horizontal transmission between adults can result in the transfer of low numbers of spores, but these spores must be transmitted to and ingested by a larva to cause a new infection.

Infection by *OE* can be debilitating and sometimes lethal to monarchs (Altizer and Oberhauser 1999; De Roode et al. 2007). In extreme cases, preadult death follows from sepsis (whole body infection) if early instar larvae ingest high numbers of parasite spores. More commonly, heavily infected butterflies die as pupae (Plate 4b, c), do not fully eclose, or are too weak to cling to their pupal cases and expand their wings, resulting in severe deformities (Plate 4d). Infected monarchs that appear physically normal can still suffer debilitating effects including shorter adult life spans, smaller body sizes, reduced mating success, and reduced flight ability (Altizer and Oberhauser 1999; Bradley and Altizer 2005; De Roode et al. 2007). Population-level impacts

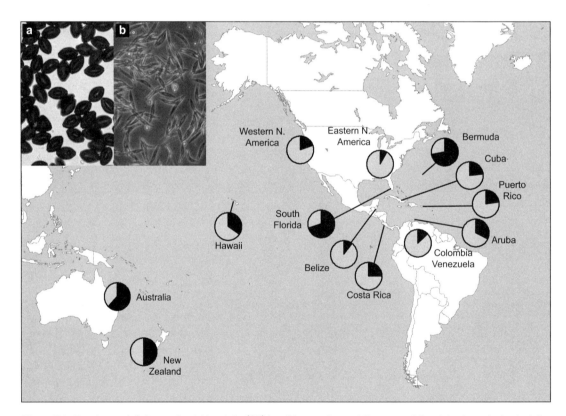

Figure 7.1. Prevalence of *Ophryocystis elektroscirrha* (OE) in wild monarch populations around the globe. In each pie chart, the black portion shows the proportion of examined monarchs that were heavily infected with OE (i.e., had > 100 spores per sample, following the nondestructive sampling technique described in Altizer et al. 2000). Years of sampling and sample sizes for each population are as follows: Aruba (2012; 93), eastern North America (1990–2011; 14,453), Belize (2011; 59), Colombia and Venezuela (1995; 47), western North America (1997–2011; 12,900), Puerto Rico (2008–2010; 218), Cuba (1995–1999; 100), Costa Rica (2008–2012; 422), Hawaii (2003–2010; 947), New Zealand (2009; 6), Australia (2005–2009; 229), South Florida (1994–2011; 731), Bermuda (2012; 29). Insets: (a) Spores of OE, shown here in a light micrograph (1000×), generally range 13–14 μm in length and ~8.5 μm in breadth. This sample was obtained from the exterior of a monarch abdomen. (b) Merozoites of OE appear as elongated cells that fill the hemolymph of an experimentally infected monarch 6 days postpupation (smaller round cells are monarch hemocytes). Neogregarines like *Ophryocystis* are distinguished from other gregarine parasites by having an extra cycle of vegetative replication, resulting in higher numbers of parasites per host and more virulent infections (Vega and Kaya 2012).

of *OE* are unquantified, but a modeling study suggested that up to 50% fewer monarchs might inhabit heavily infected populations than would be expected without the parasite present (Altizer et al. 2004).

Several features of the *OE*-monarch interaction make it relatively easy to study in both the field and laboratory. Monarchs can be readily assessed for parasite loads in several ways (Leong et al. 1992; De Roode et al. 2007). Larvae can be inoculated by feeding them precise numbers of spores transferred onto milkweed pieces, and past work has shown that negative effects of infection on monarch fitness are highly dose-dependent (De Roode et al. 2007). Inoculation studies also showed that larvae exposed to even a single parasite spore can develop heavy infections (with 10^6 or more spores on adults), demonstrating that this parasite is capable of explosive replication within the host (De Roode et al. 2007).

Given the high costs of infection, we might expect monarchs to be strongly defended against *OE*. Some monarch genotypes do show evidence of parasite resistance (De Roode and Altizer 2010; Lefèvre et al. 2011b), yet most monarchs are susceptible following experimental exposure, and infections are common in many wild populations (Figure 7.1). In this chapter, we begin by reviewing recent findings on the key mechanisms used by monarchs to defend themselves against parasite infection, including the genetic basis for variation in resistance to *OE*, determinants of monarch innate immunity, and the relevance of immune defense to *OE* infection. Next, we demonstrate how compounds monarchs derive from their milkweed host plants protect them against *OE*, and we review recent evidence that monarchs, like a handful of other animals, engage in self-medication. Monarchs are best known for their spectacular long-distance migrations in North America, and a growing body of evidence supports the idea that long-distance migration could offer protection against parasite transmission. We conclude by considering directions for future work, including the implications of monarch defenses for host-parasite evolution, and responses of the host-parasite interaction to current and future environmental change.

VARIATION IN MONARCH RESISTANCE AND TOLERANCE TO INFECTION

Natural selection should favor mechanisms that reduce infection and alleviate parasite-driven host fitness loss (Combes 2001). Like other animal hosts, insects have evolved a wide range of defense mechanisms, including genetic incompatibilities between host cell receptors and parasite molecules, cellular and humoral (extracellular) immune responses that combat parasites internally, and behaviors that reduce exposure to infection or parasite replication (Parker et al. 2011). All these defenses can be divided into three major categories based on the outcome for the host: (1) qualitative resistance (also known as anti-infection resistance or avoidance) reduces the probability that a host becomes infected; (2) quantitative resistance (also known as clearance or control) reduces parasite growth or burden upon infection; and (3) tolerance reduces fitness loss due to infection without limiting the infection itself (De Roode and Lefèvre 2012). Studies on monarchs indicate that they can use all three defense types to protect themselves against *OE*.

A growing number of studies published since 2001 show that monarchs vary genetically in both qualitative and quantitative resistance to *OE*. At the cross-population level, Altizer (2001) compared *OE* resistance in migratory monarchs derived from eastern North America (Mexico), from western North America (California), and from resident monarchs from South Florida. This study assessed resistance by inoculating larvae with known numbers of spores and measuring the resulting spore loads. Progeny from monarchs obtained from eastern North America suffered lower parasite loads (indicative of quantitative resistance) than monarchs obtained from western North America, and South Florida monarchs had similar susceptibility to those from eastern North America. More recent research has suggested that across North America, monarchs belong to a single large population based on neutral molecular genetic markers (Lyons et al. 2012), and follow-up studies that used a larger number of family lines per population (De Roode et al. 2008a; De Roode and Altizer 2010) did not find population-level differences in resistance between eastern and western monarchs (using progeny derived from monarchs sampled at their respective wintering sites). These later studies did, however, confirm that monarch families within each location varied genetically in their susceptibility to infection.

A further study asked whether monarchs also vary genetically in their tolerance to infection. Lefèvre et al. (2011b) used 19 family lines from California to quantify declines in monarch adult longevity with increasing spore load; lower reductions of life span with increasing spore load would indicate greater tolerance (Simms 2000; Stowe et al. 2000; Råberg et al. 2007; Baucom and De Roode 2011). This study showed that monarch families vary significantly in qualitative and quantitative resistance, but did not find evidence for genetic variation in tolerance. One explanation for this finding is that tolerance has become fixed in this population, as past theoretical models suggest that host tolerance to infection (which also benefits the parasite through increased transmission opportunities) should proceed to fixation (Boots and Bowers 1999; Roy and Kirchner 2000; Miller et al. 2005). Resistance, by comparison, comes with costs that might maintain both resistant and susceptible genotypes, unless pathogen prevalence is very high.

A lack of genetic variation in tolerance could also arise if monarchs derive tolerance from their diet rather than from underlying genetics. Indeed, a recent study showed that monarchs feeding on certain milkweed species as larvae obtained higher levels of tolerance in addition to resistance (Sternberg et al. 2012). Because antiparasite defense mechanisms can be costly and trade off with each other, organisms might invest in only a subset of possible defenses (Boots and Begon 1993; Kraaijeveld and Godfray 1997; Rolff and Siva-Jothy 2003; Sadd and Schmid-Hempel 2008, Lefèvre et al. 2011a). Thus, it is possible that the ability to obtain tolerance from larval food plants renders the evolution of genetic tolerance superfluous.

Many studies have shown that host resistance is determined not only by host genotypes, but also by interactions between host and parasite genotypes (e.g., Carius et al. 2001; Lambrechts et al. 2006). Indeed, host-parasite genotype interactions explained much of the variation in the proportion of monarchs that became infected and the parasite loads produced in those monarchs in two studies (De Roode et al. 2008a; De Roode and Altizer 2010). In particular, some host families were highly susceptible to infection, whereas other host families were more resistant, but to only a subset of parasite genotypes. In other words, no single host was protected against all parasite genotypes, and no single parasite infected all host genotypes well. These host-parasite genotypic interactions could explain the maintenance of variation in host resistance and parasite infectivity and growth.

Because work to date strongly supports host-parasite genetic interactions, we might also expect to find evidence for local adaptation, whereby parasites are optimally adapted to infect their local hosts, or, conversely, that hosts are optimally adapted to resist their local parasites (e.g., Lively 1989; Thrall et al. 2002). Although several studies have now asked whether local adaptation occurs in the monarch-parasite system, this does not appear to be the case across eastern and western North America and South Florida (Altizer 2001; De Roode et al. 2008b; De Roode and Altizer 2010). One explanation for the absence of local adaptation involves high levels of gene flow among populations (Lively 1999; Greischar and Koskella 2007). Indeed, recent studies provided evidence for high levels of gene flow among monarchs across all of North America (Lyons et al. 2012) and showed a high annual influx of eastern North American migrants into the South Florida breeding population (Knight and Brower 2009). Alternatively, local adaptation might further depend on the biotic environment in which hosts and parasites interact (Cory and Myers 2004; Laine 2007, 2008; Lazzaro et al. 2008). For example, it is possible that monarchs and their parasites are locally adapted only within the context of the particular milkweed host plant species with which they co-occur, an idea we describe in more detail below.

IMMUNE DEFENSE AND RESPONSE TO INFECTION

In animals, resistance is often studied in the context of immune defenses (Schmid-Hempel 2005). In monarchs and other insects, these immune defenses can take several forms, including immune cells and the activity of immune enzymes and peptides (Dunn 1990; Rincevich and Muller 1996; Lavine and Strand 2002; Vega and Kaya 2012). For example, cellular immunity in insects and other invertebrates involves hemocytes (blood cells that are functionally equivalent to white blood cells in vertebrates; Figure 7.2). Hemocytes can aid in the recognition, phagocytosis, and encapsulation of microbial pathogens (Rolff and Reynolds 2009) and can be quantified in insect blood and tissues. Another common invertebrate defense is melanization, which involves the deposition of melanin pigment through the activity of the enzyme phenoloxidase (PO), effectively producing a dark layer of material around pathogens to suffocate them or

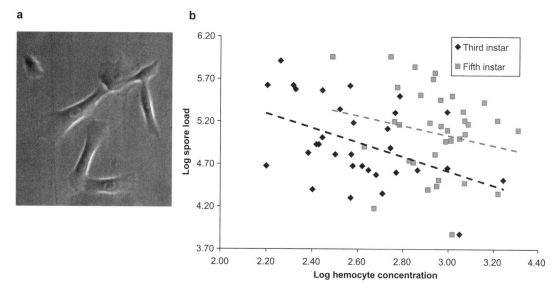

Figure 7.2. (a) Monarch hemocytes include 4 major immune cell types; plasmatocytes (shown here) are involved in aggregation and encapsulation and spread out on contact with substrates. Other hemocyte types include granulocytes as the major phagoyctic cells, oenocytoids, involved in phenoloxidase production, and spheroid cells, with functions not well known. (b) The relationship between larval hemocyte concentration and the final OE spore load of emerging adult monarchs. All monarchs were inoculated with a dose of 100 spores as third instars. Data for monarchs sampled for hemocytes at third instar (dark diamonds, black line; $R^2 = 0.20$) and at fifth instar (gray squares, gray line; $R^2 = 0.04$) show a significant negative relationship whereby greater hemocyte concentrations predict reduced parasite load (S. Altizer, J. C. de Roode, and M. Strand, unpublished data).

render them inactive (e.g., Christensen et al. 2005). Parasites that are too large to be engulfed, such as parasitoid eggs, can be walled off by aggregations of hemocytes that adhere to foreign bodies and trigger melanization, collectively called encapsulation (Rolff and Reynolds 2009). Insects also produce a variety of antimicrobial peptides that can lyse cell membranes and perform other functions; specific peptides are generally effective against broad classes of microorganisms, such as gram-negative bacteria (Hetru et al. 1998).

In monarchs, immune defenses probably confer some degree of protection against OE. As one indirect line of evidence, older larvae are known to be highly resistant to even high doses of OE spores (Altizer and Oberhauser 1999; S. Altizer, unpublished data), and past work on other insects has shown that older larvae have stronger and more mature immune defenses (Eslin and Prevost 1996). A recent experiment in monarchs showed that two measures of immunity (hemocyte concentration and phenoloxidase activity) increase sharply with larval age and almost perfectly mirror increased resistance to OE among older larvae (S. Altizer, unpublished data). In the wild, most monarchs are probably exposed to OE at early ages (first or second instars), when larvae eat spores deposited onto their egg chorion or surrounding milkweed leaf, but those that escape this early exposure might use immune defenses to effectively fight off infection.

More direct evidence for protective effects of monarch immune defense was provided by an experiment showing that larval hemocyte concentrations correlate negatively with OE spore loads in emerging adults (Figure 7.2), suggesting that hemocytes in larvae might attack OE vegetative cells or limit their within-host replication. As a result of lower parasite loads, monarchs with greater hemocyte concentrations as larvae also lived longer as adults (S. Altizer, unpublished data). Interestingly, these experimental studies showed no evidence that monarch immune defenses changed following infection by OE, as might be expected if parasite exposure induced monarchs to produce more immune cells or enzymes. Together, these results indicate that monarch defenses explain some of the variation seen in infection patterns in previous experiments and in the wild.

An open question concerns the degree to which environmental conditions, such as temperature, nutrition, crowding, and attack by other enemies (including parasitoids and predators), affect monarch immune defenses or resistance to *OE*. For example, experimental studies on other lepidopteran hosts showed that animals reared at higher densities experience reduced disease resistance and decreased time to death (Goulson and Cory 1995; Reilly and Hajek 2008). Similarly, monarchs reared at high densities experienced greater infection probability and stronger negative fitness effects of *OE* (Lindsey et al. 2009), but the researchers did not measure underlying immune defense. Past work on damselflies, flies, and grasshoppers shows that warmer temperatures can improve host resistance and lower the virulence of infection by fungi, parasitoids, and parasitic mites (Fellowes et al. 1999; Elliot et al. 2002; Robb and Forbes 2005). Indeed, many insects behaviorally raise their body temperature following infection by basking or moving to warmer areas (Adamo 1998; Elliot et al. 2002). One study in monarchs showed that larvae exposed to warmer temperatures emerged with lower *OE* parasite loads but suffered greater costs of infection than monarchs reared under cooler conditions (Lindsey 2008). Although *OE* replication appeared to decline with warmer temperatures, whether this change was caused by underlying immune defense again remains unknown. Understanding the consequences of environmental variation for host defense falls under the purview of ecological immunology, a discipline that examines the underlying causes of variation in immune function among individuals or populations (e.g., Norris and Evans 2000; Schulenburg et al. 2009). For monarchs that tend to encounter widely different environments at different locations both within and among populations, it is important to understand whether and how these conditions affect immune defense and associated infection risk.

SELF-MEDICATION AND MILKWEED USE AS A BEHAVIORAL DEFENSE

Virulence, defined as the detrimental effects of parasite infection on host fitness, and resistance (defined above) are commonly assumed to be determined by host genotype, parasite genotype, or an interaction between them. However, it is increasingly clear that host resistance to any given pathogen, and the virulence suffered by infected hosts, depend on the environment in which hosts and parasites interact (Brown et al. 2000; Mitchell et al. 2005; Lazzaro and Little 2009; Wolinska amd King 2009). This environment includes not only abiotic factors, such as temperature, but also the broader community of interacting species (Lafferty et al. 2006). Species can affect each other directly, such as through one species consuming another, or indirectly. Indirect effects occur when one species affects the performance of another species through its effects on the density or traits of a third species (Wootton 1994; Abrams 1995; Werner and Peacor 2003). For example, a parasite that lowers the abundance of an herbivore could indirectly benefit the plants on which the herbivore feeds. In addition, grazing by one herbivore can alter plant defense traits, which subsequently affects the performance of another herbivore (Van Zandt and Agrawal 2004). Such trait-mediated effects can influence tritrophic interactions among plants, herbivores, and their natural enemies (Cory and Hoover 2006), especially in cases where plant nutrition and secondary compounds alter the growth and virulence of herbivore parasites (Cory and Hoover 2006).

As herbivores, monarchs specialize on milkweed plants, mostly in the genus *Asclepias* (Ackery and Vane-Wright 1984). Milkweeds are well known to contain a variety of toxic secondary chemicals, most prominently the cardiac glycosides, often referred to as cardenolides (Malcolm 1991). Importantly, monarchs feeding on different milkweed species encounter different types and concentrations of cardenolides (Malcolm and Brower 1989; Zehnder and Hunter 2007). Cardenolides exert their toxic effects by inactivating sodium-potassium ion channels that cross the cell membranes of many eukaryotes (Malcolm and Brower 1989), and monarchs, being specialist feeders on milkweeds, have evolved sodium-potassium channels with reduced sensitivity to cardenolides (Zhu et al. 2008b; Dobler et al. 2012; Zhen et al. 2012). In fact, monarchs sequester cardenolides in their own tissues, conferring toxicity and protection against vertebrate predators (Brower et al. 1968; Brower and Fink 1985). Despite their reduced sensitivity to cardenolides, monarchs still suffer negative effects from high concentrations of cardenolides, through reduced larval survival, larval growth, and adult longevity (Zalucki et al. 2001a, 2001b; Sternberg et al. 2012; Agrawal et al., this volume, Chapter 4).

Toxic milkweeds can also protect monarchs against parasite infection. One study showed that monarchs reared on a more toxic milkweed (*Asclepias curassavica*, tropical milkweed or blood flower) were more resistant to *OE* infection and better able to tolerate infection than monarchs reared on a less toxic milkweed (*A. incarnata*, swamp milkweed; De Roode et al. 2008a). Further work showed that cardenolide differences between milkweed species were responsible for the different infection outcomes. In another experiment, De Roode and colleagues (2011b) showed that infected monarchs reared on aphid-infested *A. curassavica* plants (that produced lower concentrations of cardenolides) suffered greater parasite loads than monarchs reared on plants without aphids. Moreover, greater parasite loads in this study correlated with lower concentrations of two nonpolar cardenolides. A third experiment using 12 milkweed species with different cardenolide profiles showed that infected monarchs lived longer when reared on milkweeds with greater cardenolide concentrations (Figure 7.3a, b; Sternberg et al. 2012). The major exception was that infected monarchs reared on the high-cardenolide species *A. physocarpa* lived as short as monarchs reared on low-cardenolide species, potentially a result of toxic effects of high cardenolide levels on the monarchs themselves (Zalucki et al. 2001a, 2001b; Sternberg et al. 2012; Agrawal et al., this volume, Chapter 4).

An important question is how milkweed cardenolides increase both resistance to and tolerance of infection (Sternberg et al. 2012). One experiment suggested that the presence of toxic milkweed in monarch guts at the time of infection, but not feeding on toxic milkweed in the days following infection, reduces parasite loads in adults (De Roode et al. 2011a). These results indicate that medicinal milkweed reduces the effective number of parasites initiating an infection rather than reducing parasite growth once infection has happened (De Roode et al. 2011a). There are three possible ways in which this may happen. First, milkweed cardenolides could directly interfere with sporozoites as they emerge from *OE* spores in the monarch midgut (cf. Cory and Hoover 2006). Second, high-cardenolide milkweeds might enhance monarch midgut-based immunity, which reduces the number of sporozoites traversing the midgut wall to initiate the infection (cf. Lee et al. 2006; Povey et al. 2009; Alaux et al. 2010). Third, high-cardenolide milkweeds could provide monarchs with a midgut bacterial community that reduces parasite infection (cf. Koch and Schmid-Hempel 2011; Pinto-Tomás et al. 2011). Ongoing research will help determine which of these mechanisms is at play.

Regardless of the mechanism, infected monarchs benefit from high-cardenolide milkweeds, raising the question of whether monarchs can self-medicate by preferentially feeding on high-cardenolide milkweeds when infected. Lefèvre et al. (2010, 2012) carried out a series of experiments to determine whether monarchs can indeed do this. They first showed that infected caterpillars do not preferentially consume more toxic milkweed when given a choice. This result is not surprising, given that consumption of medicinal milkweed following infection does not reduce parasite outcomes (De Roode et al. 2011a), and early-instar caterpillars in the wild rarely have the opportunity to choose between different milkweed species (Zalucki et al. 1990). In contrast, female butterflies visit multiple patches of milkweed to oviposit (reviewed in Ladner and Altizer 2005), and therefore are more likely to be able to choose among milkweed species. Lefèvre et al. (2010, 2012) carried out host plant choice tests with female butterflies and found that infected females were twice as likely to lay eggs on the more toxic (antiparasitic) *A. curassavica* than on *A. incarnata*, whereas uninfected females had no preference (Figure 7.3c). Although infected females are unable to eliminate their own parasites and cannot prevent the passive transfer of spores to their own progeny (De Roode et al. 2009), they can preferentially lay their eggs on milkweed that reduces infection and disease in their offspring.

Self-medication was first described in primates (Wrangham and Nishida 1983; Huffman and Seifu 1989), and it has been argued that such behaviors can be displayed only by animals with strong cognitive skills (Sapolsky 1994). However, monarchs and other insects commonly use medicinal plants (DeRoode and Lefèvre 2012; De Roode et al. 2013). In addition to the monarch studies cited above, wood ants use conifer resin to lower the growth of bacterial and fungal pathogens (Christe et al. 2003; Chapuisat et al. 2007; Castella et al. 2008). Indeed, because insects are typically easier to use for manipulative experiments than vertebrates, we expect that studies in monarchs and other insects will provide many novel insights into the ecological and evolutionary consequences of animal medication.

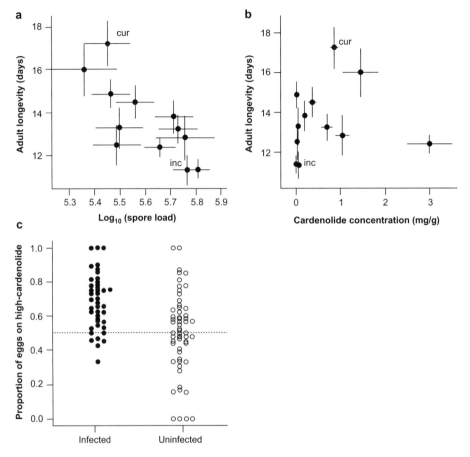

Figure 7.3. (a) Variation in parasite spore load and adult longevity experienced by monarchs reared on 12 different species of milkweeds (Sternberg et al. 2012). (b) Across 12 species of milkweeds, higher cardenolide concentrations initially increase the longevity of infected adults, but then decrease it again at very high levels of cardenolides (Sternberg et al. 2012). (c) The proportion of eggs laid on the high-cardenolide *A. curassavica* when infected and uninfected monarch butterflies are provided a two-way choice between *A. curassavica* and *A. incarnata* plants (Lefèvre et al. 2010). Data points and error bars in (a) and (b) denote species means and standard errors; data points in (c) denote individual butterflies.

LONG-DISTANCE MIGRATION AS A BEHAVIORAL DEFENSE

Long-distance migration in monarchs could serve as a different form of behavioral defense against parasite transmission. Monarchs occur worldwide (Figure 7.1) and migrate long distances in the temperate zone, but in tropical and subtropical locations such as Central America, Hawaii, and South Florida they breed year-round and presumably do not migrate long distances (Ackery and Vane-Wright 1984; Knight 1998). Even within North America, monarchs exhibit striking differences in migratory behavior despite high gene flow across locations (Pierce et al., this volume, Chapter 23). Monarchs that inhabit eastern North America undergo a spectacular two-way migration, traveling from as far north as southern Canada to overwintering sites in the mountains of central Mexico (Urquhart and Urquhart 1978; Brower 1995). Monarchs in western North America inhabit a smaller breeding range and migrate a shorter distance to overwintering sites scattered along the California coast (Nagano et al. 1993), and monarchs in South Florida breed throughout the year and do not migrate (Knight 1998).

Extensive field sampling of multiple migratory and resident monarch populations provides support for the idea that migration confers protection against *OE* infection. Initial work showed that the prevalence of *OE* varies sharply among populations (Figure 7.1) and appears to reach its highest levels in resident populations (Leong et al. 1997a; Altizer et al. 2000). Differences in *OE* infection among North American monarch populations with different migratory behaviors reflect this same pattern. Within eastern North America, fewer than 10% of monarchs sampled at the wintering sites in Mexico have been heavily infected over the past several decades. Monarchs in western North America experience low to moderate prevalence (5–30%), and resident monarchs in S. Florida experience consistently high infection prevalence (75–100%; Altizer et al. 2000; Bartel et al. 2011).

Seasonal migration likely affects at least two processes known to have consequences for parasite spread: (1) parasite transmission opportunities and (2) the survival of infected hosts (reviewed in Altizer et al. 2011). First, prolonged use of habitats allows parasite infectious stages to accumulate over time, such that migration allows animals to escape from contaminated habitats (i.e., "migratory escape"). During the time that monarchs are migrating and at overwintering sites, harsh winters and a lack of hosts could eliminate most parasites from their breeding grounds, allowing animals to return to largely disease-free conditions the following year. Long-distance migration can also lower pathogen prevalence by removing infected animals from the population (i.e., "migratory culling"). In this scenario, diseased animals are less likely to migrate long distances owing to the combined energetic costs of migration and infection.

We have found evidence that both migratory culling and migratory escape can cause variation in *OE* prevalence within and among wild monarch populations. First, a continent-scale analysis in eastern North America showed that parasite prevalence increased over time throughout the monarchs' breeding season, with highest prevalence associated with more intense habitat use and longer residency, consistent with the idea of migratory escape (Figure 7.4a; Bartel et al. 2011). Experimental monarchs infected with *OE* flew shorter distances and at reduced flight speeds (Bradley and Altizer 2005),

and field studies showed that parasite prevalence decreased as monarchs moved southward during their fall migrations (Figure 7.4b; Bartel et al. 2011), consistent with the idea of migratory culling. Parasite prevalence was also higher among butterflies sampled at the end of the breeding season than among those that reached their overwintering sites in Mexico. Together, these processes might be responsible for some of the striking differences in

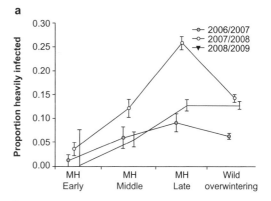

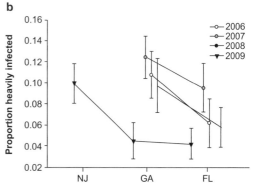

Figure 7.4. Parasitism by OE relative to migration in eastern North American monarchs. (a) Parasite prevalence across the annual migratory cycle using data from a citizen science project (Monarch Health, www.monarchparasites.org; $N = 5294$ samples) and data from wild overwintering populations in Mexico ($N = 5337$) from 2006 to 2009. For Monarch Health (MH) data, samples were excluded below 40°N latitude and for observers consistently reporting >70% infection. A migratory cycle captures data for monarchs breeding in year t and overwintering in Jan or Feb of year t+1. Definitions of early, middle, and late breeding season are as described in Bartel et al. (2011). Error bars represent standard errors. (b) Parasite prevalence among wild-caught migrating adult monarchs from three different locations (NJ, GA, and FL) along the eastern fall migratory flyway from 2006 to 2009 ($N = 1917$). Sites are arranged from northernmost (left) to southernmost (right) locations. Error bars represent standard errors. Data are reprinted from Bartel et al. 2011 with permission.

parasite prevalence reported among wild monarch populations with different migratory behaviors (Altizer et al. 2000).

An outstanding question is whether parasitism (as an ecological force) has driven the evolution of monarch migratory behavior, as would be expected if migration were a true behavioral defense. In other words, the potential benefits of escaping high infection rates by costly parasites might outweigh some of the costs of migration (energy expenditure, mortality risk, delayed reproduction), such that monarchs might migrate farther, or remain away from their breeding grounds longer, in the presence of high infection risk. Indeed, benefits of avoiding natural enemies (predators and brood parasites) have been suggested as driving greater migratory propensity in birds (McKinnon et al. 2010). On the other hand, a more likely scenario is that lower parasite risk is an incidental benefit of monarch migration, and that the migration itself evolved and is sustained for other reasons (namely the benefits of resource acquisition in the form of vast milkweed resources in North America; Malcolm 1987).

Irrespective of the underlying direction of causation, associations between migration and parasite transmission could drive evolutionary changes in pathogen virulence and host resistance. In support of this idea, laboratory studies showed that parasite isolates from the longest-distance migratory population in eastern North America were less virulent than isolates from short-distance migratory and resident populations (Altizer 2001; De Roode et al. 2008b; De Roode and Altizer 2010), suggesting that longer migration distances cull monarchs carrying more virulent parasite genotypes. However, as noted earlier in this chapter, we have not found evidence for greater resistance to infection among migratory monarchs, as might be expected if the high costs of infection favor more resistant host genotypes. It is possible that the evolution of host resistance is hindered by host-parasite genotype interactions that govern the outcome of infection (De Roode and Altizer 2010), or by selection for behavioral defenses in the form of host preferences for toxic milkweeds (described earlier in this chapter). It is also possible that migratory monarchs that face the extreme energetic demands of traveling thousands of kilometers cannot afford to invest resources in costly defenses; thus, migration itself might constrain resistance from evolving to higher levels.

SUMMARY AND FUTURE CHALLENGES

Because of their migration and use of medicinal milkweeds, monarchs and their parasites have become an important model system for studying how ecological variation alters host defenses and parasite transmission. Understanding evolutionary outcomes in this host-parasite system also offers exciting challenges for future work, especially in the context of environmental change. One future challenge is to identify conditions that favor local adaptation of hosts and parasites (Thompson 1994; Kawecki and Ebert 2004), whereby parasites are best at infecting hosts they encounter in their local environment, and hosts are best at resisting local parasites. Many studies have been done on host-parasite local adaptation, yet evidence for local adaptation remains elusive (Kaltz and Shykoff 1998; Greischar and Koskella 2007; Hoeksema and Forde 2008), including in the monarch-*OE* system (De Roode and Altizer 2010). One reason for this might be that most studies use "common garden" experiments, in which hosts and parasites are isolated from their home environment and exposed to each other under common conditions. This approach could be problematic if host resistance and parasite infectivity (the measures most often used in local adaptation studies) depend on environmental properties, such as other species like milkweeds (Cory and Hoover 2006; Wolinska and King 2009). In other words, hosts and parasites might be adapted to each other only within the natural community in which they live (Cory and Myers 2004; Laine 2007, 2008; Lazzaro et al. 2008), and most laboratory studies exclude relevant features of the environment.

Another open question involves how geographic variation in parasite risk affects the evolution of self-medication behaviors, both prophylactic (before disease occurs) and therapeutic (after disease occurs; Hart 2005). Prophylaxis is defined as a fixed strategy displayed by both infected and uninfected individuals, whereas therapeutic medication is a plastic response, displayed by infected individuals only. So far, the animal medication literature suggests that prophylactic medication is more likely to evolve when parasite risk is high and predictable, whereas therapeutic medication is more likely to evolve when parasite risk is low and unpredictable (Lozano 1991, 1998; Hart 2005), in part because the use of medicine can be costly, and uninfected animals should avoid these costs. Monarchs provide an excellent opportunity to test these predictions because parasite prevalence varies widely among populations (Figure 7.1). In particular, we expect that monarchs from migratory populations, where parasite risk is low, will display therapeutic medication, in which infected, but not uninfected, monarchs prefer to oviposit on medicinal milkweed. In contrast, populations with high parasite risk are expected to display prophylactic medication, whereby all monarchs prefer to oviposit on more toxic milkweed.

Monarchs and their parasites also provide an opportunity to study the evolution of parasite virulence. Researchers have long asked why parasites evolve and maintain the capacity to harm their hosts (Alizon et al. 2009). In other words, why do parasites, which depend on their hosts for their survival and reproduction, cause disease or even kill their hosts? Conventional wisdom has held that over evolutionary time, parasites evolve to become more benign (Burnet and White 1972), but this idea has been challenged (Anderson and May 1982). The current leading hypothesis for virulence evolution is that virulence can evolve as a byproduct of natural selection maximizing parasite transmission, which requires that parasites must replicate within their hosts, thus causing damage (Levin and Pimentel 1981; Anderson and May 1982; Bremermann and Thieme 1989; Van Baalen and Sabelis 1995; Frank 1996). This hypothesis, commonly known as the trade-off hypothesis, suggests that parasite fitness, quantified as lifetime transmission, is maximized at an intermediate level of replication, at which parasites cause a considerable amount of virulence. We tested this hypothesis using monarchs and their parasites and found support for the trade-off hypothesis, such that parasite fitness was maximized at an intermediate spore load, corresponding to a 10–20% reduction in monarch preadult survival and mating success (De Roode et al. 2008b). The next challenge is to ask how other sources of ecological variation, including long-distance migration and variation in host plant chemistry, might further shape the evolution of parasite virulence.

As a final note, the eastern North American monarch migration is now considered an endangered phenomenon (Brower and Malcolm 1991) because of deforestation of overwintering grounds (Brower et al. 2002), loss of critical breeding habitats (Pleasants and Oberhauser 2012), and potential climate-related shifts in geographic range and migration phenology (Oberhauser and Peterson 2003; Batalden et al. 2007; Ramírez et al., this volume, Chapter 13). As a species, monarchs remain common and widespread, but their great migration in North America might be at risk, especially if habitat loss at monarch wintering sites in Mexico leads to lower survival and return rates for spring migrants. At the same time, tropical milkweed (the toxic *A. curassavica*) has become a popular garden milkweed in the United States and Canada, and these plants do not die back during the winter in areas that experience milder climates, allowing for year-round breeding in the southern United States. As one possible harbinger of future change, citizen scientists have submitted dozens of reports of monarchs breeding during the winter in the coastal southeastern United States during the past 15 years (Howard et al. 2010; Batalden and Oberhauser, this volume, Chapter 19). We would expect *OE* prevalence to reach high levels in these winter breeding pockets in the southern United States, a prediction we are currently testing (Satterfield et al. 2014). In the long term, this winter breeding phenomenon could gradually replace the long-distance migratory population with a population that remains in the United States year-round, with a range expansion in the summer and contraction (to southern latitudes) in the winter. This shift in migratory ecology will likely boost parasite prevalence and could also lead to evolutionary changes in parasite virulence.

ACKNOWLEDGMENTS

Many people contributed to the work summarized in this chapter, and the authors are especially indebted to Becky Bartel, Lincoln Brower, Andrew Davis, Mark Hunter, Thierry Lefèvre, Michael Maudsley, Karen Oberhauser, and Eleanore Sternberg. Alma De Anda, Ernest Williams, Patrick Guerra, and an anonymous reviewer provided constructive comments on earlier versions of the chapter. SA was supported by NSF grant DEB-0643831 and JCdR was supported by NSF grants DEB-1019746 and DEB-1257160.

PART III

Monarchs in a Changing Climate
An Overview

KELLY R. NAIL AND KAREN S. OBERHAUSER

Monarchs, like all organisms, face the contemporary challenges of a rapidly changing climate. As ectotherms, they have limited physiological mechanisms for altering their body temperature, and they are heavily influenced by the weather conditions of their environment. Monarchs can use behavioral strategies to cope with changing environmental conditions, ranging from basking to warm their flight muscles and congregating in roosts to retain heat, to long-distance movement to seek suitable conditions. However, their capacity to cope with the challenges of altered temperature and precipitation patterns, as well as increasingly frequent extreme weather events, remains uncertain. This section explores the direct and indirect effects of climate on monarchs' development and survival at different life stages, and the ways in which climate can serve as a cue for triggering shifts in monarchs' behavior and physiology.

DIRECT EFFECTS OF CLIMATE ON MONARCHS

Development rates for monarchs, like other ectotherms, depend strongly on temperature, and extreme temperatures can directly affect monarchs of all life stages. Egg production is positively correlated with increasing temperatures (Zalucki 1981b), although extended exposure to hot temperatures probably reduces lifetime fecundity (Oberhauser 1989). The effects of increasing temperatures on monarch development rates are generally positive (Zalucki 1982), although prolonged exposure to extreme heat results in decreased adult size, increased development time, and increased mortality (York and Oberhauser 2002). In Chapter 8, Nail and colleagues show that monarchs can survive short periods of high temperatures up to 42 °C, albeit with sublethal effects. In general, the warmer temperatures predicted under climate change scenarios may allow monarchs to produce more generations over the course of a year, but if successive generations need to leave their natal grounds to avoid lethally hot temperatures (Malcolm et al. 1987; Batalden et al. 2007), summer recruitment may suffer.

Cold temperatures also affect monarch development and survival. Monarchs are commonly thought to be unable to withstand freezing temperatures, but they have more cold tolerance than might be expected. Migratory adults freeze and die at −8.2 °C, but if they are wet, they freeze at a warmer −4.7 °C, illustrating the sometimes complex interactions between climate variables (Larsen and Lee 1994). Similar freezing temperatures were found for adult monarchs at the Mexican overwintering sites (Anderson and Brower 1996). In their chapter, Nail and colleagues report that monarch eggs, larvae, and pupae can tolerate colder temperatures, with median supercooling (freezing) points ranging from −9.6 to −26.1 °C, depending on developmental stage.

Beyond temperature alone, severe weather events can also have direct negative impacts on monarchs. Brower et al. (2004) documented high mortality during an early 2002 snow storm in the overwintering colonies. In Chapter 9, Williams and Brower underscore the ways in which overwintering sites in Mexico satisfy monarch microclimate requirements (see also Calvert and Brower 1986; Anderson and Brower 1996). They describe features of the oyamel fir forests that make this area climatically suitable for monarchs, including both the protective effects of the forest canopy, and the heat retention effects that arise from monarch clustering. As Williams and Brower point out, ongoing conservation efforts are required to maintain the intact forest ecosystem that can support high numbers of monarchs; however, other researchers have indicated that climate change could render the current Mexican overwintering locations unsuitable for monarchs (Oberhauser and Peterson 2003), and effective conservation strategies must consider the effects of climate change.

On a more positive note, monarchs can often withstand thermal extremes through physiological acclimation induced by changing temperatures. For example, cold temperatures can induce rapid cold hardening in adult monarchs (Larsen and Lee 1994). Rapid cold hardening is a physiological shift that allows insects to withstand cold temperatures that would otherwise be lethal. Adult monarchs can quickly alter their physiology after exposure to cool temperatures, allowing them to withstand normally lethal cold temperatures. This cue may assist migrating monarchs that are exposed to unseasonably cold temperatures, or even developing eggs or larvae in cool spring or fall conditions. Similarly, some insects become more tolerant of hot temperatures through the expression of heat shock proteins (Neven 2000). In Chapter 8, Nail and colleagues suggest that monarch larvae were able to acclimate to extremely hot temperatures after several days of exposure; larvae exposed to these temperatures for one or two days halted development during exposure, but after four or six days of exposure they resumed development.

Adult monarchs can thermoregulate through shivering, basking, and congregating in roosts, and by using postures that minimize their exposure to the sun (Kammer 1970; Masters et al. 1988). These behaviors allow monarchs to survive extreme temperatures and even to fly and reproduce in unfavorable conditions (Wells et al. 1990). As Williams and Brower point out in Chapter 9, monarchs cluster in thermally ideal locations in their winter roosting trees. Additionally, melanism (dark coloration) on adult monarchs can enhance the amount of absorbed solar energy and their ability to thermoregulate (Masters et al. 1988; Davis et al. 2005).

Monarch larvae often move to more suitable locations on milkweed host plants or even into the surrounding leaf litter to raise or lower their temperature (pers. observ.). An example of physiological acclimation to cool or warm temperatures is variation in larval coloration; the amount of black pigment is negatively related to rearing temperatures (Solensky and Larkin 2003; Davis et al. 2004), likely allowing larvae to absorb relatively more radiant energy in cooler temperatures.

INDIRECT EFFECTS OF A CHANGING CLIMATE

Climate change can affect monarchs by altering the habitats on which they depend. The suitable area for oyamel firs in Mexico is predicted to decrease by almost 70% by 2030 and almost completely by 2090 as a result of changing temperature and precipitation regimes (Sáenz-Romero et al. 2012; Ramírez et al., this volume, Chapter 13). Even if climatic regimes in the overwintering sites remain suitable for monarchs, it is unclear whether a changed plant community could support them.

In Chapter 10, Brower and colleagues describe the importance of nectar sources for migrating monarchs traveling through Texas and northern Mexico. Addressing a single, extreme weather event (a winter storm in Mexico), Brower et al. (2004) documented potential consequences of a changing climate for overwintering monarchs. Similarly, the study reported here by Brower and colleagues documents the effects of a severe drought on native plants, and consequently monarchs, in fall 2010. Monarchs in areas with severe drought showed a 44% decline in lipid levels (fat reserves) compared with those tested in previous years, making the increased drought conditions predicted by climate change models concerning. Monarchs arriving in the Mexican overwintering colonies in fall 2010 had nearly normal lipid levels, however, suggesting they were able to compensate by finding adequate nectar in northeastern Mexico. These findings illustrate the importance of informed and broad conservation efforts to ensure that both milkweed and nectar sources are available throughout the monarchs' range.

Any changes that affect milkweed can influence monarchs as well, and as with other temperate plants (Blackman et al. 2009), the availability and quality of milkweed are affected by precipitation. For example, Batalden and Oberhauser (this volume, Chapter 19) show that native milkweed in Texas was more plentiful during autumns with high rainfall. Increased CO_2 levels cause increased plant consumption by some herbivores (Hughes and Bazzaz 1997). Increased CO_2 appears to have both positive and negative effects on milkweed defenses, increasing latex production and leaf toughness, but decreasing cardenolide expression (Vannette and Hunter 2011a). While these effects do not appear to affect milkweed consumption by monarch larvae (Vannette and Hunter 2011a), they could have long-term effects that we do not yet understand.

Finally, climate could affect monarch interactions with their natural enemies. Changing temperature or precipitation could affect parasitoid and predator abundance, and changing milkweed quality could affect the attractiveness of monarchs to their natural enemies by altering their ability to sequester chemical defenses or their own nutritional quality as prey. These topics are an important area for future work.

CLIMATE CUES

Temperature can cue changes in monarch physiology and behavior, ranging from diapause induction to flight direction. Diapause is triggered by fluctuating fall temperatures, along with senescing milkweed and decreasing day length (Goehring and Oberhauser 2002). Similarly, Guerra and Reppert (2013) found that temperature provides an important cue for adult monarchs; prolonged exposure to cold temperatures, similar to those in the Mexican overwintering sites during the winter, is required to trigger the northward flight of eastern North American migratory monarchs in the spring.

MODELING THE EFFECT OF CLIMATE ON MONARCHS

Computational models can help predict the location of suitable habitat for species under changing temperature and precipitation patterns, and researchers have used several kinds of models to understand monarch responses to climate. One type of model—an ecological niche, or correlative model—uses the current conditions at sites occupied by a species to define its ecological niche, the environmental parameters present where the organism exists. While knowledge of the organism's physiology can inform the choice of parameters considered in the model, only occurrence information is needed to build the model. Once the model is built, scientists can project the geographic distribution of the niche in the future, based on climate change models. Ecological niche modeling of monarchs in eastern North America showed that they have different habitat requirements in their overwintering and breeding ranges. These models predicted that monarchs are likely to lose wintering habitat (Oberhauser and Peterson 2003), and either gain or lose breeding habitat (Batalden et al. 2007) under climate change regimes, depending on both their own ability and that of their host plants to track climate changes.

Another type of correlative model uses time series of abundance data and various measures of climate; researchers use statistical methods to test for correlations between climate parameters and organism abundance. These correlations can then be used to predict impacts of changing climate. A recent correlative study showed that increased spring precipitation and intermediate temperatures in Texas predicted increased summer abundance of Ohio monarchs (Zipkin et al. 2012). This same model showed that Ohio climate conditions did not correlate with summer abundance. While Zipkin et al. did not overlay their findings on climate change prediction scenarios, their results could be used in this manner.

Finally, mechanistic models explicitly incorporate an organism's physiology. They allow us to construct models based on our knowledge of how changes in temperature or other key environmental variables will affect vital rates, such as birth or death. We can then predict where monarchs are and are not found currently, based on our knowledge of the spatial distribution of the parameters known to affect the species, or predict where they will or will not be able to survive in the future. In Chapter 11, Zalucki and colleagues examine the effects of both temperature and rainfall on monarch population abundance, using CLIMEX, a software program used successfully in the past to estimate areas where monarchs might establish if suitable host plants were available (Zalucki and Rochester 1999, 2004). Here, Zalucki et al. estimate monarch responses to climate variables and compare their estimates with contemporary monarch population numbers. They found no evidence that the decline in monarch numbers demonstrated by overwintering site data (Brower et al. 2011; Rendón-Salinas and Tavera-Alonso 2013) is due to the effects of climate alone; however, they note that the model accounts for climate impacts mainly in breeding habitats and not, for example, at the overwintering sites addressed by Ramírez et al. (this volume, Chapter 13).

In sum, the chapters in this section provide key empirical data that document ways in which climate affects monarchs, and an example of a modeling analysis of climate's possible effect on recent variation in monarch numbers. Together, these approaches will help us predict how monarchs could respond to future climate change. Our conservation efforts will be more effective if they utilize an understanding of monarch responses to climatic conditions, modeling approaches that incorporate these complex interactions between monarchs and climate, and a realistic understanding of future climate scenarios.

8

What's Too Hot and What's Too Cold?

Lethal and Sublethal Effects of Extreme Temperatures on Developing Monarchs

Kelly R. Nail, Rebecca V. Batalden, and Karen S. Oberhauser

We exposed immature monarchs to extreme cold and hot temperatures for varying lengths of time and quantified both sublethal (adult size and development time) and lethal effects. Larvae have upper and lower thermal limits of 42 and −20 °C. Although most larvae survived short exposure during the daytime to temperatures up to 42 °C, they suffered sublethal effects from temperatures 38 °C and higher, including smaller adult mass and slower development. Nighttime temperatures of 34 °C during periods with daytime temperatures of 38 °C resulted in lower survival, showing that respites from elevated temperatures are important in allowing monarchs to survive temperature stress. Median supercooling points (SCPs) for immature monarchs ranged from −26.1 to −9.6 °C, with eggs having the coldest SCP and third instars having the warmest SCP. Larvae appear to be freeze-intolerant, with 50% mortality not occurring until temperatures fall below each stage's respective SCP; however, eggs seem to be chill-intolerant, with mortality occurring before their SCP. Interestingly, third instars were most susceptible to both cold and heat stress. These findings can help inform future modeling and conservation efforts for monarchs throughout their life cycle.

INTRODUCTION

Monarchs are affected by climate in their wintering sites (Oberhauser and Peterson 2003; Zalucki and Rochester 2004; Ramírez et al., this volume, Chapter 13; Williams and Brower, this volume, Chapter 9) and during the breeding season (Zalucki and Rochester 2004; Batalden et al. 2007), when extensive rain or prolonged cool and cloudy conditions can reduce egg laying and increase development time, and prolonged hot or dry spells can reduce adult life span and fecundity (Zalucki 1981b; Masters et al. 1988; Masters 1993). Zalucki and Rochester (2004) predicted fluctuations in monarch abundance in the eastern North American population due to the effects of climate on their phenology and fecundity (but see Zalucki et al., this volume, Chapter 11). Climate can also influence milkweed abundance and quality (Zalucki and Kitching 1982b; Zalucki and Rochester 2004).

Work on upper thermal limits of monarchs has included both modeling and empirical studies. Ecological niche models based on occurrence data from the Monarch Larva Monitoring Project (MLMP; see Oberhauser et al., this volume, Chapter 2) predict a marked northward summer range shift for eastern North American monarchs under climate change scenarios (Batalden et al. 2007), driven largely by increasing temperatures. MLMP data show no monarch presence above a monthly mean temperature of 30 °C (all temperatures in this paper refer to degrees Celsius). Continuous exposure to temperatures of 36° causes significant larval mortality (Zalucki 1982b), while single or repeated 12-hour pulses of 36° increase development time without increasing mortality (York and Oberhauser 2002). This finding suggests that low nighttime temperatures are important, yet observed climate change and future predictions indicate increased frequency of higher nighttime lows (IPCC 2007).

To our knowledge, no studies have examined the lethal and sublethal impacts of temperatures above 36° or explored the effects of elevated nighttime temperatures on immature monarchs. It is possible they can survive warmer or cooler temperatures than those used to build the models. If monarchs can withstand conditions different from those they currently inhabit, it is possible that published models (Oberhauser and Peterson 2003; Zalucki and Rochester 2004; Batalden et. al. 2007) do not reflect their actual thermal tolerance and that they will be able to survive the severe changes predicted by climate models. Additionally, none of the niche models incorporate selection on character traits such as thermal tolerance and resulting evolution.

Contrary to the typical strategy of overwintering in Mexico, some monarchs breed in the southern United States throughout the winter, although the proportion of breeding vs. diapausing individuals is not known (Howard et al. 2010; Batalden and Oberhauser, this volume, Chapter 19). However, breeding during the winter poses risks; winter temperatures in the southern United States can fall below freezing. While adult monarchs can leave areas with unsuitable temperatures, immature monarchs may be exposed to cold and possibly lethal temperatures.

The lower *development* threshold for monarch larvae is 11–12° (Zalucki 1982b), but lower *lethal* thresholds for immature stages are poorly understood. Although little is known about immature monarch cold tolerance, much work has been done on insect cold tolerance in general. Insects are classified into three categories of cold temperature-dependent mortality: chill-intolerant, freeze-intolerant, or freeze-tolerant (Lee 2010). Chill-intolerant insects experience mortality before freezing. Freeze-intolerant, or freeze-avoidant, insects survive until their body freezes (i.e., ice nucleation occurs). Finally, freeze-tolerant insects can survive ice formation within a limited temperature range. The classification of the cold-tolerance strategy of monarchs is not known for any life stage. Adult monarchs do not survive internal ice formation (Larsen and Lee 1994), so they may be freeze-intolerant. Although the supercooling points of adult monarchs is known (−8° if they are dry and −4° if wet; Anderson and Brower 1996), no previous work has assessed freezing points of immature monarchs or cold-mediated lethal and sublethal effects.

METHODS, RESULTS, AND DISCUSSION

In a series of four experiments, we investigated lethal and sublethal impacts of elevated and lowered temperatures on immature monarch survival and development. Experiment 1 determined the upper physiological limits of larval development. Experiment 2 tested the effect of high nighttime temperatures combined with a daytime temperature that caused sublethal impacts (increased development time and decreased size), but no lethal impacts. Experiment 3 measured the supercooling points (SCPs, subzero temperature at which intracellular fluid freezes) of immature monarchs. Finally, experiment 4 determined the lower lethal temperatures of immature monarchs.

In all four experiments, wild-caught monarchs laid eggs on *Asclepias curassavica* in a greenhouse. For experiments 1 and 2, immature monarchs were kept in Percival growth chambers (LD 12:12 h photoperiod; 30°:25°), except when they were exposed to treatment conditions. In experiments 3 and 4, they were kept in the same growth chambers (LD 15.5:8.5 h photoperiod; 22°:20°) for at least 12 hours for eggs and 48 hours for all other stages, before being exposed to cold temperatures during a specific stage. After cold exposure, these monarchs were returned to the same growth chamber conditions for the duration of their development. The pre-exposure photoperiod and temperatures in experiments 3 and 4 simulated conditions experienced during winters in the southern United States. In experiments 1, 3, and 4, individuals were kept throughout their development in separate 500 ml plastic deli containers with ventilation holes in the lids. In experiment 2, they were kept in petri dishes until they became fifth instars and then moved to the deli containers to pupate. In all experiments, rearing containers were cleaned and larvae given fresh milkweed (*A. syriaca*) daily.

Experiment 1: Upper temperature limits of larval development

Experiment 1 was completed in two sequential rounds; round 1 treatment temperatures were 38° and 40°, and round 2, 42° and 44°. All experimental larvae were offspring of wild-caught individuals collected from first-generation monarchs in St. Paul, MN, in June 2007.

Within 12 hours of hatching, individuals were placed in rearing containers and randomly assigned to an experimental group (stage, temperature, and duration of heat exposure). We exposed larvae within 24 hours of hatching or molting into first, third, or fifth instars to 12-hour pulses of 38°, 40°, 42°, or 44° over periods of 1, 2, 4, or 6 days. They were always returned to nighttime temperatures of 25°. Sample sizes ranged from 17 to 20 for each treatment group, including the control, which was kept at 30° during the day and 25° at night.

We observed the larvae daily, tracking mortality and development. Using 12° as the lower threshold temperature, we measured development time in days and degree days (Zalucki 1982b). We assessed two additional effects of heat exposure: the ability to pupate without falling and adult size. If individuals fell from the lids of their containers when they attempted to pupate, we recorded the event and used thread and tape to affix the cremaster back to the lid. Adult size was measured as mass and right forewing length (i.e., distance from wing base to apex). We measured each individual's mass after its wings dried but before it had fed.

All monarchs exposed to 44° for any length of time died before adulthood; of 234 individuals exposed to 42°, only 16 survived to be adults. Therefore, we could not compare sublethal effects across treatments at these temperatures. Mortality did not differ between control groups in rounds 1 and 2 (Figure 8.1; 0.125 mortality for both groups, Fisher's exact test, $P = 1.00$), so we compared mortality across all treatments.

Survival varied with temperature, treatment timing, and treatment duration (Figure 8.1), with higher survival at lower temperatures and shorter treatment duration. Individuals exposed as third instars were less likely to survive to adulthood, particularly after 6 days of exposure, with 70% and 20% surviving at 38° and 40°, respectively. Only 53% of fifth instars survived to adulthood after 6 days exposure to 40°. Overall mortality increased with exposure to higher temperatures; 18% died at 38°, 28% at 40°, and 93.3% at 42°.

Of 451 individuals that survived to pupate in all treatments, 143 dropped from the lids of their containers during or just after pupation. Under natural conditions, falling is likely to result in death, either directly from the fall or from another source, such as predation. Here, the distance to the floor of the container

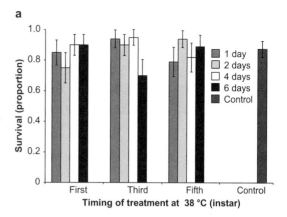

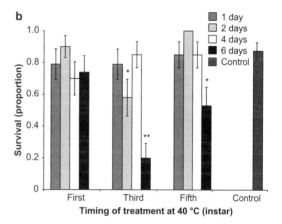

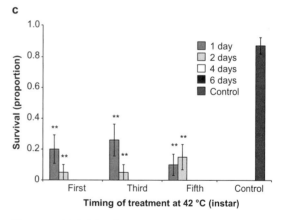

Figure 8.1. Probability of survival by treatment timing and duration for (a) 38 °C, (b) 40 °C, and (c) 42 °C. Asterisks (* and **) indicate treatment combination is significantly different from control with $P < 0.05$ and $P < 0.01$, respectively. Error bars represent standard error. No monarchs exposed to 44 °C for any length of time survived.

was small, and when we taped pupae back onto the container lid, subsequent survival was 94%. Consequently, this potentially lethal effect of exposure to heat stress was not represented in our mortality estimates. The proportion of individuals that fell did not depend on temperature, but individuals exposed as fifth instars were more likely to fall at durations of 4 or 6 days, with 58% and 72% falling, respectively (Figure 8.2). Individuals exposed as third instars also showed elevated risk of falling, with 33%, 38%, and 46% falling if exposed for 1, 2, or 4 days, respectively. The lack of effect for third instars in the 6-day treatment may be due to the high mortality (and consequent low sample size) of this group. There was no difference in pupation ability whether the individual pupated while still in the treatment or after returning to control conditions ($\chi^2 = 0.218$, df = 1, 63, $P = 0.64$).

Temperature, but not the duration or timing of treatment, affected adult mass. Mass decreased with increasing temperature, and monarchs exposed to temperatures above 38° were lighter than controls. Wing length in the treatment groups did not differ from controls (Table 8.1).

When we assumed that development rates continue to increase with increasing temperature, development time to adult, measured in degree days, was longer in nearly all treatments compared to control (Figure 8.3a, b). This effect was not as strong for fifth instars as it was for first and third instars. However, insects have developmental maxima as well as minima, so we know that our assumption of continually increasing development rates with increasing temperature is inaccurate. Thus, we recalculated degree days to exclude time exposed to elevated daytime temperatures, assuming that development ceased while monarchs were in the heat treatments. The adjusted degree-day totals (Figure 8.3c, d) reflect all the time spent in control conditions but only the 12 nighttime hours each day (at 25°) spent under treatment conditions. This assumption worked fairly well for some treatments, resulting in degree-day development times similar to the controls.

It is possible that some individuals acclimated to the hot temperatures, and started to develop under these conditions. With the assumption that development ceases in hot conditions, acclimation could explain the fact that some individuals with longer exposure times spent fewer degree days to develop than control individuals. At 40°, fifth instars exposed for 4 or 6 days used significantly fewer adjusted degree days to develop than controls. The fact that this was not true for first and third instars suggests that younger larvae were less able to acclimate to high temperatures (Figure 8.3d). Many insects produce heat-shock proteins (HSPs) in response to exposure to high temperatures, including other Lepidoptera (Fittinghoff and Riddiford 1990; Sakano et al. 2006). These HSPs may then in turn increase thermotolerance at temperatures previously unsuitable for development (Neven 2000).

We determined when the developmental lags occurred by detailed tracking of individuals (Table 8.2; note that only the 40° treatment is shown in

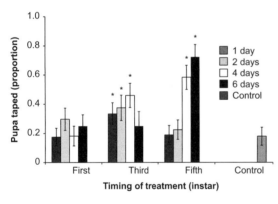

Figure 8.2. Probability that pupae fell by treatment timing and duration. Asterisk (*) indicates treatment combination is significantly different from control with $P < 0.05$. Error bars represent standard error.

Table 8.1. Adult mass and right wing length by treatment temperature in Experiment 1

Temp (°C)	Male mass (g) (SE)	Female mass (g) (SE)	Male RWL (mm) (SE)	Female RWL (mm) (SE)
38	0.59 (0.006)[a]	0.55 (0.007)[a]	52.4 (0.22)[a]	52.0 (0.23)[a]
40	0.57 (0.007)[b]	0.52 (0.007)[b]	51.7 (0.24)[b]	51.2 (0.24)[b]
42	0.52 (0.025)[c]	0.49 (0.023)[b]	50.0 (0.86)[b]	50.8 (0.83)[ab]
Control	0.61 (0.02)[a]	0.57 (0.02)[a]	51.2 (0.64)[ab]	51.3 (0.55)[ab]

Note: Different letters indicate treatments that are significantly different within columns (ANOVA, Tukey LSD tests, $P < 0.05$).

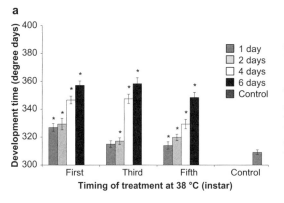
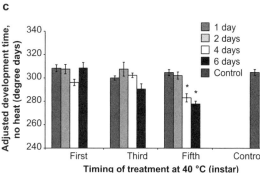
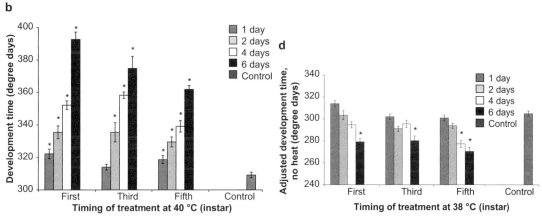

Figure 8.3. Development time, measured in degree days, by treatment timing (first, third, or fifth instar) and duration at (a) 38 °C and (b) 40 °C. Adjusted development time, measured by excluding time spent at elevated temperatures, by treatment timing and duration at (c) 38 °C and (d) 40 °C. Asterisk (*) indicates treatment combination is significantly different from control with $P < 0.05$. Error bars represent standard error.

this table for illustrative purposes). All individuals exposed to elevated temperatures as first instars took longer to become second instars than control larvae. At 38°, their average time as second instars was not significantly different from control, even though individuals in the 4 or 6 day exposure groups were still exposed to elevated temperatures as second instars. At 40°, their average development times as second and third instars were longer than the control only for larvae exposed for 6 days.

Third instars exposed to 38° did not slow development, but those exposed to 40° took longer than control individuals to become fourth instars (Table 8.2). Their development times as fourth instars were not significantly different from control, except for those exposed to 40° for 4 and possibly 6 days.

For individuals exposed to elevated temperatures as fifth instars, 38° did not slow their development, but all those exposed to 40° spent longer as fifth instars than the control group (Table 8.2). Fifth instars exposed to 40° for 6 days spent more time as pupae than the controls.

Experiment 2: Effects of high nighttime temperatures

Experiment 2 was completed in three sequential rounds. Each round tested a different treatment temperature, which differed from the control during the day and night, rather than only during the day as in experiment 1. Daytime treatment temperature was 38° for all replications and 30°, 32°, or 34° at night in rounds 1, 2, and 3,

Table 8.2. Development time (followed by sample size) for each larval instar by each factorial combination of treatment, timing, and duration

Timing (instar)	Duration (days)	1st	2nd	3rd	4th	5th	Pupa	Total
Control	Control	2.10 (40) [a]	1.72 (39) [a]	1.87 (39) [a]	2.42 (38) [abcde]	3.74 (39) [a]	8.10 (36) [ab]	19.94 (36) [a]
1	1	2.71 (17) [b]	1.71 (17) [a]	2.00 (16) [ab]	2.19 (16) [abc]	3.75 (16) [a]	8.07 (15) [abc]	20.47 (15) [abcd]
	2	2.83 (18) [b]	2.12 (17) [a]	1.95 (19) [ab]	2.26 (19) [abcd]	4.00 (19) [abcd]	8.00 (18) [abcd]	21.00 (18) [cdefg]
	4	2.80 (15) [b]	2.29 (14) [ab]	2.00 (14) [ab]	2.29 (14) [abcdef]	4.00 (14) [abcde]	8.00 (14) [abcd]	21.43 (14) [defg]
	6	2.84 (19) [b]	2.89 (18) [b]	2.69 (15) [b]	2.67 (15) [abcdef]	4.13 (15) [abcde]	8.00 (15) [abcd]	23.40 (15) [h]
3	1	–	–	2.61 (17) [b]	1.82 (17) [a]	4.12 (17) [abcde]	8.20 (15) [a]	19.93 (15) [ab]
	2	–	–	2.53 (13) [ab]	2.77 (13) [abcdef]	4.00 (12) [abcde]	8.18 (11) [a]	21.00 (11) [bcdefg]
	4	–	–	2.61 (19) [b]	3.21 (19) [f]	4.22 (18) [abcde]	7.88 (17) [bcd]	21.82 (17) [g]
	6	–	–	2.59 (13) [b]	3.23 (13) [ef]	4.67 (6) [abcdef]	7.75 (4) [bcd]	22.25 (4) [efgh]
5	1	–	–	–	–	4.56 (18) [bcdef]	8.18 (17) [a]	20.24 (17) [abc]
	2	–	–	–	–	5.05 (20) [f]	7.85 (20) [cd]	20.65 (20) [abcde]
	4	–	–	–	–	5.21 (19) [f]	8.00 (17) [abcd]	20.59 (17) [abcde]
	6	–	–	–	–	5.31 (13) [f]	8.67 (12) [e]	21.42 (12) [defg]

Notes: Only the 40° treatment is shown, for illustrative purposes. Shaded values represent time larvae spent in the heat treatment. Times are shown as mean number of days for each instar. Treatment combinations followed by different letters are significantly different within columns (ANOVA, Tukey LSD tests, $P < 0.05$).

respectively. We used the same factorial design for treatment timing and duration as in experiment 1, and measured the same lethal and sublethal indicators of heat stress. Larvae for round 1 were F1 offspring of wild-caught individuals collected in St. Paul, MN, and western Wisconsin in June 2008. Rounds 2 and 3 individuals were F2 offspring of these individuals.

Control mortality did not differ among rounds ($\chi^2 = 1.875$, df = 2, 57, $P = 0.39$), so our analysis of mortality includes comparisons across all rounds. Control development time, however, was different across rounds, so we cannot compare this response variable between rounds (ANOVA $F = 3.61$, df = 2, 45, $P = 0.04$). Round 1 was conducted in midsummer, while rounds 2 and 3 occurred into late fall; thus, milkweed quality may have contributed to the observed changes in development time. Alternatively, differences between F1 and F2 generations could have affected development time.

Across all treatments, mortality increased with length of exposure to high nighttime temperatures longer than 2 days, with 72% and 68% survival after 4 and 6 days, respectively, compared with 81% survival for the control group (Figure 8.4a). At nighttime temperatures of 34°, only 3.8% of larvae exposed as first instars survived to adulthood (Figure 8.4b), but exposure at other ages or temperatures did not affect survival.

Experiment 3: Supercooling points of eggs and larvae

We determined SCPs of individual eggs; first, third, and fifth instars; and pupae in a –80° freezer. We used contact thermocouple telemetry and lowered the temperature by approximately 1 °C/min using a standardized foam-insulated box (Carillo et al. 2004). Fifth instars were held in filter paper capsules to ensure constant contact with the thermocouple, whereas firsts and thirds were attached to the probe using high vacuum grease. Temperatures were recorded 10 times per second by a multichannel data logger. SCP was recorded as the lowest temperature reached before the observed increase in temperature (latent heat of fusion) indicating a state change of the intercellular fluid from liquid to solid. Sample sizes ranged from 10 to 14 individuals for each developmental stage tested. All experimental individuals were offspring of wild-caught larvae collected in Minnesota and western Wisconsin in summer 2011.

All life stages had median SCPs well below 0°, and no individuals within any of the treatment groups froze at temperatures warmer than –4° (Figure 8.5). Eggs had the lowest median SCP of –26.1°, followed by pupae at –17.5°. First instars had a median SCP of –12.4°, and third and fifth instars, –9.6° and –10.3°, respectively. All individuals were lowered to temperatures well below their SCP (below –70°) and there was no survival after removal from the freezer.

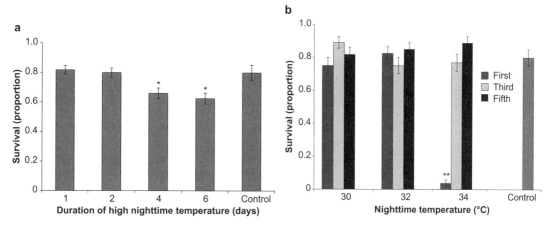

Figure 8.4. Probability of surviving elevated nighttime temperatures when daytime temperature was 38 °C by (a) treatment duration and (b) temperature and timing. Asterisks (* and **) indicate treatment combination is significantly different from control with $P<0.05$ and $P<0.01$, respectively. Error bars represent standard error.

Experiment 4: Lower lethal temperatures of eggs and larvae

We measured lower lethal temperatures (LT_{50}, the temperature at which 50% of individuals die from the effects of cold temperature) for eggs and first- and third-instar larvae, using offspring of wild-caught individuals collected in Minnesota and western Wisconsin in summer 2011. We cooled monarchs in groups of 10 (eggs and first instars) or 5 (third instars) at a rate of approximately 0.3 °C/min and then immediately removed them when the desired temperature was reached (Carillo et al. 2004). We tested three (eggs and first instars) or five (third instars) groups at each temperature; thus total sample sizes at each temperature were 30 (eggs and first instars) and 25 (third instars). We assessed their response to several temperatures, beginning at 0° and decreasing temperatures at intervals of 5°, with the minimum temperature for each age class at least 10° below its SCP. We warmed monarchs gradually and assessed survival by recording the number hatched (eggs) or alive after one day (larvae); larvae were considered alive if they were observed moving or if they moved in response to a tactile stimulus. We also raised monarchs to eclosion to assess postexposure survival. Our response variable was the proportion of each group that survived; we then used logistic regression on these proportions to determine the LT_{50}.

Some survival was recorded below the median SCP for each life stage, but no survival occurred

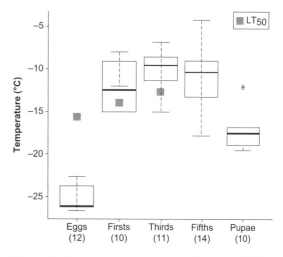

Figure 8.5. Box plot of observed supercooling points (SCPs) along with LT_{50} for respective age classes. Line inside each box represents the median SCP, with the box representing the first to third quartile of SCP data. Dashed lines outside the box represent the range of data, with outliers represented by open circles (outliers are data points more than 1.5 times the interquartile range). Sample sizes are listed in parentheses. LT_{50} for eggs and first and second instars shown for comparison.

below the lowest recorded SCP for the respective life stage (Figure 8.6). The temperature predicted to be lethal for half the population (with standard error) is $-15.6° \pm 1.2$ for eggs, $-14.0° \pm 0.76$ for first instars, and $-12.7° \pm 0.56$ for third instars (Figure 8.6). Of the initial 359 immature monarchs that hatched from eggs or survived the initial freezing as larvae or pupae, only 85 survived to eclosion, including only 23% of the control group, so we

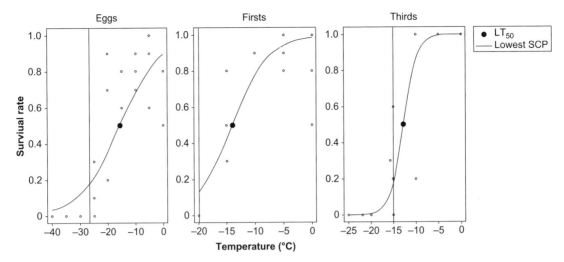

Figure 8.6. Lower lethal temperatures for monarch eggs and first- and third-instar larvae. Logistic regression fitted lines used to calculate LT_{50} for each stage: Eggs = $1/(1+e^{-(2.136 + 0.137 * temperature)})$; Firsts = $1/(1+e^{-(4.406 + 0.314 * temperature)})$; Thirds = $1/(1+e^{-(8.509 + 0.669 * temperature)})$. Vertical lines on each graph represent the lowest recorded SCP for that life stage; closed dots represent the predicted temperature at which half the larvae would die (LT_{50}).xs

were unable to compare sublethal effects of cold treatments.

CONCLUSION

These experiments provide additional information on immature monarch tolerance of extremely high and low temperatures. Larval mortality increased substantially between 40 and 42 °C, regardless of the timing or duration of exposure. This suggests a physiological limit between 40 and 42°. No larvae survived any exposure to 44°. Third instars showed higher mortality than either first or fifth instars in the heat stress experiments, particularly when exposed to 40 °C for 6 days, suggesting that third instars are more susceptible to heat stress. Similarly, York and Oberhauser (2002) found that mortality was higher for third than for first or fifth instars when exposed to 12-hour pulses of 36°. Third instars that survived the initial heat exposure were less able to pupate successfully after being removed to control conditions, indicating a long-term effect of exposure to high temperatures. We cannot explain the increased susceptibility of third instars but suspect that development during this stage is particularly vulnerable to thermal stresses.

Fifth instars exposed for 6 days to 40° also had higher mortality than controls, possibly because individuals attempted to pupate while exposed to treatment conditions, or because of increased physiological stress as a result of preparing to pupate coupled with heat stress. Fifth instars exposed to elevated temperatures for 4 or 6 days were also more likely to fall as they pupated.

There were also sublethal effects of heat stress. Individuals exposed to higher temperatures weighed less, which could result in decreased male mating success (Solensky and Oberhauser 2009a), fecundity (Oberhauser 1997), and survival during migration (Masters et al. 1988; Arango 1996; Van Hook 1996; Alonso-Mejía et al. 1997). There may also be direct effects of exposure to high temperatures on these fitness correlates; it would be interesting, for example, to measure effects on fecundity. Additionally, overall development time increased with exposure to elevated temperatures, which could lead to increased risk of predation during this vulnerable stage (Zalucki and Kitching 1982b; Oberhauser et al. 2001, this volume, Chapter 6).

High temperatures affected development time in ways that are not easily explained by degree-day calculations; this is expected when individuals are exposed to supraoptimal temperatures. When total degree days were recalculated to exclude daytime hours at temperatures above 38°, the total degree days needed for development did not differ from control individuals for most treatment combinations, suggesting that development stopped during exposure to elevated temperatures. This assumption was less accurate for longer exposure times, suggesting that after a certain amount of time under unfavorable conditions, monarchs no longer confined development to more favorable nighttime hours. This appeared to be true for all larvae exposed to 38° for 6 days, but only for fifth instars when exposed to 40°. Larvae may be acclimating to higher temperatures, or the rate of development may be determined by a balance between the risks of delayed development and the risks of development at elevated temperatures.

Developmental delays largely occurred while individuals were exposed to heat stress; when they returned to control conditions, their development returned to a normal pace. Larvae appeared to try to escape the hot temperatures; we observed first instars apparently seeking shade under the container labels and older larvae under leaves or the filter paper at the bottom of their container. Individuals also ate less, if at all, under high temperatures. These behaviors were recorded only anecdotally but suggest behavioral changes in response to elevated temperatures, as observed by Serratore et al. (2012) in a field study.

Increased nighttime temperatures also pose a risk. Monarchs exposed constantly to 36° die before adulthood, while repeated 12-hour pulses of 36° do not increase mortality (York and Oberhauser 2002). These findings, coupled with our data showing some survivorship up to 42° when nighttime temperatures were 25°, indicate that a decrease in temperature at night is necessary to cope with extreme temperatures. When the temperature dropped only to 34° at night, first instars suffered substantial mortality, possibly caused by desiccation, since first instars have the highest ratio of surface area to body mass. Dead first instars exhibited signs of dehydration, but because observations were made only once a day, it is unknown whether they desiccated before dying or postmortem. Nighttime temperatures of 34° or lower did not lead to increased mortality for third or fifth instars, suggesting that any decrease in temperature from daytime to nighttime is beneficial in the short term, as long as temperatures drop below the 36° threshold (Zalucki 1982b; York and Oberhauser 2002).

While monarchs can survive short exposures to temperatures above the previously assessed limit of 36°, individuals exposed to higher temperatures were more likely to fall during pupation, weighed less, and developed more slowly. Ecological niche models do not predict monarchs present in areas with a mean monthly temperature above 30° (Batalden et al. 2007), although other models predict monarchs present, but doing poorly, at this temperature (Zalucki and Rochester 2004). Even though monarchs can survive temperatures above 30°, potentially lethal temperatures (40° or higher) as well as temperatures that cause sublethal impacts (38° or higher) are possible within a month with a mean temperature of 30°.

In the cold tolerance experiment, no larvae survived exposure to −20°, but many survived at −10°. For eggs, no hatching was recorded after exposure to −30°, but many eggs hatched after exposure to temperatures as low as −20°. Monarch eggs appear to be chill-intolerant, as their LT_{50} (−15.6°) is warmer than the median SCP (−26.1°). Conversely, first- and third-instar larvae appear to be freeze-intolerant, with half the monarchs dying at or near their respective SCPs. These findings show that monarchs can survive brief periods of cold, as much as 22° to 37° below the stage-specific developmental zeroes reported by Zalucki (1982b). However, the high mortality rates in the control group (reared in cooler, late-fall Texas temperatures) suggest that extended periods at temperatures warmer than the SCP and LT_{50}s may also be lethal; hence, mortality of both mature and immature monarchs resulting from extended exposure to nonlethal cold conditions should be tested with future experiments.

Species can respond to climate change in three general ways: movement, adaptation, or extirpation (Holt 1990). Mobile species may be able to track their ecological niches as the climate changes; there is evidence that some European and North American butterflies have done so (Parmesan et al. 1999; Parmesan and Yohe 2003; Crozier 2004a; Breed et al. 2012). According to ecological niche models, in the future, monarchs will need to move northward from their current range in June and July, and then return southward in August to track the conditions they currently use for reproduction (Batalden et. al. 2007). Currently, only the spring generation appears to move northward before laying eggs (Journey North 2013); during most of the summer, monarchs remain in approximately the same geographic range (MLMP 2013).

It is unclear whether monarch summer generations will respond to suboptimal (i.e., too hot) conditions with movement. If, under a changed climate, an individual monarch survives a stressful temperature regime as a larva, it might attain greater fitness by relocating as an adult, perhaps moving north before laying eggs to promote offspring fitness. Such movement occurs in the spring generation, but the mechanisms that prompt migration out of the southern United States are unclear. If monarchs are to track their moving thermal niche (the area with a range of temperatures in which they can survive) throughout the summer, they will probably need to respond to different cues from those that currently trigger spring or fall migration.

Climate models predict that the overwintering grounds in Mexico may soon be unsuitable for monarchs (Oberhauser and Peterson 2003) or oyamel fir trees (Ramírez et al., this volume, Chapter 13), indicating that the eastern North American monarch population may require different overwintering habitat. Whether monarchs can successfully overwinter in other areas depends in part on their being able to survive the colder temperatures and different habitats present in areas such as the southern United States. The absolute minimum temperatures present in the southern United States are warmer than the LT_{50} for all stages tested, indicating that cold temperatures are unlikely to be a limiting factor for monarchs overwintering there, although we have measured cold tolerance only of monarchs and not their host plants. There may be time for a frozen host plant to regenerate leaves in time to provide food for an egg or possibly a first instar; however, this is unlikely for a later instar. On the other hand, pupae exposed to freezing temperatures do not depend on fast regeneration of milkweed leaves. Thus, the risks imposed by freezing temperatures are likely to vary with different monarch stages, and understanding the effects of freezing on milkweed plants, including leaf regeneration times, will help us interpret the results reported here.

ACKNOWLEDGMENTS

We thank the University of Minnesota Monarch Lab for their help, especially G. Bowers, E. Nimmer, W. Caldwell, K. Duhn, and N. Howard; and S. Kells, M. Eaton, and R. Venette for their use of equipment and their sharing of cold tolerance knowledge. Ernest Williams, Myron Zalucki, and Leslie Ries provided valuable comments on an earlier version of the manuscript. This material is based on work supported by the National Science Foundation (DBS 0710343 to K. Oberhauser and R. Batalden, ISE 0917450 to K. Oberhauser, Graduate Fellowships to K. Nail and R. Batalden) as well as a DeWind Award to K. Nail from The Xerces Society.

9

Microclimatic Protection of Overwintering Monarchs Provided by Mexico's High-Elevation Oyamel Fir Forests

A Review

Ernest H. Williams and Lincoln P. Brower

Monarchs survive the winter by taking advantage of microclimatic features of the high-elevation fir forest in central Mexico. The forest canopy serves as an insulating layer that moderates temperature extremes and shields the overwintering colonies from excessive solar radiation, rain, hail, snow, and wind; in these ways, the canopy protects the butterflies from freezing and from exhaustion of their lipid reserves. Monarchs attain additional temperature and humidity benefits in their dense bough and trunk clusters. Furthermore, the warmest part of the forest is at an intermediate height above the ground, which is where monarchs form most clusters. In combination, these microclimatic features provide protection for overwintering monarchs: blanket-like thermal insulation, umbrella-like shield against precipitation, windbreak against disruptive winds, hot-water bottle–like buffering by tree trunks against thermal extremes, and a vertical location that reduces the likelihood of freezing. All these features illustrate the importance of a dense, undisturbed forest for overwintering monarchs and show that thinning the forest increases risks for their overwinter survival.

LOCATION OF OVERWINTERING COLONIES

Images of the spectacular overwintering clusters in Mexico emphasize the extraordinary character of the monarch's migratory phenomenon (Plate 6). Each fall, monarchs from central and eastern North America leave their summer breeding range because, having evolved from tropical ancestors, they cannot survive the deep freezes of temperate-zone winters (Ackery and Vane-Wright 1984; Larsen and Lee 1994; Brower 1996a). They migrate to volcanic highlands in central Mexico where they find a narrow range of specific climatic requirements and cluster in precise locations where they benefit from microclimatic features of the surrounding forest. Microclimate refers to small-scale patterns of temperature and humidity that are affected by vegetation, solar radiation, wind direction and velocity, precipitation, evapotranspiration, and radiation (Geiger et al. 2009). The response of the butterflies to the combined effects of these microclimate determinants is key to understanding their overwintering survival.

Monarch overwintering sites have been reported on 12 separate mountain ranges between 19° and 20° north latitude and between 100°50′ and 99°40′ west longitude in the Transverse Neovolcanic Belt of central Mexico. Most are found from 2900 to 3300 m elevation (Bojórquez-Tapia et al. 2003; Slayback and Brower 2007; Slayback et al. 2007). Recent aerial surveys found no other mountain ranges supporting colonies, although a small colony was reported on the western slope of Popocatepetl, just southeast of Mexico City, during the 1978–1979 (W. Calvert, pers. comm. 1979) and 2001–2002 overwintering seasons (Cevallos 2002).

It is cold at high elevations on these mountains; the lapse rate, the decline in minimum daily temperature with elevation, has been measured in the Monarch Butterfly Biosphere Reserve as −4.4 °C/500 m (Weiss 2005). Monarch overwintering colonies form characteristically below ridge crests, most likely to escape high-elevation winds, and near arroyos with streams or seeps (small springs) as sources to which the butterflies periodically fly to drink (Calvert and Brower 1986; Calvert et al. 1989; Bojórquez-Tapia et al. 2003). The southwesterly aspect provides more afternoon sun, enabling them to fly to water and also to re-form their clusters after being blown down by storms. Adiabatic cooling of rising air masses that have come over the Pacific Ocean provides periodic moisture to these slopes (Calvert and Brower 1986; Calvert et al. 1989). Although the mountains lie in a summer wet season fog belt, the winters are dry (Manzanilla 1974; Brower 1995), so availability of water becomes critical as the dry season advances. Steep slopes also allow cold air to drain from ridge crests down to cold pockets below, leaving the mid-slope slightly warmer (Weiss 2005).

The boreal-like forests at these elevations are dominated by the oyamel fir, *Abies religiosa*, although other species are also present, including the Mexican cypress *Cupressus lindleyi* (= *C. lusitanica*) (Earle 2011), the Mexican pine *Pinus pseudostrobus*, and some broadleaf trees, especially oaks and alders (Manzanilla 1974; Brower et al. 1977; Calvert et al. 1989). This high-altitude area is a relict ecosystem that had a wider distribution during glacial times (Manzanilla 1974; Snook 1993; Brower 1999). A diverse understory includes numerous herbaceous species, including shrubs of *Senecio* and *Eupatorium* spp., with mosses and lichens on the forest floor (Brower et al. 1977; Calvert and Brower 1986; Núñez and Garcia 1993; Snook 1993; García-Serrano et al. 2004; Cornejo-Tenorio and Ibarra-Manríquez 2008). Increased vegetation, both canopy and understory, leads to greater moderation of environmental conditions (Geiger et al. 2009). The mosaic of vegetation creates microclimatic variation by absorbing and reradiating heat energy.

Overall, monarchs encounter a protective microclimatic envelope produced by the interplay of the physical features of elevation, slope, exposure, and water, and the biological features created by the fir forest ecosystem (Brower 1996a). Temperature and precipitation are key features. The sites must be cold enough to lower the consumption of the monarchs' lipid reserves and keep them in reproductive diapause (Barker and Herman 1976; Tauber et al. 1986; James 1993), but not so cold that they freeze to death. At the same time, the surrounding environment must stay warm enough to allow butterflies to fly when needed, as when clusters are disrupted by storms, and wet enough to provide nearby drinking sources and to reduce the likelihood of fires (Brower 1985a; Calvert and Brower 1986; Brower and Missrie 1998). In the pages that follow, we examine these effects in detail. The importance of microclimate is shown by the butterflies themselves in the locations where they aggregate. We know that the colonies move downslope as spring approaches and re-form where they are partially exposed to sun and where water is available nearby (Calvert and Brower 1986), but we do not know the environmental cues that cause them to move.

PROTECTIVE EFFECTS OF FOREST CANOPY

Thermal insulation

The insulating effect of the forest, with temperatures moderated under the canopy, has long been emphasized (Calvert and Brower 1981; Calvert et al. 1982; Alonso-Mejía et al. 1997; Brower 1999; Geiger et al. 2009). The butterflies avoid clearings, where radiational heat loss leads to nocturnal temperatures dropping as low as −11 °C (Manzanilla 1974); instead, they aggregate under the protective canopy of dense forests (Plate 6), where the temperature rarely falls to freezing. The denser the forest, the stronger the insulating effect. By comparing forests of different densities, Calvert et al. (1982, 1984) reported that a decrease in density of 100 trees/ha correlated with a lowering of minimum temperatures by 0.37 to 0.53 °C. A similar conclusion was reached by Brower et al. (2011), who found that a thinned forest (34% open canopy) was colder on average by 0.33 °C than a denser part of the same forest (13% open canopy). Although monarchs do become somewhat cold hardened (Larsen and Lee 1994), they succumb to hard freezes, especially when wet (Anderson and Brower 1996), so protection from extreme cold is necessary.

Not only must the butterflies be protected from freezing at nighttime, but they must also remain cool enough during the day to slow the usage of their energy reserves. Stored lipids made from nectar consumed during the fall migration (Brower et al. 2006; this volume, Chapter 10) provide the butterflies with energy to maintain their basal metabolism and allow flights to drink water during the long winter, as well as to fuel the beginning of the spring remigration. Although monarchs attempt to feed from both *Senecio* and *Eupatorium* spp. that flower at the overwintering sites, these resources are not sufficient to replenish lipid reserves prior to migration (Brower 1985a). Like all biochemical processes, the metabolic rate of monarchs depends on body temperature, and as ectotherms, whose body temperature is influenced by the surrounding environment, they burn lipids at a rate influenced by ambient temperatures. In the heat of the open sun during the day, they would use up their energy supply too quickly; estimates are that active butterflies at 25 °C would exhaust all lipids in less than 40 days of the 150-day winter (Masters et al. 1988). As Weiss (2005) noted, higher temperatures also increase water deficits by reducing relative humidity, so the butterflies would have to fly more to rehydrate, thus burning even more of their limited fuel reserves. Furthermore, reproductive diapause can be broken by high temperatures (Barker and Herman 1976; Tauber et al. 1986; James 1993), and spring migration may begin too early. These factors all point to the need to avoid excessive warmth.

In summary, a dense canopy provides thermal stability (Brower et al. 1985a; Geiger et al. 2009), with a balance of cold to reduce lipid loss and maintain reproductive torpor but with sufficient warmth to avoid freezing and allow flight to water. Thus, overwintering survival is facilitated by the blanket-like insulating capacity of the vegetative cover.

Shield from precipitation

A second important feature of an intact canopy is that it provides an umbrella-like cover that deflects precipitation from falling directly on the butterfly clusters (Anderson and Brower 1996). The canopy also protects the butterflies from dislodgement by snow. Butterflies wetted by rain, dew, or snow are killed at higher subfreezing temperatures than dry butterflies (Larsen and Lee 1994; Anderson and Brower 1996): 50% of wetted butterflies froze at −4.2 °C whereas 50% of dry butterflies remained alive down to −7.7 °C (Anderson and Brower 1996). Temperatures of a few degrees below zero are regularly encountered in the overwintering colonies, so shielding from precipitation is a critical protective feature of a full canopy.

Windbreak

The forest canopy also serves as a windbreak, lessening the frequency with which butterflies are blown off branches and onto the ground. High winds can dislodge tens of thousands of butterflies (Calvert and Brower 1986), and on the ground, they are more susceptible to wetting from dew and subsequent freezing (Brower 1999), as well as to predation by mice (Brower et al. 1985a; Glendinning et al. 1988; Glendinning and Brower 1990). When cold and stranded on the forest floor, monarchs may crawl onto the understory vegetation to escape the ground; they can crawl at 4.6 °C, shiver at 8.0 °C to elevate thoracic temperature, and fly at 13.0 °C (Alonso-Mejía et al. 1993). If blown to the forest floor into shady areas without direct insolation, they may not escape the cold zone next to the ground, where the probability of freezing is higher. Furthermore, winds can be drying and thus increase the need for drinking, which requires energetically expensive flights to find water (Brower 1999). Firs are effective windbreaks, however, and coniferous forests in general provide even more resistance to wind than do deciduous forests (Geiger et al. 2009).

MICROCLIMATIC EFFECTS UNDER THE FOREST CANOPY

Bough clusters

Classic images of overwintering monarchs show conifer branches draped in dense layers of butterflies, with only a few twigs and needles apparent through a nearly continuous layer of orange and black wings (Plate 6). An immediate question is, why do monarchs cluster in such large numbers; is there an advantage to being part of dense clusters on tree boughs?

Most bough clusters are found on oyamel fir, the most abundant tree species in the high-elevation overwintering colonies, but clusters also form on smooth-bark Mexican pine and Mexican white cedar or cypress. While fir branches are needle-dense and drape in a way that may facilitate the formation of dense aggregations, we do not know if the architecture of firs actually provides the best substrate for clustering (Brower 1999). Oyamel firs appear to form the climax community at these elevations, but abundant cedars, upon which monarchs also densely cluster, are found on the south slope of Cerro Pelón and appear to represent a fire-induced subclimax community.

Studies of microclimate inside and immediately outside bough clusters by Brower et al. (2008) showed slight microclimatic moderation for those butterflies inside each cluster. This conclusion was drawn from gently lowering dowels with attached temperature/humidity loggers (hygrochron iButtons, Maxim Integrated Products) into clusters. For each cluster, one iButton was located inside and another remained just outside the cluster. During the day, based on hourly measurements from 1200 to 1700 hours, the inside of the clusters averaged from 0.2 to 0.6 °C cooler, with relative humidity 1–2% higher than the air just outside the cluster. The difference at night was minimal, with the average of hourly temperatures inside the clusters only 0–0.2 °C warmer from midnight to 0800 hours. These measurements were taken during moderate weather, and the differentials between the inside and outside may be greater during occasional temperature drops and more extreme conditions. Whether these differences are enough to affect monarch lipid usage or survival is uncertain. Another effect is that butterflies inside and on the bottom of the bough clusters are better protected from wetting than those on the outside (Anderson and Brower 1996).

Trunk clusters

Monarchs cluster densely in the overwintering colonies not only on boughs but also on tree trunks, where they sometimes aggregate so tightly that the trees themselves are no longer visible (Plate 6). When clusters start forming in November, the butterflies first settle on the outer branches of the trees, but by January they have also packed onto the trunks (Calvert and Brower 1986). Trunk clusters are generally 5–15 m above the ground (Brower et al. 2011), but sometimes extend to the base of the tree, as happens after a storm when butterflies have been knocked down to the forest floor and begun to crawl back up (Brower et al. 2011). The presence of dense aggregations on tree trunks raises the question of whether roosting on trunks gives the butterflies any microclimatic advantages.

Trunks do provide protection against freezing. Calvert et al. (1983) found that following a severe 1981 winter storm in the Sierra Chincua colony, 78% of monarchs on the tree trunks remained alive, versus 56% in boughs. Brower et al. (2009) reported a similar result after a 1992 storm in the Herrada colony, with 43% survival on tree trunks and only 5% survival on boughs. Thermal buffering by living tree trunks is expected because their high heat capacity leads to slower warming up and cooling down than the surrounding air (Brower et al. 2009; Geiger et al. 2009). The difference can be seen by examining measurements of specific heat, which is the amount of heat needed to raise a unit of mass 1°C. The specific heat of a living tree, which is essentially wet wood, falls between that of dried wood at 1.2 J/g-K (Simpson and TenWolde 1999) and that of water at 4.2 J/g-K (CRC 1992). Analysis of five species of firs (Simpson and TenWolde 1999) showed that the average moisture content ranged from 25% to 49% for heartwood and 53% to 63% for sapwood. When our research team bored into trees in the Sierra Chincua colony, water squirted out of 12 of 60 oyamels, confirming how wet they are. The above data indicate that the specific heat of a living tree is at least 2.0 J/g-K, about 1000 times greater than the specific heat of an equivalent volume of air, which is 0.0013 J/cm^3-K (1 g of wood is approximately 1 cm^3). Thus, trunk surfaces cool and warm slowly compared with the surrounding air. The small size of insects leads to rapid heat conformity with their surroundings (Casey 1992), so those clustered on tree trunks of high thermal mass should receive significant thermal moderation through both conduction and convection.

To compare the temperature of tree trunks and the immediately surrounding air, Brower et al. (2009) placed paired iButtons on oyamel fir trunks, with one of each pair directly on the bark and the other near the first iButton but on a dowel extending 3 cm away from the trunk. The experiment was replicated in two different colonies with a total of 22 trees for at least 27 days. The results showed not only significant moderation of temperature variation by the trunks, but also that larger trunks had a greater buffering effect. At night, smaller trees (mean = 38.7 cm dbh, diameter at breast height) were 0.8–1.5 °C warmer than ambient, while larger trees (mean = 72.0 cm dbh) remained 1.1–2.2 °C warmer. During the day, smaller trees were 0.8–0.9 °C cooler than ambient, while the large trees were 1.0–2.0 °C cooler. All measurements of trunks and surrounding air were significantly different, and three of four comparisons of smaller and larger trees yielded significant differences. As would be expected from temperature changes within masses of wet wood, larger trees provide greater microclimatic buffering. In addition, relative humidity remains higher in the surrounding air near the trunks during the day because of the cooler temperatures (Geiger et al. 2009).

The greater protection provided by larger trees raises the question of what these forests must have looked like centuries ago, before the beginning of logging. The overwintering forests currently comprise trees smaller on average than they used to be. Calvert (2004b) described the average dbh of trees in the Sierra Chincua colony in 1979 to be 31 cm, and Brower et al. (2009) reported a similar mean of 32.8 cm in 1985. In the early 2000s, Keiman and Franco (2004) measured trees from 13 plots in the Sierra Chincua, with mean dbh values ranging from 1.9 to 39.2 cm, and only 2 plots had means above 30 cm. Recently, Brower et al. (2009) chose 11 of the largest oyamel firs available for an analysis of trunk microclimate, and the largest tree had a dbh of 91.8 cm. In contrast, oyamel fir diameters of up to 2 m (and heights of 50 m) have been reported in the past (Loock 1950; Earle 2011), and one of us (LPB) has found large stumps up to 1.8 m diameter on the south face of Cerro Pelón. Furthermore, Manzanilla (1974) reported that in an old-growth oyamel fir forest west of the overwintering colonies, 15% of 331 trees had a diameter greater than 80 cm, with the largest at 2 m, and some trees were at least 272 yrs old. Logging over the years has reduced the average size of trees in the colonies, and the largest are gone completely.

Calvert (2004b) estimated that in the Sierra Chincua colony during the winter of 1985–1986, approximately 10% of the butterflies were in trunk clusters and 90% in bough clusters. Would a higher percentage of butterflies have aggregated on trunks rather than boughs centuries ago in uncut old-growth forest with large trees? Large trunks provide more space for clustering, and the buffering of temperature changes would be greater. Compared with trees with an average diameter of 55 cm, which was the average diameter of the 22 large trees studied by Brower et al. (2009), trees of 2 m diameter would provide 3.6 times as much surface area for roosting and 13 times as much volume for thermal buffering. This increase results from changes in diameter only, but larger trees would also have been taller and provided an even more expanded surface and buffering capacity. Strong microclimatic buffering provided by the trunks of such large trees would enhance protection from large fluctuations in temperature, i.e., an increased hot-water bottle effect. We hypothesize that, in pre-Columbian forests, a higher percentage of overwintering monarchs would have been in trunk clusters than is the case currently, and the butterflies would have therefore benefited from enhanced protection against freezing as well as reduced consumption of lipid reserves.

Vertical profile

Whether on boughs or on trunks, most clusters occur at heights intermediate between the forest floor and the canopy (Brower et al. 1977; Keiman and Franco 2004). Measurements in 2008 of 18 bough clusters in the Sierra Chincua, where the canopy ranged from 20 to 30 m high, showed that clusters ranged in height above the ground from a lower average of 5.9 ± 2.5 m to an upper average of 15.4 ± 2.4 m (Brower et al. 2011). Similar height measurements had been found some years earlier for trunk clusters (Calvert and Brower 1986). This persistent pattern of clusters forming at intermediate heights is determined to some degree by the architecture of the trees but suggests that the butterflies benefit from avoiding both the ground and the canopy.

There is good reason for the butterflies to avoid the forest floor and the tree crowns. The temperature in a montane forest remains coldest near the ground, which is a deep heat sink, and dew formation is also more likely there (Geiger et al. 2009); in addition, radiational cooling in the tree crowns leads to rapid cooling at night, as well as greater exposure to wind, rain, hail, snow, and dew. Measurements through the vertical profile of the oyamel forest have shown that the nightly minimum temperature remains higher at intermediate heights than at the ground or in the canopy (Figure 9.1). The strongly curved thermal profile at night becomes uniform during the day as both air and ground surface warm (Geiger et al. 2009; Brower et al. 2011). With reduced insolation, cloudy periods produce lower daytime temperatures but higher nighttime minimum temperatures throughout the vertical profile. In all conditions, however, intermediate heights are warmer at night than those near the ground or in the canopy.

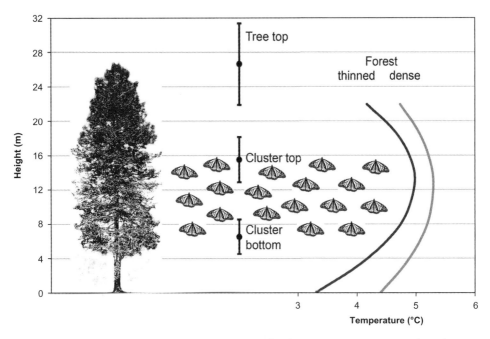

Figure 9.1. The relationship between nighttime vertical temperatures (right) and monarch cluster heights (center) compared with an oyamel fir 27 m tall (left). The clusters averaged from 6.4 m above the ground up to 15.5 m, more than 10 m below the canopy, with most monarchs concentrated at middle levels (black height bars show mean ± SD). The curved lines on the right show nighttime vertical temperature profiles from the ground to 22 m on two oyamel firs, one in a thinned forest area and the other in a denser area. The midsections of the vertical profiles are about 1–2 °C warmer than both the ground and the upper canopy. Reprinted from Brower et al. 2011 by permission of The Lepidopterists' Society.

Calvert and Brower (1981) found monarchs crawling up available foliage when they had been blown or knocked to the cold ground. Butterflies caught on the ground at night are more likely to freeze, so getting onto vegetation is important; in experimental studies, those butterflies that crawled at least 30–40 cm above the forest floor escaped freezing mortality even at very low ambient temperatures (Calvert and Cohen 1983; Alonso-Mejía and Arellano-Guillermo 1992; Alonso-Mejía et al. 1997). This means that the presence of understory vegetation is a component of the complex microclimate environment to which monarchs respond. The understory remains warmer and drier than surrounding open areas, and even monarchs that do not get off the ground are better off in understory vegetation than being more exposed (Calvert et al. 1986); nevertheless, they are more subject to freezing if they remain within 1 m of the ground.

Important microclimatic differences occur across vertical transects of the forest. At the warmest height (12 m) above the ground, nighttime temperatures were approximately 0.5 °C colder in areas where the forest had been thinned than in the denser forest, and the thinned forest was even colder at both lower and higher heights (Brower et al. 2011). The effect of forest density is also apparent in the daily variation in temperature at each height, with the denser forest providing greater buffering capacity and experiencing less variability. In all these cases, however, the best protection from freezing is found at intermediate heights beneath the canopy, and this is the height at which dense butterfly clusters form on boughs and trunks.

Although primordial forests with trees up to 60 m in height no longer exist in the Monarch Butterfly Biosphere Reserve, they would have provided an expanded zone of modified microclimate under the canopy. In the Sierra Chincua, where the majority of our research has been done, most of the oyamel trees do not exceed 30 m in height (Brower et al. 2011); among 62 trees measured during the 1984–1985 overwintering season, the maximum trunk diameter in the colony area was 65 cm, and the oldest tree was 84 years old (Brower et al. 2009). In contrast, in mostly undisturbed forests in the states of Mexico, Morelos, and Puebla (east of Michoacán), Manzanilla (1974) determined that some oyamels were more than 140 cm in diameter, more than 61 m in height, and more than 200 years old. Manzanilla's forests were also nearly pure (323 of 327) oyamel stands. Thus, the maximum heights, diameters, and ages of the oyamels in the undisturbed forest were more than twice those in the current Chincua forests. We do not know how the primordial monarch overwintering phenomenon differed from that in the current largely degraded forests.

Behavior

In addition to their choices of where to cluster, monarchs can respond to microclimatic conditions with thermoregulatory behaviors. When cold and stranded on the ground, they can shiver to increase thoracic temperatures enough to enable flight (Kammer 1970), but this succeeds only when they can raise their temperatures above 13 °C. When too warm in the intense sun of high elevations, monarchs may move to shade or take sun-minimizing postures (Masters et al. 1988). Another thermoregulatory behavior is that of flying up from the clusters to glide in cold air. While flight may seem energetically expensive, this behavior costs less energy than remaining flightless and overheated in a warm cluster; in flight, the thorax is warm from muscular usage, but the abdomen cools to ambient (Masters et al. 1988).

IMPORTANCE OF FOREST QUALITY

A protective layer of dense forest serves to shield overwintering monarchs from freezing cold, excessive warmth, precipitation, and wind. Forest cover is destroyed by clear-cut logging and degraded by moderate tree thinning, forest fragmentation, and fires set to clear the land for agriculture (Manzanilla 1974; Williams et al. 2007; Ramírez et al., this volume, Chapter 13). Where the forest has been degraded, the butterflies are exposed to greater climatic variation. A thinned forest has reduced insulative capacity, allowing greater radiational cooling at night (Calvert et al. 1982; Brower et al. 2011), which leads to lower minimum temperatures. In addition, greater daytime warming, higher wind speeds, and reduced buffering in the absence of large trees threaten microclimatic integrity. In response, the butterflies may experience increased desiccation, greater burning of stored lipids leading to starvation, insufficient fuel for spring migration, and early departure from overwintering colonies (Brower 1999; Brower et al. 2009). Partial thinning can also degrade habitat quality of a larger area because of subcanopy air circulation (Weiss 2005; Geiger et al. 2009). Additional threats to the fir forest are climate change (Oberhauser and Peterson 2003; Sáenz-Romero et al. 2012) and bark beetle and mistletoe infestations that may lead to an increase in forest thinning (Vázquez 2009).

The orientation of North American mountain ranges allows arctic air masses to dip down into Mexico (Brower 1999), bringing large winter storms that can buffet the monarch colonies and drop to lethally freezing temperatures. These weather events highlight the need for a protective canopy. Storms in 1981–1982 that penetrated the forest led to 40% mortality (Calvert et al. 1983, 1984), and an even more severe storm in January 2002 was estimated to have killed 75% of the butterflies in two colonies and upward of 500 million butterflies across the overwintering region (Brower et al. 2004). In addition to their impacts on clustering monarchs, storms lead to mudslides and soil erosion, degrading the habitat (Aridjis 2004; Brower et al. 2010). Historical records suggest that storms intense enough to kill a large percentage of butterflies—those that produce wetting and are followed by severe drops in temperature—occur about once a decade. Because the impact of these storms can be so severe, and because their impacts are exacerbated by thinned forests, it is imperative that mature forests are protected and that tree regeneration, including natural reseeding, is encouraged in degraded areas (Keiman and Franco 2004).

Despite the conspicuous value of the forest canopy for protecting the butterflies, degradation of the forest within the overwintering area has continued. From 1971 to 1999, 44% of conserved forest was degraded (Brower et al. 2002). Between 1999 and 2008, heavy illegal logging occurred (Simmon et al. 2008), and although the rate of illegal deforestation has decreased over the past four years (Navarrete et al. 2011; Ramirez et al., this volume, Chapter 13), salvage logging of trees downed by storms (Anon. 2011) continues to degrade the forest in the legally protected Monarch Butterfly Biosphere Reserve (Brower, pers. observ.). In addition, rising temperatures and different precipitation patterns caused by climate change will alter the forest in ways that may severely reduce the umbrella and blanket effect of the oyamel fir forest (Allen et al. 2010; Flores-Nieves et al. 2011). The overwintering area is equal to less than 0.01% of the summer breeding area (Brower 1999), a ratio that illustrates the importance of protecting this small area of concentrated usage. Since 1995, the size of the total monarch overwintering population has declined significantly, a decline that follows both the loss and degradation of the oyamel overwintering habitat and increasing loss of breeding habitat (Brower et al. 2012a, 2012b; Pleasants and Oberhauser 2012). In our opinion, continued logging is having a devastating effect. The combined evidence from these microclimate studies indicates that an intact fir forest ecosystem is key to the survival of overwintering monarch butterflies in Mexico.

ACKNOWLEDGMENTS

We thank the collaborators in our three consecutive microclimate studies: Linda S. Fink, M. Isabel Ramírez, Raul R. Zubieta, Daniel A. Slayback, M. I. Limon García, P. Gier, J. A. Lear, T. Van Hook, S. B. Weiss, W. H. Calvert, and W. Zuchowski, and we acknowledge Jeremy Greenwood for translating documents for us. We also thank Linda Fink, Cuauhtémoc Sáenz-Romero, Ben Slager, and Karen Oberhauser for critically reading the manuscript and providing thoughtful suggestions for improvement. We are grateful to the personnel of La Reserva de la Biosfera Mariposa Monarca for facilitating our research over the past several years. Support was provided by National Science Foundation grant DEB-0415340 to Sweet Briar College, with Lincoln Brower and Linda Fink as principal investigators, the October Hill Foundation, and the Monarch Butterfly Fund. EHW was supported by the Christian A. Johnson Professorship Fund at Hamilton College.

10

Effect of the 2010–2011 Drought on the Lipid Content of Monarchs Migrating through Texas to Overwintering Sites in Mexico

Lincoln P. Brower, Linda S. Fink, Ridlon J. Kiphart, Victoria Pocius, Raúl R. Zubieta, and M. Isabel Ramírez

Nectar sources in Texas and northern Mexico allow monarch butterflies to accumulate lipid reserves that support them while overwintering in Mexico. In 2010–2011 this area had the worst drought on record, raising concern that limited nectar would reduce the butterflies' lipid reserves. In October 2011, at the peak of the fall migration through central Texas, we collected monarchs nectaring on frostweed (*Verbesina virginica*) at four sites for lipid analysis. As predicted, monarchs' lipid levels were significantly below normal. In contrast, monarchs in an irrigated field of gayfeather (*Liatris mucronata*) contained normal lipid titers. Because of the water stress across Texas, we predicted that monarchs would arrive at overwintering sites in November with lower lipids than normal. In fact, monarchs in two colonies had normal lipid levels. We hypothesize that there must be areas in eastern Mexico where monarchs had access to copious nectar. An alternative explanation is that monarchs leaving Texas with low reserves died before reaching the overwintering sites. The first hypothesis seems more likely and, if correct, argues for the importance of conserving floral corridors in eastern Mexico.

INTRODUCTION

Most of the energy that adult butterflies use to fuel their activity is obtained from flower nectar (Robertson 1928; Tooker et al. 2002; Brower et al. 2006). Nectar contains sugar that butterflies can convert to lipids, store in their abdominal fat body, and mobilize as needed (Brown and Chippendale 1974; Heinrich 1983; Brower 1985a; Nation 2002). Monarchs' utilization of lipids is dynamic, because their energetic needs and the availability of nectar change during the spring and summer breeding seasons, the fall migration, the five-month overwintering in Mexico, and the subsequent spring remigration into the Gulf Coast states (Brower et al. 2006).

Monarchs arrive at the Mexican wintering sites in November with high lipid levels. Nectar sources during the overwintering season are scarce, and the butterflies must survive the winter on lipid reserves stored during the fall migration (Brower 1977, 1985a, 1999; Alonso-Mejía et al. 1997); butterflies that arrive in Mexico with low initial reserves risk starvation. Fall migrants collected in the northern United States have moderate, highly variable lipid levels, while those collected in Texas and northern Mexico have substantially higher, and still highly variable, lipid levels (Brower 1985a; Brower et al. 2006). This pattern indicates that the butterflies accumulate most of their lipid reserves in the southern United States and northern Mexico, and therefore that fall nectar resources in this region must be crucial for the butterflies' winter survival.

The Texas Hill Country is an uplifted Cretaceous limestone formation that occurs on the eastern portion of the Edwards Plateau and runs southward through Central Texas. The vegetation includes a mix of evergreen savannas and upland deciduous communities growing on thin soils, and moist river bottomland communities on deeper soils. The area is renowned for the diversity and abundance of its fall wildflowers (Enquist 1987; Wrede 2010). Calvert (1998) noted the importance of its riparian areas for roosting and nectaring fall migrant monarchs. Observations since 2002 suggest that nine of the most important fall nectar sources for the migrants are in the Asteraceae: *Ageratina (Eupatorium) havanensis, Baccharis neglecta, Conoclinium (Eupatorium) coelestinum, Helianthus maximiliani, Liatris mucronata, Solidago canadensis, S. nemoralis, Verbesina encelioides* and *V. virginica* (Kiphart, pers. observ.; Enquist 1987; Diggs et al. 1999; Lady Bird Johnson Wildflower Center Native Plant Database; USDA Plants Data Base). Of these, *Verbesina virginica* (frostweed), which grows extensively in the understory of riverine forests, appears to be the most important (Kiphart, pers. observ.).

From October 2010 through September 2011, Texas and adjoining states entered a period of dry weather; by 31 October 2011, most Texas counties had reached the status of *exceptional drought* (National Weather Service 2011; Nielsen-Gammon 2011; Stengle 2011; NOAA National Climatic Data Center 2012). This was the driest 12-month period in Texas since 1895, when instrumental records began (NOAA National Climatic Data Center 2013). Further evidence of the drought intensity comes from an analysis of tree rings dating back to 1550, indicating just one other year (1789) in which Texas had such severe conditions (Dawson 2011). The drought extended through south-central Texas into the northern Mexico states of Chihuahua and Coahuila. The drought in the five northern states of Mexico was also the worst since the Mexican government began recording rainfall 70 years ago (Salazar and Rodriguez 2011).

Most monarchs migrating through Texas on their way into Mexico in the fall pass through a central flyway that crosses the Texas Hill Country (Calvert and Wagner 1999; Howard and Davis 2009; Quinn 2011). Thus, in the fall of 2011, most of the eastern North American monarch population had to pass through a highly water-stressed region.

The research reported here addressed three questions: (1) What was the impact of the drought on native fall nectar sources along the central Texas flyway? (2) Did the effect of the drought on the flora prevent monarchs from adequately building up their lipid reserves? (3) Did monarchs arrive at the overwintering sites in Mexico with lower lipid stores than in other years?

To answer these questions, we observed the effects of the drought on the flora and on the region's rivers in October 2011. We collected fall migrant monarchs during these observations, and overwintering monarchs at two sites in Mexico in November 2011 (Figure 10.1), for lipid analyses. In March 2012, after the winter and spring rains, we reexamined the flora and river flows. We predicted that both the Texas and overwintering samples from Mexico would have low lipid reserves compared to years with moderate to normal rainfall.

LIPID LEVELS IN ADULT MONARCH BUTTERFLIES IN SUMMER, FALL, AND EARLY WINTER

To put the 2011 data in perspective, we first review the geographic and temporal patterns of monarch lipid accumulation for years prior to the drought. Using data published in Brower (1985a), Alonso-Mejía et al. (1997), Brower et al. (2006), plus additional unpublished data, this summary groups together samples that were previously presented separately. The locations of the collection sites are indicated in Figure 10.1. Descriptive statistics and the individual samples included in each group are presented in Table 10.1, with histograms of the data in Figure 10.2.

Freshly eclosed butterflies (Group 1)

A baseline value of 29 mg of lipid (range 5–57 mg) was found in 45 freshly eclosed butterflies. These butterflies were collected as wild chrysalids in Plainfield, Massachusetts (MA) on 15 September 1977 and 15–23 September 1979. They eclosed in our laboratory, had no access to food, and were killed 24–48 hours after eclosion. These data indicate that only small amounts of lipids are available from larval feeding.

Summer breeding butterflies (Group 2)

Summer adults were captured in open fields in MA and Virginia (VA) while they were nectaring, ovipositing, and courting. The collections were made in Hampshire County MA from 21 July to 16 August 1979; and Sweet Briar VA from 23 to 30 August 1999 and 7 August to 2 September 2001. Their mean lipid content was 18 mg (range 0–92 mg, $N = 413$). This

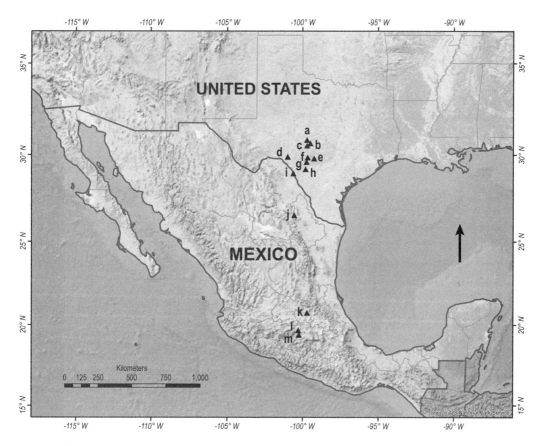

Figure 10.1. Texas and Mexico collecting locations of monarchs that listed in Table 10.1 and shown in Figure 10.2. Sites are listed from north to south. (a) Whispering Water Ranch, TX (Group 9b, 30°49.793′N 99°42.874′W, elevation 604 m) (b) Lucky Boy Ranch, TX (9a, 30°38.374′N 99°29.323′W, 504 m) (c) Native American Seed Co., TX (10, 30°30.892′N 99°42.591′W, 515 m) (d) Devils River, TX (5b, 29°53.204′N 100°59.414′W, 446 m) (e) Medina River floodplain, TX (9c, 29°47.736′N 99°14.331′W, 426 m) (f) Frio River floodplain, TX (29°50.990′N 99°40.360′W, 570 m) (g) Garner State Park floodplain, TX (4d, 9d, 29°35.603′N 99°44.590′W, 444 m) (h) Uvalde, TX (4a, 4b, 4c, 29°11.642′N 99°48.417′W, 276 m) (i) Quemado, TX (5a, 28°57′N 100°37′W, 245 m) (j) Bustamante Canyon, Mexico (6, 26°32.928′N 100°33.651′W, 504 m) (k) San Javier, Queretaro, Mexico (7, 20°44.861′N 99°42.935′W, 2240 m) (l) Sierra Chincua, Mexico (11a, 19°40′29.92″N 100°17′52.93″W, 3311 m) (m) Cerro Pelon, Mexico (11b, 19°23′21.81″N 100°15′32.68″W, 3249 m).

average lipid content, the lowest we found in any phase of the annual migration cycle, is significantly lower than in the freshly eclosed group ($t_{65} = 8.8$, $P < 0.0001$). The skewed distribution (see Figure 10.2.2), together with the low mean, indicates that summer breeding butterflies are using up the small reserves carried over from larval feeding. Further evidence that breeding monarchs draw down reserves, rather than accumulating them, is the fact that within the VA samples, breeding butterflies with worn wings had significantly lower lipid content than breeding butterflies with better wing condition (Brower et al. 2006).

Fall migrants, north of Texas (Group 3)

The 890 migrant butterflies collected in New England, the mid-Atlantic, and the Midwest had significantly higher lipid levels than summer breeders ($t_{1056} = 13.6$, $P < 0.0001$). Their mean lipid content was 34 mg (range 3–169 mg). The monarchs in Group 3 were collected in Amherst MA (3–27 September 1979), Cape May and Beach Haven NJ (29–30 September 1979), and Lawrence KS (23 September 1979); and Sweet Briar VA (between 31 August and 1 November of 1998–2001). All Group 3 butterflies were collected while nectaring, with the exception of

Table 10.1. Lipid content (mg per butterfly) of summer, fall, and early overwintering monarch butterflies

Group no.		Samples	Years	Inclusive dates	N	Mean	SD	Min	Max
Samples before 2011									
1		Freshly eclosed, MA (chrysalid dates)			45	29	10	5	57
	1a	MA 1977	1977	15 Sept	32	29	9	17	57
	1b	MA 1979	1979	15–23 Sept	13	29	14	5	53
2		Summer breeding, MA and VA			413	18	12	0	92
	2a	MA	1979	21 July–16 Aug	229	19	13	3	80
	2b	VA	1999, 2001	23 Aug–2 Sept	184	17	12	0	92
3		Fall migrants, MA, NJ, KS 1979; VA 1998–2001			890	34	28	3	169
	3a	Massachusetts, Amherst	1979	3–27 Sept	137	26	15	4	76
	3b	New Jersey, Cape May, nectaring on *Solidago*	1979	29 Sept	36	21	17	5	100
	3c	New Jersey, Beach Haven, sitting on bushes	1979	30 Sept	14	61	30	5	107
	3d	Kansas, Lawrence	1979	23 Sept	122	18	15	3	96
	3e	Virginia, Sweet Briar	1998–2001	31 Aug–1 Nov	581	40	31	4	169
4		Fall migrants, TX Hill Country			182	71	39	7	226
	4a	Uvalde, Henderson's pecan orchard	1982	12 Oct	59	77	43	24	226
	4b	Uvalde, roosting	1982	16 Oct	40	59	35	7	172
	4c	Uvalde, Leona Park	1982	19 Oct	40	62	29	14	151
	4d	Garner State Park	1994	10 Oct	43	84	40	15	161
5		Fall migrants, TX-Mexico border			111	85	45	14	192
	5a	Quemado TX	1982	26 Oct	39	52	23	14	113
	5b	Dolan Creek, Devil's River Conservation Area, TX	1994	11 Oct	72	103	44	18	192
6		Fall migrants, northeastern Mexico, Bustamante Canyon			149	98	54	6	234
	6a	Bustamante Canyon	1976	30 Oct	47	126	44	39	234
	6b	Bustamante Canyon	2006	19 Oct	102	85	54	6	221

7		Fall migrants, close to wintering sites	1977	31 Oct	101	149	43	8	281
8		Mexico overwintering November			705	128	41	20	237
	8a	Sierra Chincua	1981	7 Nov	32	126	40	59	202
	8b	Sierra Chincua	1982	9 Nov	174	142	36	54	227
	8c	Sierra Chincua, Chivati, Contepec, Cerro Pelon, Rosario	1993	8–18 Nov	449	121	142	20	237
	8d	Sierra Chincua	2004	2 Nov	50	145	41	40	231
2011 samples									
9		TX Hill Country, nectaring on *Verbesina virginica*			135	40	29	5	138
	9a	Lucky Boy Ranch, Mason, Mason Co.	2011	11 Oct	6	15	12	5	39
	9b	Whispering Water Ranch, Menard, Menard Co.	2011	11 Oct	40	38	24	13	121
	9c	Medina River floodplain, Bandera Co.	2011	12 Oct	48	56	28	7	105
	9d	Garner State Park oxbow, Uvalde Co.	2011	13 Oct	41	28	28	5	138
10		TX Hill Country, nectaring in *Liatris mucronata* nursery, Native American Seed Co., Junction, Kimble Co.	2011	11 Oct	36	80	42	15	174
11		Mexico overwintering samples November 2011			199	134	49	36	262
	11a	Sierra Chincua	2011	17 Nov	99	122	53	36	262
	11b	Cerro Pelon	2011	18 Nov	100	146	42	48	253

Note: Statistics are given for each of the 11 group totals, followed by those for the individual collections comprising the groups.

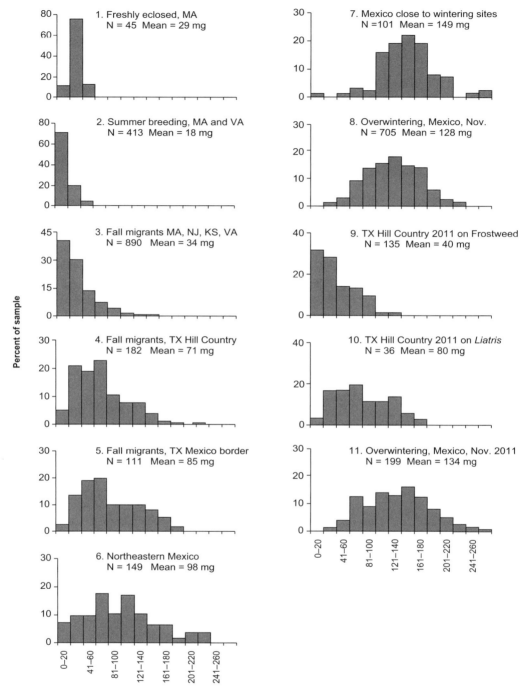

Figure 10.2. Histograms of lipid contents of butterflies in Groups 1–11 (Table 10.1). Left column: 2596 adult monarch butterflies collected prior to 2011 (1) as chrysalids and (2) as adults during the summer breeding season; and as adults (3) during the fall migration south to Virginia, (4) during the fall migration in the Texas Hill Country, (5) near the Texas-Mexico border, (6) along the Sierra Madre in northeastern Mexico, (7) immediately north of the overwintering sites, and (8) at the overwintering sites shortly after arrival in November. Right column: butterflies captured in fall 2011 (9) as migrants while nectaring on *Verbesina virginica* in four sites in the Texas Hill Country during the height of the drought; (10) as fall migrants while nectaring on *Liatris mucronata* in a well-watered commercial garden in the Texas Hill Country during the height of the drought; and (11) in mid-November shortly after arriving at their overwintering sites on Cerro Pelon and the Sierra Chincua.

the 14 butterflies in Beach Haven NJ, collected while "sitting on bushes" (Walford 1980).

Fall migrants, Texas (Groups 4 and 5)

We collected monarchs during 1982 and 1994 in two areas of the central flyway in Texas. The June to October (5 month) cumulative rainfall in 1994 was 13.6 inches, close to the 50-year (1962–2011) average of 14.4 inches; the 1982 rainfall, 9.8 inches, was the fourth lowest value (NOAA National Climatic Data Center 2013). In 2011, for comparison, the rainfall for June through October was 5.9 inches. The central Hill Country sample (Group 4) includes 182 monarchs collected by W. H. Calvert in October 1982 and 1994, in and adjacent to Uvalde and Garner State Park. Butterflies were collected from clusters in late afternoon and early evening. The mean lipid content of these migrants was 71 mg (range 7–226 mg). This is more than twice the amount measured in the northern migrants (Group 3). That fall migrants store larger amounts of lipids is also indicated by the shift from a skewed distribution (Groups 2 and 3) toward a normal distribution.

In the western part of the Texas Hill Country near the Rio Grande and the Texas-Mexico border (Group 5), Calvert collected clustering monarchs in Quemado, Maverick County in October 1982, and Brower and Calvert collected clustering monarchs along Dolan Creek in the Devils River Conservation Area, Val Verde County, in October 1994. These 111 butterflies had a mean lipid content of 85 mg (range 14–192 mg).

Fall migrants, northeastern Mexico (Group 6)

In northeastern Mexico along the eastern side of the Sierra Madre Occidental, Calvert collected roosting butterflies in Bustamante Canyon, Nuevo Leon on 30 October 1976. Brower and Zubieta collected from several roosts in the same site on 19 October 2006. The mean lipid content of the 149 butterflies was 98 mg (range 6–234 mg), continuing the trend of increasing lipid content.

Fall migrants, approaching the wintering sites (Group 7)

Roosting monarchs were collected by Calvert on 31 October 1977 in San Javier, Mexico, 141 km NNE of Angangueo, the heart of the overwintering region. The mean lipid content of 101 butterflies was 149 mg (range 8–281 mg). This group had the highest lipid content we have measured.

Early-season wintering butterflies, Mexico (Group 8)

We analyzed the lipids in 705 clustering butterflies collected in November 1981, 1982, 1993, and 2004 from overwintering sites in the Monarch Butterfly Biosphere Reserve. The butterflies were collected primarily in Sierra Chincua but also in Sierra Chivati-Huacal, Cerro Altimirano (Contepec), Cerro Pelon, and Sierra Campanario (Rosario). Their mean lipid content was 128 mg (range 20–237 mg), higher than the Bustamante samples (Group 6), but lower than the San Javier sample (Group 7).

Together, these data show that migrant monarchs add to their lipid reserves between the northeastern United States and Texas, and continue adding lipids as they fly south from Texas to the overwintering sites (Figure 10.1). Lipid levels in the Texas and northeastern Mexico samples (Groups 4–6) are significantly higher than in the northeastern U.S. migrants (Group 3); and lipid levels in the overwintering region samples (Groups 7 and 8) are significantly higher than in Texas and northeastern Mexico ($F_{5,387} = 627$, $P < 0.0001$; ANOVA and Tukey HSD test, JMP 9.0.2).

2011–2012 OBSERVATIONS AND DATA COLLECTION: METHODS

The geographic coordinates and elevations of observation and collection sites in Texas and Mexico (Figure 10.1) were determined with Garmin GPS units and verified on Google Earth (accessed September–October 2012).

Fall 2011 observations and collections in Texas

Texas Monarch Watch butterfly counts from 1993 to 1995 (Calvert and Wagner 1999) and 2000 to 2004 (Quinn 2011), and the 2002–2011 butterfly roost database assembled by Journey North (Howard and Davis 2009; Howard 2011, 2012), indicate that the fall migration through the central Texas flyway at 30°N peaks between 4 and 17 October. On this basis, we made collections and observations on

11–13 October 2011. We drove approximately 650 miles through 9 counties (Bandera, Gillespie, Kendall, Kerr, Kimble, Mason, Menard, Real, and Uvalde) in the Texas Hill Country. In addition to collecting monarchs from five locations, we qualitatively assessed the effects of the drought on several rivers and streams and on the flora along roadsides, on several ranches, and in Garner State Park.

Assisted by Texas citizen scientist colleagues, we netted monarchs in four riparian sites where the butterflies were nectaring on or flying near *Verbesina virginica*, and in one commercial garden where the butterflies were nectaring on *Liatris mucronata*. On a scale of 1 (pristine) to 5 (faded and tattered) (Malcolm et al. 1993), 89% of the 171 monarchs we collected scored ≤ 2, indicating that the butterflies were migrants. We put each captured butterfly into a glassine envelope, and then into a plastic bag inside an insulated bag containing ice packs. We froze the butterflies in late afternoon to early evening of the day of collection, and transported them on 14 October in the insulated bag to Virginia where they were kept frozen until processed for lipids in spring 2012.

Late fall 2011 collections in Mexico

One Mexican overwintering sample ($N = 99$) was collected from Sierra Chincua on 17 November 2011, and came from four clusters near the headwaters of the Arroyo Hondo. The second sample ($N = 100$) came from Cerro Pelon on 18 November 2011, from four clusters in the El Puerto colony above the Llano de los Tres Gobernadores. The butterflies were put into glassine envelopes and kept on ice until they could be stored in a freezer on the evening of collection, and then transported on ice to Virginia in March 2012.

Lipid analyses

Lipid analyses of the 2011 butterflies were carried out at Sweet Briar College. The procedure, described in Brower et al. (2006), includes drying the butterflies, weighing them, extracting the lipids in petroleum ether, and then weighing the extracted lipid. The data are presented as mg of lipid per butterfly. Since our previous analyses found no significant difference between the lipid contents of male and female monarchs, the data for the sexes are combined. To put these lipid data in perspective, the wet weight of adult monarch butterflies ranges 300–900 mg, dry weight, 125–350 mg, and the weight of lipids, 0–296 mg (Brower et al. 2006).

Statistical tests

Analysis of variance and t-tests were carried out using JMP (9.0.2, SAS Institute 2010). Samples were tested for normality, and $\log_{10}$ transformations were applied to skewed samples to produce normal distributions. Multiple comparisons among means used the Tukey-Kramer honestly significant difference test, with an alpha level of 0.05. We report Welch ANOVA values for samples with unequal variances.

2011–2012 OBSERVATIONS AND DATA COLLECTION: RESULTS

General observations in the Texas Hill Country during and after the drought

While driving through the Texas Hill Country in October 2011, a period in which the weather was generally clear and hot, we saw very few monarchs along the roadsides and no evidence of their migrating close to the ground or in the sky. As is characteristic of the Texas Hill Country during the late summer and fall, highly localized and intense rains occurred in the week before our visit. For example, 3–5 inches fell on 8–9 October near Bergheim in Kendall County, and we noted occasional patches of greening and blooming flowers as we traversed the Hill Country. Overall, however, this part of Texas was desiccated and most of the native wildflowers, herbaceous plants and trees growing along roadsides, in pastures, and adjacent to dried riverbeds were severely stressed (Figure 10.3 and Plate 7).

The rivers were at or near historically low flow levels. The Medina River, for example, (Figure 10.3) is a typical Hill Country stream of crystal clear water flowing over limestone outcroppings, with bald cypress (*Taxodium distichum*) lining the banks (Texas Parks and Wildlife Department 2012). From October 1982 through February 2011, its mean monthly flow at the Bandera gauging station ranged from 1.01 to 7871 cubic feet per second (cfs) (USGS 2012a). For just the month of October, the mean monthly flow for 1982–2010 ranged from 7.2 to 630 cfs, with an overall October average

Figure 10.3. Medina River, two photos taken at the same location. (a) 12 October 2011, during the drought. Kiphart is standing in the dry riverbed, immediately adjacent to our Medina River collecting site (Color Plate 7c). (b) 31 March 2012, after the flow had recovered. Frequent floods and droughts, and highly variable river flows, are characteristic of the Hill Country. Photos by Lincoln Brower.

of 109 cfs. Starting in March 2011, the river began setting new low monthly records. By July 2011, its mean monthly flow had dropped to 0.33 cfs, and at the time of our October visit (Figure 10.3a), the river had ceased flowing completely: the flow was 0 cfs every day from August through November 2011 (USGS 2012b). The contrasts in the river flows and vegetation between fall 2011 and spring 2012 were dramatic. On our return to the Medina River on 31 March 2012 (Figure 10.3b), its flow was about 62 cfs, and in the adjacent forest, flood debris from spring rains had accumulated to a height of about 4 m.

The Texas fall 2011 collections

On 11 October, we visited Lucky Boy Ranch, Whispering Water Ranch, and the Native American Seed Company; on 12 October, the Medina River floodplain in Bandera County and the Frio River Floodplain in Real County; and on 13 October, the Frio River floodplain in Garner State Park in Uvalde County. Lucky Boy Ranch is located about 20 miles southwest of Mason on the Mason–Kimble County line, on a bluff overlooking the Llano River. The river had very low flow, with severe drought conditions surrounding it (Plate 7b). Ranch owner Monika Maeckle stated that the *V. virginica* and other composites growing along the river edge were about half their normal heights. Ashe juniper trees (*Juniperus ashei*) growing among the Texas live oaks (*Quercus virginiana* var. *fusiformis*) on the moderately steep slope above the river's floodplain were brown from extreme desiccation. Subsequent surveys in adjacent counties by the Texas Forest Service reported that hundreds of millions of trees, including junipers, were severely affected or killed by the drought (Edgar and Carraway 2011), and Maeckle (pers. comm., 2 May 2012) reported that many of the junipers had died (Plate 7b) (see also Stengle 2011). The trees along the far bank of the river include sycamores (*Platanus americanus*) and native pecans (*Carya illinoinensis*). Fall migrant monarchs frequently form clusters in these trees (Maeckle, pers. comm., 2 May 2012), but during our midday visit we only found monarchs, in very low numbers, nectaring on drought-stressed *V. virginica* growing along the edges of the river. From about 11:30 a.m. to 12:30 p.m., we netted four males and two females. The butterfly abdomens were thin, indicative of low titers of lipid.

Whispering Water Ranch is 12 miles south of Menard, in Menard County, immediately downstream from the origin of Elm Creek that flows northeast into the San Saba River. Ranch owner Carolyn Dippel said that in the second week of September a wave of migratory monarchs passed through, and that the week before our visit, large clusters had formed on chinkapin oak trees (*Quercus muehlenbergii*) along the river. Monarchs were more abundant here than at Lucky Boy Ranch; from about 1:15 to 2:45 p.m., we netted 27 males and 14 females on *V. virginica* growing along the edge of Elm Creek adjacent to the oaks.

The Native American Seed Company site contained a 0.5 ha cultivated plot of gayfeather (*Liatris mucronata*) located in the floodplain of the Llano River, in Junction, Kimble County, Texas. The area surrounding the site was extremely dry, but company manager Emily Neiman told us that the plot had been irrigated. Several hundred *Liatris* were in excellent condition and blooming profusely. Nearby, in an area that had not been irrigated, large patches of *V. virginica* plants had died back. Monarchs were not abundant, but from about 4:30 to 5:00 p.m., we netted 19 males and 17 females nectaring on and flying among the *Liatris*.

The Medina River floodplain site (Figure 10.3c) is adjacent to the Medina Garden Nursery, in Bandera County, Texas. The collection site (~1.6 ha) included native vegetation with a mature native pecan overstory, bordering the bone-dry Medina River (Figure 10.3a). Notwithstanding the drought, the floodplain soil was moist and there was a lush understory of *V. virginica* (Plate 7c). Monarchs were moderately abundant here, and from 10:30 a.m. to about 1:00 p.m., three of us netted 35 male and 14 female monarchs nectaring on the *Verbesina*.

The Frio River floodplain contains an extensive riverine habitat at the H. E. Butt Foundation Camps in Leakey, Real County about 28 km NNE and upstream of Garner State Park. At this site we drove along the flat limestone riverbed from 3:30 to 4:30 p.m. and explored the riverbanks and tributary edges. The areas back from the river edges were extremely dry, and the *V. virginica* and other nectar sources were drought-stressed. We saw no monarchs.

The Garner State Park floodplain site is also on the Frio River, but in Uvalde County (Plate 7d). From mid-July through 20 September 2011, except for a short pulse to 35 cfs on 13 August, the river flow at a point 13 km downstream was 0 cfs (USGS 2012c). The habitat was in an old dry oxbow, and had a pecan tree overstory. The *V. virginica* understory was severely drought-stressed and the plants were about a half meter shorter than those in the Medina River floodplain (compare Plates 7c and d). From 11:45 a.m. to about 2:45 p.m., three of us netted 29 males and 13 females that were nectaring.

Lipid levels of Texas and Mexico butterflies

The 135 butterflies in the four samples collected from *V. virginica* (Group 9, subgroups 9a–9d in Table 10.1) contained 5–138 mg of lipid (mean = 40 mg). The butterflies collected on Whispering Water Ranch and the Medina River floodplain (Groups 9b and 9c) had significantly higher lipid content than butterflies collected at Lucky Boy Ranch and Garner State Park (Groups 9a and 9d) (ANOVA $F_{3,131}$ = 14.95, $P < 0.0001$).

The 36 butterflies collected on *L. mucronata* in the irrigated field at the Native American Seed Company (Group 10) had higher lipid stores than the butterflies collected on *V. virginica*. These *Liatris* butterflies contained a mean of 80 mg of lipid (range 15–174 mg), significantly higher than the *V. virginica* butterflies (Group 10 vs. Group 9, t_{69} = 6.5, $P < 0.0001$).

The November 2011 samples from the two Mexican overwintering sites had much higher lipid levels than any samples from the United States (mean = 134 mg, range 36–262 mg, Group 11; Group 11 vs. Group 10, t_{233} = 6.2, $P < 0.0001$). The butterflies from Cerro Pelon had significantly higher lipid levels than the butterflies from Sierra Chincua (Pelon mean = 146 mg, range 48–253 mg; Chincua mean = 122 mg, range 36–262 mg; t_{186} = 3.55, $P = 0.0005$).

Comparison of 2011 samples with previous samples

The four samples of Texas migrants collected on *V. virginica* in 2011 (Group 9) had significantly lower lipid levels (40 mg) than the Texas migrants collected in 1982 and 1994 (Group 4, 71 mg; Table 10.1 and Figure 10.2, t_{236} = 8.54, $P < 0.0001$). In contrast, the lipid content of the 2011 *Liatris* butterflies (80 mg) was not significantly different from the earlier Texas samples (Group 10 vs. Group 4, t_{216} = 0.99, $P = 0.16$).

Thus, by comparing the lipids of butterflies collected in 2011 with butterflies collected in the same area of Texas in 1982 and 1994, we conclude that the drought resulted in a 44% drop in the butterflies' lipid reserves. That this reduction was due to the stressed floral resources is strongly supported by what fortuitously amounted to a control sample, of butterflies able to nectar on watered *L. mucronata*.

Our prediction that the butterflies arriving at the overwintering sites in 2011 would be deficient in lipids, however, was not borne out by the data. Lipid levels in Cerro Pelon and Sierra Chincua butterflies collected in mid-November 2011 (Group 11) were not significantly different from previous November collections (Group 8) (t_{283} = 1.54, $P = 0.13$). Thus, the monarchs arriving at the overwintering sites had normal lipid titers.

DISCUSSION

We interpret our results, of low lipid titers in October 2011 Texas migrants and normal lipid titers in November 2011 overwintering monarch butterflies, as indicating (1) a negative effect of the 2011 drought on the condition of migrating monarch butterflies; and (2) the importance of floral resources in northern Mexico for fall migrants.

The strongest evidence that the drought had a negative effect on the lipid titers of the Texas butterflies was the fact that the butterflies from the irrigated *Liatris* garden contained significantly more lipid than the water-stressed *V. virginica* samples. Because the monarchs were collected between the late morning and late afternoon, then kept alive for several hours before they were killed, they had the opportunity at each site both to feed and to convert much of the sugar from this feeding into lipids. Sugar moves out of butterfly crops quickly: in laboratory experiments, crops released 5–30 μl of sucrose solutions within 1–2 hours in both monarchs (Cherry 2006) and *Vanessa cardui* (Hainsworth et al. 1991). In the fat body, the conversion of sugars into lipids is even more rapid: Brown and Chippendale (1974) injected radioactively labeled glucose into the abdomens of migratory monarch butterflies and determined that radioactive lipids (diglycerides and triglycerides) appeared within 6 minutes. Thus, the higher lipid content of butterflies collected on the *Liatris* indicates that this patch of flowers was a better nectar source than the patches of *V. virginica*.

We did not measure the volume or composition of the nectar on the two flower species. Given that nectar varies widely in chemical composition and volume (Baker and Baker 1982), some of the difference between butterflies collected on *V. virginica* and *L. mucronata* might be due to differences between plant species. Evidence that the drought itself was responsible for the low lipid titers, however, comes from the fact that the *Verbesina virginica* was not equally stressed at all collection sites. The butterflies collected at the sites where *V. virginica* was less stressed (Whispering Water Ranch and Medina River floodplain) had higher lipids than those at the more stressed sites (Lucky Boy Ranch and Garner State Park).

Alternative explanations for the low lipids in fall 2011 Texas monarchs must be considered. One possibility is that we missed the peak of the fall migration, and thus collected stragglers that were in poorer condition than the main wave of butterflies. This was not the case: Journey North's animated weekly reports indicate that the October 2011 timing was normal (Howard 2012), and we were there during the migration peak.

Another possibility results from the fact that all 2011 Texas butterflies were collected while nectaring, but the butterflies in the earlier Texas samples (Groups 4 and 5) were collected from roosts. We had planned to collect both roosting and nectaring monarchs in 2011, but we were constrained to midday observations on clear, warm days, after overnight clusters break up. Since Brower et al. (2006) determined that nectaring monarch butterflies generally have lower lipid levels than nearby roosting butterflies, could this explain why the 2011 Texas lipid levels were so low? Two samples collected in Austin, Texas, in October 1979 argue against this. Brower et al. (2006) included descriptive statistics for these two samples, which are not included in Group 4 of this paper because we no longer have the data from individual butterflies: 42 roosting monarchs had a mean lipid content of 112 mg, and 66 monarchs nectaring on a wild, native *Liatris* species had a mean lipid content of 100 mg. The 1979 nectaring sample had higher lipid levels than our 2011 nectaring samples, and the 12 mg difference between the 1979 roosting and nectaring butterflies is much smaller than the difference between our 2011 nectaring butterflies and the 1982 and 1994 Texas roosting samples. We cannot rule out the possibility that the collection difference partly explains the difference in lipid levels between earlier Texas samples and 2011, but it is unlikely to be the primary factor.

How did the monarchs arrive at the Chincua and Pelon overwintering sites with normal lipid levels? One hypothesis is that many of the butterflies migrating through the central flyway died of starvation or desiccation and did not reach the overwintering sites, while those butterflies migrating along the Texas coastal flyway had greater access to nectar, increased their lipid stores, and made it safely to the overwintering area. This explanation is unlikely because the extreme drought affected both flyways (North American Drought Monitor map for 31 October 2011, NOAA National Climatic Data Center 2012).

Is it possible that a high proportion of the monarchs passing through both Texas flyways starved before reaching the overwintering sites, and that only monarchs with high lipid stores survived the entire trip? Although the 2011–12 overwintering population in Mexico was, in fact, small (Brower et al. 2012a), indirect evidence suggests that high starvation among the migrants was unlikely. Powered flight is energetically expensive, but when monarchs encounter appropriate weather conditions they are able to soar and glide for long distances with minimal energy expenditure (Gibo and Pallett 1979; Calvert 2001). Brower et al. (2006) used this fact plus other data to conclude that the energy stored in lipids is more important for butterflies' winter survival than for fueling the migration itself. This suggests that starvation en route was unlikely to be a major risk for butterflies during the peak of the migration, although we cannot rule it out.

It is possible that the combined stress of dry environmental conditions and low lipid stores led to high monarch mortality in 2011 by disrupting the migrants' water balance. Flying insects have high rates of water loss, but they generate large quantities of metabolic water when they oxidize lipids as fuel (Chapman 1998). A monarch flying across a dry landscape will have difficulty finding nectar or free water, and if it has insufficient lipids to burn as fuel, its water loss might be lethal. In addition, Gibo and McCurdy (1993) suggested that monarchs with low lipid stores might use water as ballast, because the butterflies' soaring ability is affected by their center of gravity. If this is, in fact, true, then butterflies with the lowest lipid stores would be the least efficient at conserving energy through soaring flight. In the overwintering areas, masses of monarchs drink water, and it is assumed that maintaining their water balance is an important winter challenge (Calvert and Brower 1986; Brower 1999). Monarchs' water budget during both the migration and the overwintering season merits further study.

Even if starvation and/or desiccation increased the mortality of southern migrants above normal, this would not explain the recovery of lipid levels found in the overwintering butterflies. Ninety-five percent of the Texas butterflies had lipid levels lower than the median lipid level (133 mg) of the Mexican sample (Figure 10.2, histograms 9 and 10 versus histogram 11). That is, only a small percentage of the Texas butterflies had as much lipid as the average Mexican butterflies. Even if we assumed survival of only the 25% of the Texas butterflies with the highest lipid levels (>71 mg for Groups 9 and 10 combined), the lipid levels in these butterflies are still significantly lower than in the butterflies from Mexico (Wilcoxon test, $P < 0.0001$).

We propose, therefore, that the butterflies encountered areas with sufficient rainfall and adequate floral resources somewhere along the Sierra Madre range in eastern Mexico to build their lipids back up to normal levels. Supporting this hypothesis is the fact that the drought moderated to the southeast in Mexico (NOAA National Climatic Data Center 2012) where the monarchs shift their flight direction from southwest to southeast, following the Sierra Madre Occidental (Calvert 2001). Brower (1985a) and Brower et al. (2006) focused on the significant increase in butterflies' lipid stores between the northern United States and Texas, and did not look closely for a geographic pattern in Texas and northern Mexico. Though not statistically significant, Figure 10.2 shows a trend of increasing lipids in samples from the Texas Hill Country (Group 4) to the Texas-Mexico border (Group 5) to northeastern Mexico (Group 6); the increase in lipid stores for samples between northeastern Mexico (Group 6) and the wintering area (Groups 7 and 8) is statistically significant. This pattern draws attention to the importance of nectar resources all along the monarchs' migratory route, so that during dry years those regions with higher rainfall may produce sufficient nectar to compensate for regions in drought conditions.

In contrast to the importance of key sites for migrating shorebirds, Brower et al. (2006) concluded that specific stopover sites did not seem crucial for monarchs. Our 2011 field collections invite a reevaluation of this statement. Within Texas, the numerous floodplains and riverine forests are crucial nectaring habitats, and these are probably especially important during years of drought.

Our findings make a compelling case to search for areas in Mexico where abundant fall floral resources must occur, and ensure their protection. Brower et al. (2006) were concerned that changes in nectar resources due to massive habitat modification might be affecting monarchs' ability to store sufficient lipids to survive the overwintering period, and noted that sufficient data were not available to ascertain this effect. The lacuna in our knowledge of locations where monarchs build up their lipid reserves as they migrate through northern Mexico offers an opportunity for research with important application to the conservation of the migratory phenomenon of the monarch butterfly.

ACKNOWLEDGMENTS

For hospitality and access to their lands, we are grateful to Monika Maeckle and Bob Rivard (Lucky Boy Ranch), Carolyn and Chipper Dippel (Whispering Water Ranch), Emily Neiman and George Cates (Native American Seed Co.), William Collins (H. E. Butt Foundation Free Camps), Rick Meyers (Garner State Park), and Ernesto Carino and Ysmael Espinosa (Medina Garden Nursery). We thank Patty Leslie-Pasztor, Mary Kennedy, Mike Quinn, and Jenny Singleton for field assistance, and Carol Cullar for information on the monarch migration through Quemado TX. For the 1977–2006 lipid analyses of several thousand monarchs, we are grateful to past technicians and students, most especially Sue Swartz, Peter Walford, Alfonso Alonso-Mejía, Tonya Van Hook, Chris Kisiel, Donna McLaughlin and Cara Cherry. We also thank William Calvert for several of the earlier Texas and Mexico monarch collections. James L. Nation provided insight into lipid synthesis by insects. Rebecca V. Batalden and Kelly Nail provided helpful comments on this paper. We thank the Dirección General de Vida Silvestre, SEMARNAT Mexico, for butterfly collecting and export permits. This study and data synthesis were supported by NSF Award 0949650 to LPB and LSF and by The October Hill Foundation.

11

Estimating the Climate Signal in Monarch Population Decline

No Direct Evidence for an Impact of Climate Change?

Myron P. Zalucki, Lincoln P. Brower, Stephen B. Malcolm, and Benjamin H. Slager

Using a CLIMEX model that estimates the seasonal distribution of the abundance of monarch butterflies anywhere in the world, we estimate the effect of climate on long-term population dynamics in North America. We use daily maximum and minimum temperatures and rainfall from 1970 to 2010 for 25 locations that cover the breeding range across eastern North America to generate a series of abundance indices. Although there is considerable variation in the population indices over this period, no systematic trend due to climate on monarch breeding populations was detected. We conclude that the observed decline in overwintering populations is due to other factors; most likely forest degradation at overwintering locations and the loss of milkweed populations in breeding habitats caused by widespread agricultural use of herbicides and genetically engineered soybean and corn crops. However, climate impacts on mortality during spring and autumn migrations and at overwintering sites cannot be ruled out with our model.

INTRODUCTION

Each year, monarch butterflies east of the Rocky Mountains of North America migrate in autumn to aggregate in about a dozen overwintering locations in the mountains of central Mexico (Brower et al. 2012a). Since the early 1990s, the eastern population of North American monarchs has declined, based on annual measures of the combined areas of all known overwintering populations in Mexico from 1994–95 through the 2010–11 season (Brower et al. 2012a). The population did not recover in the subsequent two years, being 2.89 and 1.19 ha, respectively (Rendón-Salinas et al. 2013).

A number of threatening processes (sensu Caughley 1994) may be responsible for the decline, including logging in and adjacent to forested overwintering sites (Brower et al. 2002, 2009), widespread agricultural use of herbicides that have reduced the abundance of milkweeds, *Asclepias* spp., in breeding areas (Zalucki and Lammers 2010; Pleasants and Oberhauser 2012), loss of breeding habitat (CEC 2008), possibly presence of Bt maize (Losey et al. 1999; Jesse and Obrycki 2000, 2004; but see discussion), as well as the general impact of climate change on monarchs and the ecosystems across which they move during their annual migration (e.g., Oberhauser and Petersen 2003; Batalden et al. 2007; Sutherst et al. 2011).

Climate exerts a major influence on the distribution and abundance of insects, and monarchs are no exception (Zalucki and Rochester 2004). Climate influences monarchs directly by affecting developmental rates and immature survival (Zalucki 1982; York and Oberhauser 2002; Nail et al., this volume, Chapter 8), adult reproduction (Zalucki 1981b), and adult survival (Anderson and Brower 1996); and indirectly by affecting host plant resources for adults and immatures (Zalucki and Rochester 1999) as well as interactions with parasitoids and predators (Koch et al. 2005).

The influence of climate on insect populations is generally inferred by collecting a long time series of abundance data and correlating these with various measures of climate (see e.g., Maelzer and Zalucki 1999; Zalucki and Furlong 2005). The climate measures are many, arbitrary, and varied, from simple means of minimum and maximum temperatures and precipitation over various periods of time to temperature in the hottest month, range of temperature in the coldest quartile, and so forth. Zipkin et al. (2012) recently applied such a regression-based approach to a monarch population accounting for the effects of climate during different stages of the annual remigration. We take a different approach here and use a model of climate suitability for monarchs to estimate their potential geographical distribution worldwide. We then use that model to estimate temporal variation in climate suitability at individual localities, given long-series climate data for that location. The method is described in Zalucki and Furlong (2008, 2011) and Li et al. (2012), but after a brief discussion of monarch movement, we include details below for those not familiar with the method.

MONARCH MOVEMENT PATTERNS

The monarch's population ecology in North America is unique in that virtually the entire eastern population migrates to and overwinters at a small number of locations on 12–13 mountains in an area smaller than 1800 km² in the neovolcanic highlands of central Mexico (Calvert and Brower 1986; Brower 1995; Hobson et al. 1999; Brower et al. 2002; Slayback et al. 2007). Migrant monarchs coalesce to form compact roosting sites where the density of butterflies in trees is approximately 50 million per hectare (Calvert 2004b; Brower et al. 2004). The total area of these sites has been estimated each winter since 1994–95 and is interpreted as representing the combined success of the spring-summer breeding in the United States and Canada, and the migration south (Brower et al. 2012a).

In spring, surviving individuals return to the coastal states of the Gulf of Mexico and establish a fresh spring generation that then continues the journey north across the eastern United States to southern Canada, thus reestablishing the summer breeding range (Cockrell et al. 1993; Malcolm et al. 1993; Miller et al. 2011). A realistic mechanistic-based model for monarchs in North America would be spatially explicit and represent the landscape as a mixture of breeding and nonbreeding habitats in a GIS framework (e.g., Zalucki and Rochester 2004; Feddeman et al. 2004). Monarch population dynamics can also be modeled in both time and space using relatively simple discrete equation formalism, with some overlapping of generations (Yakubu et al. 2004; Zalucki et al. unpublished) (Figure 11.1).

Based on known monarch population biology, there are about four generations during the spring and summer each year (Malcolm et al. 1987), the last of which becomes both the overwintering population in Mexico and the spring remigrants that produce the first spring generation, G_1 (Figure 11.1). These subsequently give rise to the second to fourth generations (G_2, G_3, G_4) in the summer, in the north-central and northeastern United States and southern Canada. The relatively long life span of adult monarchs means that generations overlap in time and space (Figure 11.1), and it is likely that some of G_3 and all of G_4 migrate to spend the winter at sites in Mexico. More than 90% of overwintering monarchs have fed as larvae on the northern milkweed species *A. syriaca* (Malcolm et al. 1993), demonstrating that most monarchs that reach Mexico come from the geographic range of this host. Subsequent stable isotope studies have supported this conclusion (Wassenaar and Hobson 1998). Migration mortality occurs on the journey south to Mexico and only a proportion survive, say JS_i in the ith year. The monarchs that survive the winter in Mexico continue the cycle the following year (Figure 11.1). We are interested here in the size of the population that becomes established at the overwintering sites in Mexico, defined as population $G_{W,i}$. Estimates of $G_{W,i}$ are based on annual censuses of the total area of all combined overwintering colonies, conducted in December or January when the butterflies have coalesced (Brower et al. 2011). This annual cycle of migration and overwintering has been listed as an endangered biological phenomenon (Brower and Malcolm 1991).

Here we use a climate model to estimate spring-summer breeding success; we label this ($G_1:G_4$), which is the outcome of four overlapping generations in space and time (G_1 to G_4 in Figure 11.1). As is apparent (Figure 11.1), we have data for one variable ($G_{W,i}$), a model to estimate ($G_1:G_4$) (see Equation 1), and three additional unknown survivals that can be influenced by climate: the journey south, JS_i; overwintering survival, OWS_i; and the journey north, JN_i to recolonize the southern United States. Based on the above we can write the balance

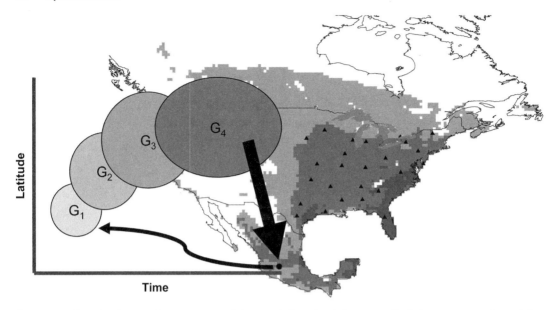

Figure 11.1. Spatial and temporal representation of the annual life cycle of the eastern North American monarch population; G1–G4 represent the overlapping spring-summer breeding generations; arrows show the autumn migration south to overwintering sites in Mexico and the migration north each spring. Locations of the 25 sites (triangles) used to represent the temporal effects of climate (see Table 11.1 for details). Shading indicates annual growth index (GI_A) from the CLIMEX model (see text for details): lightest shade = 5–15, middle shade = 16–30, darkest shade > 30.

equation for the size of the overwintering population from one year, $G_{W,i}$ to the next, $G_{W,i+1}$ as a product of four key processes:

$$G_{W, i+1} = G_{W,i} * OWS_i * JN_i * (G_1{:}G_4)_i * JS_i \qquad (1)$$

SIMULATING CLIMATE EFFECTS ON GEOGRAPHICAL DISTRIBUTION AND ABUNDANCE OVER TIME

Bioclimatic modeling has a long pedigree in ecology (e.g., Petersen 2003) and a number of methods are available for fitting climate envelopes. These methods vary in both the climate variables used and the way in which envelopes are defined. Some use standard statistical techniques, such as generalized linear modeling, and various permutations: GLM, GAM, GARP, GRASP (for a list of acronyms, see the glossary at the end of this chapter), ensembles of regression trees, or neural networks (Thuiller 2003; Worner and Gevrey 2006). Others use computer programs that incorporate specialized algorithms including HABITAT (Walker and Cocks 1991), BIOCLIM (Busby 1991) and its various incarnations (BIOCLIM/BIOMAP and now ANUCLIM), DOMAIN (Carpenter et al. 1993), MaxEnt (Phillips et al. 2006; Elith et al. 2011), and others. At their core, all these multivariate statistical techniques use distributional data to infer the climatic factors that most likely account for the species' range and are essentially regression-based approaches. These methods attempt to determine the "climate envelope," a set of environmental covariates (usually based on meteorological variables) that encompass the locations where a species currently exists, or was known to exist from collection records. The chosen covariates tend to be arbitrary and not necessarily related to biological processes that determine abundance and distribution. In addition, they may have a temporal resolution, for example temperature in the hottest month or range of temperatures in the coldest quartile, that does not accord with the scale at which species grow or die. As many collection records come from sites without meteorological recording stations, the various methods usually rely on interpolated climate surfaces.

The ability of these statistical and rule-based approaches to model species distributions and make meaningful predictions has been criticized (Kriticos and Randall 2001; Sutherst 2003; Webber et al. 2011; although see Robertson et al. 2003; van

Plate 1. Children participating in monarch environmental education projects. Clockwise from top left: (a) Monarchs Without Borders (Canada). (b) Monarchs in the Classroom, Driven to Discover (U.S.). (c) Monarchs Across Georgia (U.S.). (d) Monarch Butterfly Fund workshops (Mexico). Photos by Michel Tremblay (by permission of Insectarium de Montréal), Mike Fitzloff, Kim Bailey, and Eneida Montesiños, respectively. (Chapter 1)

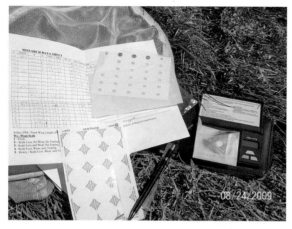

Plate 2. Monarch citizen scientists. Clockwise from top left: (a) Monitoring tools for MLMP, Monarch Health, Monarch Watch, and individual research on monarch mass collected by Denny Brooks, Michigan. (b) Students observing monarch eggs on schoolyard milkweed plant for the MLMP. (c) MLMP volunteer Ilse Gebhard with monarch rearing operation for MLMP and Monarch Health parasitism studies. (d) Monarch with Monarch Watch tag. Photos by Denny Brooks, Elizabeth Howard, Russ Schipper, and Orley Taylor, respectively. (Chapter 2)

Plate 3. Many organisms, from insects to vertebrates, consume milkweed. (a, top) Fifth-instar monarch larva consuming common milkweed (*Asclepias syriaca*). (b, bottom) Rabbit consuming green milkweed (*A. viridiflora*). Photos by Ellen Woods and Tom Collins, respectively. (Chapter 4)

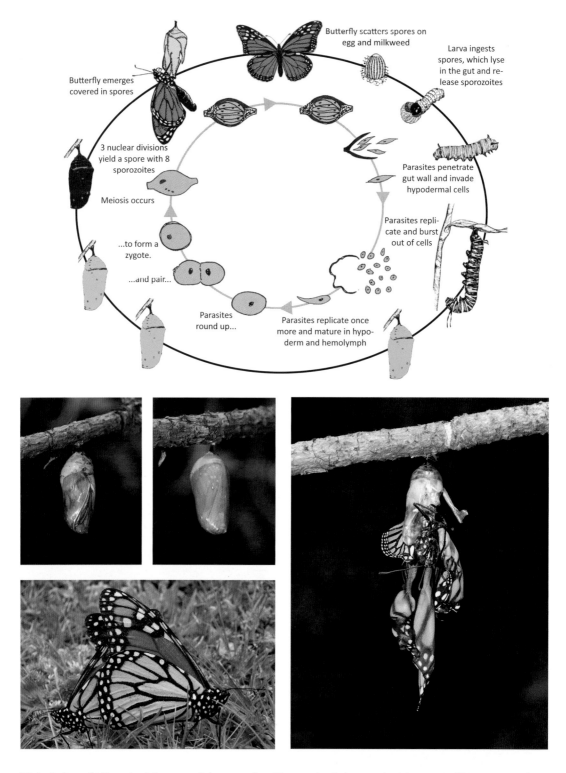

Plate 4. (a, top) Life cycle of the monarch (outer oval) and its parasite *Ophryocystis elektroscirrha* (OE; inner circle). Clockwise from left: Negative effects of heavy parasite infection include pupal death (compare (b) heavily infected pupa with (c) uninfected pupa). (d) Deformities of the adult butterfly that prevent eclosion. (e) Reduced mating ability (photo illustrates a healthy mating pair). Life cycle art and diagram by S. Altizer, J. C. de Roode, and Kristen Kuda; photos by J. C. de Roode. (Chapter 7)

Plate 5. Monarch caterpillar strategies for avoiding milkweed defenses. Clockwise from top left: (a) First instar grazing leaf hairs. (b) First instar cutting protected circle to avoid latex. (c) Fifth instar notching leaf to cut off flow of latex. (d) Fifth instar eating notched leaf. Photos by Ellen Woods (a, c, d) and Anurag Agrawal (b). (Chapter 5)

Plate 6. Monarch overwintering clusters in the Sierra Chincua colony. Clockwise from top left: (a) On hanging oyamel fir branches (8 January 2007). (b) On oyamel tree trunks (5 February 2008). (c) Inside the forest (6 February 2008). Photos by Ernest Williams. (Chapter 9)

Plate 7. From top: (a) Severely grazed ranch near Menard, Texas, showing the effects of the drought. The ground beneath the live oaks is devoid of grass and wildflowers (11 October 2011; 30°49.632′N 99°43.805′W; elevation 632 m). (b) Floodplain of the low flowing Llano River. The effects of the drought are visible on the grasses and shrubs in the foreground and on the dying ashe juniper trees above the riverbank in the background. Lucky Boy Ranch, near Mason, TX, 11 October 2011. (c) Collection site in the Medina River floodplain, Bandera County, TX, immediately adjacent to the riverbed shown in Figure 10.3. Because the river floods frequently here, the soil was moist and the understory *Verbesina virginica* beneath the pecan overstory was lush (12 October 2011). (d) Collection site in the desiccated oxbow floodplain adjacent to the Frio River, Garner State Park, Uvalde County, TX. The habitat is geologically older and drier than the Medina River floodplain. *V. virginica* in the understory was severely desiccated and the nectaring monarchs were exceedingly difficult to net (13 October 2011). Photos by Lincoln Brower. (Chapter 10)

Plate 8. Monarch oviposition. Clockwise from top: (a) On common milkweed (*Asclepias syriaca*). (b) On swamp milkweed (*A. incarnata*). (c) In the lab on tropical milkweed (*A. curassavica*). Photos by Candy Sarikonda, Pat Davis, and J. C. de Roode, respectively.

Plate 9 South American *Asclepias* species found within the range of *D. erippus* in Bolivia and Argentina. Habitat descriptions and estimates of altitudinal ranges are based on field observations (photos by Steve Malcolm, except *A. candida*). (a, top left) *A. barjoniifolia*, frequent >1500 m in Bolivia south to NW Argentina in dry habitats (7 April 2010, 5 km NW Zudañez, east of Sucre, Bolivia). (b, top right) *A. boliviensis*, frequent >2500 m in southern Peru, Bolivia, and NW Argentina, by rivers (14 October 2009, Huaytú, near Buena Vista, Santa Cruz, Bolivia). (c, middle left) *A. candida*, uncommon >1000 m, spring ephemeral in eastern Bolivia, western Paraguay, western Brazil (29 October 2009, near Santiago de Chiquitos, Bolivia; courtesy of Darwin Initiative Project 16-004, Conservation of the Cerrados of Eastern Bolivia). (d, middle right) *A. curassavica*, abundant >2500 m, occurs commonly everywhere, especially near water (15 October 2009, east of Samaipata, Bolivia). (e, bottom left) *A. mellodora*, infrequent >1000 m, spring ephemeral in lowland grasslands of Bolivia, Brazil, Paraguay, and Argentina (19 October 2009, Warnes, north of Santa Cruz, Bolivia). (f, bottom right) *A. pilgeriana* (= *A. flava*), uncommon >2000 m in spring and early summer near Villa Serrano, east of Sucre, Bolivia, and near Tucumán, Argentina (18 October 2009, near Villa Serrano, Bolivia). (Chapter 20)

Plate 10. *Danaus erippus*, the southern monarch. Clockwise from top: (a) Male nectaring at *Eupatorium arnottianum* by the Rio Grande at El Siambón, Tucumán, Argentina. (b) Female basking below roost site by the Rio Grande. (c) Fifth-instar larva feeding on *Asclepias barjoniifolia* at 2800 m altitude near Sucre, Bolivia. Photos by Steve Malcolm. (Chapter 20)

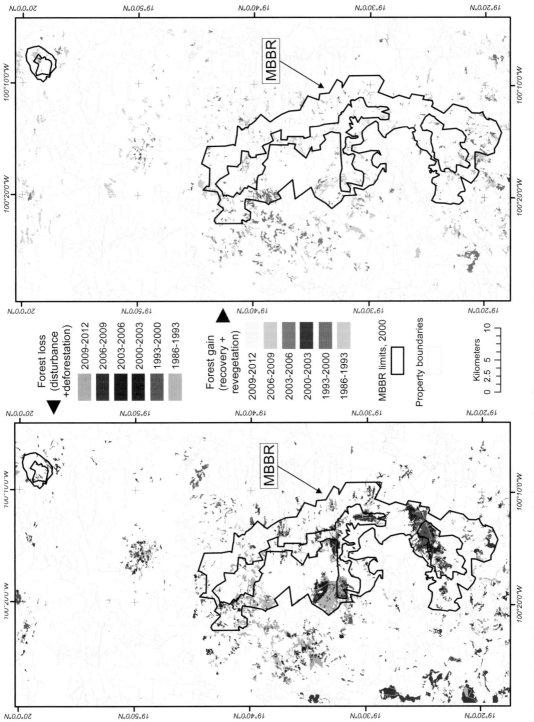

Plate 11. Spatial distribution of forest loss (left) and forest gain (right) from 1986 to 2012 in the MBBR and its area of influence. MBBR limits for the year 2000 are shown; the inner line shows the core zone, and the outer line the buffer zone. Note that forest changes are associated with land ownership boundaries, indicated by lighter lines. (Chapter 13)

Plate 12. *Abies religiosa* trees. Clockwise from top left: (a) With signs of decay on the upper part of a crown. (b) Partially defoliated (right tree). (c) With most of the crown defoliated (extreme left tree). (d) Entirely dead (center tree), at the El Rosario Sanctuary, Monarch Butterfly Biosphere Reserve, Michoacán, Mexico (March 2011, photos by Cuauhtémoc Sáenz). (Chapter 13)

Plate 13. Monarch habitat conservation. (a, top) Thinning the seedling density in a seed production field of a showy milkweed (*Asclepias speciosa*) at Hedgerow Farms, Yolo County, California. (b, bottom) Summer burns in Oklahoma result in fresh growth for native *A. viridis*. Photos by John Anderson and Kristen Baum, respectively. (Chapters 16 and 17)

Plate 14. Many insects eat monarchs. These natural enemies include parasitoids and predators. (a, top) A yellow jacket, *Vespula* sp., consuming a monarch larva. (b, bottom) The spine-shouldered stink bug, *Podisus* sp., which secretes a digestive enzyme into caterpillars, allowing it to suck the resulting liquid. Photos by Ellen Woods and Anurag Agrawal, respectively. (Chapter 6)

Plate 15. Monarchs nectaring on (a, top) showy and (b, bottom) common milkweed (*Asclepias speciosa* and *A. syriaca*). Photos by John Anderson and Candy Sarikonda, respectively.

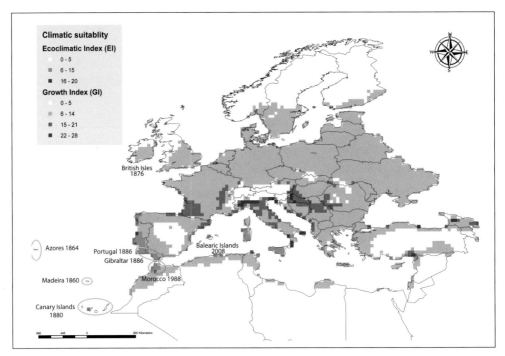

Plate 16. Monarchs in Europe. (a, top) Dates of first sightings of monarchs and climatic suitability for year-round monarch breeding (purple to blue) and seasonal breeding (shades of red) (see Chapter 11 for more details on the model). Areas with permanent breeding populations (blue rings) seem to have established many years after the first recorded sightings and establishment of milkweeds. Note that predictions are on a coarse scale and local microclimates may be more suitable. (b, bottom) Fifth-instar monarchs defoliating *Gomphocarpus fruticosus*, a milkweed species introduced into Spain from Africa during the eighteenth century. *A. curassavica* is also used by monarchs in Spain and was introduced from Central America, also during the eighteenth century. Photo by Juan Fernández-Haeger. (Chapter 22)

Klinken et al. 2003). An alternative approach is to incorporate explicit measures of the physiological processes that interact with climate and thereby influence the organism's distribution and abundance (see for example Crozier and Dwyer 2006; Kearney and Porter 2009; Chown et al. 2010). Here, we have chosen to utilize the software program CLIMEX (Maywald and Sutherst 1991; Sutherst and Maywald 2005) because it is a readily accessible modeling package that integrates physiological processes into simple population modeling, and hence produces results that are more likely to be biologically meaningful.

CLIMEX is normally used to generate the distribution of potential abundance maps for species as part of biological control programs and pest risk analysis, and it has been applied to a variety of plants, animals, and pathogens (e.g., Sutherst et al. 1989; Yonow and Sutherst 1998; Matsuki et al. 2001; Kriticos et al. 2003; Wharton and Kriticos 2004; Yonow et al. 2004; Sutherst and Maywald 2005; Watt et al. 2009). However, it has not yet been used widely to model a species' population changes over time, although this should be a useful application of the method (Zalucki and Furlong 2005; Zalucki and van Klinken 2006; Sims-Chilton et al. 2010; Li et al. 2012).

The rationale for using CLIMEX has been described many times, and we will not repeat the arguments in detail here; however, see e.g., Yonow and Sutherst (1998) and papers above. Briefly, CLIMEX calculates a weekly growth index (GI_W) scaled between 0 and 1, analogous but not equivalent to a population growth rate, which describes the suitability of that week for the species population at a location. The growth index is a product of temperature (TI) and moisture (MI) indices. For both TI and MI, a range of conditions of temperature and moisture are maximally suitable (that is to say, have a value of 1). On either side of the optimum range, suitability decreases as temperatures and moisture change beyond the optimum range. Above some upper thresholds and below some lower thresholds, "growth" ceases (namely, suitability of both TI and MI is 0). The annual growth index, GI_A, is given by $GI_A = \Sigma\ GI_W/52$, where 52 is the number of weeks in a year over which GI_W is summed. This arrangement accords with the ecological Law of Tolerance (Shelford 1963) and the ecological Law of the Minimum (van der Ploeg et al. 1999).

A measure of the relative suitability of a location for species' persistence is summarized in a single annual ecoclimatic index (EI), which is normally scaled to 100: EI = 100* GI_A*SI*SX; SI includes four cumulative stress indices, describing the species' response to the extremes of cold (CS = cold stress), heat (HS), dry (DS), and wet (WS) conditions, and SX, if needed, includes interactions between extremes of these conditions; cold-dry (CDS), cold-wet (CWS), hot-dry (HDS), and hot-wet (HWS) stresses. The stress indices (SI and SX) are accumulated weekly at specified rates whenever conditions exceed a specified threshold. The thresholds and rates constitute the parameter set that relates the biology or physiology of the organism being modeled to climate (temperature and moisture) effects.

Indices are calculated weekly using long-term monthly average maximum and minimum temperatures, rainfall, and humidity values as inputs for each location. For temporal variation at a site, monthly or daily data can be used. Areas with positive values for EI are suitable for species persistence; the larger the value of EI, the more suitable the location. Generally an EI of less than 5–10 would be considered marginal for species persistence year-round at a site or location. The values for parameters (thresholds and rates) that describe the effect of temperature (TI) and moisture (MI) on "growth" or suitability, and the various stress indices and their interactions can be estimated from laboratory or field studies (Zalucki and van Klinken 2006; De Villiers et al. 2013). Unknown or poorly measured values are estimated by an iterative procedure that involves comparing the modeled distribution with the known distribution, and adjusting parameter values in an attempt to achieve concordance. Comparing the modeled and observed species distribution in an area *not* used for the initial fitting procedure (e.g., another continent) can test the generality of the model (Maywald and Sutherst 1991; Webber et al. 2011).

We developed a CLIMEX model for monarchs based on Australian seasonal distributional data and rearing experiments (Zalucki 1981b, 1982) and used it successfully to describe the seasonal geographical distribution of monarchs in North America (Zalucki and Rochester 2004) as well as to estimate potential distribution elsewhere (Zalucki and Rochester 1999) (Figure 11.2a, b). Monarchs are now established in Spain and elsewhere in the Mediterranean region (Fernández-Haeger et al., this volume, Chapter 22)

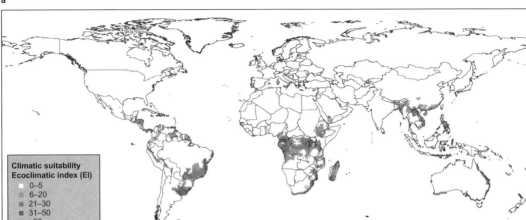

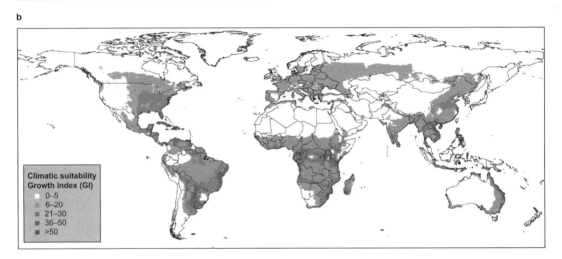

Figure 11.2. (a) Ecoclimatic index (EI) and (b) annual growth index (GI_A) generated by the CLIMEX model for monarchs. EI denotes areas where monarchs could breed year-round; GI_A indicates the potential seasonal expansion in range where they could migrate and breed if suitable milkweed hosts are present. See Zalucki and Rochester (1999) for more details. EI and GI_A are quite reasonable for North and South America (the latter for the closely related *Danaus erippus*) and Australia, where monarchs were introduced. The model indicates that southern Spain, Portugal, and northern Morocco are climatically suitable, and since invasive milkweed has established, monarchs have been introduced and also established (Fernández-Haeger et al., this volume, Chapter 22).

where our model indicated they could establish if introduced along with suitable hosts (Figure 11.2a, b); therefore, the model appears to represent monarchs' breeding climatic niche.

If climate is the main determinant of sites where a species can persist, we might expect a strong influence of climate variability or weather on the temporal variation in abundance of that species at a given location (see Zalucki and Furlong 2005; Zalucki and van Klinken 2006; Lawson et al. 2010; Sims-Chilton et al. 2010; Li et al. 2012). If we have the long-term daily weather data for a location, we can use the estimated species response to climate variables, namely our CLIMEX model, to infer the variation in suitability (fluctuations in the various indices) at the location over time, based on the observed climate record (Zalucki and Furlong 2005; Zalucki and van Klinken 2006; Li et al. 2012). Essentially, we generate a model of likely seasonal population indices based on climate alone, using information on a species' geographic distribution to derive its responses to climatic conditions, i.e., its climatic niche.

Our aim here is to investigate whether the decline in overwintering populations is related to our model estimates of spring-summer breeding success. We used two indices, the standard CLIMEX annual growth index, GI_A, and one we derive (see below).

METHODS

The CLIMEX model for monarchs (Zalucki and Rochester 2004) was run over time using climate data for 25 locations that cover the breeding range across North America (Table 11.1, Figure 11.1). The climate data span 41 years, from 1970 to 2010, and were kindly supplied by Janis Lentz from the National Climate Data Center. Missing data were interpolated temporally. In nearly all cases, missing data were runs of 1–5 days and ranged from 0.01% (St Louis) to 0.8% (Kalamazoo) (mean over all sites = 0.18%) over the period of interest. Temporal projections in CLIMEX were made using daily maximum and minimum temperatures (°C) and daily rainfall (mm) data as input.

We used the CLIMEX model to generate two indices as estimates of monarch abundance over the spring-summer breeding period (Figure 11.1, all of G1:G4 combined) for each year over the 41 years for each site. The simplest index of abundance uses the annual growth index (GI_A) for each site that CLIMEX calculates and outputs as a matter of course. The second index is based on the GI_W values (also provided as standard CLIMEX output when using daily meteorological input data). For this index, we allowed monarchs to lay an egg at each site as soon as

Table 11.1. Locations (latitude, longitude) used in generating expected monarch population size based on climate alone

Location	State	Lat	Long	First Date[a]	Main Body[b]	W[c]	Mean GIA[d]	SD	Min	Max
Ocala	FL	29.16	−82.18	15 Mar	1 Apr	1	34.6	4.36	26	42
San Antonio	TX	29.42	−98.48	19 Mar	1 Apr	1	13.6	6.07	1	29
Lake Charles	LA	30.23	−93.22	19 Mar	6 Apr	1	25.8	6.62	9	40
Waco	TX	31.57	−97.80	29 Mar	10 Apr	1	14.0	4.95	7	28
Jackson	MS	32.30	−90.18	29 Mar	15 Apr	1	19.5	5.79	10	34
Montgomery	AL	32.37	−86.32	3 Apr	15 Apr	1	19.0	5.78	8	33
Athens	GA	33.91	−83.32	8 Apr	24 Apr	1	18.1	6.07	7	34
Huntsville	AL	34.73	−86.59	13 Apr	24 Apr	1	19.3	5.65	8	31
Little Rock	AR	34.75	−92.29	13 Apr	28 Apr	1	17.4	3.85	10	25
Tulsa	OK	36.02	−95.90	17 Apr	3 May	5	16.8	5.00	8	29
Lynchburg	VA	37.41	−79.14	27 Apr	7 May	5	15.9	5.32	8	29
Louisville	KY	38.25	−85.76	27 Apr	12 May	5	17.1	5.29	8	32
St Louis	MO	38.60	−90.20	2 May	12 May	5	15.0	5.42	6	27
Olathe	KS	38.91	−94.78	2 May	16 May	5	16.7	5.25	6	25
Cape May	NJ	38.94	−74.91	2 May	16 May	5	15.8	5.14	7	26
Zanesville	OH	39.95	−82.00	7 May	21 May	5	14.2	4.06	6	23
Peoria	IL	40.74	−89.61	12 May	25 May	5	13.8	5.44	4	27
Wilkes-Barre	PA	41.24	−75.88	12 May	25 May	5	12.5	3.41	6	20
Buffalo	NY	42.84	−78.19	21 May	3 June	5	10.7	2.50	6	16
Grand Rapids	MI	42.96	−85.67	21 May	3 June	5	11.9	3.42	6	19
Germantown	WI	43.23	−88.11	21 May	3 June	5	10.5	3.03	4	16
Sioux Falls	SD	43.56	−96.73	26 May	3 June	5	9.3	3.51	2	22
Toronto	ON	43.79	−79.23	26 May	8 June	5	8.6	2.82	4	15
Burlington	VT	44.48	−73.21	31 May	8 June	5	10.7	2.75	6	17
Minneapolis	MN	45.07	−93.38	31 May	12 June	5	15.6	2.68	10	21

Note: Shading denotes central sites

[a] Date of expected first oviposition. Oviposition before 1 May: latitude = 0.11 (oviposition date) + 27.88; oviposition from 1 May and south of 41°40′N: latitude = 0.47 (oviposition date) + 9.52; oviposition north of 41°40′N: latitude = 0.07 (oviposition date) + 38.59. Oviposition date is days from 15 March for the main cohort of monarchs determined from egg peaks. Data collected 1981–1983.
[b] Date of arrival of the main body of migrants based on Cockrell et al. (1993). Data collected 1981–1983.
[c] Weights used to generate approximately 10% of monarchs from southern locations (lat <35.0°N) and 90% from the central (lat 35.0–42.0°N) and northern (lat >42.0°N) sites.
[d] Mean annual growth index (GIA) and its standard deviation (SD) for each site over the 41 years.

GI_W was positive. We calculated the "return" on each week's egg lay, PGR_i, by calculating the product of GI_W adjusted by weekly stress indices from the week of laying, i, to the week of adult emergence, approximately 340 degree days (dd) later (see Zalucki 1982), $i+340dd$ for that cohort, and we undertook this calculation for all weeks for each site, as follows:

$$\text{week} = i + 340dd$$
$$PGR_i = \prod_{\text{week}=i} GI_i, \quad \text{for week } i = 1, 2, \ldots, 52 \quad (2)$$

CLIMEX provides degree days for each week based on the species' physiological development requirements, which are known for monarchs (Zalucki 1982). Eggs are laid every week until GI_W goes to 0 at the end of the year, and the return on egg laying can be summed over the year (which is equal to all generations at a location) in order to generate an index of abundance for that site for each year. Note that if GI_i for any week i goes to zero before a cohort is complete, namely somewhere in the interval [i, $i+340dd$], then all generated cohorts affected by the event will effectively go extinct, and PGR_i equals 0 for that period. Essentially, we treat adjusted GI_W as a weekly survival rate for the calculation of this index. The estimated index of abundance for a year is then the average across all sites, or sites are weighted depending on location (Table 11.1, and see below) according to their purported relative contribution to overwintering populations (Malcolm et al. 1993). We have broadly classified sites as southern, central, and northern based on latitude (see Table 11.1) but weighted northern and central sites equally for this analysis. Essentially, southern sites are downplayed.

RESULTS

For the annual growth index (GI_A), the mean over the 41 years across all sites is 16 (SE = 0.30). Not unexpectedly, quite a bit of variation can be observed among sites and years (Table 11.1). The lowest GI_A value in one year was 1 for San Antonio, Texas, and the highest value was 42 for Ocala, Florida. Generally GI_A is higher for southern sites than for central and northern sites, as the season is shorter at higher latitudes (Table 11.1). Overall the poorest year was 1988 (GI_A averaged across all sites = 11.4) and the best was 2004 (GI = 20.5). Since monarchs supposedly come mainly from the central and northern regions, we can weight these sites more than others (Table 11.1) and generate a weighted average series (Figure 11.3). The weightings are arbitrary, and assume that the central and northern areas will contribute approximately 90% of monarchs, as suggested by Brower and Malcolm (1991) and confirmed by Wassenaar and Hobson (1998). Weighting does make some, but not large, changes to the series. In fact, we ran various weightings for central and northern sites and the changes are small (analysis not shown), suggesting climatic conditions are similar over large areas, and better areas elsewhere in the range compensate for poorly suitable areas.

Comparing the weighted GI_A and the second method of calculating an abundance index, PGR, which uses GI_W and development times due to seasonal temperatures, we get the same general picture (Figure 11.3), but with more variation over time. The lowest year is still 1988. The highest values for PGR in order occur in 1981, 1973, and 2004. For weighted GI_A the order is 2004, 1973 and 1975. Not surprisingly, the two methods of calculating an index of abundance are correlated ($r^2 = 0.60$, $n = 40$, $P < 0.0001$).

For neither method is there any indication of a systematic trend in climate suitability for monarchs as estimated by our two indices (see Figure 11.3,

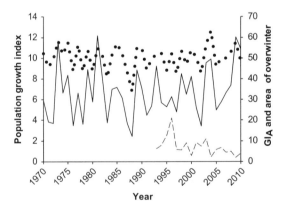

Figure 11.3. Modeled monarch spring-summer breeding population growth potential based on annual average population growth index, PGR (see text for details), weighted by purported monarch regional productivity (solid line) for 25 sites in eastern North America, with weighted annual GI_A (dotted line) for comparison (scale for the latter is 0–70). Area of overwintering (ha) in Mexico also shown (dashed line, from mid-1990s, same scale as GI_A).

slope of regression of GI_A with time = 0.024, r^2 = 0.0024; slope for PGR = 0.043, r^2 = 0.0416) from 1970 to 2010. If anything, the climate may have improved, whereas the area of overwintering in Mexico has declined (Figure 11.3; Brower et al. 2012a).

Some weak tests of the temporal model

Cockrell et al. (1993) presented data on the timing of monarch recolonization across North America in the spring (Table 11.1). The average date that GI_W first goes positive at a site (Figure 11.4a) is not a good indicator of return migration for the main body of migrants. This finding is perhaps not unexpected because weather conditions can be highly variable in the spring, and conditions at locations from which monarchs are coming will influence arrival times. On the other hand, the mean date of first positive return on eggs, namely the average date when PGR goes positive for a site (Figure 11.4b), is a good indicator. Essentially, the main body of monarch remigration (sensu Cockrell et al. 1993) appears to be timed to maximize population growth.

As expected, climate change trends are reflected in the date GI_W first goes positive, based on three sites running south to north on about the same longitudes: Little Rock (AR), Olathe (KS), and Minneapolis (MN). The date GI_W first goes positive may be getting earlier, particularly in the south and center (Figure 11.5a, Little Rock shows a decline of 1.4 days every 10 years, Olathe, a decline of 2.3 days every 10 years, but Minneapolis an increase of 0.3 days every 10 years, although none of the regressions are significant). The date of first positive return at this stage is more stable (Figure 11.5b, with a decline of 1.4 days for Little Rock, a decline of 0.14 days for Olathe, and an increase of 0.4 days for Minneapolis; again, none of these trends are statistically significant), perhaps reflecting the effect of springs that start off favorably but then deteriorate because of late winter storms.

Pleasants and Oberhauser (2012) estimated monarch egg production in the Midwest from 1999 to 2010, essentially for what we have called G3 and G4. They showed that overwintering areas in Mexico correlated with these estimates. Treating PGR as a measure of returns on eggs, we multiplied our index by their estimate of egg production. The area of overwintering populations were correlated with this measure (Figure 11.6), but we did not improve on their correlation.

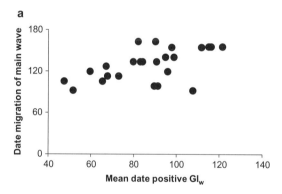
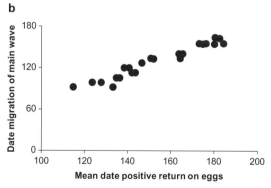

Figure 11.4. Date of migration by the main wave of spring migrants (from Cockrell et al. 1993) plotted against (a) the mean date of first positive GI_W for each of 25 sites across eastern North America, and (b) the mean date of first positive return on eggs or PGR_i. Date is in days from 1 Jan.

DISCUSSION

The two weighted indices suggest that if climate alone determines the size of the recorded monarch overwintering area, then it should not have declined, but it has (Brower et al. 2012a). Climate could still be partially responsible for the decline; overwintering survival, and survival during the northward spring migration and the southward autumn migration may be influenced by climate in ways not accounted for by our models.

Overwintering survival can be dramatically affected by winter storms (Brower et al. 2004). Less dramatic, but perhaps more important, is the role played by intact forest in monarch survival at overwintering sites through this period. Intact forests

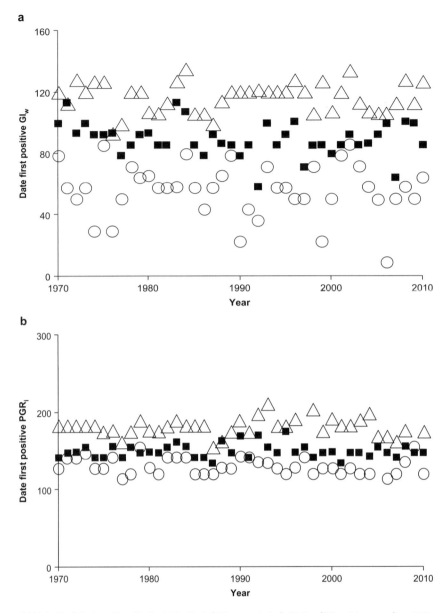

Figure 11.5. (a) Date (day) first positive GI_w for Little Rock (AR, open circles), Olathe (KS, solid squares), and Minneapolis (MN, open triangles) over time. (b) Date first positive return on eggs (PGR_i) for the same sites. Date is in days from 1 Jan (see text for regression slopes).

and large trees provide a blanket, umbrella, and "hot-water bottle" that effectively ameliorate extreme conditions (Brower et al. 2008, 2009; Williams and Brower, this volume, Chapter 9). As the oyamel forest continues to be degraded (Brower et al. 2002), survival may well have decreased, particularly if climate at the site is changing as well.

Monarchs migrating both north and south are more likely to survive the journey with suitable wind speeds and directions (Gibo 1986). In addition, nectar resources are crucial sources of lipids necessary for migration and overwintering survival (Masters et al. 1988; Brower et al. 2006, this volume, Chapter 10) and climate is likely to influence the

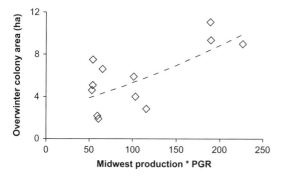

Figure 11.6. Overwintering colony area of monarchs in Mexico as a function of Midwest egg productivity (from Pleasants and Oberhauser 2012) and population growth index (PGR) for sites in the Midwest (shaded in Table 11.1). The line is fitted to square-root transformed area ($\sqrt{\text{Area}} = 0.0067 \cdot X + 1.635$, $r^2 = 0.3727$, $P = 0.021$).

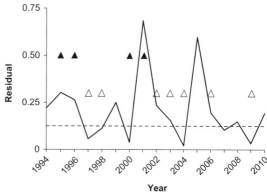

Figure 11.7. Combined effects of overwintering survival, and survival during the migrations north and south inferred as the residual of Equation 1 (see Equation 3 in text) over time. The inference is based on the assumption that the area of overwintering in Mexico in successive years is a reasonable estimate of population size and that the index of abundance for spring-summer breeding success due to climate is estimated by PGR. The expected value for R_i assuming only 50% survive each of the contributing processes is also shown (dashed line). The incidence of severe (solid triangles) and moderate (open triangles) monarch killing storms at overwintering sites in Mexico also shown (taken from Fink et al. in prep).

availability of nectar (e.g., Brower et al., this volume, Chapter 10).

Using our model, we can rearrange Equation 1 to examine the combined effect of these processes and call it the residual R_i, which is a product of overwintering survival, and the proportion surviving the migrations north and south:

$$R_i = OWS_i * JN_i * JS_i = G_{W,i+1} / G_{W,i} / PGR_i \quad (3)$$

Since each of the processes is a proportion, the product should be bounded between 0 and 1 and usually low. If, on average, the chance of surviving each process were 50%, then R_i would equal 0.125 (Figure 11.7). We are essentially predicting the variation in R_i, which could reflect the effect of climate (or other factors) on these processes; however, again no systematic trend is seen in these values, although they fluctuate, suggesting good and bad years (Figure 11.7). A systematic trend might be expected if climate change were responsible, for example, because of gradual warming. Interestingly, many of the very low values for R_i coincide, by and large, with winter storms (Figure 11.7), which are known to reduce monarch winter survival in Mexico (Calvert et al. 1983). The one exception was in 2002 when a severe storm was recorded, but R_i is high.

It is possible that our model is projecting at too coarse a scale to illustrate what could be divergent effects of climate on local monarch abundance. For example, Zipkin et al. (2012) studied 90 sites in Ohio and showed that when the available growing degree days were above average for each site, it was the coolest sites that were expected to have the greatest monarch abundances. This effect became null for sites with mild temperatures and was actually reversed at the warmest sites; nevertheless, our model did not detect a systematic trend in monarch abundance that might have been expected if climate change were responsible, for example, due to gradual warming.

So what could be responsible for the decline in overwintering populations? What are the likely threatening processes? Conditions for breeding, overwintering in Mexico, or migrating may have deteriorated, or, more likely, a combination of the three has caused the decline. The analysis above suggests that climate for the breeding spring-summer generations over the last 41 years has not been a problem.

As a result of increased herbicide use in agroecosystems as a consequence of genetically engineered glyphosate-resistant soybean, the abundance of milkweeds has declined dramatically (Pleasants and Oberhauser 2012), which could impact egg laying or potential reproduction (Zalucki and Lammers 2010). In addition, the widespread cultivation of Bt-expressing crops such as maize may also influence monarch survivorship negatively on the remaining milkweed in agricultural habitats. The Bt issue has been mired in controversy from the time it was first reported that the pollen from some Bt transformations of maize was toxic to monarch larvae (Losey et al. 1999). Although a series of papers published in PNAS (e.g., Sears et al. 2001) showed the risk to monarchs from Bt (defined as probability of exposure multiplied by a toxic effect) was small, none of these papers put that risk into a population ecological context, such as by using a Population Viability Analysis (Zalucki et al. in preparation).

Reduced numbers of monarchs in Mexico could be an effect of forest degradation at overwintering sites, but our CLIMEX model does not address this issue. Similarly, empirical analyses to determine whether climate change effects are impacting overwintering monarchs in the deteriorating forests of Mexico have not been completed, but see Oberhauser and Peterson (2003), Sáenz-Romero et al. (2012), and Ramírez et al. (this volume, Chapter 13) for modeled impacts.

Climate change can also have subtle effects. Warm latitude insects will go a little further poleward each year, as has been widely reported (Hill et al. 2011), but conditions are also becoming suitable earlier each year. However, the remigration data over the period of 1997–2012 in the Journey North database shed no light on the question of whether the spring remigration is occurring earlier (A. Davis, pers. comm.). Although the effect of climate on the returns on these eggs has remained relatively stable, there are likely to be interaction effects on host plant availability and suitability. Plants will senesce faster and become unsuitable sooner. Host plants have a strong influence on early stage survival in monarchs (Zalucki et al. 2001b, 2012), but the effect of climate on host plant quality and availability has yet to be addressed in this system. As temperatures continue to increase and rainfall patterns change, the consequences for relatively common species, such as monarchs, could be very strongly negative. As Brower (2001) suggested, monarchs may well be the "canary in the cornfield" after all.

The big challenge in sorting out the processes affecting monarch decline will be data, both lack of long-term historic data and the acquisition of relevant current data. At present, only the population in Mexico over the winter is well estimated. It is essential that monitoring data across the monarchs' spring-summer range be collected in a consistent fashion and integrated into estimates of abundance. This collection and monitoring would reduce the number of unknowns we have to ponder. The many monarch citizen science programs (Oberhauser et al., this volume, Chapter 2) offer the potential to contribute to this outcome. Estimating migration survival will not be straightforward.

GLOSSARY OF ACRONYMS

All the acronyms below refer to approaches or software packages that attempt to describe the environmental variables (usually climate) associated with where a species is found and so "predict" where else it might occur. Most are variations on a theme and essentially describe the set of attributes (usually climate) associated with a species' location data.

BIOCLIM/ANUCLIM: Software packages that use climate pattern-matching with a minimum bounding rectangle (MBR) to describe the climatic envelope of a species and to predict its occurrence

CLIMEX: A software package that differs from those above in that it is a process-oriented model describing species response to climatic variables, and predicting relative climatic suitability for a species at selected locations and the relative climatic similarity between different locations

DOMAIN: A software package that matches climate pattern among locations using a point-to-point similarity index

GARP (Genetic Algorithm for Rule-set Production): Generates environment-description rules using machine-learning techniques based on point location data, where plants and animals have been collected from herbaria or museums, and both climatic and nonclimatic data

GLIM/GAM (Generalized Linear Interactive Modeling/Generalized Additive Models): A very general set of statistical procedures for fitting species response functions or regression models to survey data so as to predict the probability of occurrence of species at a fine scale

GRASP (Generalized Regression Analysis and Spatial Prediction): A regression modeling approach used to establish relationships between a response variable (e.g., species presence) and a set of spatial predictors (location climate data)

HABITAT: A software package that aims to circumscribe the environmental envelope of a species (a combination of environmental variables where it lives) and so predict the environments in which it may be present

MaxEnt (Maximum Entropy modeling): A machine learning technique based on the distribution of maximum entropy to predict species distribution

ACKNOWLEDGMENTS

The authors are particularly grateful to Janis Lentz for tracking down the climate data, and to Karen Oberhauser, Kelly Nail, Darren Kriticos, and an anonymous reviewer for comments on the manuscript.

PART IV

Conserving North American Monarch Butterflies

An Overview

LINCOLN P. BROWER AND LINDA S. FINK

MONARCHS AT RISK

The eastern North American monarch population is on a statistically significant decline (Brower et al. 2012a; Rendón-Salinas and Tavera-Alonso 2013) and ominously, in Chapter 12 in this section, Jepsen and Black also report an 84% decline in the average number of monarchs per overwintering site in the western population from 1997 to 2011. The monarch is a resilient species with high reproductive potential and the population declines can be reversed, but only if additional conservation strategies are enacted quickly.

Illegal logging in Mexican overwintering sites, until recently the largest individual threat to the monarch phenomenon, has been reduced in recent years (WWF 2012b; Vidal et al. 2014; Ramírez et al., this volume, Chapter 13). That the population continues to decrease, as manifested by the area occupied by wintering monarchs, indicates that multiple factors are responsible. Other stressors addressed in this section include forest microclimate degradation in Mexico through continuing low intensity logging (Ramírez et al., this volume, Chapter 13), and milkweed habitat loss in the United States (Pleasants, this volume, Chapter 14). Extreme weather events and long-term climate change are also addressed by Ramírez and colleagues (see also Brower et al., this volume, Chapter 10; Williams and Brower, this volume, Chapter 9). It is clear that monarchs face anthropogenic habitat deterioration in their breeding areas, along migration routes, and in their overwintering areas, and that long-term persistence of the migration will depend on expanded conservation efforts at all points of their annual cycle. In this section, we document threats to monarchs and actions to address these threats.

HABITAT LOSS AND DEGRADATION

Pleasants and Oberhauser (2012; Pleasants, this volume, Chapter 14) document a dramatic loss of milkweeds from agricultural areas in Iowa between 1999 and 2012, and they project this to regional losses throughout the U.S. Midwest. Through the rapid adoption of genetically modified glyphosate-resistant corn and soybeans, midwestern farmers can use glyphosate herbicide to eliminate almost all unwanted plants from agricultural fields, including milkweeds and nectar sources. Prior to the extensive extermination of milkweeds by glyphosate, 92% of monarchs overwintering in Mexico fed on *Asclepias syriaca* (Malcolm et al. 1993), at least 50% of these butterflies originated in the midwestern corn belt (Wassenaar and Hobson 1998), and a majority were estimated to originate in agricultural fields (Oberhauser et al. 2001; Pleasants and Oberhauser 2012). The swift and extensive disappearance of milkweeds from agricultural fields is clearly a major factor in the eastern monarch population's decline. In Chapter 12, Jepsen and Black suggest that milkweed losses will also prove to be one factor responsible for the decline of the western population, although we need more data to document the distribution and abundance of milkweeds in the western United States.

Herbicides and roadside management policies that eliminate milkweeds also eliminate other nectar sources, yet there has been no systematic documentation of the extent to which nectar loss is occurring, nor any determination of how it affects energy balance in migratory monarchs. Monarchs consume nectar all along their migration route to fuel their flight, but their need for nectar is most acute in Texas and northern Mexico, where they accumulate the majority of the reserves that will support them through the winter (Brower et al. 2006, this volume, Chapter 10). Decreases in nectar availability through this region may be contributing to monarchs' decline.

Dozens of the 458 known overwintering sites in California were lost to development in the twentieth century; because many of these sites have not been systematically monitored, the rate of recent losses is not known (Jepsen and Black, this volume, Chapter 12). The Xerces Society, in collaboration with other groups, is creating a database of California overwintering sites to fill this knowledge gap. In Mexico, one overwintering area has been destroyed by illegal logging. Monarch colonies were recorded in the Lomas de Aparacio area in the southern Cerro Campanario between 1996–97 and 2006–07 (and probably occurred there earlier); satellite images taken in March 2004 and February 2008 document clear-cutting of the area between these two dates (Simmon et al. 2008).

In Mexico, small-scale tree removal is continuing in the core and buffer zones of the Monarch Butterfly Biosphere Reserve (MBBR) (Ramírez et al., this volume, Chapter 13; Vidal et al. 2014). This affects the microclimate (Weiss 1998; Williams and Brower, this volume, Chapter 9), and thus the rate at which the butterflies metabolize lipid reserves (Masters et al. 1988) and their ability to maintain their water balance and survive storms (Anderson and Brower 1996). Diversion of water and unsustainable ecotourism also degrade overwintering habitats. In Mexico during the 2012–2013 season, colonies were found in only 9 of 17 sites monitored by WWF-Mexico (Rendón-Salinas and Tavera-Alonso 2013). Almost all these sites have had variable occupancy over the years, but the absence of butterflies from so many sites may indicate that the changed forest conditions are unacceptable. More detailed measurements correlating forest condition, microclimate, and butterfly occupancy would allow this hypothesis to be tested.

In Chapter 15, Martínez-Torres and colleagues review the importance of fire control policy in the oyamel fir and pine forests in which the butterflies roost in Mexico. Fires burn up to several hundred hectares per year in the MBBR (CEC 2008) and occur generally during the dry season from March through July. Most fires result from clearing land for agriculture and some are accidental, but others are set deliberately in local conflicts or to obscure illegal logging. Martínez-Torres et al. point out that until 2012, government agencies were not collecting systematic data on MBBR fires and that the response to fires by government and local groups has not been coordinated. Additionally, the fire management tactics practiced in the reserve have not been based on ecological principles.

Clearly, further research is needed to determine the role of fire in maintaining the quality of the forests for monarchs, and the development of an ecologically sound fire management policy for the MBBR should be a high priority. The negative ecological effects of salvage logging (Lindenmayer and Noss 2006), for example, might justify its prohibition following fires and other natural disturbances such as storms and bark beetle infestations (Rodriguez 2009).

CLIMATE CHANGE AND SEVERE WEATHER

In the western and midwestern United States and Texas, monarch populations have been negatively affected by severe droughts (Brower et al., this volume, Chapter 10; Jepsen and Black, this volume, Chapter 12). In Mexico, killing storms with cold, clear nights following heavy precipitation have led to high mortality of wetted butterflies (Calvert et al. 1983; Anderson and Brower 1996; Brower et al. 2004, 2009). On a short time scale, individual weather events such as these cannot be ascribed to climate change, and there is not, at this point, evidence that mortality attributable to drought or winter storms has been increasing.

Over the longer time scale, however, models predict that climate change in Mexico will have strong, negative effects on overwintering monarchs because the MBBR's suitability will diminish both for the butterflies and for the oyamel fir, *Abies religiosa* (Oberhauser and Peterson 2003; Sáenz-Romero et al. 2012; Ramírez et al., this volume, Chapter 13). Using ecological niche modeling, Oberhauser and Peterson (2003) predicted that an increased frequency of storm-related mass mortality would render the current overwintering areas less suitable. The resolution of their climate maps is 0.5° × 0.5°, and the entire winter range of the monarchs falls in a single pixel. Finer-scale projections of winter climate might modify this prediction, given the highly variable nature of storms across mountainous terrain.

Sáenz-Romero et al. (2012) and Ramírez et al. (this volume, Chapter 13) used Random Forest classification models (Lawler et al. 2006) to predict the future distribution of oyamel firs. Projecting forward to 2090, Sáenz-Romero et al. (2012) predicted a 96.5% reduction in climatically suitable habitat for the oyamel fir. Similarly, in their chapter in this section, Ramírez and colleagues' models predict that by 2090 the climate in the entire MBBR may become unsuitable for oyamel firs. They propose that assisted migration of oyamel firs into areas outside the MBBR, where the future climate will be more suitable, be considered as one management strategy to provide future habitat for monarchs. Sáenz-Romero et al. (2012) also propose experimental plantings within the MBBR of conifers with phenotypic similarities to oyamels but with physiological tolerances suited to the projected climate.

In the face of global climate change, bioclimatic models are increasingly important components of conservation planning, but the diversity of modeling approaches can lead to substantially different range predictions (Heikkinen et al. 2006; Lawler et al. 2006). As a result, the future distributions of monarch butterflies and oyamel fir trees will continue to be uncertain.

TAKING ACTION TO REVERSE MONARCH POPULATION DECLINES

Monarch scientists and enthusiasts have been concerned about milkweed losses in the U.S. Midwest for almost 30 years (Brower 1985b), and momentum is building for milkweed protection and planting by private landowners (Diffendorfer et al. 2013).

Substantial recovery of the lost milkweed acreage, however, will require participation by the stakeholders who develop policies for the management of large land areas, including agribusiness, transportation departments, public lands managers, and federal and state soil conservation agencies (see Shahani et al., this volume, Chapter 3).

In Chapter 16, Borders and Lee-Mäder describe Project Milkweed, a collaborative initiative between The Xerces Society, plant nurseries, USDA-NRCS, and other pollinator conservation groups to increase the availability of region-specific native milkweed seeds for large plantings. Project Milkweed started in 2010, and seeds from a number of the milkweed species have not yet been harvested and planted. The success of this project will depend on the growers' ability to maintain consistent production of commercial quantities of local milkweeds, landowners' ability to establish self-sustaining milkweed populations, and The Xerces Society and its partners' ability to foster a demand for milkweeds in large-scale plantings.

Land management strategies also affect milkweed availability. In Chapter 17, Baum and Mueller suggest that appropriately timed mowing or burning can increase milkweed abundance and quality. They compared the seasonal availability of *Asclepias viridis* in several Oklahoma sites under different disturbance regimes. Removal of aboveground growth in midsummer, by either burning or mowing, enhanced late-summer regrowth, which was then available for late-breeding monarchs. Similar effects of mowing regimes on *A. syriaca* quality were demonstrated by Fischer (forthcoming) in southern New York State. Experiments with additional milkweed species at different latitudes are needed to develop regionally appropriate management recommendations; timing of management actions will be critical and vary by habitat and latitude.

In Mexico, it is very likely that the current boundaries of the MBBR are insufficient for the long-term continuation of the monarch migration. Five of six wintering areas outside the MBBR (Slayback and Brower 2007) do not have the same level of protection as areas within the Reserve, yet the probability of a viable number of butterflies surviving severe winter storms is higher when individual colonies are spread over a broader geographic area. In addition, climate models (Oberhauser and Peterson 2003; Sáenz-Romero et al. 2012; Ramírez et al., this volume, Chapter 13) make a pressing case for reforestation to create new habitats that will be suitable in the future. The current boundaries of the MBBR were established in 2000 to replace boundaries set in 1986, based on 14 years of data, discussion and compromise (Bojórquez-Tapia et al. 2003; Missrie 2004). Another 14 years have passed, and expanding protection of suitable forests beyond the boundaries of the MBBR is warranted.

The monarch butterfly is a symbol of trinational cooperation among Canada, the United States, and Mexico, and there is a strong commitment by scientists, citizen scientists, government agencies, and NGOs to protect the migratory phenomenon (CEC 2008; Shahani et al., this volume, Chapter 3). Despite this commitment, we conclude with a painful acknowledgement: the unparalleled migration and overwintering of the monarch butterfly are threatened (Monarch E.S.A. petition 2014). If additional conservation measures are not taken quickly, we may lose a unique phenomenon. The cause will be failure of stewardship and the incessant human usurpation of the resources needed by other organisms that have the right to share this planet.

12

Understanding and Conserving the Western North American Monarch Population

Sarina Jepsen and Scott Hoffman Black

A volunteer citizen monitoring effort in California reveals that since 1997, the abundance of overwintering monarchs has declined by 50% from the 17-year average. Although climatic variation is probably important as a driving factor in monarch population dynamics, the loss or degradation of overwintering and breeding habitat may also influence monarch abundance. The Xerces Society is working to understand the status and distribution of monarch overwintering and breeding habitat in the western United States. To achieve this goal, we have created a geographic and informational database of overwintering sites to inform site conservation, implemented a monitoring program to fill gaps in our knowledge of overwintering monarch status and distribution, and developed a habitat assessment protocol to evaluate the condition of overwintering habitat. We are currently collecting and mapping breeding habitat observations. The Society is working with partners to conserve overwintering sites on the California coast by developing and disseminating guidance to influence site management, and reviewing the legal status of monarchs to aid advocates in the protection of key sites.

INTRODUCTION

While most people are familiar with the monarchs that migrate thousands of miles from all over eastern North America to overwinter in central Mexico, the monarchs of western North America receive far less attention. Every fall, hundreds of thousands of monarchs, and at times more than a million, arrive at forested groves along the California coast and aggregate en masse. Throughout this chapter, we refer to these seasonal aggregations as overwintering sites, although in some cases these aggregations are much more temporary than the term "overwintering" implies.

A long-term citizen monitoring effort, the Western Monarch Thanksgiving Count (WMTC) (Monroe et al. 2013; Oberhauser et al., this volume, Chapter 2), provides annual estimates of the number of overwintering monarchs at approximately 100 California coastal sites and has revealed declines in monarch abundance since 1997 (Figure 12.1). While climatic factors such as rainfall and drought may explain much of the variation in the annual abundance of overwintering western monarchs (Frey et al. 2003b; Stevens and Frey 2004, 2010), loss or alteration of breeding and wintering habitat is also likely to inhibit the ability of monarchs to survive and reproduce.

The Xerces Society is working to understand the status and distribution of monarchs and their habitat in the western United States. We aim to promote western monarch overwintering habitat conservation by influencing management and advocating for the protection of overwintering sites. We are also working to restore western monarch breeding habitat by encouraging the mass production of regionally appropriate native milkweed seed for large-scale habitat restoration projects (Borders and Lee-Mäder, this volume, Chapter 16). Here, we provide background on our current understanding of the status of western monarchs and detail our approach to conserving the monarch and its overwintering

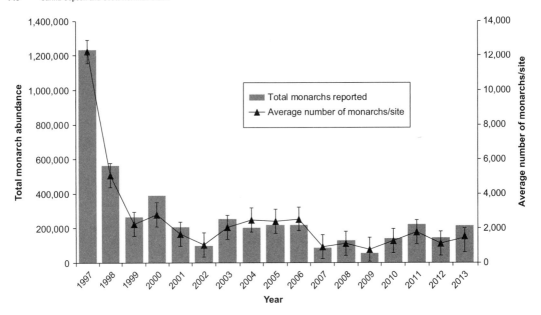

Figure 12.1. Western Monarch Thanksgiving Count total and average abundance estimates with standard error of the means at 63 overwintering sites, 1997–2013 (Monroe et al. 2014).

phenomenon. This approach includes four key strategies: (1) Working to understand the status and distribution of western monarchs' overwintering habitat through creation of a database of overwintering sites, development of a monitoring project to fill gaps in that database, outreach to recruit and train volunteer monitors to participate in the WMTC, and initiation of a habitat assessment protocol for overwintering sites; (2) Creating and updating monarch habitat management guidelines, developing site specific management plans, and providing workshops for overwintering site managers; (3) Compiling and mapping known and potential natal areas to document the current distribution of monarch breeding habitat in the western United States; and (4) Collaborating with partners at Lewis and Clark Law School to analyze existing laws and regulations pertaining to monarchs and overwintering habitat, in order to understand their legal status and provide tools to advocates who seek to protect these habitats.

STATUS AND DISTRIBUTION OF OVERWINTERING WESTERN MONARCHS

Monarchs have historically aggregated en masse in the fall and winter at numerous sites scattered along 620 miles of the California coast from northern Mendocino County to as far south as Baja California, Mexico (Lane 1993; Leong et al. 2004) (Figure 12.2). Smaller aggregations of monarchs consisting of tens to hundreds of butterflies have been reported from Arizona and southeastern California (CNDDB 2012; Xerces 2012d; Monroe et al. 2013,). An estimated 458 monarch cluster locations are known (Xerces 2012d), including sites that host small or large monarch clusters; are current, historic, or of unknown status; and host monarchs very temporarily (transitory) or throughout much of the fall and winter (climax).

Prior to monitoring efforts that began in the 1980s, the historic distribution and size of the western monarch population was largely unknown. Few records exist of monarchs until the mid-1900s, although early accounts can be found of overwintering masses of monarchs from Monterey, California in 1869 and 1873, and from Santa Cruz in 1888 (Lane 1993; Brower 1995).

Monarch abundance at western overwintering sites can vary dramatically from year to year (Figures 12.1, 12.4), and historic estimates of the overall overwintering population size range from 1 to 10 million (Nagano and Lane 1985; Nagano and Freese 1987). Leong et al. (2004) used data from the

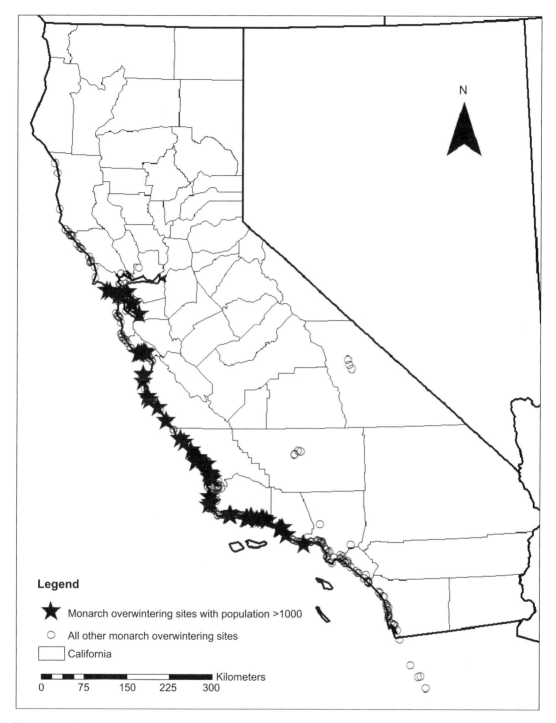

Figure 12.2. Distribution of current and historic monarch overwintering sites in California. Black stars represent sites that have hosted more than 1000 monarchs in the past decade (Xerces 2013).

California Natural Diversity Database (CNDDB) from 1990 to 2000 to estimate the maximum number of overwintering monarchs for a single season to be more than 2.3 million. Since 1997, a dedicated team of volunteer citizen scientists estimate the number of monarchs aggregating at California overwintering sites each year as part of the WMTC (Monroe et al. 2013; Oberhauser et al., this volume, Chapter 2). This effort is coordinated by Mía Monroe, Candace Fallon, Dennis Frey, and Shawna Stevens. While the WMTC includes 261 sites, only a subset (83–163 sites) are monitored in any given year. Fifteen sites have been consistently monitored every year since 1997; the total numbers of monarchs observed in these sites each year are illustrated in Figure 12.3. Whether we consider data from all monitored sites, or only from the sites that have been monitored consistently, the same pattern is apparent. The total number of monarchs counted in the most recent WMTC is approximately 211,000, an order of magnitude below the 1990–2000 maximum, although these counts underestimate the total population size, since not all sites are monitored.

A comparison of abundance estimates from the WMTC at overwintering sites in 1997 and 2013 show a decline of approximately 90% in the average number of monarchs per site. The total number of monarchs reported in 1997 surpassed 1.2 million, which is high for the period of 1997–2013, but at the low end of the range of historic estimates of the overall western monarch population (Leong et al. 2004). The decline from 1997 to 1999 was particularly dramatic; the average number of monarchs per site in 1999 was only 18.5% of the average in 1997 (Figure 12.1).

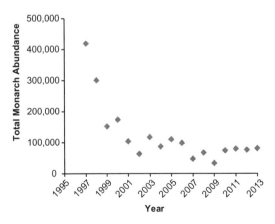

Figure 12.3. Estimated monarch abundance at 15 overwintering sites monitored every year since 1997 (Monroe et al. 2013).

An early estimate by Tuskes and Brower (1978), using a mark-release-recapture method, concluded that a maximum of 90,000 monarchs roosted at Natural Bridges State Park in Santa Cruz County, 1975–1976. Currently, the largest known site is Pismo Beach in San Luis Obispo County, with an estimated 30,293 monarchs in 2013 (Figure 12.4). Only 31 of the 163 sites monitored in the 2013 WMTC hosted more than 1000 monarchs (Monroe et al. 2013).

Anecdotal accounts exist of dramatic western monarch population fluctuations from the 1990s (Figure 12.4). The winter of 1994 showed a dramatic decline in monarch numbers at some sites; only 1500 monarchs were counted at Pismo Beach when 150,000 monarchs had been previously recorded at that site (Monarch Newsletter 1994 as cited in Meitner 1995). Brower noted a virtual absence of monarchs at overwintering sites in 1994, followed by a reappearance in 1995 attributed to a westward shift of winds and possible influx of immigrants from Mexico (Brower and Pyle 2004). Long-term mark-release-recapture studies by Dennis Frey and Monarch Alert at one site in San Luis Obispo County (Pismo Beach State Park) from 1990 to 2009 reveal enormous population fluctuations, and Villablanca (2010) predicts that the population at that site will gradually go extinct.

The factors that influence western monarch population dynamics are still largely unknown; however, Stevens and Frey (2010) examined WMTC data for the time period 1998–2007 and suggest that increasing drought conditions were the most likely cause of changes in monarch population size; the severity of the drought in key monarch breeding states (California, Arizona, Nevada, and Oregon) explained the variation in monarch abundance during that time period.

In addition to climatic factors, we assume that the western monarch population also may be influenced by the loss of milkweed breeding habitat and changes in the amount and quality of overwintering sites, but the relative contribution of each of these factors has not been studied. For the monarchs in eastern North America, the loss of milkweed breeding habitat, especially from agricultural fields due to the increasing use of herbicide-tolerant crops, has contributed to the population decline (Pleasants and Oberhauser 2012). Other potential threats to western monarchs include disease, parasitism, and predation; pesticide

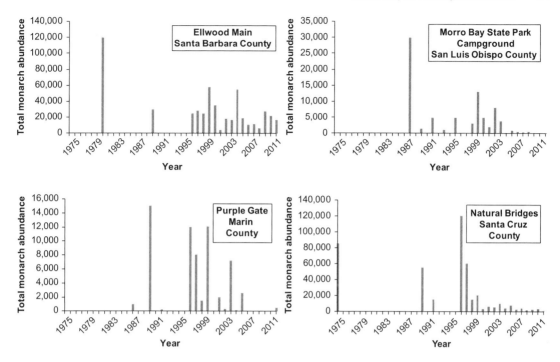

Figure 12.4. Monarch population estimates from 1 November to 15 December at four sites in California: Ellwood Main (Santa Barbara County), Morro Bay State Park Campground (San Luis Obispo County), Purple Gate (Marin County), and Natural Bridges (Santa Cruz County) (Monroe et al. 2013; Xerces 2013).

use in breeding and overwintering habitat; climate change; and overcollection.

To better understand the status and distribution of western monarchs at their overwintering habitat, in 2010, we began to compile multiple sets of existing data. Prior to this project, the most comprehensive western monarch dataset was the CNDDB, housed by the California Department of Fish and Wildlife and created from data gathered by researchers who visited and evaluated 333 monarch roosting sites from 1984 to 2000. This dataset represented a comprehensive picture of monarch distribution and site-specific abundance at California overwintering sites; however, many existing data sources, such as the WMTC dataset from 1997 to 2013 and many other published and unpublished reports, are not included in the CNDDB.

We worked to update the CNDDB and incorporate all available information about California overwintering monarch distribution, abundance, population status, and habitat into a single informational and spatial database, and to make that information freely available to researchers and conservationists. We used information from the CNDDB, the WMTC, published literature and unpublished reports, and personal communication with numerous western monarch researchers; state, county, and local land managers; agency employees; and others. Geographic data were revised based on site visits and conversations with monarch researchers and WMTC volunteers with site-specific knowledge to more clearly illustrate the boundaries of current and historic monarch groves.

The resulting data set includes 478 current or historic monarch roosting locations, with the earliest records dating from 1970. These data include locations of trees used as climax sites, transitory sites, and sites with too little data to characterize as either. Only 83 of the sites have hosted more than 1000 monarchs in the past decade (Figure 12.2); 62 have been extirpated or possibly extirpated due to development, tree trimming, natural factors (such as storm or fire), and other factors; and the status of 231 sites is largely unknown because they have not been monitored in more than 10 years. The database provides a comprehensive picture of individual sites, from completed survey work and resulting population estimates, to habitat characteristics, and to information such as landownership or driving directions.

This comprehensive Xerces database can be used to identify knowledge gaps and prioritize research or conservation efforts appropriately. It also serves as a useful tool to assess the potential impact of housing developments or other projects on monarch overwintering habitat. For example, in 2011, an overwintering site in San Luis Obispo County was proposed for development, and the database was queried to quickly provide information to elected officials on the historic and current level of monarch occupancy of that site. The geographic component of the database allows users to determine the proximity of overwintering sites to proposed projects, and to the boundaries of the Coastal Zone. The Coastal Zone is an area generally about 914 meters inland from the Pacific Ocean (as defined in the Coastal Zone Management Act) and is important because sites within the Coastal Zone can receive greater protection for their biological resources than sites outside of it. Understanding the boundaries of overwintering sites and the number of sites owned by agencies and other entities can help to prioritize outreach to these agencies to encourage monarch habitat conservation. Ultimately, we aim to move this database online and make all nonsensitive data publicly available.

In 2011, we initiated a monitoring project to fill gaps in our understanding of the distribution, population status, and habitat condition of overwintering monarchs. We selected 120 sites based on historic population size, time since last monitoring effort (sites that had not been visited in 10 years or more were prioritized), and public landownership. Each site was monitored to assess site occupancy and estimate the number of aggregating monarchs, evaluate habitat condition, and delineate site boundaries.

Although WMTC citizen monitors routinely visit approximately 100 monarch overwintering sites each year, data on the type or quality of the habitat are not routinely collected. To better understand the condition of monarch overwintering habitat, we developed a habitat assessment protocol that contains attributes that are considered important to monarchs, including microclimate (temperature, humidity, and wind conditions within and outside the site), topography, water source, presence and type of nectar species available, and tree species. The protocol also guides the collection of information on any disturbances or habitat alterations within the site and surrounding habitat, such as diseased trees, pesticide use, or high levels of human visitation. This tool will enable Xerces staff, WMTC volunteers, and others to assess threats to sites and, if used regularly, to document changes over time. In addition, the data will add to our knowledge of monarch overwintering habitat requirements and can inform management and restoration. This habitat assessment protocol is being distributed to current WMTC volunteers and used in workshops to train future volunteers.

CONSERVATION AND MANAGEMENT OF MONARCH OVERWINTERING HABITAT

Pyle and Monroe (2004) suggest that the most vulnerable element of the monarch annual cycle is the overwintering stage. One reason for their vulnerability perhaps lies in the fact that so many individuals are concentrated in a very small area. Overwintering monarchs have very specific microclimatic requirements, such as protection from wind and storms, absence of freezing temperatures, exposure to dappled sunlight, and presence of high humidity (Chaplin and Wells 1982; Calvert et al. 1983; Masters et al. 1988; Anderson and Brower 1996; Leong 1999; Williams and Brower, this volume, Chapter 9). Fall or winter blooming flowers that provide monarchs with nectar may be important to maintain lipid reserves required for winter survival and the spring migration (Tuskes and Brower 1978).

Habitat change in the areas of the monarch overwintering sites is not new; the historical composition of coastal vegetation in California included groves of native trees that presumably hosted dense monarch aggregations (Lane 1984, 1993). Many of these groves have been gone for decades. Monarch overwintering habitat in California continues to be directly threatened by urban development and, to a lesser extent, agricultural development. Habitat alterations, such as tree trimming or tree removal, or natural factors such as fire, severe storms, or disease or senescence of trees, can alter the structure and microclimate of an overwintering site and reduce its suitability for monarchs (Sakai and Calvert 1991).

More than two decades ago, a statewide report documented the loss or destruction of 38 monarch overwintering sites, including 16 lost to housing developments (Sakai and Calvert 1991). Eleven of these sites were lost in the period 1985–1991; the remaining 27 sites were lost prior to 1985 (Sakai and Calvert 1991). In the 1990s, housing developments displaced 11 additional monarch overwintering sites (Meade 1999). The Xerces Society database currently lists 62 sites that have likely been made unsuitable for monarchs, but many of those localities need to be monitored to determine whether monarchs have returned, and the condition of the habitat needs to be assessed. At present, two overwintering sites in San Luis Obispo County are slated for housing developments (S. Jepsen, pers. observ.). Anecdotal reports also suggest that overwintering sites have been lost as a result of tree cutting or trimming (Sakai and Calvert 1991), or that the monarch population has declined after tree trimming (but see Villablanca 2010).

At present, most overwintering sites are dominated by non-native blue gum *Eucalyptus* (*E. globulus*) or red river gum *Eucalyptus* (*E. camaldulensis*), although many sites also contain native trees such as Monterey pine (*Pinus radiata*), Monterey cypress (*Cupressus macrocarpa*), and western sycamore (*Platanus racemosa*) (Xerces 2012d). *Eucalyptus* are exotic invasive species that were introduced into California from Australia in 1853 (Butterfield 1935) and have been shown to reduce biodiversity (Bossard et al. 2000). *Eucalyptus* removal is highlighted as a restoration goal for some natural areas (IELP 2012), and conflicts can emerge between monarch habitat conservation and *Eucalyptus* removal efforts.

Many monarch overwintering sites contain aging or diseased trees. For example, Monterey pine is affected by pitch canker (*Fusarium circinatum*), a fungus that causes swollen lesions that girdle branches, trunks, and exposed roots. The disease was first observed in California in Santa Cruz County in 1986; in the decades following, it spread to 18 coastal counties (Wikler et al. 2003). As aging or diseased trees lose limbs or die, sites can become less suitable for monarchs and pose a public safety hazard. To ameliorate safety hazards, land managers prune aging or diseased trees, and they may do so without consulting experts who could provide advice on the monarch's habitat requirements.

Recent studies suggest that monarchs do not prefer *Eucalyptus* trees. They use native tree species more than would be expected, given the low density of native trees relative to *Eucalyptus* in many overwintering groves (Griffiths 2012); however, restoration of overwintering sites with native tree species can take decades because many of California's native conifers are relatively slow-growing. Lane (1993) recommends that removal of *Eucalyptus* be done in phases while native trees are planted so that viable monarch habitat will be continually present.

Active management of overwintering sites is the best way to maintain or restore habitat for aggregating monarchs. In 1993, monarch researchers and conservationists completed a set of conservation and management guidelines (Bell et al. 1993). Since then, our understanding of monarch habitat requirements and grove restoration has advanced. Working with Dr. Stuart Weiss and others, we are updating these guidelines to include more up-to-date information on overwintering habitat requirements and management recommendations that include defining boundaries; monitoring monarchs and habitat conditions; assessing hazardous trees; and trimming, removing, or planting trees or flowers for nectar sources. After a peer review, these guidelines will be publicly available and will serve as an important tool for land managers and private landowners who are responsible for monarch overwintering sites.

In addition to general guidelines, we recommend that all significant overwintering sites have a site-specific management plan that provides guidance on tree trimming, hazardous tree removal, tree planting, trail placement, and augmentation of nectar sources. Monarch researchers (Stuart Weiss, Kingston Leong, John Dayton, and others) work with land managers to develop site-specific plans, and their approach and experience continue to inform the development of these guidelines. We are currently working with the U.S. Forest Service to develop a site-specific management plan for an overwintering site on the Los Padres National Forest in Monterey County.

Finally, we conduct outreach to agencies and individuals who manage overwintering sites to provide information on monarch habitat management. In recent years, the Society has partnered with Monarch Alert to provide workshops to land managers about monarch habitat requirements, habitat management, and population monitoring.

WESTERN MONARCH BREEDING HABITAT

Loss of milkweed habitat is considered a primary factor threatening monarchs in the eastern United

States (Pleasants and Oberhauser 2012), but the contribution of milkweed abundance, distribution, and quality to monarch population dynamics in the western United States has not been studied. Identifying the current distribution of milkweed that is used by monarchs is the first step to conserving breeding habitat.

Although numerous anecdotal observations of monarch larvae on wild milkweed exist from monarch enthusiasts, lepidopterists, and wildlife professionals, the most important monarch breeding areas in the western United States have not yet been identified. Stevens and Frey (2010) compared the distribution of milkweed to variation in climate suitability and proposed maps of the most likely areas in the western United States to support monarch breeding (including nearly all of California and Arizona, most of western and southern Nevada, and scattered areas of Utah, Oregon, southern Washington, and western Idaho). Pyle (1999, this volume, Chapter 21) reports monarchs breeding across Idaho as well as in many areas of eastern Washington.

To better understand the current distribution of monarch breeding habitat in the western United States, we created identification guides for milkweed species found in western states. Then, we developed and distributed a questionnaire to targeted audiences soliciting information about milkweed and its use by monarchs. The survey and identification guides have been distributed to state and federal resource agency staff, natural resource consultants, members of native plant societies, botanists, lepidopterists, and monarch enthusiasts. The responses are being mapped, along with existing records of milkweed and monarch larvae from herbaria, literature, unpublished reports, and personal communication with knowledgeable individuals. This map of monarch breeding areas will be used to inform breeding habitat assessments and conservation.

LEGAL STATUS AND PROTECTION OF WESTERN MONARCHS

In 1987, the California state legislature acknowledged the significance of preserving monarch overwintering sites and passed Assembly Bill #1671 (Bell et al. 1993). The following year, Californians passed a Parks and Wildlife Bond Act, Proposition 70, which allocated $2 million for monarch habitat site acquisition (Allen and Snow 1993). Despite these early legislative efforts, monarchs and their habitat receive little legal protection in many cases, and the protection they do receive is not widely understood.

To better understand the legal status of monarchs and their overwintering sites in California and to provide tools to advocates who seek to protect them, we collaborated with the International Environmental Law Project (IELP) at Lewis and Clark Law School to analyze relevant existing laws and policies: federal law and management plans relating to national parks, national forests, and federal military bases; laws and management plans for California state parks; city and county ordinances applicable to public and private land; and Local Coastal Programs (LCPs) developed by cities and counties to protect biological resources under the authority of the California Coastal Act (IELP 2012). We developed model legislation for municipalities that wish to protect monarchs and monarch habitat (IELP 2012), based on existing ordinances that provide strong protection for monarchs or their habitat. This analysis has also informed efforts of advocates working to protect monarch overwintering habitat. In addition to the model legislation, our work with the IELP addressed three questions, summarized below.

Does legislation protect monarch overwintering sites from irreversible damage?

The level of protection afforded to monarch overwintering habitat depends extensively on landownership and location. Overwintering habitat on land managed by most federal or state agencies is generally protected from development but not from alterations that can occur as part of routine tree maintenance activities. On military land, however, the protection of overwintering habitat is less certain; in some cases, management plans protect monarch roost sites only when it is consistent with the mission of the agency.

Monarch habitat in parks managed by cities or counties is generally protected from development by prohibitions against vegetation removal, although habitat is generally not protected from development on land they manage that is not in parks. Monarch habitat on any city or county land is generally not protected from inadvertent alteration due to routine management activities, such as tree trimming. Some cities have adopted monarch-specific guidelines for pruning trees, but they apply only when monarchs are present. These ordinances do not provide meaningful protection, since alterations at any time of the year can potentially make habitat unsuitable for monarchs.

Notable exceptions include Santa Cruz County and the cities of Goleta (in Santa Barbara County) and Capitola (in Santa Cruz County). Santa Cruz County requires a permit and consultation with a qualified monarch expert prior to tree removal and trimming in monarch groves. Goleta and Capitola prohibit vegetation removal at overwintering sites and expressly consider and prevent impacts to monarchs in any development activity. Goleta contains the Ellwood sites, which are among the largest in California. The city has an integrated general plan and Local Coastal Program that designates overwintering sites as Environmentally Sensitive Habitat Areas (ESHAs), thus providing protection "against any significant disruption of habitat values" (IELP 2012). Prior to any development where there is probable cause to believe that monarch overwintering habitat may exist, a site-specific study must be conducted; if it is found that monarchs use that site, the site and a 50-foot buffer are protected from development.

Capitola prohibits the removal of "monarch trees" year-round, except when there are safety concerns, and then removal is permitted only after consultation with a monarch expert. Capitola also requires that microclimatic data be gathered before and three years after any construction project that occurs at one of the monarch overwintering sites within the city to determine whether the project resulted in microclimatic changes to the overwintering site.

The California Coastal Act and the associated Local Coastal Programs developed by individual cities and counties may provide the strongest protection for some overwintering sites. The California Coastal Act implements the Federal Coastal Zone Management Act, meant to encourage protection and enhancement of coastal zones, and conservation of terrestrial and marine species and their habitats. The California Coastal Act applies to habitat and species within the Coastal Zone. Coastal counties and cities are required to develop Local Coastal Programs that must be approved by the California Coastal Commission. Once a city or county LCP is approved, its regulations apply to all developments in the Zone. Otherwise, the California Coastal Act applies. Some LCPs include specific protection for monarch overwintering sites during the winter; for example, Santa Barbara County designates specific overwintering sites as ESHAs; however, most LCPs do not.

Additional protection for monarch habitat within California's Coastal Zone could be obtained if the California Coastal Commission designated as ESHAs all significant overwintering sites that do not currently fall into existing, approved LCPs. Alternatively, an amendment to the California Coastal Act could designate all significant monarch overwintering sites as ESHAs. Some protection of monarch habitat may be obtained under the California Environmental Quality Act (CEQA). This law requires that environmental impacts be considered and alternatives proposed if a project has the potential to substantially reduce the population or range of a wildlife species, or substantially reduce its habitat. In cases where a state, county, or municipal project could harm monarchs or monarch habitat, an Environmental Impact Review is required under the CEQA. This review must assess the project's environmental impacts, identify mitigation measures for significant effects, and consider project alternatives.

Does legislation protect monarchs from overcollection?

Collection of monarchs in California is not tracked, so the scale of this practice is unknown. However, clusters at roost sites may be vulnerable to overcollection because the butterflies aggregate in such high densities. An active commercial monarch production industry exists in the western United States, and monarchs are probably collected on a regular basis from the wild to augment breeding stock, although a representative of the International Butterfly Breeders Association reports that monarchs are collected in the spring, rather than the winter, for this purpose (Dale McClung, pers. comm.). Federal (nonmilitary) and state agencies generally require permits to collect monarchs on land they manage, and there is limited access to land managed by the military. Cities and counties generally do not require permits for monarch collection, although a few city ordinances—such as one in Pacific Grove—ban the collection of monarchs. Monarchs on private land are not protected from overcollection. The California Department of Fish and Wildlife currently provides no protection to monarchs, although it may be within their mission to do so.

Does legislation require management of monarch habitat through enhancement, restoration, or other means?

The trees in many overwintering sites are old or diseased, and they require active management to maintain conditions both appropriate for monarchs and safe for visiting humans. In 2004, a limb from a diseased tree within the Pacific Grove monarch sanctuary fell on a visitor and killed her. Her family subsequently sued the city and was awarded a settlement of $1 million (Chawkins 2010). Very few jurisdictions require restoration or active management of monarch overwintering habitat. One notable exception is Santa Cruz County, which requires restoration of monarch habitat in its LCP.

CONCLUSION

California hosts the only large monarch overwintering aggregations in the United States, yet this unique phenomenon may be threatened. The Xerces Society is working with partners to better understand the population status and distribution of overwintering monarchs in the West, the condition of overwintering habitat, and the distribution of breeding habitat. We are trying to better understand fall and winter monarch habitat conditions and requirements to help land managers conserve, enhance, and restore overwintering sites, and are developing tools to help advocates protect monarchs and their habitat.

Although efforts to understand, protect, and restore monarch overwintering habitat are underway, important questions remain that, when answered, can better inform conservation efforts. The role played by loss of overwintering and breeding habitat in monarch population dynamics in the western United States is largely unknown, on both broad and site-specific scales. While something is known about habitat requirements of monarchs at California overwintering sites, an increased understanding of the factors that limit their survival could contribute to habitat restoration efforts by directly informing management actions that benefit monarchs. Furthermore, a solid understanding of the location of monarch breeding habitat in the western United States, coupled with increased knowledge of the habitat characteristics most important to monarchs, will ultimately allow conservationists and land managers to target specific breeding locations for conservation and enhancement measures.

ACKNOWLEDGMENTS

The Monarch Joint Venture, Bay and Paul Foundation, Hind Foundation, Strong Foundation, and Xerces Society members have funded our western monarch conservation work since 2010.

We appreciate the assistance provided by Candace Fallon, Katy Gray, and Ashley Minnerath in the preparation of this chapter. We are grateful to Mía Monroe, Carly Voight, Jen Zarnoch, Brianna Borders, Dennis Frey, Jessica Griffiths, Francis Villablanca, Stuart Weiss, Kingston Leong, Walt Sakai, Dan Meade, John Dayton, Chris Nagano, Jim Chu, and numerous Western Monarch Thanksgiving Count volunteers for their contributions to the conservation work described in this chapter. Karen Oberhauser, Bob Pyle, and John Pleasants reviewed the chapter and provided invaluable comments.

13

Threats to the Availability of Overwintering Habitat in the Monarch Butterfly Biosphere Reserve

Land Use and Climate Change

M. Isabel Ramírez, Cuauhtémoc Sáenz-Romero, Gerald Rehfeldt, and Lidia Salas-Canela

The availability of overwintering sites in the Monarch Butterfly Biosphere Reserve (MBBR) depends on the conservation of oyamel fir (*Abies religiosa*) forests. These forests are threatened by land cover changes and by the decoupling of oyamel populations from the climate to which they are adapted, because of climate change. From 1986 to 2012, a total of 4300 hectares of conserved oyamel fir forest were lost or disturbed (8% of the MBBR), mostly as a result of anthropogenic activities. The average annual loss rates varied from 0 to 2.4%. The most intense disturbances occurred right after the expansion of the protected area, peaking from 2000 to 2003. Decoupling of suitable climate and oyamel populations is predicted to reduce the area of suitable climatic habitat for oyamels by 50% by 2030, and completely eliminate it by 2090. There are forest damage by pests and disease, likely related to drought stress brought on by climate change. Future availability of the oyamel fir forests requires avoiding further land cover change, and finding alternatives that mitigate the impacts of climate change, such as planting oyamel in sites in which suitable climate is predicted for the future.

INTRODUCTION

Monarchs and oyamel firs

Monarch populations overwintering in Mexico prefer cool and humid, well-protected mountainous habitats. Fred Urquhart (1976, p. 173) emphasized the association between climate and the insect's selection of forests in the Transversal Volcanic System (TVS) in central Mexico: "I'm convinced that the monarch's selection of the Sierra Madre [actually the TVS] for overwintering is no random choice. Butterflies are poikilotherms, that is, creatures that adjust their body temperatures to the ambient air. At this 9000-feet elevation, winter temperatures hover from just below freezing to just above. Ideal for monarchs! Inactivated by the chill, they burn up almost none of the reserve fat they'll need on their northward flight." The high-elevation location of the overwintering sites results in the cool, humid climate that promotes monarch survival in many ways (summarized by Williams and Brower, this volume, Chapter 9).

In terms of habitat, overwintering monarchs prefer forests dominated by the conifer *Abies religiosa* (sacred fir or oyamel fir), a large tree with a dense crown resembling that of the Douglas fir (*Pseudotsuga menziesii*). Oyamel firs are restricted to high elevations (2400–3600 m) in the TVS (lat 19–20°N), where the highest mountains of Mexico are located. Here, overwintering monarchs are sheltered by dense, mature populations of *A. religiosa* in the Monarch Butterfly Biosphere Reserve (MBBR), located between the states of Michoacán and México in central Mexico (Anderson and Brower 1996). The butterflies minimize mortality during cold and sometimes rainy winter nights by clustering in densely packed colonies, thus taking advantage of the umbrella, blanket, and hot-water bottle effects provided by the forest canopy and branches and trunks of oyamel firs (Anderson and Brower

1996; Brower et al. 2009; Williams and Brower, this volume, Chapter 9). Consequently, in order to protect monarch overwintering, we must also protect mature oyamel stands within the MBBR and in other sites where monarch colonies are established (Slayback et al. 2007).

Oyamel fir management: Conservation, harvest, and restoration

The area hosting overwintering monarchs in Mexico is a Natural Protected Area (NPA) that falls under the UNESCO category of Biosphere Reserve (Man and Biosphere Program, UNESCO 2010). Mexican environmental legislation (Ley General de Equilibrio Ecológico y Protección al Ambiente: LGEEPA) defines two types of zones in Biosphere Reserves:

1. The core zone that guarantees the preservation of ecosystem components and the environmental services they provide, in which activities are restricted to research and low-scale ecotourism; and
2. The buffer zone, protecting the core zone through research, environmental education, and application of sustainable land use management (LGEEPA 2007).

The TVS oyamel forests have a long history of use and extraction, beginning in the sixteenth century when indigenous Otomí and Mazahua people arrived in the region. By the late nineteenth century, mining companies had obtained large tracts of forestlands, extracting wood for fuel, mines, and railway construction. After the Mexican Revolution ended in 1920, these forestlands were divided into communal properties under the legislative communal land-use categories of Ejidos and Indigenous Communities; however, from 1920 to 2000, governmental forest policies were ineffective in leading the productive capabilities of communal land managers toward sustainable forest management (Boyer 2005). During that time, logging bans were in place, but a lack of enforcement infrastructure led to high levels of illegal logging. At the same time, logging concessions were common, despite their negative impact on forest structure and composition (Merino and Hernández 2004).

In 1986, an area of 16,110 ha was set aside as an NPA, with a logging ban established within the 4490 ha core zone (DOF 1986). In 2000, the federal government expanded the reserve to 56,259 ha and established a logging ban across 13,551 ha of the expanded core zone (DOF 2000). In 2002, nongovernmental organizations (NGOs) and federal and state Mexican environmental authorities created the Monarch Butterfly Conservation Fund (Missrie and Nelson 2007) to compensate forest owners within the core zone in exchange for their renunciation of existing logging licenses. An additional payment was granted to forest owners to support conservation activities within the core zone, irrespective of their possession of logging licenses (Missrie and Nelson 2007; Honey-Rosés et al. 2009).

After the 1986 decree, government and nongovernment agencies promoted actions aimed at regaining forest cover, and these activities were further increased after the 2000 decree. Reforestation has been the main forest conservation activity, followed by soil conservation, fencing, and maintenance efforts; however, the absence of any systematic record or monitoring strategy to assess the effectiveness of these efforts is unfortunate. Between 2003 and 2009, nearly 700 ha of recovering forest stands were identified within the core zone, but it is unknown whether the recovery resulted from reforestation or natural regeneration (Venegas-Pérez et al. 2011).

One factor that encouraged the revision of the 1986 MBBR decree was its ineffectiveness in protecting the ecological integrity of oyamel stands and the monarch colonies they shelter (Brower et al. 2002; Ramírez et al. 2003). Most of the forest disturbance was caused by illegal logging and (to a lesser extent) by authorized timber harvest, forest fires, domestic timber extraction, agricultural clearing, and natural forest loss (Honey-Rosés 2009a). From 1993 to 2006, forest disturbance was detected in 5239 ha of the reserve area located in the state of Michoacán, 61% of which was explained by illegal logging, 33% by extemporaneous authorized extractions (extraction of a volume of commercial wood previously approved by authorities, but after the authorized time frame had expired), and only 6% by authorized logging executed during approved periods (Navarrete et al. 2011).

Pressures on oyamel fir stands by climate change

Conservation and reforestation efforts to mitigate forest disturbance problems inside the MBBR have been attempted by land users, reserve authorities, several governmental agencies (from municipal to state and federal levels), and a number of NGOs; taking the projected negative effects of climate change into consideration, existing efforts might not be enough to preserve monarchs' overwintering habitat. A study using seven climatic models with scenarios of greenhouse-gas emissions predicted a temperature increase of 3.7 °C and a precipitation decrease of 18.2% as average for Mexico by the end of the twenty-first century (Sáenz-Romero et al. 2010). These changes imply that trees will suffer increased heat and drought stress, and thus reduce natural defenses against insects and disease. Recent studies in many parts of the world document the massive death of trees from causes related to stress induced by climate change, such as seen in *Pinus edulis* (pinyon pine) in the southwestern United States (Breshears et al. 2005), *Populus tremuloides* (aspen) in the Rocky Mountains in the United States (Worrall et al. 2008) and Canada (Hogg et al. 2002), and several other species worldwide (Allen et al. 2010).

The process of forest decline due to climate change may eventually affect the Mexican oyamel forests. The nearly exclusive preference by overwintering monarchs for the shelter offered by these forests and their microclimatic features poses a challenge to monarchs' continued use of the MBBR. Researchers inside and outside the MBBR have observed a recent increase in the number of oyamel trees with signs of dieback and defoliation, likely caused by drought stress associated with climate change (Plate 12) (Flores-Nieves et al. 2011).

The objectives of this study are to (1) assess the reduction of oyamel fir habitat caused by land cover changes since the first MBBR protection decree was issued in 1986, (2) estimate the current and future climate in oyamel forests inside the MBBR in order to reveal the degree of current and future decoupling between fir populations and their suitable climate, and (3) suggest priority sites for reforestation, taking into account both contemporary and predicted climatic conditions.

METHODS

Forest cover change from 1986 to 2012

To identify changes in forest area and general condition between 1986 and 2012, we used a map series generated for a long-term land cover monitoring project of the MBBR and its surroundings. These maps show land cover for the years 1986, 1993, 2000, 2003, and 2006, and were generated at a 1:75,000 scale. We used a systematic method based on visual interpretation of multiresolution imagery and high-resolution photographs, as well as field validation of imagery observations. The study area comprises the 56,259 ha of the MBBR plus 286,993 ha surrounding it. The procedure for generating the maps is described by Ramírez and Zubieta (2005), Ramírez et al. (2007), and Navarrete et al. (2011).

To update the map series for 2009 and 2012, we used Landsat ETM+ images 27/46–47 from March 2009 and January 2012 (USGS 2012d), SPOT images from January 2009 (ERMEX 2009) and January 2012 (ERMEX 2012), and very-high resolution GeoEye images from March 2010 (ORBIMAGE Inc. 2010). On-screen visual interpretation and map analysis were carried out using ArcGis 9.0 (ESRI 2004).

In an initial analysis, we extracted the polygons of dense forest cover (canopy cover > 70%, indicative of well-preserved forests) from all maps. We then calculated the dense forest net change within the MBBR core and buffer zones and their surroundings. The rate of transformation of dense forest was estimated using the formula proposed by the FAO (1996):

$$t = \left(1 - \frac{S_1 - S_2}{S_1}\right)^{1/n} - 1$$

where: t = rate of transformation; S_1 = dense forest surface at time 1; S_2 = dense forest surface at time 2; and n = number of years between times 1 and 2. The result was multiplied by 100 to obtain the percent rate of transformation.

In the next analysis, we overlaid six pairs of maps (1986–1993, 1993–2000, 2000–2003, 2003–2006, 2006–2009, 2009–2012) to identify land cover change processes. We constructed transition matrices for each period and identified the following four land cover change processes (sensu Ramírez et al. 2003; Velázquez et al. 2003): (1) Deforestation: elimination of forest cover implying a transformation from forest to nonforest land uses; (2) Disturbance: reduction in the percent cover of the forest canopy to levels <70%; (3) Recovery: restoration of dense woodland on previously disturbed areas; and 4) Revegetation:

establishment of secondary vegetation as a result of farmland abandonment (Chokkalingam and DeJong 2001; FAO 2001). We quantified and mapped both positive (recovery and revegetation) and negative (deforestation and disturbance) processes.

Contemporary and future climate for sites with presence of oyamel firs

To assess the potential changes in climate for oyamel fir sites in the MBBR, we obtained climate estimates (Sáenz-Romero et al. 2010; Crookston 2012) for 42 field sites used by Ramírez (2001) and 9 used by Sáenz-Romero et al. (2012) from the Mexican National Forest Inventory. We considered climate trends from the contemporary period (defined as the average of the 1961–1990 climate normals), the average of the decade surrounding 2030 (average of the 2026–2035 climate), and the decades surrounding 2060 and 2090 for mean annual temperature and precipitation. Future climates are presented as the mean for three General Circulation Models (GCM) and two greenhouse-gas emission scenarios for each GCM (3 GCM × 2 emission scenarios = 6 model scenarios; Crookston 2012).

The GCM and emission scenarios used were the following: (1) Canadian Center for Climate Modeling and Analysis (CCC) for emission scenarios A2 and B1; (2) Hadley Centre (HAD) for emission scenarios A2 and B2; and (3) Geophysical Fluid Dynamics Laboratory (GFD) for emission scenarios A2 and B2. In general, the A2 scenario reflects unrestrained carbon emissions (a "business as usual" or "pessimistic" scenario), while the B1 and B2 scenarios incorporate a substantial reduction of emissions resulting from fast technological change from fossil-fuel energy generation to zero-emission technologies ("optimistic" scenarios). The methodological details, downscaling techniques, and spline modeling are described in detail by Rehfeldt (2006) and Sáenz-Romero et al. (2010); for an explanation of emission scenarios see IPCC (2000). Weather station monthly records, climatic splines, grids, and instructions for interrogating the spline climatic model are available at Crookston (2012).

We also use the Random Forests classification tree of Sáenz-Romero et al. (2012) to map the habitat within and surrounding the MBBR to be suitable for the fir in the contemporary climate and for the decade surrounding 2030. Briefly, this technique develops a bioclimatic model that predicts what is the climatic space (combination of intervals of values of a set of climatic variables found relevant) where a given species can occur (named suitable climatic habitat), by comparing the climate of current locations where the species occurs against the climate in which a species is not found. This suitable climatic habitat can be viewed as the climatic niche of the species, the geographic region with a unique combination of climatic variables that allow a species to persist (Rehfeldt et al. 2006). Because this model was developed for predicting the presence-absence of oyamel fir throughout its geographic distribution, we were uncertain about its ability to represent the portion of the distribution specifically in the MBBR. To provide assurance that the model was suitable for our purposes, we used it to predict the probability of the contemporary climate being suitable for oyamel fir in the 42 sites containing oyamel fir visited by Ramírez (2001). Because these sites were not used to build the model, this exercise can be considered one of validation. In interpreting predictions by a bioclimatic model, it is important to realize that the model predicts where the climate may be suitable for a species, but not necessarily where the species actually occurs. Typically, the predicted geographic space encompasses an area larger than the actual distribution, largely because a species can be absent in suitable climatic space for other reasons, such as limitations of seed dispersion, soil conditions, or competition (Zonneveld et al. 2009).

In making predictions from a Random Forest algorithm (R software, version 2.15.1; R Development Core Team 2004), "votes" are cast as to whether the climate of an observation should be suitable. For the oyamel fir model, the algorithm casts 2500 "votes." In our validation exercise, the 42 oyamel sites received on average 94% of the votes. Because a simple majority of the votes is commonly used as a threshold for determining climate suitability, these results support the validity of the model and imply that it is suited to practical use at the scale of MBBR.

Priority sites for reforestation

Priority sites for reforestation were selected by superimposing a grid of the forest cover in 2012 dominated by oyamel fir over the grids of the predicted probability of the presence of suitable climatic habitat for oyamel trees in 2030 (grids developed by

Sáenz-Romero et al. 2012). This allowed us to identify sites in which the following two conditions are met: (1) oyamel fir forest is absent in 2012 as a result of deforestation and other human-related or natural disturbances over the past 26 years, and (2) the climate is likely (probability > 0.50) to be suitable for oyamel fir trees in the year 2030. We chose 2030 as the target horizon to realign natural oyamel populations to the climate for which their genotypes are adapted.

RESULTS

Forest loss from 1986 to 2012

In 1986, conserved forest (canopy cover ≥ 70%) comprised 69,640 ha within the study area, with just more than half (36,670 ha) located in the mountainous area included within the MBBR boundaries established in 2000. The other half was distributed in the surrounding area, also called the area of influence. Since 1986, conserved forest area has decreased at rates between 100 and 1000 ha per year, with a total affected area of 6900 ha. More than 4300 ha of this affected area were oyamel forests, with 43% in the core, 56% in the buffer, and only 1% in the area of influence.

Both positive and negative changes have occurred within the core, buffer, and area of influence zones of the MBBR (Figure 13.1). An increase in the surface area of forests was observed during the period 1986–1993, almost all within the area of influence, but 920 hectares of oyamel fir forest suffered some degree of perturbation during this period. The 2000–2003 period showed the largest loss of oyamel forest cover, coinciding with the decree expanding the reserve. During these three years, almost 3000 ha of forests were affected; two-thirds of these were inside the MBBR, where 1000 ha were oyamel fir forest (Figure 13.1). Between 2003 and 2006, and even more so between 2006 and 2009, the loss of dense forest cover was reduced, but not eliminated, in the entire study area; however, extraordinarily strong storms at the beginning of 2010 caused landslides and mudflows that affected 74 ha of forest within the reserve (Carranza et al. 2010). In addition, strong winds knocked down numerous trees, which were later removed (Figure 13.2). As a result, from 2009 to 2012, 620 ha of oyamel fir–dense forest were partially or totally cleared.

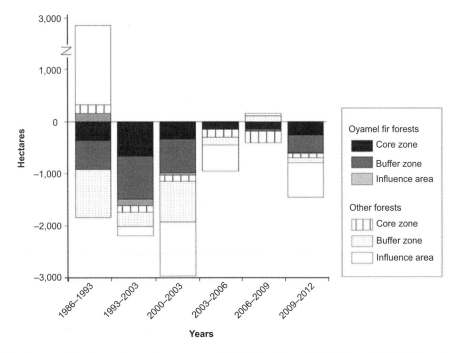

Figure 13.1. Conserved forest cover change (canopy cover ≥70%) by forest type within the Monarch Butterfly Biosphere Reserve, core and buffer zones, and its area of influence.

Figure 13.2. a) Exposed roots of trees felled by storms during 2010 (circles) and logs cut for their extraction. b) Truck inside the core zone transporting logs with a sign stating, "Extraction of trees felled by meteorological phenomena authorized with permit No. DFMARNAT/0713/2010."

Between 2000 and 2003, the annual loss rate of oyamel fir forest peaked in the core (2% per year) and buffer (2.4% per year) zones (Table 13.1). This rate decreased from 2003 to 2009 and then increased between 2009 and 2012.

During all periods, the dominant process of change was the loss of dense tree cover through disturbance, which occurred both within the MBBR and in surrounding areas (Table 13.2). The opposite process, recovery, has also been extensive; however, the area in which dense forest has been reestablished is just a little more than half the disturbed area (22,165 ha disturbed vs. 13,324 ha reestablished, Table 13.2). Regarding the other opposing processes of deforestation and revegetation, deforestation has occurred on a surface twice as large as

Table 13.1. Average annual rates (%) of conserved forest loss (negative values) or gain (positive values) from 1986 to 2012 in the core, buffer, and influence zones of the MBBR

Conserved forests	1986–1993	1993–2000	2000–2003	2003–2006	2006–2009	2009–2012
Oyamel fir forest						
Core zone	−0.8	−1.6	−2.0	−0.8	−0.9	−1.7
Buffer zone	−0.8	−1.2	−2.4	−0.1	−0.2	−1.3
Area of influence	0.9	−0.7	−0.6	0.0	0.1	−0.2
Other forests						
Core zone	0.6	−0.5	−1.0	−1.4	−2.0	−0.8
Buffer zone	−0.9	−0.3	−1.9	−0.4	0.3	−0.3
Area of influence	1.0	0.0	−1.0	−0.5	0.0	−0.7

Table 13.2. Forest cover change processes within the MBBR core and buffer zones and its area of influence, 1986–2012

MBBR zones	Deforestation (ha)	Disturbance (ha)	Recovery (ha)	Revegetation (ha)
1986–1993				
Core	3	826	637	12
Buffer	155	2496	1263	101
Influence	346	2409	6003	107
1993–2000				
Core	13	1330	201	8
Buffer	168	2148	930	58
Influence	341	2675	2663	203
2000–2003				
Core	17	784	26	
Buffer	88	1950	55	4
Influence	343	1764	202	2
2003–2006				
Core	17	358	39	
Buffer	29	275	116	5
Influence	56	675	186	15
2006–2009				
Core	83	456	152	36
Buffer	86	341	481	140
Influence	1415	1864	350	477
2009–2012				
Core	13	427	18	3
Buffer	40	520	3	4
Influence	3	868	0	0
TOTAL 1986–2012	3217	22,165	13,324	1176

revegetation (3217 ha vs. 1176 ha, respectively); in both cases, the largest areas involved are in the area of influence surrounding the MBBR (Table 13.2). The temporal and spatial patterns of the different processes indicate that they are associated with land property boundaries, and hence with community decisions that lead to forest preservation or disturbance (Plate 11).

Contemporary and future climate suitable for the presence of oyamel fir trees

Estimates of contemporary mean annual temperature and precipitation (obtained from the spline climatic model) averaged across the 52 MBBR sites that include oyamel firs were 11.1 °C and 1023 mm, respectively. On average, future predictions of the six model-emission scenarios anticipated an increase of mean annual temperature of 1.4, 2.2, and 3.7 °C for the decades centered in the years 2030, 2060, and 2090, respectively (Figure 13.3a). The variation among model scenarios is due mostly to greenhouse-gas emission scenarios rather than to differences between different General Circulation Models (GCM). For example, the annual temperature increment for 2090 predicted for the GCM Geophysical Fluid Dynamics (optimistic emission scenario B1) is only 2.3 °C, but the most pessimistic one (the Hadley GCM scenario A2) is 5.0 °C.

Precipitation is predicted to decrease, although there is high variation among model-scenarios (Figure 13.3b). The averages of the six model-emission scenarios of the predicted percent change in precipitation compared with the current climate (average from 1961–1990) were −7.3%, −8.1%, and −14.4% for 2030, 2060, and 2090, respectively. For 2090, the Hadley B2 scenario shows a change of only −3.8%, but an extraordinarily large (and potentially devastating) change of −40.7% was predicted for the Canadian A2 scenario.

The decoupling of oyamel trees from their suitable climate range inside the MBBR is evident when mean average temperature is plotted against mean annual precipitation in the 52 sites using both contemporary and year 2090 predicted climate (averaging estimations across the six model scenarios) (Figure 13.4).

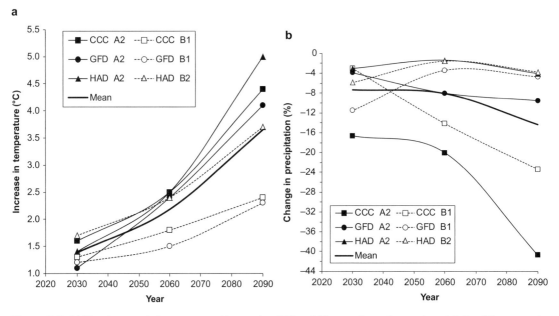

Figure 13.3. (a) Mean increment of average annual temperature (°C), and (b) mean change in annual precipitation (%), compared with contemporary climate (1961–1990), averaged from 52 study sites inside the MBBR, using global climate models from the Canadian Center for Climate Modeling and Analysis (CCC, greenhouse-gas emission scenarios A2 and B1), Hadley Center (HAD, scenarios A2 and B2), and Geophysical Fluid Dynamics Laboratory (GFD, scenarios A2 and B1), for decades centered in 2030, 2060, and 2090.

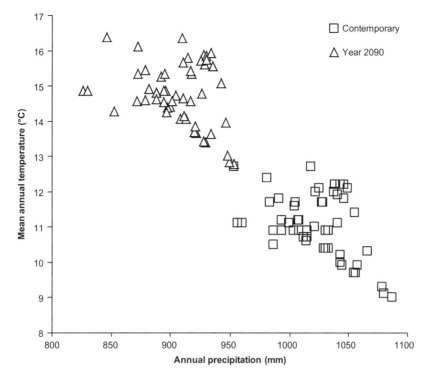

Figure 13.4. Mean annual temperature plotted against mean annual precipitation for the contemporary climate (1961–1990) and for the decade surrounding the year 2090, as estimated from the average of six emission scenario projections, for 52 study sites inside the MBBR.

Priority sites for reforestation

Superimposing the predicted geographic distribution of suitable oyamel climatic habitat by the year 2030 over contemporary oyamel forest cover indicates that several sites to the east of the MBBR will maintain a climate suitable for the species (Figure 13.5). Currently, most of this area is devoted to farmland, offering an opportunity to work with local owners to revert the area to forests and address multiple environmental goals, including monarch habitat conservation, carbon sequestration, water capture, and soil erosion management.

DISCUSSION

The creation of the NPA in 1986 aggravated logging within the reserve (Hoth 1995; Brower et al. 2002), and evidence mounts of a cause and effect explanation. The favored reason is that lack of both adequate planning and consultation with landowners, as well as lack of a comprehensive strategy before implementation (Merino and Hernández 2004; Ibarra 2011), resulted in "panic deforestation" (Fernando Rosete, pers. comm. 2007). Trees were cut quickly, before federal authorities began enforcing the decree. It was expected that the change of legal status to Biosphere Reserve in the year 2000, which expanded the legally protected area, would reverse the process of degradation; however, our data show that the most intense disturbance in the MBBR took place over the first three years following this change, from 2000 to 2003. Several authors argue that extensive disturbances often result from social conflict (Honey-Rosés 2009a; Honey-Rosés 2009b; Alatorre 2007) while others argue that the public policies for forest management and conservation have not been linked to the needs and capabilities of communities, thus encouraging conflicts inside and outside communities (Merino and Hernández 2004; Ibarra 2011). As a result, nearly 60% of disturbance is due to illegal logging and more than 30% to authorized forestry practices that

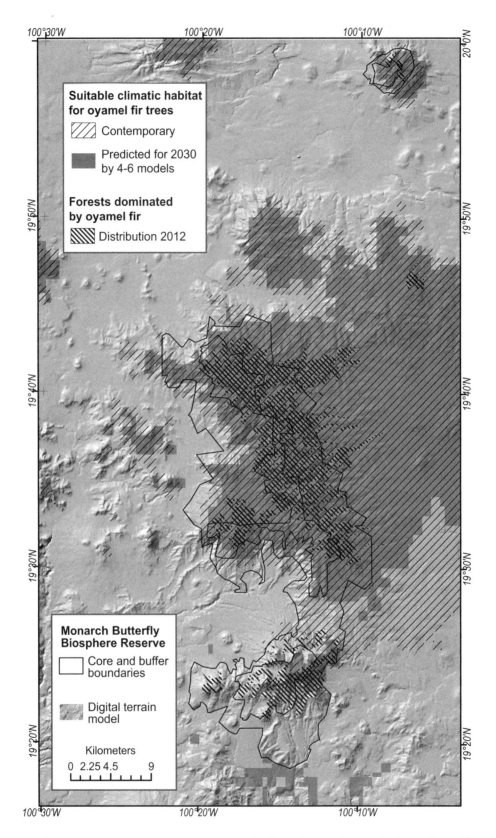

Figure 13.5. Contemporary and predicted suitable climate habitat for *Abies religiosa* and current distribution of oyamel fir as dominant forests. Predicted suitable habitat for 2030 is consensus among 4 or more model scenarios with a probability of presence of oyamel fir > 0.5.

do not strictly follow the management program (Navarrete et al. 2011).

The threat of disturbance in the MBBR is common to mountainous areas that host monarch colonies outside the reserve (Slayback et al. 2007). Franco et al. (2006) applied a methodology comparable to ours in a study in the Nevado de Toluca National Park, documenting an annual disturbance rate of 0.8% between 1972 and 2000. In the Izta-Popo National Park, Galicia and García-Romero (2007) estimated an average deforestation rate of 0.35% from 1970 to 2000. Such disturbances, on top of possible habitat transformation due to climate change, threaten the survival of monarch migration in Mexico (Oberhauser and Peterson 2003). Hence, there is a real need to monitor and analyze both land use change and climatic change in ecologically important areas (Slaymaker 2001).

Decoupling of oyamel firs from the climate for which they are adapted

The predicted changes in temperature and precipitation for current oyamel habitat over most decades are comparable to those predicted for Mexico on average (Sáenz-Romero et al. 2010), with the exception of the very strong decrease in precipitation predicted by the Canadian A2 scenario for the MBBR (−40.7 %, Figure 13.3b). The latter decrease is much higher than the one predicted by the same model scenario over all of Mexico (−28.5 %; Sáenz-Romero et al. 2010). In all models, the contemporary and the predicted climates do not overlap in climatic space by the end of the century; by year 2090 the area with climate suitable for oyamels is likely to be absent in the MBBR (Figure 13.4 and Saénz-Romero et al. 2012).

The mechanistic causes of the effects of climate change on the oyamel forests are likely to be severe stress on the trees, caused by a combination of increased temperature and decreased precipitation. This stress will make them more susceptible to insect attacks and diseases, and eventual death (Plate 12). The belief is widespread that pests and disease kill trees (Anon. 2009; Martínez 2009), but it is likely that such causes are secondary. The likely primary cause is climate-related stress, which lowers tree defenses, leading to pest outbreaks (Sturrock et al. 2011).

Oberhauser and Peterson (2003) studied the potential impacts of climate change on the microclimatic requirements of overwintering monarchs, and thus focused on winter climate (January means). Their climate models predicted little change in winter temperature, and an increase in precipitation. These results are not in conflict with ours, because we modeled climate using all 12 monthly means and included 2 General Circulation Models that they did not consider (Canadian, which predicts an important decrease in precipitation, and Geophysical Fluid Dynamics).

Priority sites for reforestation

Areas that do not currently contain oyamels but are predicted to have suitable climatic habitat for oyamel fir trees in the year 2030 should be priorities for oyamel fir reforestation (Figure 13.5). These areas are distributed mainly to the east of the MBBR in the state of México. Considering that suitable climatic habitat for *A. religiosa* is predicted to occur at progressively higher altitudes, we suggest that reforestation at the established priority sites use seeds collected from sites approximately 300 m lower in elevation. This strategy of shifting seed movement upward would compensate for climate change between the 1961–1990 period (contemporary) and projections for 2030 (for a detailed account of the 300 m estimate, see Sáenz-Romero et al. 2012).

The decreasing area of the predicted suitable climatic habitat inside the MBBR forces us to consider extension of the NPA into new areas where oyamels are expected to be viable. MBBR core zone boundaries were delimited based on the historical presence of overwintering monarch colonies (Bojórquez-Tapia et al. 2003). Oyamel reforestation efforts outside the MBBR that aim to realign the trees with their predicted climatic habitat would also aim to create future overwintering sites, but this action does not guarantee that monarchs will arrive in those places; we are unsure whether the navigation system of monarchs is flexible enough to change a genetically programmed destination (Zhan et al. 2011); however, we can hope that monarchs selected the sites because of the microclimate found in oyamel forests rather than the presence of the tree itself. If these conditions can be offered by other large-sized forest tree species, the chance that monarchs will "accept" the change of overwinter locations would be greater, even if oyamels are absent.

Either way, we encourage an active forest management strategy that aims to preserve and create novel locations of dense populations of healthy oyamel trees for sheltering overwintering monarchs in a future climate. This is surely a better option than merely witnessing the decline of the forests caused by anthropogenic stresses like land use and by climate change. Doing nothing to accommodate climate change is perhaps the worst option.

ACKNOWLEDGMENTS

Financial support was provided by Monarch Butterfly Fund; also to MIR and LSC by PAPIIT-UNAM (grant IN301411); and to CSR by CONACYT–State of Michoacán (grant 2009-127128), University of Michoacán scientific research grant, and SEP-PIFI-2009 fund. LSC received a scholarship from CONACYT. We thank Nicholas Crookston (USDA-Forest Service, Moscow, ID) for technical support; Karen Oberhauser, Myron Zalucki, and an anonymous reviewer for helpful suggestions on a previous version of the document; Dan Slayback (Science Systems & Applications, NASA) for providing GeoEye 2010 images; Jorge Carranza (CONANP-SEMARNAT) and José Antonio Navarrete (CIGA-UNAM) for providing SPOT 2009 and SPOT 2012 imagery, respectively.

14

Monarch Butterflies and Agriculture

John M. Pleasants

More than a decade ago, agricultural fields in the U.S. Upper Midwest accounted for a high proportion of milkweeds and a large fraction of new monarchs produced during the summer months. During the past 15 years, two types of genetically modified (GM) crop plants, Bt corn and glyphosate-tolerant corn and soybeans, have rapidly dominated agricultural acreage in the Midwest. Initial concerns about the consumption of Bt corn pollen by monarchs were laid to rest but then replaced by newer concerns about glyphosate-tolerant crops. The use of glyphosate has virtually eliminated milkweeds from agricultural fields. From 1999 to 2012, the amount of milkweed on the Midwest landscape declined by 64% and Midwest monarch production declined by 88%. Over the same period, the number of overwintering monarchs declined by 72%. Milkweeds on Conservation Reserve Program land have now become an important resource for monarchs. I show that herbicide kills monarch eggs and larvae by defoliating plants. I also examine several reasons for the observed high density of monarch eggs on agricultural milkweeds and estimate the potential milkweed resource in presettlement times.

INTRODUCTION

The presence of milkweeds (*Asclepias syriaca*) in agricultural fields has concerned farmers and weed scientists because of their impact on crop yield (Bhowmik 1994). What was not appreciated until recently was the importance of milkweeds in agricultural fields for monarchs. This realization resulted from several studies addressing the question of whether Bt corn poses a threat to monarchs. Hartzler and Buhler (2000) measured milkweed densities in different habitats in Iowa in 1999. Oberhauser et al. (2001) and Pleasants and Oberhauser (2012) used these densities and the amount of area occupied by different habitats to estimate that in 1999, approximately half the milkweeds in the Upper Midwest were in agricultural fields. Additionally, they found a higher per-plant density of monarch eggs on milkweeds in agricultural fields than on milkweeds in nonagricultural habitats, and they estimated that about 80% of all monarchs in the Midwest fed as larvae on milkweeds in agricultural fields. Because approximately half of all monarchs that overwintered in Mexico in the late 1990s came from the corn belt region of the U.S. Midwest (Wassenaar and Hobson 1998), it is clear that milkweeds in agricultural fields are important for overall monarch production; in fact, over the last decade, the estimated number of monarchs produced in the Midwest correlated positively with the number of monarchs in the overwintering colonies in Mexico (Pleasants and Oberhauser 2012).

In 1996, two genetically modified (GM) crop plants with the potential to affect monarchs became available. One of these, Bt corn, contains a bacterial gene that encodes a crystal toxin protein; when incorporated into plant tissues, this protein inhibits feeding by lepidopteran larvae and targets the European corn borer in particular. The second GM crop, glyphosate-tolerant (brand name Roundup Ready) soybeans, contains a gene from a bacterium that confers tolerance to the herbicide glyphosate. Corn varieties with glyphosate tolerance have also been developed, and this trait is sometime stacked (combined into the same plant line) with the Bt trait.

GM crops and monarchs: Bt corn

In 1999, a laboratory study raised concerns about the potential negative effect of Bt corn pollen blown onto milkweed leaves and consumed by monarch larvae (Losey et al. 1999). At that time, 26% of corn acres were planted with Bt corn (USDA, Economic Research Service 2012). Subsequent field studies focused on the temporal and spatial overlap between monarch larvae and Bt corn pollen (Oberhauser et al. 2001), the amounts of Bt pollen to which monarchs might be exposed (Pleasants et al. 2001), and the sensitivity of the larvae to different amounts of Bt corn pollen on milkweed leaves both in the lab (Hellmich et al. 2001) and in the field (Stanley-Horn et al. 2001). Because corn pollen does not travel very far from fields (Pleasants et al. 2001; Kawashima et al. 2004), the primary danger of Bt corn pollen would be to monarch larvae feeding on milkweeds in or immediately adjacent to cornfields. Sears et al. (2001) combined all risk factors from these studies and concluded that the risk to monarchs was very small, primarily because the chance of encountering a pollen density on milkweed leaves that would reduce growth and survival was very low. In a follow-up study, Dively et al. (2004) examined monarch larvae exposed to naturally occurring Bt corn pollen in the field; they estimated that Bt pollen could result in 0.6% mortality among the monarch population as a whole, a result that could mean several hundred thousand negatively affected individuals, assuming a starting population in excess of 100 million (Garcia-Serrano et al. 2004). Their conclusion was based on a lower percentage of Bt corn plantings than exists today, suggesting that the risk today might be greater, but it was also calculated on the basis of the higher density of milkweed in agricultural fields in about 2000. The amount of milkweed in agricultural fields has been drastically reduced since that time by glyphosate herbicide (see below), eliminating concerns about Bt corn pollen.

GM crops and monarchs: Glyphosate-tolerant corn and soybeans

Weed control practices in agricultural fields that affect milkweeds will likely affect monarchs. Historically, weed control was accomplished primarily by cultivation or tillage, but since the 1940s the application of herbicides has become an important component. Traditional herbicide use (prior to crop emergence) produced only moderate control of milkweed, such that in the 1970s and 1980s, milkweed infestation in agricultural fields was viewed to be increasing, with 10.5 million ha infested in the north-central states (Martin and Burnside 1980). The more recent adoption of no-till farming practices has made milkweed even more difficult to control (Buhler et al. 1994; Yenish et al. 1997).

Glyphosate is a potent herbicide to which milkweeds are susceptible (Bhowmik 1994; Pline et al. 2000), but because it has a detrimental effect on conventional crop plants, it could not be applied after crop emergence until the development of GM glyphosate-tolerant crops. Glyphosate-tolerant (Roundup Ready) soybeans were introduced in 1996 and had reached a 93% adoption level by 2012. Corn with the glyphosate-tolerant trait was introduced in 1998 and had reached a 73% adoption level by 2012 (Figure 14.1) (USDA, Economic Research Service 2012). Glyphosate use in soybeans went from 2.2 million kg in 1994 to 41.7 million kg in 2006 (the last year for which data are available and when adoption of glyphosate-tolerant soybeans was 89%), and glyphosate use in corn went from 2.0 million kg in 2000 to 28.5 million kg in 2010 when the adoption level was 70% (USDA, National Agricultural Statistics Service 2011).

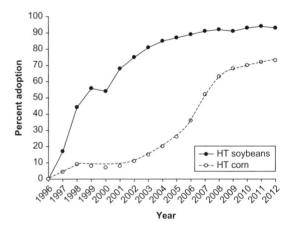

Figure 14.1. Percent adoption of corn and soybeans with the herbicide (glyphosate) tolerant (HT) trait since introduction, based on data from USDA, ERS (2012). Some varieties of HT corn also have the Bt trait (stacked) while some do not.

EFFECT OF HERBICIDE USE ON MILKWEED PLANTS

I began monitoring plots of milkweeds in six agricultural fields, between 10 and 51 ha in size, near Ames, Iowa, in 2000 and continued monitoring those plots for several years, using GPS to relocate them (see Pleasants and Oberhauser 2012 for details). This period covered the transition from nonglyphosate herbicide use to glyphosate herbicide use in conjunction with increased adoption of Roundup Ready soybeans, and later Roundup Ready corn, and thus provided an opportunity to compare the response of milkweeds to these two weed control treatments. From 2000 to 2003, I monitored each of the sites weekly for monarch activity during the study periods. In each plot (of approximately 20 m^2 in size), every milkweed stem was counted and inspected for monarch eggs and larvae, which were identified to instar. From 2004 to 2009, monarch activity was not monitored but the number of milkweed stems was counted once each year in early August.

In general, whether nonglyphosate or glyphosate herbicide was used, the leaves on treated milkweed plants turned yellow within a few days of application, and then often turned brown, dried up, and fell off within a week. When nonglyphosate herbicide was used, some plants did not recover and the entire stem turned brown, but many plants sprouted new branches from leaf axils, and they appeared fully recovered in 2–3 weeks. When glyphosate herbicide was used, most plants were killed; those that survived had little if any resprouting from leaf axils.

For cornfields, the effect of nonglyphosate herbicide application on milkweed plants typically involved a reduction in the number of stems 1–2 weeks after herbicide application, followed by an increase in stems over the next 1–2 weeks to the same level or greater than just before application (Figure 14.2a). The reduction in stem number was due to stem death or apparent death, and the subsequent increase was due to stems that recovered and produced large side branches that were counted as new stems, along with the appearance of new stems by vegetative reproduction. The effect of glyphosate herbicide use in cornfields could not be evaluated because this herbicide was not applied in the cornfields examined here. In soybean fields treated with nonglyphosate herbicide, the milkweeds recovered, as seen with corn, but in fields treated with glyphosate there was no recovery and only a small number of stems survived (Figure 14.2b).

Besides exploring the effect of herbicide use within a season on the number of milkweed stems, I also examined between-year changes in number of milkweed stems. No seedling recruitment occurred in any of the fields I examined, such that between-year changes were due to stems arising from previously existing plants. I compared the number of milkweed stems in each field near the end of one season (first week in August) with the number of stems at the same time the next year after an intervening herbicide application (Table 14.1). To examine the effect of different herbicide treatments, I used a standard least-squares linear model with treatment ("type of crop (corn or soybean)" and "herbicide" combination) and "year" as main effects, and "site" and "site by plot" as random effects. The treatment effect was significant, but the year effect was not, so the model was rerun with the treatment effect only. The treatment effect (crop and herbicide combination) was statistically significant ($F = 28.4$, df = 3,14, $P < 0.001$) and all four treatments were different from each other (Tukey HSD at 0.05 level). For soybeans, the greatest decline in milkweed numbers occurred in fields sprayed with glyphosate, whereas milkweed numbers declined proportionately less in fields sprayed with nonglyphosate herbicide (Table 14.1). For cornfields sprayed with nonglyphosate herbicide, the decline in milkweed was quite low, and in the two fields where corn was planted in consecutive years, milkweed numbers actually increased (Table 14.1). For comparison, the number of milkweed stems in nonagricultural plots did not change appreciably from one year to the next (Pleasants unpublished data).

Based on these results, we can speculate about milkweed populations in agricultural fields prior to glyphosate use. The data are lacking to consider how tillage alone has affected milkweed populations, but for the majority of fields that formerly used nonglyphosate herbicide, we would expect a moderate negative effect on milkweed populations, with a greater negative effect in soybean fields than in cornfields. With corn-soybean rotation, the corn year would have allowed the milkweed population to decline less rapidly; however, with the introduction of glyphosate-tolerant soybeans and the consequent use of glyphosate herbicide, the decline of milkweeds in soybean fields would have accelerated.

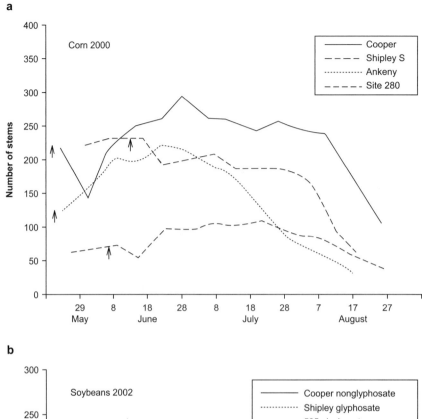
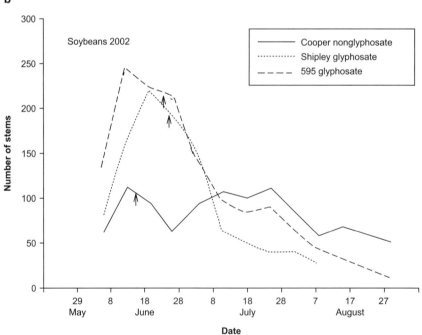

Figure 14.2. Milkweed stem numbers in fields before and after herbicide application. Dates when herbicide was applied to each site are indicated by arrows. (a) Change in the number of milkweed stems in cornfields at 4 sites in 2000. All sites were treated with a nonglyphosate herbicide. The Ankeny site showed no reduction in numbers after application, probably because the application was ineffective for unknown reasons. (b) Change in the number of milkweed stems in soybean fields at 3 sites in 2002. Two fields, Shipley North and Site 595, were sprayed with glyphosate herbicide whereas the Cooper site was sprayed with a nonglyphosate herbicide. Cases where it appears that milkweed numbers were declining before herbicide application are an artifact of application occurring between weekly sampling intervals.

Table 14.1. Percent change in number of milkweed stems for type of crop field and type of herbicide used

Site and year	Nonglyphosate			Glyphosate
	Corn	Corn[a]	Beans	Beans
Shipley N.				
2001	−40.1			
2002	−37.3			−90.0
2003		25.0		−74.7
Site 595				
2001	−16.4			
2002				−85.1
Cooper's				
2001	−14.3		−41.8	
2002	−9.4		−67.8	
2003	−17.3		−22.2	
2004	9.5		−40.3	
2005	9.3	43.5		
Airport Rd.				
2009				−87.7
Average	−14.5	34.3	−43.0	−84.4

[a] Field also in corn the previous year, all other fields in crop rotation

After Roundup Ready corn was introduced, every year would have produced a significant decline in milkweeds, resulting in a precipitous decline in milkweed numbers. All the plots I examined beginning in 2000, comprising a total of about 1000 stems, were devoid of milkweeds by 2008 after experiencing a combination of glyphosate and nonglyphosate use over the years. A similar loss of milkweeds was noted in Minnesota fields (Oberhauser unpublished data).

EFFECT OF HERBICIDE USE ON MONARCHS

Herbicide application, either glyphosate or nonglyphosate, can have two potential effects on monarch populations. First, monarch eggs on milkweed plants at the time of herbicide application will die if defoliation occurs, and larvae will die if they are unable to move to unaffected plants. Second, when milkweeds disappear because they are killed by herbicide, it reduces the milkweed resource available to monarchs, thus potentially reducing the size of the monarch population that can be supported.

Figure 14.3 shows monarch population egg phenology (average eggs per milkweed stem in corn and soybean fields) over four seasons at study sites near Ames, Iowa. The shapes of the phenology curves for larvae per stem are very similar, although the position is shifted to the right. The phenology curves show the two main generations of monarchs that occur in Iowa and the Midwest in general, with one generation of eggs laid in early summer (May–June) and a second (and possibly third overlapping) generation of eggs laid in July and August (Figure 14.3). In the cornfields in my study, nonglyphosate herbicide was applied from 21 May to 8 June (C ng in Figure 14.3), coinciding with first-generation eggs and larvae. Today, the majority of corn is glyphosate-tolerant. The time frame for the first application of glyphosate herbicide in glyphosate-tolerant cornfields is from 5 May to 30 May and a second application is often made 14–21 days later, as early as 19 May and as late as 20 June (C g in Figure 14.3) (ISU Weed Specialist Micheal Owen, pers. comm.). This coincides with the first monarch generation. Thus, both nonglyphosate and glyphosate herbicide application in corn appear to closely overlap the first generation of eggs and larvae and could result in monarch mortality.

In the soybean fields examined here, nonglyphosate herbicide was applied between 15 June and 23 June (B ng in Figure 14.3). Very few eggs or larvae were present at this time in three of the four years, but first-generation eggs and larvae were still present in 2003. Glyphosate herbicide was applied in late June, a period of low numbers of eggs and larvae at all sites. However, the potential time frame for the application of glyphosate herbicide to soybeans is 20 May to 10 June, with a second application sometimes two or three weeks later, as early as 3 June and as late as 1 July (B g in Figure 14.3) (Micheal Owen, pers. comm.). Thus, as in corn, the timing of glyphosate herbicide application potentially overlaps the first generation of monarchs, resulting in mortality.

I could directly document monarch deaths resulting from the defoliation caused by herbicide in 2000, 2002, and 2003, when I monitored monarch activity. I measured the survival of 13 cohorts of eggs that were present just before herbicide application based on the number of second-instar (L2) larvae present in the next census following herbicide treatment. I used cohorts in fields in which either nonglyphosate or glyphosate herbicide was used because both produced an initial defoliation. I compared their survival with survival of 13 cohorts that had not experienced herbicide treatment, choosing treatment and non-treatment cohorts to control for the type of crop field and weekly interval during the season. The herbicide-treatment cohorts were from 6 cornfields and 7 soybean fields, and the

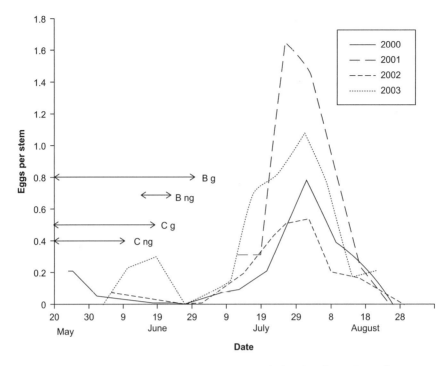

Figure 14.3. Dates when herbicide is likely to be sprayed relative to monarch phenology. B = soybeans, C = corn, ng = nonglyphosate herbicide, g = glyphosate herbicide. Monarch phenologies for 4 seasons are shown, each based on average eggs per stem over all study sites per year.

non-treatment cohorts were from 8 cornfields and 5 soybean fields. For each sampling interval from 23 May to 10 July, there were as many treatment cohorts as non-treatment cohorts. For the 157 eggs in the herbicide-treatment cohorts, 3 L2 larvae were found in the next census (1.9%). For the 144 eggs in the non-treatment cohorts, 25 L2 larvae were found in the next census (17.4%). The difference is statistically significant ($P < 0.001$, one-tailed Fisher's exact test based on the prediction that survival would be lower with herbicide application).

Pleasants and Oberhauser (2012) examined the effect of glyphosate herbicide use on milkweed and monarch populations in the U.S. Upper Midwest from 1999 to 2010. We used two surveys of milkweed densities in Iowa, one from 1999 (Hartzler and Buhler 2000) and the other from 2009 (Hartzler 2010) to gauge how milkweed densities had changed. These densities were converted to the area occupied by milkweeds in those habitats across Iowa using USDA land-use information. I have extended that analysis through 2012 and corrected a mistake in Pleasants and Oberhauser (2012). In that paper we used for our analyses the milkweed densities for different habitats given in the tables of Hartzler and Buhler (2000) and Hartzler (2010); however, those densities were only for sample plots that had milkweeds. To calculate the overall density of milkweeds, we should also have included the zeroes from plots that had no milkweeds. The table in Hartzler (2010) shows that the proportion of crop fields infested with milkweeds went from 51% in 1999 to 8% in 2009 and that milkweed density in infested fields dropped from 23 (average in corn and soybeans) to 5 m^2/ha. Thus the overall crop-field milkweed density should be 11.73 m^2/ha for 1999 and 0.4 m^2/ha for 2009, a decline of 96.5%. I have now redone tables 1 and 3 from Pleasants and Oberhauser (2012) with the corrected milkweed density values for different habitats. Based on this analysis I estimate that between 1999 and 2012, 98.7% of milkweeds in agricultural fields were eliminated, and that there has been a 64% reduction in milkweeds on the Iowa landscape, 72% of the loss occurring in agricultural fields. This loss is probably representative of the loss of milkweeds throughout the agriculturally intense landscapes of the Midwest as a whole (Pleasants and Oberhauser 2012); however, because monarch egg densities are about 4 times higher per milkweed stem in agricultural fields relative to nonagricultural habitats (Pleasants and Oberhauser 2012), the loss of agricultural milkweeds represents an even greater loss of potential resource than the 64% reduction in milkweeds would indicate.

Yearly monarch production in the Midwest can be estimated by combining an indicator of the number of milkweed stems on the landscape each year (density in m²/ha) with the number of monarch eggs per milkweed stem, obtained from the Monarch Larva Monitoring Project (MLMP 2013; see Pleasants and Oberhauser 2012 for assumptions used in this estimate). Figure 14.4 shows Midwest monarch production estimates for 1999 through 2012. Based on the regression equation it is estimated that an 88% reduction in Midwest monarch production has occurred over this period. A decline is also seen in the size of the overwintering population (Brower et al. 2012a). Based on the overwintering numbers from 1999 to 2012 (Rendón-Salinas and Tavera-Alonso 2013), I performed a regression that indicated a population size decline of 72%. The similarity in the percent decline in Midwest monarch production and the percent decline in the size of the overwintering population suggests the two are related. In addition, Pleasants and Oberhauser (2012) found a strong correlation between Midwest production of monarchs in each year and the size of

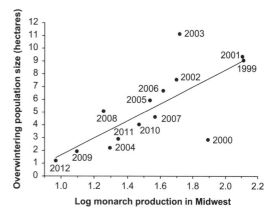

Figure 14.5. Monarch production in the Midwest vs. size of the overwintering population. Regression statistics: $F_{1,13} = 14.89$, $P = 0.002$, $r^2 = 0.55$, $y = -5.02 + 6.67x$.

the overwintering population, suggesting a causal relationship between loss of milkweeds, declining midwestern monarch production, and a decline in the overall monarch population. Extending that analysis through 2012 further strengthens the correlation (Figure 14.5). While other factors such as loss of wintering habitat and persistent drought in Texas probably have contributed to the monarch population decline, the loss of milkweeds in the Midwest appears to be the primary cause.

MILKWEED ABUNDANCE PRIOR TO AGRICULTURE

Extensive agricultural fields in the Midwest were not present before European settlement, so it is of interest to speculate on the amount of milkweed that might have been available to monarchs in pre-settlement times and whether the size of the monarch population that could be supported then was larger or smaller than today. No historical records exist that would provide such information, so we must rely on indirect estimates. Prairie remnants contain a vestige of pre-settlement vegetation, and the abundance of milkweed on these sites could be used to estimate pre-settlement abundance. White (1983) surveyed a number of prairie remnants in Iowa and quantified the percent cover of all plant species present according to plant community type. I used soil types as a proxy for community types (White 1983) and estimated the area occupied by soil types in Iowa (Oschwald et al. 1965). I then used

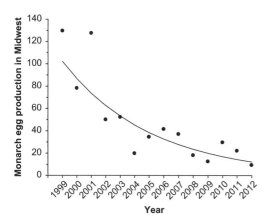

Figure 14.4. Estimates of monarch numbers produced in the Midwest since 1999. The units for egg production are relative; they are the product of milkweed density in amount of cover (m² per hectare) and eggs per milkweed plant (see Pleasants and Oberhauser 2012 for details). The regression line is based on the log of production: $F_{1,13} = 34.35$, $P = 0.00008$, $r^2 = 0.74$, $y = 2.08 - 0.071x$ where $x = 1$ for 1999.

information on the percent cover of milkweed species in each community type to estimate milkweed cover values for the area of the state that was formerly prairie and is now in farmland.

White (1983) found three milkweed species in the remnant prairies: *A. viridiflora, A. verticillata,* and *A. tuberosa* (note the absence of *A. syriaca*). I estimated the overall percent cover of these three species as follows: *A. viridiflora* 0.11%; *A. verticillata* 0.22%; and *A. tuberosa* 0.32% (combined cover 0.65%). These data can be compared with estimates of the percent cover of the most common milkweed on the landscape, *A. syriaca,* using data from Pleasants and Oberhauser (2012), Hartzler and Buhler (2000), and Hartzler (2010). Milkweed is found in agricultural fields, CRP (Conservation Reserve Program) land, pastures, and roadsides. The percent cover of *A. syriaca* has decreased in all these habitats, except along roadsides, since 1999. In agricultural fields this reduction has been caused by glyphosate use; in the other habitats the reasons are less clear (Pleasants and Oberhauser 2012). In agricultural fields, percent cover of *A. syriaca* was 0.11% in 1999 but had declined to 0.0015% by 2012. The percent cover on CRP land has gone from 1.42% to 0.71% and on pastures, 0.04% to 0.02%. For roadsides, the percent cover has remained fairly constant at 0.76% (Hartzler 2010). Combining percent cover information with the amounts of land in each of these categories on the Iowa landscape, I estimate that the overall percent cover of *A. syriaca* went from 0.194% in 1999 to 0.068% in 2012.

It appears that the milkweed percent cover in 1999 was about a third of that based on historical prairie remnants, and current percent cover is about 10% of historical estimates. Interpreting these data with regard to the availability of milkweed resources for monarchs and the size of the monarch population during "pre-European settlement" versus "today" requires some caution. The data from prairie remnants may not be representative of the presettlement prairie vegetation in Iowa. The nine remnants surveyed by White (1983) include the largest ones in the state (68–97 ha), but the locations do not provide a representative coverage of the state. Another complicating factor is that in terms of their resource potential for monarchs, it is difficult to compare *A. viridiflora, A. verticillata,* and *A. tuberosa,* the three milkweed species found in the vegetation surveys of remnants, with *A. syriaca,* the predominant species used by monarchs today (Seiber et al. 1986). Monarchs lay eggs on *A. viridiflora, A. verticillata,* and *A. tuberosa* and larvae can complete their lives on these species (Orley Taylor, pers. comm.), but they might differ from *A. syriaca* in their attractiveness to ovipositing females and larval survival (Ladner and Altizer 2005) owing to differences in cardenolide and latex content (Zalucki et al. 1990; Zalucki and Malcolm 1999; Sternberg et al. 2012; Agrawal et al., this volume, Chapter 4). We also know that *A. syriaca* plants in agricultural fields have higher egg densities than *A. syriaca* plants in nonagricultural areas.

A. syriaca undoubtedly increased in abundance after European settlement as a result of the disturbance associated with agriculture. The primary pre-settlement habitat for midwestern *A. syriaca* was probably excavated soil around badger mounds or other animal burrows; it is rare in undisturbed prairie (Platt 1975) but is well adapted for colonizing disturbed areas (Wilbur 1976). In the eastern United States, milkweed abundance is certainly higher today than it was pre-European settlement as a result of clearing of forests for agriculture and habitation, and creation of disturbance habitat (Brower 1995); thus, milkweeds have likely been added in the eastern United States, while their numbers in the Midwest have remained similar or more recently decreased. How abundant were monarchs in pre- or early settlement times? Brower (1995) points out anecdotal observations from the 1850s and 1860s indicating large numbers of fall migrating monarchs, but there is no way to quantify these observations to compare them with current monarch numbers.

EXPLANATIONS FOR HIGHER EGG DENSITY ON AGRICULTURAL MILKWEEDS

One reason for the importance of agricultural milkweeds to monarchs is the higher density of eggs per milkweed stem in agricultural fields compared with stems in nonagricultural habitats. Several potential explanations can be given for the difference in egg density that are not mutually exclusive: egg survival might be higher in agricultural habitats because of less predation or more favorable microclimate; ovipositing monarchs might be more attracted from a distance to milkweeds in agricultural fields; or females might be more likely to lay eggs on plants in agricultural fields once they find them. With regard to egg survival, Oberhauser et al. (2001) found that early-instar survival was higher in corn than in nonagricultural habitats (survival in soybeans was not systematically examined). With regard to overall attraction, Zalucki and Suzuki (1987) found that monarch females preferred ovipositing in small patches and on isolated plants, similar to what is found in agricultural fields (J. Pleasants, pers. observ.). In addition, the ability to find milkweed plants in monoculture crop fields could be greater than in mixed vegetation habitats because of the lower structural and chemical diversity of the vegetation (Floater and Zalucki 2000; Jactel et al. 2011).

With regard to the greater likelihood of oviposition on agricultural milkweeds once the plants are found, females might be responding to higher leaf quality. Milkweed plants in cornfields are shaded by the corn canopy, as are the lower halves of plants in soybean fields. Agrawal et al. (2012a) found that in shaded habitats *A. syriaca* leaves were larger, less tough, and had lower cardenolide content and lower induced latex production, possibly increasing their quality for monarch larvae (Oyeyele and Zalucki 1990). Also, egg densities on milkweeds with young or resprouted leaves tend to be higher than on old leaves (Zalucki and Kitching 1982a), so resprouting after nonglyphosate herbicide application may also be partly responsible for the higher egg densities on milkweeds in agricultural fields. Some butterflies oviposit more on leaves with higher nitrogen (Letourneau and Fox 1989; Prudic et al. 2005), although a study involving monarchs found that leaf nitrogen concentration did not influence oviposition (Oyeyele and Zalucki 1990), and plant nitrogen levels do not appear to affect monarch growth, since larvae are able to compensate for lower plant nitrogen by eating more (Lavoie and Oberhauser 2004). In a preliminary study in late July 2002, I analyzed nitrogen levels of leaves of milkweed plants growing in one cornfield, one soybean field, and several nonagricultural areas (two roadsides and one pasture) (9–12 leaves per site) using a CE Flash 1112 autoanalyzer (CE Elantech, Inc.). There were significant differences in percent nitrogen among the corn and soybean fields and the nonagricultural sites ($F = 29.29$, df = 2,51, $P < 0.001$). There was a significantly higher percent nitrogen in leaves of milkweeds from soybean fields (mean − 4.8%) than cornfields (mean = 3.9%) and both were higher than nonagricultural areas (mean = 2.7%) (t-tests significant for all with $P < 0.006$ or smaller; significance based on Bonferroni corrected alpha value criterion of 0.017). Although Oyeyele and Zalucki (1990) did not find an oviposition preference with leaf nitrogen, their highest observed nitrogen level was below the mean levels for corn and soybeans.

CONCLUSIONS

Agriculture has clearly played, and continues to play, a significant role in monarch population biology, largely through its effect on the abundance of milkweeds. Historically agriculture in the eastern United States increased the abundance of *A. syriaca* over its original abundance, and agriculture in the Midwest also increased the abundance of *A. syriaca*, compensating, or perhaps more than compensating, for the loss of other milkweed species in the native prairie that it replaced. Within the last decade, however, the use of glyphosate herbicide, made possible by the widespread adoption of genetically modified corn and soybeans, has drastically reduced milkweed abundance in the Midwest. Milkweed abundance sets a ceiling on the size of the monarch population that can be supported, and that ceiling has been lowered substantially. Year-to-year variation in monarch population size is influenced by many factors, most notably weather conditions, but that variation plays out under declining food resources, as seen in Figure 14.4. The overall result is a decline in the size of the monarch population (Brower et al. 2012a; Pleasants and Oberhauser 2012), and this lower ceiling reduces the extent to which the monarch population can rebound from adverse events.

With the disappearance of milkweeds from agricultural fields, milkweeds present in other habitats become more important for monarch populations. The habitat of greatest importance is CRP land, potential agricultural land that has been set aside and planted with a cover of grasses and forbs. In Iowa, 59% of available milkweeds are now found in CRP land. This land is part of the agricultural system. The amount of CRP land is determined by government policies and the relative economic value of taking the government subsidy versus growing crops. In 2012 the amount of CRP land for the Midwestern states (North and South Dakota,

Nebraska, Kansas, Minnesota, Iowa, Missouri, Wisconsin, Illinois, Michigan, Indiana, Ohio) had declined by 1.56 million hectares from its high in 2007 of 7.12 million hectares (USDA, Conservation Programs 2013). With the lowering of the ceiling brought about by the disappearance of milkweed in crop fields, it is important that further lowering not occur through the loss of CRP land. It is possible, however, to enhance the value of CRP land for monarchs as well as other species such as pollinators through appropriate management practices (e.g., Baum and Mueller, this volume, Chapter 17). The next most important habitat is roadsides, accounting for 36% of milkweeds in Iowa. Department of Transportation practices, such as the timing of mowing and herbicide spraying, need to be examined for their effect on monarch production.

ACKNOWLEDGMENTS

I thank Karen Oberhauser, Chip Taylor, and Lincoln Brower for discussions and data sharing. The data on monarch egg densities used in the analyses of the relationship between overwintering numbers and monarch production in the U.S. Upper Midwest were collected by volunteers in the Monarch Larva Monitoring Project.

15

Fires and Fire Management in the Monarch Butterfly Biosphere Reserve

Héctor Martínez-Torres, Mariana Cantú-Fernández, M. Isabel Ramírez, and Diego R. Pérez-Salicrup

Fire is a common disturbance that affects most terrestrial ecosystems and has become a high priority in forest management in Mexico. Current practices of fire management, based mainly on firefighting and fire suppression, have had negative effects on many forests worldwide. To avoid such effects, it is important to provide forest management alternatives based on both sound ecological research and social feasibility. In this chapter we document the forest fires that occurred during 2012 in the Monarch Butterfly Biosphere Reserve (MBBR) and evaluate current knowledge of fire regimes in monarch overwintering forests. Based on information and interviews with the authorities responsible for fighting fires in the MBBR in the states of Michoacán and México, we present preliminary summaries of current uses of fire by local inhabitants of the reserve, and describe the current forest management strategy followed by authorities. We identify important gaps in our understanding of the role of fires in the reserve, and provide recommendations for new fire management strategies for local authorities, forest managers, and conservation organizations.

INTRODUCTION

Since the discovery in the 1970s of the monarch migration phenomenon from the United States and Canada to the mountains of central Mexico (Urquhart 1976), forests of the overwintering sites have been the target of international conservation attention. Significant efforts have been made to preserve these forests by federal, state, and municipal authorities; environmental groups and other nongovernmental organizations (NGOs); researchers; local inhabitants; and other stakeholders. One of the most important events in the efforts to secure the integrity of these forests was the decree that established a Natural Protected Area (NPA) in 1980, after which two more decrees were issued, resulting in today's Monarch Butterfly Biosphere Reserve (MBBR). Today, the MBBR covers 56,259 ha, with 13,552 ha in three core zones where only conservation, environmental education, and research activities are allowed. The remaining 47,707 ha are part of the two buffer zones where forest management, centered on timber and resin extraction, is permitted. Recently the MBBR was designated a World Heritage Site (DOF 1980, 1986, 2000, 2009; UNESCO 2013).

Conservation efforts in the MBBR are based on the premise that well-preserved forests are needed to maintain the monarch populations that overwinter year after year; therefore, the paucity of research on the effects of human and natural disturbances on these forests is surprising (Rendón-Salinas et al. 2007). Research on disturbances both natural and caused by humans, as well as their effects on forest structure and biological composition, is very important. It is necessary to generate concrete recommendations to guide forest management and conservation practices, and to ensure those recommendations are compatible with a forest structure and species composition that supports monarch colonies and, at the same time, meets forest users' demands.

One of the most important disturbances that occurs in coniferous forests worldwide is fire (Pyne 1996; Rowell and Moore 1999). Fires have influenced terrestrial ecosystems for millions of years, and for many ecosystems, fire is a fundamental process that triggers regeneration of the arboreal community (Agee 1993; Pausas and Keeley 2009).The effects of fires differ drastically across ecosystems, and ecosystems might differ substantially in their dependence on fire as an ecosystem process. Fires, like other disturbances, can be characterized for a given ecosystem in terms of their return intervals (frequency), the energy they liberate (intensity), their impact on vegetation (severity), their spatial extent (area affected), the season of the year in which they occur (timing), the spatial pattern they present (patchiness), and their synergies with other known natural or human disturbances. The ranges of values in which these variables naturally occur in a given ecosystem constitute the Natural Fire Regime (Agee 1993; Jardel 2010). Fire has also been used by humans for hundreds of thousands of years, an interaction that has created an extensive body of empirical knowledge among some cultural groups, who have used and manipulated fire in sophisticated ways for usages as diverse as agriculture, livestock production, and even forestry (Pyne 1996, 2010; Pausas and Keeley 2009).

It is important to remember that fire can cause social, economic, and ecological disasters. As a consequence of those negative effects, the general perception of fires was mostly negative during most of the twentieth century (Pyne 2010), and that perception can explain, at least in the United States, the development of fire policies focused on suppression (Egan 2009). Perhaps the most important outcome of these policies was the suppression of natural fire regimes, particularly in western coniferous forests, which had negative consequences for forest management that are still being felt. One potential negative consequence of fire suppression is the accumulation of forest fuels that can promote catastrophic fire events, with intensities and severities much higher than those to which species might be adapted. Another consequence is the modification of some ecosystem functions provided by naturally occurring fires, such as seed scarification and the opening of adequate clearings where tree seedlings can establish.

To avoid catastrophic fires and promote a fire management strategy that can be incorporated into a strong forest management scheme, it is fundamental to understand the natural fire regimes in the dominant forest types within the MBBR. At the same time, it is important to understand the fire risks associated with activities conducted by inhabitants of the 63 legally established settlements within the reserve. These activities include income-generating activities such as forest management and extensive (as opposed to intensive) livestock production, and activities meant for the direct support of families, such as small-scale agriculture and the gathering of fruits, mushrooms, and firewood.

Here, we document and synthesize information on natural fire regimes and fire management in the dominant coniferous forests in the MBBR. We address four specific questions: (1) What do we know about fire regimes in coniferous forests in the MBBR? (2) What is the importance of fires as perceived by local authorities? (3) How do local people currently use fire? and (4) What is the institutional response to wildfires? Answers to these questions should help us identify research priorities that will improve our understanding of the role of fire in the reserve.

METHODS

We conducted a literature review of natural fire regimes in the different coniferous forest types in the MBBR. Natural fire regimes can be described in relation to variables such as average return interval, severity, intensity, affected area, and synergies with other disturbances (Agee 1993). Few studies have been conducted of fire within the MBBR, so we used data generated for similar forests, particularly for those dominated by *Pinus* spp. (Minnich et al. 2000; Park 2003; Rodríguez-Trejo and Fulé 2003; Jardel et al. 2004) and by *Abies* spp. (Ángeles-Cervantes and López-Mata 2009).

We also searched for relevant information on fires in the reserve in scientific publications, technical reports, and government files. We searched the online Database on Fire Regimes and Fuels for Mexican Terrestrial Ecosystems (www.zotero.org/fuego_mex), a compilation resulting from a systematic search conducted by the National Autonomous University of Mexico, University of Guadalajara, and University of Washington. We also searched Thompson Reuters Web of Science and Google Scholar using the key words fire, *Pinus*, *Abies religiosa*, and Monarch Butterfly Biosphere Reserve. These last searches were conducted in August 2012. We requested information about fires that occurred in 2012 from the Natural Protected Areas National Commission (CONANP-MBBR; for a list of acronyms, see the glossary at the end of this chapter) and in the offices of the National Forestry Commission in the states of Michoacán and México (CONAFOR-Michoacán and CONAFOR-Estado de México). With these data, we were able to identify and visit 14 sites that burned in the MBBR in 2012, and we conducted a general evaluation of the size, intensity, severity, and source of ignition of these fires. Data on the size of fires were generated by walking around the fire boundaries, making a polygon with a GPS unit, and then estimating the area. Intensity was inferred from apparent flame length, as estimated from charred tree trunks and burnt vegetation. Fire intensity can be divided into surface fires (flames < 1 m), understory fires (flames 1–3 m), and crown fires (flames > 3 m) (Agee 1993; Sugihara et al. 2006). Severity was also estimated by visually evaluating vegetation mortality. Low severity implied that only herbaceous plants died, while tree mortality would indicate high severity. The source of ignition was informed by members of the firefighting brigades (see below). We also inquired about fires for previous years but found that information on fires was not systematically collected until 2012.

We conducted four semistructured interviews to understand the importance of fires in the perception of federal, state, and municipal authorities with an interest in the MBBR, and to identify the institutional response to fires. Our interviewees included the fire manager of the MBBR (Ing. Cesar Torres), the former head of the Michoacán State Forestry Commission's (COFOM) East Region IV (Ing. Estanislao Esquivel), the fire chief of the Forest Protection Agency of the State of México (PROBOSQUE; Ing. José Mendez), and the Fire Department chiefs of CONAFOR-Michoacán (Ing. Javier González) and CONAFOR-México (Ing. Martín Tapia). We also obtained information from open interviews with the person in charge of coordinating the fire department of CONAFOR-Michoacán East Region (Ing. Edilberto Sánchez), and the person in charge of the permanent firefighting brigade of CONANP-Michoacán in Zitácuaro (Mr. Ubaldo García). Finally, we attended the Sixth Gathering of the Restoration and Forest Protection East IV Committee, in which the headmaster of COFOM's East Region IV (Ing. Efraín Sánchez) explained the activities they conduct in regard to forest fires.

RESULTS

Natural and managed fire regimes

We found no literature sources that directly evaluate any of the components of natural fire regimes in coniferous forests of the MBBR, but we found information on similar vegetation types in other sites in Mexico. We also found anecdotal information about the effects of fires on forest structure, the use of fire for income-generating and self-sustenance activities by inhabitants of the MBBR, and about the ignition of fires as an action to intimidate other local communities (Table 15.1).

Coniferous forests in the reserve can be divided into pine- and fir-dominated forests (Cornejo-Tenorio et al. 2003, Giménez-Azcárate et al. 2003). Pine forests are dominated by *Pinus pseudostrobus* and *P. hartwegii* (smooth bark Mexican and Hartweg's pines, respectively). Fir forests are dominated by *Abies religiosa* (oyamel fir), although other pines and *Cupressus lusitanica* (Mexican cypress) trees also occur in fir forests. From studies conducted at other sites in Mexico, we infer that these kinds of forests have contrasting natural fire regimes. Pine forests have fire regimes characterized by return intervals of 5–10 years, with fires of low severity (i.e., low mortality of the dominant canopy vegetation as a result of the fire event) and low intensity (i.e., superficial fires with insufficient energy to ignite trees, though some tree bases might show charred bark), which usually cover 1–10 ha (Jardel et al. 2009, 2010; Fulé and Covington 1998). Fires in fir forests, on the other hand, have longer return intervals (>40 years), with high severity (i.e., many canopy trees die after a fire, creating gaps for a regenerating tree cohort) and high intensity (i.e., energy liberated by the fire ignites trees, either from the base, which are passive crown fires, or from one tree crown to the next, active crown fires), covering areas greater than 10 ha (Ángeles-Cervantes and López-Mata 2009). This contrast between fire regimes for pine- versus fir-dominated coniferous forests suggests that these forests should have different fire management strategies. It is particularly important to avoid severe and intense wildfires that could destroy the *Abies* forests, where the wintering

Table 15.1. Number of sources for each category of information found in the literature, and number and sources of anecdotal information

Information category	No. of references or interviewees	Reference source or interviewee's title
Fire in vegetation types similar to those at MBBR	9	Ángeles-Cervantes and López-Mata 2009; Fulé and Covington 1998; Jardel et al. 2004, 2009; Minnich et al. 2000; Park 2003; Rodríguez-Trejo and Fulé 2003; Rodríguez-Trejo and Myers 2010; Rodríguez-Trejo et al. 2011
Forest management in the MBBR	18	Brenner 2009; Brower et al. 2002; Byers 2004; Hoth 1995; Honey-Rosés 2009b; Honey-Rosés et al. 2009, 2011; Ibarra 2011; Martin 2002; Merino 1997; Merino and Hernández 2004; Navarrete et al 2011; Ramírez et al 2003; Rendón-Salinas et al. 2007; Sáenz et al. 2005; Sigala and Campos 2001; Tucker 2004; WWF 2004
Fire related laws and rules that apply to MBBR	6	CONAFOR 2012; CONANP 2001, 2011; Jardel 2010; Jardel et al. 2009; NOM-015-SEMARNAT/SAGARPA-2007
Anecdotal information	8	Fire manager of the MBBR (Ing. Cesar Torres); former head of Michoacán State Forest Commission's (COFOM) East Region IV (Ing. Estanislao Esquivel); chief of the Forest Protection Agency of the State of México (PROBOSQUE; Ing. José Mendez); Fire Department chief of CONAFOR-Michoacán (Ing. Javier González) and CONAFOR-State of México (Ing. Martín Tapia); Fire Department coordinator of CONAFOR-Michoacán East Region (Ing. Edilberto Sánchez); head of permanent firefighting brigade of CONANP-Michoacán in Zitácuaro (Mr. Ubaldo García); and headmaster of COFOM's East Region IV (Ing. Efraín Sánchez)

sites are located; however, a suppression policy could be counterproductive in the long term.

We found no information on the use of fire and fire management within the MBBR; however, documents about the social, political, and economic conflicts associated with forest management in the reserve often report fire as an issue of importance, although in most references only anecdotally. These studies analyze the processes by which different localities make common decisions, and the way they have historically managed the ecosystems, including how public policies have restricted their access to forest resources since the MBBR was created (Merino and Hernández 2004).

2012 fire season in the MBBR

According to the information provided by CONANP-MBBR and CONAFOR-Michoacán, 45 forest fires occurred inside the reserve during the 2012 season, from January to June. They covered a total area of 176.6 ha, or 0.31% of the total reserve area. Of these, 18 fires were in the state of México and affected 30.5 ha (40% of the fires and 17% of the burned area). In Michoacán, 27 forest fires affected 146.1 ha (60% of the fires and 83% of the burned area) (Table 15.2, Figure 15.1).

The 2012 fires affected areas in eight different municipalities. San José del Rincón had the highest fire incidence (11 fires), followed by Zitácuaro (10 fires) and Ocampo (9 fires). The smallest wildfire covered 0.4 ha and the largest, 41.4 ha. Five fires occurred inside the core zone and 38 in the buffer zone, and two could not be appropriately located in either one of these zones (i.e., data from the brigades were not accurate enough).

In regard to fire severity, we could gather information only for Michoacán, where all fires had low severities, i.e., fire-associated tree mortality was none to low. Sixteen fires affected pine forests (total area=107.5 ha), and 11 affected fir forests (total area=38.6 ha). Ten different ignition causes were reported (Table 15.2). The most frequent ignition sources were agricultural activities (10 fires, 22%), followed by campfires (8 fires, 18%), and forestry activities (7 fires, 16%). The remaining fires were ignited by miscellaneous human activities such as garbage burning and tourist campfires. It is interesting to point out that only one fire was reportedly caused by a lightning bolt (Table15.2).

Table 15.2. 2012 MBBR forest fire locations, extents, and causes

	State[a]	Municipality	Property	MBBR Zone	Ha	Fire Causes[b]
1	Mex	Temascalcingo	Cerrito de Cárdenas	Buffer	2.00	ND
2	Mex	San José del Rincón	Ejido El Depósito	Buffer	0.50	I
3	Mex	San José del Rincón	El Depósito	Buffer	1.00	II
4	Mex	San José del Rincón	Ejido la Mesa	Core	5.00	IX
5	Mex	San José del Rincón	Ejido La Trampa	Buffer	1.00	II
6	Mex	San José del Rincón	Ejido La Trampa	Buffer	0.50	II
7	Mex	San José del Rincón	Ejido La Trampa	Buffer	1.50	II
8	Mex	San José del Rincón	Ejido San Joaquín la Milla	Buffer	3.00	II
9	Mex	San José del Rincón	Ejido La Esperanza	Buffer	0.50	II
10	Mex	San José del Rincón	Ejido Jerónimo Lamilla	Buffer	3.00	II
11	Mex	San José del Rincón	Ejido El Depósito	Buffer	1.00	I
12	Mex	San José del Rincón	Ejido San Juan Palo Seco	Buffer	0.50	I
13	Mex	Villa de Allende	Ejido El Aventurero	Buffer	2.00	VI
14	Mex	Villa de Allende	Ejido El Aventurero	Buffer	1.00	I
15	Mex	Villa de Allende	Ejido El Aventurero	Buffer	1.00	ND
16	Mex	Villa de Allende	Ejido San José Villa de Allende	Buffer	1.00	ND
17	Mex	Villa de Allende	Ejido Cuesta del Carmen	Buffer	2.00	ND
18	Mex	Donato Guerra	Ejido El Capulín	Buffer	4.00	IV
19	Mich	Ocampo	Ejido Asoleadero	Buffer	1.0	X
20	Mich	Zitácuaro	C.I. Crecencio Morales	Buffer	2.5	III
21	Mich	Angangueo	Ejido El Rosario	Buffer	2	VI
22	Mich	Angangueo	Ejido El Rosario	Buffer	3.5	VI
23	Mich	Ocampo	Ejido Hervidero y Plancha	Buffer	1.00	V
24	Mich	Zitácuaro	C.I. Crecencio Morales	Buffer	0.5	III
25	Mich	Ocampo	Ejido Asoleadero	Buffer	1.5	X
26	Mich	Zitácuaro	C.I. Crecencio Morales	Buffer	2.0	VIII
27	Mich	Zitácuaro	C.I. Crecencio Morales	Unidentified	2.0	VIII
28	Mich	Ocampo	Ejido Ocampo	Buffer	6.0	V
29	Mich	Angangueo	Ejido Angangueo	Buffer	6.0	V
30	Mich	Zitácuaro	C.I. Crecencio Morales	Core	1.5	VIII
31	Mich	Zitácuaro	C.I. San Juan Zitácuaro	Buffer	3.0	VIII
32	Mich	Zitácuaro	Ejido Crescencio Morales	Core	6.0	VI
33	Mich	Senguio	P.P. Rancho de Guadalupe	Buffer	34.95	V
34	Mich	Ocampo	Ejido El Rosario	Buffer	0.96	V
35	Mich	Ocampo	C.I. San Cristobal	Buffer	41.42	I
36	Mich	Ocampo	C.I. San Cristobal	Unidentified	0.41	I
37	Mich	Angangueo	Ejido Rondanillas	Buffer	3.0	I
38	Mich	Zitácuaro	C.I. Crecencio Morales	Buffer	1.25	V
39	Mich	Ocampo	Ejido Asoleadero	Buffer	3.0	VII
40	Mich	Ocampo	Ejido El Rosario	Buffer	2.00	I
41	Mich	Angangueo	Ejido El Rosario	Buffer	1.00	V
42	Mich	Zitácuaro	C.I. Crecencio Morales	Buffer	1.00	I
43	Mich	Zitácuaro	Ejido Crecencio Morales	Core	15.1	VIII
44	Mich	Angangueo	Ejido Angangueo	Buffer	1.5	I
45	Mich	Angangueo	Ejido El Rosario	Core	2.00	V

[a] Mex = México; Mich = Michoacán
[b] I = agricultural activities; II = forestry activities; III = poachers; IV = lightning; V = campfires; VI = tourist campfires; VII = smokers; VIII = intentional; IX = garbage burning; X = other productive activities; ND = not determined

Interviews with the authorities

According to information collected during interviews about fire management and forest fires, three different types of actions were implemented by institutions: (1) fire prevention, (2) direct fire suppression through firefighting, and (3) monitoring and restoration of burned sites. Fire prevention is carried out in three main ways: (1) organization and communication between

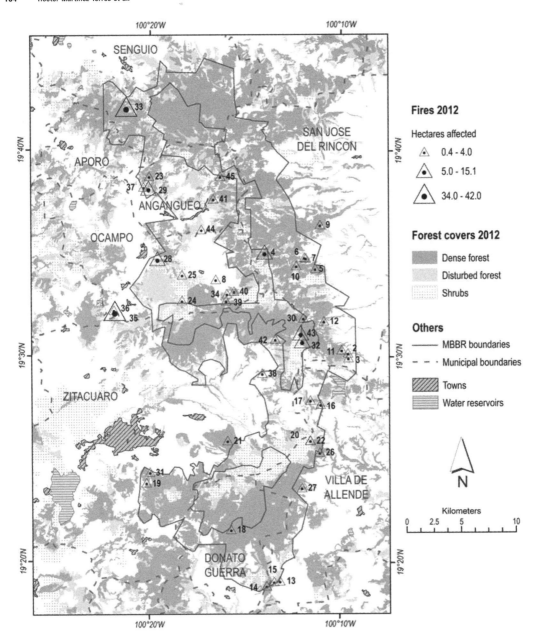

Figure 15.1. Forest fires in 2012 within the MBBR in the states of Michoacán and México. One fire, which took place in Cerro Altamirano, in the northern part of the MBBR, is not shown as it would force a change in scale that would make the remaining 44 fires illegible. Some fires appear outside MBBR boundaries; however, the brigades report that they occurred within the reserve, so the locations might have been incorrectly determined from faulty GPS readings by the brigade.

different government sectors, (2) providing of information on the causes and consequences of forest fires to inhabitants and visitors, and (3) physical actions.

All interviewees agreed that there is good communication between federal and state institutions: CONAFOR-Michoacán and CONANP-MBBR (federal); COFOM (state of Michoacán) and PRO-BOSQUE (state of México). They also reported adequate communication among municipal authorities within the reserve. We did notice, however, that in Michoacán, COFOM follows many initiatives proposed by CONAFOR, but in the state of México PROBOSQUE takes the initiative. Institutional communication between COFOM and PROBOSQUE is usually mediated by CONAFOR and CONANP-MBBR.

Interviewees in Michoacán pointed out that the Michoacán East Forest Regional Council is a key institution in fire response and management in the East Region of Michoacán. COFOM's East Region IV coordinates this council. Within its structure, a Fire Committee coordinates the actions of different institutions to prevent and fight forest fires. This council is also in charge of evaluating firefighting operations once fires are put out, with the aim of identifying mechanisms to improve the efficiency of firefighting. Key institutions that participate in the council include CONANP-MBBR, environmental NGOs with local presence, members of the forest service sector, and representatives from the indigenous communities and *ejidos*. Ejido is a communal land tenure type in Mexico, where each independent member might use a patch of land for personal benefit, but there are also common areas which are managed by the whole ejido community.

Much of the response to fires depends on brigades in charge of fighting and preventing fires. Each brigade is formed by a variable number of people, usually between 5 and 10. In the Michoacán portion of the reserve there are currently six kinds of fire brigades, all with different characteristics (Table 15.3). Every year the number of brigades changes, depending on local interest and the resources available to the ejidos and indigenous communities, municipalities, COFOM, CONAFOR, CONANP, NGOs, and private companies. CONAFOR-Michoacán and COFOM provide material support (tools and appropriate clothing) and training to the brigades, but this support must be formally requested, because cases have arisen in which the benefits granted by the institutions were used for other purposes.

Table 15.3. Classification of prevention and firefighting brigades that work within the MBBR influence area

Brigade type	Time of functioning in a year	Supporting institutions	Support from CONAFOR, COFOM, or CONANP	Support from other institutions	Resources owned by the workers	No. of brigades in 2012
Permanent	12 months	CONAFOR-Michoacán and CONAFOR- State of Mexico	Salary, training, tools, clothing, communication devices, transportation	None	None	2
Temporary	2 or 3 months	COFOM, COFOM-UMAF, COFOM-Tlalpujahua, CONANP-RBMM, PROBOSQUE	Salary, training, tools, clothing, communication devices, transportation	None	Not identified	8
Independent and temporary	2 or 3 months	Municipalities, forest associations (i.e., APFOMAC)	Training, tools	Salary, clothing, transportation	Not identified	10
From communities or ejidos with training	Not identified	Community or ejido from where the fire fighters come	Training	Money	Clothing, tools, transportation	Not identified
From communities or ejidos without training	Not identified	Community or ejido from where the fire fighters come	None	Not identified	Not identified	Not identified
Volunteers	Whenever a fire presents inside their community	Community or ejido from where the fire fighters come	None	None	Not identified	Not identified

Outreach and education

Authorities of the MBBR follow a national strategy against wildfires. This campaign, run by CONAFOR, includes brochures and workshops to warn people about the dangers of forest fires and the need to report them. NGOs and state and municipal governments have also organized efforts to promote a culture of fire prevention.

Another fire prevention strategy is the Official Mexican Norm (NOM-015-SEMARNAT/SAGARPA-2007), which provides mechanisms by which agricultural and forest management fires should be conducted. Because most of the fires in 2012 started as the result of agricultural activities, this strategy is very important. The Department of Agriculture, Livestock, Rural Development, Fisheries and Food (SAGARPA) is primarily responsible for disseminating this policy through its technicians and developers. CONAFOR provides training and outreach materials, and COFOM also promotes NOM-015; however, all interviewees agreed that NOM-015 has not been functional or successfully applied in the MBBR region. They said that it would be unrealistic to monitor each individual agricultural burn conducted by each farmer in the reserve, and to expect inhabitants to change their practices overnight. Hence, they point out the need for a sustained program to promote agricultural burns that do not cause forest fires, and to inform residents about the benefits of complying with NOM-015.

Prevention efforts in natural forests consist of repairing firebreaks and reducing forest fuels. These activities are conducted by some of the brigades outside the fire season. It is noteworthy that within the MBBR buffer zone, prescribed fires to reduce fuels are allowed, but this strategy has not been used as a fire management tool. Fuel reduction in the reserve is done mostly by extraction of downed woody debris and standing dead or infected trees. In the case of the former, downed woody debris might then be used for domestic purposes, while the latter, depending on quality, might be sold as round wood.

Firefighting

As a federal protected area, the MBBR receives priority in terms of firefighting. When a fire occurs within the reserve, human resources (brigades) and materials (vehicles, tools, communication equipment, and even helicopters, if necessary) are immediately brought to the site, sometimes at the expense of other fires outside the MBBR.

The first stage of firefighting is detection. This happens through citizen calls (free of charge) to the emergency phone number that CONAFOR promotes at the federal level. The states of México and Michoacán also provide a free number to inform about fires. Detection can also take place from observation towers, seven in the state of Michoacán (which are run by COFOM) and one in the state of México, that function during the fire season (January to June). One person per tower works during the day, although no one stays in a tower at night. The opening of these towers depends on the availability of funds from Michoacán and México state governments.

The second stage is communication within and between the brigades and their movement to fire sites; however, only some brigades are able to move to different parts of the reserve (Table 15.3). Additionally, a lack of resources makes it impossible to maintain the communication systems 24 hours a day, even during the critical fire season.

The last stage of firefighting is the direct attack on the fire front. Brigades use various techniques to fight fires, depending on intensity, terrain, and atmospheric conditions. Combat can be performed directly, by extinguishing the fire, or indirectly, by making firebreaks (i.e., reducing forest fuels). The brigades' experience and knowledge of the local terrain is essential. Usually, several brigades coordinate to fight a single intense or large fire, regardless of where it started or why. For safety reasons, the brigades fight fires only during the day, with very few exceptions.

The brigades must remain at the burned site performing control activities until there is no chance that the fire will reignite. In addition, starting in 2012, the brigade chief must complete a report with details of the firefighting and the characteristics of the fire.

Postfire: Restoration and monitoring

Each agency is responsible for collecting information from the fire brigades it supports and that fought the fire. Starting in 2012, CONAFOR established forms that must be completed with details characterizing the fire: location, site description, photographs, information about how the fire was detected, time spent in mobilization and actual combat, general data about the affected site, the equipment used, the type of fire, environmental impacts, and economic losses.

None of the institutions formally monitor sites where the fires took place, so long-term fire effects cannot be evaluated from official data. Soil erosion prevention activities and reforestation are commonly carried out, usually by the owners of the burned sites. These owners usually apply for financial or material support from government agencies or to NGOs. Only in cases of a very severe fire do COFOM or CONANP-MBBR take the initiative; however, little or no monitoring is done of the short- or long-term success of these actions.

DISCUSSION

MBBR forest fire regimes

Although we found no published studies specifically addressing fire regimes in the MBBR, sparse and anecdotal information allowed us to begin to identify some characteristics of fire regimes in these forests. We think the most valuable information will be generated from the reports filed by brigades and, of course, from their empirical learning gained from fighting forest fires. Data are also available from various federal, state, and even municipal institutions covering several years; however, the heterogeneity of these data makes it difficult to assess their quality. Therefore, we used data only from the 2012 fire season, as data from previous years are not necessarily accurate or consistent with current data.

The basic information collected from the fires allows us to make some estimates of the fire regime in the MBBR forests, such as size, intensity, severity, and seasonality. Fires in 2012 covered, on average, 3.9 ha, a relatively small area; furthermore, most of them were of low to moderate intensity according to the descriptions of Agee (1993) and Sugihara et al. (2006), since most were superficial and the height of the flame in the line of fire was generally less than 2.4 meters. Severity can be evaluated from tree mortality. While this parameter is not recorded on the forms filled out by brigades after fires, and thus we cannot know the severity of all of the fires, the fire sites we managed to visit indicated low severities. Finally, in regard to seasonality, fires occur mainly during the dry season, from March through July, with occasional fires starting in January and February.

It is impossible to distinguish a significant difference between fires in the pine forests and those in fir forests using only data from 2012. It is essential to provide adequate monitoring of fire behavior in these different ecosystems, given the differences in return interval, intensity, and severity between them (Ángeles-Cervantes and López-Mata 2009; Rodríguez-Trejo and Myers 2010.)

Relevance of MBBR fires

To understand the relative importance of MBBR fires, we can compare reserve fires with fires from the entire country of Mexico and from the states of México and Michoacán, using data from CONAFOR (2012). In 2012, through the third week in September, the national average fire size was 48.59 ha, while the state averages for Michoacán and México were 15.65 ha and 2.6 ha, respectively; fires affected 0.17%, 0.23%, and 0.11% of Mexico, Michoacán, and the state of México, respectively (CONAFOR 2012; INEGI 2012). The average size of fires in the MBBR was 3.9 ha, and the affected area was 0.31% of the reserve; thus, the proportion of land affected in the MBBR is higher than that of the country or the states in which the reserve lies, but it is important to consider that virtually all types of vegetation in the reserve are prone to fires. Fires in the MBBR are smaller, on average, than those in the entire nation or within Michoacán, but not those in the state of México, suggesting that forest firefighting might be more efficient in the state of México.

In regard to social perception, forest fires are perceived as one of the most important problems within the MBBR (CONANP 2001; Rendón-Salinas et al. 2007). We do not know the perception of fires in this region historically, but we have information about the history and management of this forest land from 1930 to 2000, and about how social relationships and institutional organization have changed during this period. As a result of the Mexican Revolution, an intense land reform took place in Mexico after 1930. Large landholdings (e.g., haciendas) present in the region were subsequently subdivided and, together with the remaining federal lands, either granted to ejidos or reinstalled to indigenous communities (*comunidades indígenas*). These ejidos and indigenous communities engaged in agricultural and forest management activities; thus, when the MBBR was finally decreed, most of the land where it stands had already been granted to local inhabitants, who saw conservation policies developed for the newly created reserve as intromissions into their livelihoods (Merino 1997; Martín 2002; Merino and Hernández 2004; Ibarra 2011).This created tension, and fire became one method of expressing land ownership by local communities. As reported for other sites in Mexico, this use of fire might have aided in the perception that local inhabitants degrade forest resources with fire, and it might have decreased interest in learning traditional ways of managing fire (Mathews 2003).

Three activities may influence people's perception of forest fires today. First and second, Payments for Environmental Services (PES) and tourism (Romo 1999; Granct and Fonfrède 2005; Brenner 2006; Orozco et al. 2008; Honey-Rosés et al. 2009; Esquivel et al. 2011) are both likely to encourage fire suppression. The third one, and perhaps the most common theme among the literature we reviewed, is illegal logging and deforestation in the MBBR, recognized by government entities (CONANP 2001; Sáenz et al. 2005), NGOs (WWF 2004; Rendón-Salinas et al. 2007), academia (Brower et al. 2002; Ramírez et al. 2003; Honey-Rosés et al. 2011; Navarrete et al. 2011), and local people (Honey-Rosés 2009b). According to our interviewees, forest fires might be intentionally started to justify land use change or to remove evidence of illegal logging. Hence the relevance of fires should not be seen as a phenomenon outside the socioeconomic and historical context of the region.

Local fire management and institutional response

In this work, intended as a first evaluation of fire management, we address published information and the visions of the academic sector and the government institutions in charge of firefighting; however, in the context of the MBBR, local inhabitants and their role should clearly be incorporated into all phases of a fire management plan, including an understanding of how fire is managed by local people, and how that knowledge can be incorporated into such a management plan (Jardel 2010; Rodríguez-Trejo et al. 2011). This research should consider the implications of establishing the reserve on land previously given to local farmers by the Mexican government, and the development of new local organizations interested in promoting sustainable forest management (SEMARNAP 1998; Sigala and Campos 2001; Byers 2004; Tucker 2004; Venegas 2010).

We identified three fire management strategies of government institutions: prevention, which involves organization and communication between institutions; firefighting; and restoration and monitoring of the burned sites. We note that most strategies reflect a vision and policy of fire suppression, but information is so scarce that we cannot evaluate whether this policy might eventually lead to counterproductive long-term effects, as have been reported for some forests in the United States (Mathews 2003; Jardel 2010; Pyne 2010; Rodríguez-Trejo et al. 2011). It is therefore essential to understand the natural fire regimes of coniferous forests in the MBBR, and to evaluate whether current firefighting policies suppress them.

In terms of prevention, firefighting brigades are the center of organization and action. Their tasks consist primarily of establishing firebreaks and other prevention activities, but except for CONAFOR permanent brigades, all brigades are active only during the fire season, and thus do not participate in many prevention activities. The lack of funds makes it difficult for them to stay active throughout the year, although some municipal governments use their own workers as part of the brigades, thus making the brigade available for the whole year. Volunteer brigades of communities and ejidos engage in the care of their forests constantly, but it is essential to provide economic and technical resources to communities to promote efficiency in prevention efforts.

Information outreach and educational programs are clearly biased in favor of fire prevention. Local inhabitants perceive that, because they are settled within a Natural Protected Area, the MBBR authorities have more institutional mechanisms to fine or punish those who damage the environment. We also think dissemination and implementation of NOM-015-SEMARNAT/SAGARPA-2007, setting norms for the use of fire in agriculture and forestry activities, have been poor. Under this scenario, MBBR authorities could develop an outreach and education campaign to inform inhabitants about natural fire regimes and adequate management of fires to promote long-term forest maintenance. Some governmental and nongovernmental institutions in the MBBR are already making efforts to learn about fire management from the experience of other protected areas. For example, in the Manantlán (state of Jalisco) and La Sepultura (state of Chiapas) Biosphere Reserves, fire management programs incorporating local expertise have been implemented for more than 10 years. It is also noteworthy that National Forestry authorities have recently changed their approach to fire management, moving toward a management scenario, and away from compulsory suppression (Rodríguez-Trejo et al. 2011).

We know very little about the traditional use of fire in the MBBR, but from the information available about the causes of fires in 2012, the people interviewed, and the MBBR management plan (CONANP 2001), it seems that the use of fire by local inhabitants is varied and not always intended to improve forest structure. Agricultural activities are the main cause of fires, so it is necessary to work more closely with the agricultural sector, since to this date, dissemination of NOM-015 is poor. Above all, NOM-015 should be presented and explained in the context of forest types and land uses in the MBBR. Intentionally caused fires, either as a product of conflict between rural settlements or to promote land use change, require special consideration. Other causes of fires, such as their use as an expression of protest (Hoth 1995) or for firewood (Merino 1997; Brenner 2009) should also be considered. For example, gathering firewood could be synergetic with fuel management, which is a viable mechanism of controlling fire danger and fire hazard. Finally, authorities of the MBBR should apply the new recommendations on fire management outlined by CONANP (2011), and include them in the new Reserve management plan.

GLOSSARY OF ACRONYMS

COFOM (Michoacán Forestry Commission): The institution that regulates activities concerning the ecosystems of Michoacán, dependent on state power.
CONAFOR (National Forestry Commission): The Mexican institution that generates plans and programs to develop and strengthen forest activities, dependent on executive power.
CONANP (National Commission of Natural Protected Areas): The Mexican institution in charge of guarding federal natural protected areas, dependent on executive power.
DOF (Diario Oficial de la Federación): The official letter of the Mexican Government. Any new law is official only after it is published in DOF.
INEGI (National Institute of Statistics and Geography): The Mexican institution in charge of generating information related to economics, population, and geography.
MBBR (Monarch Butterfly Biosphere Reserve): Natural Protected Area between Michoacán and México, created to protect the monarch butterfly migration phenomenon.
NOM (Official Norm of Mexico): Obligatory regulations that establish procedures and specifications for conducting actions. Official Norms must always derive from particular legislation.
PROBOSQUE: The institution that regulates activities concerning forest management in the state of México, dependent of the state's Ministry of the Environment.
SAGARPA (Ministry of Agriculture, Livestock, Rural Development, Fisheries and Food): The Mexican institution that makes policies for primary activities concerned with food production, dependent on executive power.

ACKNOWLEDGMENTS

We would like to thank our interviewees as well as Alternare A.C., Espacio Autónomo A.C., and Biocenosis A.C. We also thank Karen Oberhauser, Lincoln Brower, Patrick Guerra, and an anonymous reviewer for their helpful comments and corrections. This research was made possible by funds from CONACYT project 154434. HLMT and MCF received funds from Monarch Butterfly Fund to attend a meeting where portions of this work were presented. HLMT acknowledges support from Posgrado en Ciencias Biológicas, Universidad Nacional Autónoma de México.

16

Project Milkweed

A Strategy for Monarch Habitat Conservation

Brianna Borders and Eric Lee-Mäder

The loss of milkweed plants from the monarch butterfly's breeding range in North America is believed to be a major cause of recent monarch population declines. The restoration of native milkweeds is critical for monarch recruitment, but a scarcity of native milkweed seed in many regions of the United States limits opportunities to include the plants in habitat restoration efforts. To address this seed shortage, The Xerces Society launched Project Milkweed, in collaboration with the native seed industry, the USDA-NRCS Plant Materials Program, and community partners, to produce new sources of milkweed seed in key areas of the United States where seed had not been reliably available: Florida, Texas, the Southwest, the Great Basin, and California. In addition to increasing seed availability, Xerces is raising awareness about the wildlife value of milkweeds, promoting the inclusion of milkweeds in nationwide pollinator conservation efforts, and expanding markets for milkweed seed. Our goal is to implement a model program for milkweed restoration that contributes to North American monarch habitat conservation by increasing the regionwide abundance of milkweed host plants.

INTRODUCTION

Milkweeds (*Asclepias* spp.) are essential for monarch reproduction, and the loss of these plants from the monarch's North American breeding range is believed to be a significant factor contributing to monarch population declines (CEC 2008; Brower et al. 2012a; Pleasants and Oberhauser 2012). Agricultural intensification, suburban development, and the use of mowing and herbicides to control roadside vegetation have decreased the availability of milkweed (CEC 2008). A scarcity of milkweeds can lead to increased search time by gravid female monarchs (Zalucki and Lammers 2010) and decreased larval survival because of resource competition (Flockhart et al. 2012). Given these impacts on monarch fecundity when their larval host plants are limited, increasing the availability of milkweeds is an important component of monarch conservation.

Though milkweed loss at the landscape level often cannot be quantified because of limited data on historical milkweed abundance and distribution, declines in milkweed density resulting from the expansion of glyphosate-resistant corn and soybean crops have been well documented in the Midwest (Pleasants, this volume, Chapter 14). Between 1999 and 2009, Hartzler (2010) documented a 90% decline of common milkweed (*A. syriaca*) in Iowa corn and soybean fields. Pleasants and Oberhauser (2012) estimated a 58% decline of milkweed density in the landscape of the Midwest between 1999 and 2010, with a corresponding 81% decline in monarch production in the region.

To help offset the loss of monarch breeding habitat, The Xerces Society for Invertebrate Conservation, Monarch Watch, Monarch Joint Venture, and other organizations have engaged land managers, agencies, organizations, home gardeners, and individual butterfly enthusiasts to plant milkweeds for monarchs in the United States; however, commercial availability of native milkweed seed has been limited in many regions. This seed scarcity can prevent the use of milkweed in native plant restoration efforts or result in the planting of species that are not regionally native. To overcome this barrier, in 2010 we launched Project Milkweed, an initiative to increase native milkweed seed availability and promote milkweed conservation. With this main focus, The Xerces Society began working with regionally based partners to produce new sources of milkweed seed in key areas of the United States where seed had not been reliably available: Florida, Texas, New Mexico, Arizona, the Great Basin, and California. Here, we describe the scope of our seed production program and our concurrent efforts to raise awareness about the wildlife value of milkweeds and encourage their inclusion in habitat restoration plantings.

BACKGROUND

The Xerces Society is an international nonprofit organization that conducts advocacy, education, habitat restoration, and applied research to conserve invertebrates. Xerces' involvement in monarch conservation began in 1983, with a focus on overwintering habitat conservation in Mexico. At present, Xerces is engaged in the conservation and management of western monarch overwintering habitat (Jepsen and Black, this volume, Chapter 12). Project Milkweed is part of a national Pollinator Conservation Program that implements habitat restoration for native bees, butterflies, and other beneficial insects, and conducts educational outreach to a broad audience. We define "milkweed restoration" as increasing the abundance of milkweeds in a given area to provide resources for monarchs and other beneficial insects, and "habitat restoration" as enhancing habitat quality for monarchs and other wildlife by providing larval host plants and pollen and nectar sources.

To date, native milkweed seed has been reliably available in only a few regions of the United States. Availability has been greatest in the Midwest and Great Lakes regions, where several seed companies and producers offer "source-identified" milkweed seed for which the origin and ecotype are documented. Additionally, source-identified seed is available from at least one producer each in the Southeast, mid-Atlantic, and Northeast regions. Across all these regions, common milkweed (*A. syriaca*), butterfly milkweed (*A. tuberosa*), and swamp milkweed (*A. incarnata*) are the most widely available species, and their seed can be dependably purchased by the pound. A second tier of species for which seed is sometimes available by the pound include showy milkweed (*A. speciosa*), prairie milkweed (*A. sullivantii*), whorled milkweed (*A. verticillata*), and green milkweed (*A. viridis*). Because seed of these species is readily available, they can often be included in seed mixes utilized for revegetation and habitat restoration. In contrast, availability in other regions has typically been limited to small quantities of wild-collected seed, with supplies fluctuating from year to year.

While wild-collected seed can meet small-scale demand, it is rarely available in the quantities needed for large-scale restoration, and it can be prohibitively expensive compared with commercially produced seed. In regions where seed supplies cannot meet the existing demand, customers and restoration practitioners may ultimately purchase seed from distant sources. Although no significant data sets are available to inform the extent to which individual milkweed populations of a given species are genetically differentiated or uniquely adapted to their environment, using regionally native plant materials, to the extent possible, will help maximize plant adaptation and minimize artificially elevated human-driven gene flow among wild populations.

INCREASING MILKWEED SEED AVAILABILITY

Establishing partnerships

Early in our program, we approached many native seed producers and plant growers to begin a dialogue about milkweed production. Some companies were hesitant to bring new species into production for which propagation techniques are unknown, or when a market for the seed is not guaranteed. To address these concerns, Xerces offered technical support, funding to help offset production costs or purchase supplies, and in some cases, the foundation seed or transplants needed to initiate production. Through this approach, we successfully established key partnerships with various seed producers, community groups, and the U.S. Department of Agriculture's Natural Resources Conservation Service (USDA-NRCS) Plant Materials Program (which includes a network of regionally focused Plant Materials Centers).

Species selection and seed collection

To select species for seed production and restoration value in each region, we used five "ideal" criteria: (1) known value as a monarch larval host plant, (2) broad distribution across the target region, (3) adaptability to a range of habitat conditions, (4) low potential to become "weedy" and spread aggressively outside their planted areas, and (5) minimal toxicity to livestock. All target species are known monarch host plants, and they meet more of the remaining four criteria than other candidate species in their respective regions.

Following species selection, we worked to obtain the regionally sourced seed needed to initiate production efforts. Since the necessary foundation seed was not available from commercial or governmental sources (e.g., the USDA's National Plant Germplasm System), we arranged to have the seed wild-collected by volunteer and contract collectors. A wild seed collection effort entails identifying suitable source populations, acquiring the necessary permission for land access and seed collection, and making appropriately timed visits to harvest seed (Bureau of Land Management 2012). Obtaining wild-collected seed can be an uncertain endeavor. Because of the wind-aided dispersal of milkweed seeds, the window of opportunity for seed collection is narrow. On any given seed collection visit, only a fraction of the seedpods are ready for collection, and multiple visits are often required to obtain the target amount of seed. Additionally, it is important not to deplete wild populations by over-harvesting seed (Bureau of Land Management 2012). Under drought conditions, plants may not produce any seed, and collection efforts may be delayed until at least the following year. Also, populations that are being monitored for seed collection may be unexpectedly mowed or treated with herbicide before a harvest can be made.

Establishing seed production fields

Xerces' seed producer partners established milkweed seed production fields up to one acre in size, by either direct seeding or transplanting. The planting method used depended on the quantity of foundation seed and the availability of mechanized planting equipment. Native North American milkweeds are perennials; while some species (e.g., *A. fascicularis*, *A. subulata*) flower and produce seed within one year of being established from seed, others (e.g., *A. asperula*, *A. speciosa*, *A. viridis*) typically require more than one year of growth (pers. comm. with personnel from Hedgerow Farms, Native American Seed, and Arizona Western College). The establishment and maintenance of production fields requires continuous weed control, application of irrigation water as needed, and management of occasional pest and disease outbreaks. Weed growth in milkweed fields was controlled with a combination of mechanical methods (e.g., hand removal, hoeing, and shallow cultivation between planted rows) and herbicide application (Plate 13). Although milkweeds contain toxic chemicals that help protect them against herbivores (Malcolm 1991), a suite of specialist insects are adapted to feed on various milkweed plant parts (Agrawal et al., this volume, Chapter 4), including the sap-feeding oleander aphid (*Aphis nerii*), the seed-feeding small and large milkweed bugs (*Lygaeus kalmii*, *Oncopeltus fasciatus*), the leaf-feeding blue milkweed beetle (*Chrysochus cobaltinus*), swamp milkweed leaf beetle (*Labidomera clivicollis*), long-horn beetles (*Tetraopes* spp.), and the milkweed stem weevil (*Rhyssomatus lineaticollis*). Milkweed crops are also attractive to ovipositing queen (*Danaus gilippus*) and monarch butterflies. On occasion, the presence of butterfly caterpillars in milkweed field stands has resulted in severe, localized defoliation and complicated options for controlling outbreaks of other insects because of the growers' commitment to protecting caterpillars from pesticides. Finally, some milkweed crops are occasionally infected by fungal pathogens; as these have occurred, the pathogens have been identified by local extension plant pathologists and outbreaks managed accordingly.

To maximize the efficiency of seed harvesting and postharvest seed processing, we are developing mechanized protocols. Since most of the milkweed species used in this project have not previously been propagated for mass seed production, this project is an opportunity to develop and document new production methods, from seed germination through seed processing. To make this information widely available, we are developing written guidelines for the native seed industry and monarch conservationists.

REGIONAL ACCOMPLISHMENTS

Florida and Texas

Spring breeding in the Gulf States is well documented (Cockrell et al. 1993; Malcolm et al. 1993; Howard and Davis 2004), and monarchs produced

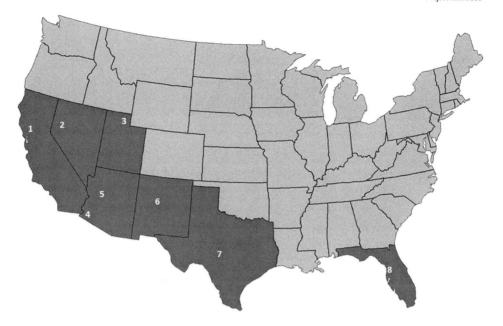

Figure 16.1. Project partner locations within the target regions of the Xerces Society's milkweed seed production initiative. (1) Hedgerow Farms, Winters, California, (2) USDA-NRCS Great Basin Plant Materials Center, Fallon, Nevada, (3) Private milkweed grower, Smithfield, Utah, (4) Arizona Western College, Yuma, (5) Painted Lady Vineyard, Skull Valley, Arizona, (6) USDA-NRCS Los Lunas Plant Materials Center, Los Lunas, New Mexico, (7) Native American Seed, Junction, Texas, (8) USDA-NRCS Brooksville Plant Materials Center, Brooksville, Florida.

at these southern latitudes play an important role in recolonizing the northern breeding range (Malcolm et al. 1993; Zipkin et al. 2012; Ries et al., this volume, Chapter 24). In Texas, we partnered with Native American Seed, one of the largest native seed producers in the state, to produce seed of antelope horns (*A. asperula* subsp. *capricornu*) and green milkweed (*A. viridis*), two of the main host plants for spring migrating monarchs in the southern United States (Lynch and Martin 1987, 1993; Martin and Lynch 1988). Using these species in restoration efforts will significantly benefit spring migrants, and monarchs also use these milkweeds in Texas and other southern Great Plains states during late August and September (Calvert 1999; Baum and Mueller, this volume, Chapter 17). The aboveground growth of *A. asperula* and *A. viridis* typically senesces during summer, but with adequate seasonal rainfall their foliage can persist into the fall (Batalden and Oberhauser, this volume, Chapter 19). Other work has demonstrated that fall regrowth of *A. viridis* can be stimulated by prescribed burning or mowing during summer (Baum and Sharber 2012; Baum and Mueller, this volume, Chapter 17).

In Florida, we are working with the USDA-NRCS Plant Materials Center in Brooksville to produce seed of pinewoods milkweed (*A. humistrata*), the main host plant used by spring breeding monarchs in the state (Lynch and Martin 1987; Malcolm et al. 1987; Cockrell et al. 1993). We also launched a small-scale production effort of aquatic milkweed (*A. perennis*) in cooperation with the Plant Materials Center and are preparing to initiate production of a southeastern ecotype of butterfly milkweed (*A. tuberosa*) in collaboration with Ernst Conservation Seeds. Information about growing Florida native milkweeds is scarce, and providing propagation guidelines for these species will help engage seed producers and plant nurseries in their production. The availability of native plant materials, particularly seed, is limited in Florida, and restoration practitioners must often utilize materials sourced from outside the state. This work will help fill that void, and greater seed availability could also increase the production of live milkweed transplants, for which anecdotal reports suggest there is significant demand.

California

In comparison with eastern monarchs, the movement patterns of western monarchs and the location and extent of their natal grounds are less well documented (Dingle et al. 2005; Stevens and Frey 2010; Pyle, this volume, Chapter 21). When winter aggregations of monarchs on the California coast disperse during the spring, some monarchs breed along the coastline (Wenner and Harris 1993), but many fly inland in search of milkweed (Nagano et al. 1993). The availability of California milkweeds during spring is therefore crucial for the production of first-generation monarchs. Because monarchs are present in California year-round, milkweed availability during summer and early fall is also important for monarch breeding success. The Central Valley has been subject to extensive land conversion for urban and suburban development and agriculture, and our work to restore milkweed to agricultural lands across the Central Valley will help offset the loss of monarch breeding habitat in the region.

We are working with Hedgerow Farms, a large-scale native seed producer, to produce seed of narrowleaf milkweed (*A. fascicularis*), showy milkweed (*A. speciosa*), purple milkweed (*A. cordifolia*), and Indian milkweed (*A. eriocarpa*). Hedgerow Farms currently has more than one acre each of narrowleaf and showy milkweed in production (Plate 13), and over the course of four growing seasons, more than 500 pounds of seed have been produced. Source-identified California narrowleaf milkweed seed has not previously been available by the pound, and showy milkweed seed previously available through other seed vendors has been sourced almost exclusively from the Midwest and Central United States. Through this partnership, we are making significant strides in producing new sources of California-sourced milkweed seed.

The Great Basin

Monarchs occur in the Great Basin, particularly Nevada, throughout the summer and into the fall (Dingle et al. 2005). Large portions of the Great Basin have become degraded as a result of frequent wildfire and an increasing prevalence of invasive weeds (Miller and Narayanan 2008; Sheley et al. 2008). To face these environmental challenges, the Bureau of Land Management and the U.S. Forest Service initiated a large-scale program to produce native plant materials and use them in rehabilitating and restoring degraded and disturbed lands.

The NRCS Great Basin Plant Materials Center (Fallon, NV) hosts large on-site milkweed populations and helped to provide the necessary foundation seed to initiate seed production in the region. We are collaborating with an individual farmer to produce Nevada-sourced narrowleaf milkweed and showy milkweed seed. Both species have a broad distribution across Nevada and the Great Basin, and seeds produced through this partnership will contribute to regional restoration efforts.

Arizona and New Mexico

Monarchs occur in Arizona throughout most of the year, with peak numbers observed during fall (Southwest Monarch Study, unpublished data). Both breeding (Funk 1968; Gail Morris, unpublished data) and nonreproductive (Gail Morris, unpublished data) monarchs have been documented in Arizona during winter. Monarch adults tagged in Arizona have been recovered at overwintering sites in both California and Mexico (Southwest Monarch Study 2013a). We have worked with Arizona Western College and the Painted Lady Vineyard to produce seed of rush milkweed (*A. subulata*) and spider milkweed (*A. asperula* subsp. *asperula*), respectively. Rush milkweed is a desert-adapted, evergreen species that hosts both monarchs and queen butterflies across its range. With sufficient rainfall, it can host caterpillars during both spring and fall. Spider milkweed is mainly a spring host plant, but it can reemerge after a midsummer senescence for a second growth cycle in response to summer rains (Southwest Monarch Study 2013a).

The population dynamics and migratory behavior of monarchs in New Mexico have not been well documented and warrant future study. We are working with the USDA-NRCS Plant Materials Center in Los Lunas, New Mexico, to produce seed of showy milkweed, broadleaf milkweed (*A. latifolia*), and antelope horns (*A. asperula* subsp. *capricornu*). Following this initial seed increase effort, we will engage a commercial grower in further expanding the production of these species. Supplies of native seed are limited in the Southwest and can be difficult to obtain for restoration and revegetation projects. These partnerships are producing the first available sources of bulk milkweed seed in the region, and our marketing and outreach campaigns are generating interest in using these species in restoration efforts.

Figure 16.2. A milkweed (*A. asperula*, *A. latifolia*, and *A. speciosa*) seed production field established at the USDA-NRCS Plant Materials Center in Los Lunas, New Mexico. Photo by David Dreesen, USDA-NRCS.

PROMOTING MILKWEED CONSERVATION

In addition to increasing milkweed seed availability in target regions, we are promoting the wildlife value of milkweeds, encouraging the inclusion of milkweeds in nationwide pollinator conservation efforts, and building markets for milkweed seed.

Project Milkweed is part of The Xerces Society's Pollinator Conservation Program, which provides training to farmers, agricultural professionals, land managers, natural resource specialists, gardeners, and educators on protecting, managing, and restoring habitat for insect pollinators and other beneficial insects. In 2012, Pollinator Conservation Program staff conducted more than 100 workshops or trainings in 35 states for an audience of more than 6000 individuals. Milkweeds are featured prominently in outreach to those audiences, and we have incorporated milkweed planting recommendations into habitat restoration and revegetation projects nationwide. We have also delivered publications and workshop content that include information on milkweed and monarchs to state and federal natural resource agencies, transportation agencies, and nongovernmental environmental organizations.

A key partner in our pollinator conservation efforts is the USDA's Natural Resources Conservation Service (NRCS), which provides financial and technical assistance to landowners and agricultural producers to help manage and conserve natural resources (USDA-NRCS 2013). Many NRCS voluntary conservation programs provide opportunities to conserve and create pollinator habitat. We provide technical support to NRCS staff (and their landowner-clients) and promote the inclusion of milkweeds in NRCS-supported pollinator conservation efforts through developing seed mixes and planting plans, providing workshops to the NRCS and their clients, and regional milkweed plant guides (Xerces Society 2013) for use by NRCS field planners. In addition to working with the NRCS Plant Materials Program on milkweed seed production, we have partnered with the agency to conduct pollinator habitat field trials nationwide.

When promoting milkweed restoration, we provide information about species' native ranges and guidance on which species are commercially available and appropriate for planting within various regions. In all instances, we promote planting milkweeds that are native to the United States, rather than introduced species such as tropical milkweed (*A. curassavica*), swan plant (*A. fruticosa*), and balloon plant (*A. physocarpa*). Of these non-native species, tropical milkweed is by far the most widely available, and it is frequently planted in gardens for its showy flowers, attractiveness to monarchs, and ease of establishment; however, there is preliminary evidence that the presence of tropical milkweed in the United States may disrupt monarchs' migratory cycle and increase transmission and virulence of the protozoan parasite *Ophryocystis elektroscirrha* (Batalden and Oberhauser, this volume, Chapter 19; Altizer and de Roode, this volume, Chapter 7). Thus, some leading monarch scientists assert that the planting of native milkweeds will better promote monarch health and sustain their natural migratory cycle.

By raising awareness about the need for milkweed restoration, we are helping expand markets for milkweed seed, which incentivizes seed producers. We are also providing direct marketing support to project partners, for example, by producing and distributing brochures that promote planting milkweed and advertise seed availability. As new sources of seed become available, we have made regionally targeted announcements to Xerces Society members, natural resource agencies, conservation districts, and nongovernmental conservation organizations.

As monarch and milkweed conservation outreach campaigns continue, markets for milkweed seed and plants will continue to expand. Either seed or transplants can be utilized for small-scale plantings (e.g., 5000 square feet or smaller) but for large-scale revegetation efforts, seeding is typically a more economical and efficient approach than transplanting. As milkweed seed becomes increasingly available, more opportunities will arise to incorporate milkweeds into large-scale plantings implemented by transportation agencies, resource conservation districts, and natural resource agencies.

SUMMARY

The Xerces Society's Project Milkweed represents a successful model for combining public outreach, seed production, and monarch-specific efforts within a larger nationwide discourse about pollinator conservation. We are contributing to North American monarch habitat conservation both by helping to offset habitat loss and by engaging a broad cross-section of stakeholders such as the native seed industry and the USDA-NRCS. As a result of our partnerships with these groups, they now routinely reference monarch butterflies in their broader discussions of pollinator conservation. While we have made important progress, widespread land conversion continues apace, and the rate of milkweed habitat loss almost certainly exceeds the rate of milkweed establishment resulting from all combined restoration activities. Conservation actions of a broad, national scope are needed to realistically offset habitat loss and ensure monarch breeding success in North America. Our work is helping to lay the foundation to make that possible.

ACKNOWLEDGMENTS

We thank project partners Hedgerow Farms, Native American Seed, the Painted Lady Vineyard, Arizona Western College, and the USDA-NRCS Plant Materials Centers in Florida, New Mexico, and the Great Basin. We thank the Desert Botanical Garden, Oecohort LLC, Chris Parisi, Bob Sivinski, Gail Haggard, Linda Kennedy, Eric Roussel, Laura Merrill, Tom Stewart, and Tatia Veltkamp for providing foundation seed. Greenheart Farms produced an essential crop of spider milkweed transplants. Gail Morris, Chip Taylor, Robert Michael Pyle, and an anonymous reviewer provided helpful comments on the manuscript. Funding was provided by a national USDA-NRCS Conservation Innovation Grant, Hind Foundation, SeaWorld and Busch Gardens Conservation Fund, Disney Worldwide Conservation Fund, The Elizabeth Ordway Dunn Foundation, The William H. and Mattie Wattis Harris Foundation, the Monarch Joint Venture, Turner Foundation Inc., The McCune Charitable Foundation, and Xerces Society members. We also thank Ernst Conservation Seeds.

17

Grassland and Roadside Management Practices Affect Milkweed Abundance and Opportunities for Monarch Recruitment

Kristen A. Baum and Elisha K. Mueller

Monarchs reproduce in the southern Great Plains (a region typically defined as encompassing parts of Texas, Oklahoma, New Mexico, Kansas, Colorado, and Nebraska) in the spring and early summer. They are absent from this region during midsummer (especially July), but some return in August and September and reproduce prior to fall migration; therefore, milkweed availability in this region is important for monarchs during spring and fall migration. We studied milkweed in the portion of the southern Great Plains in north-central Oklahoma within rangelands managed with prescribed fire, prairies managed with annual mowing, and roadsides mowed several times per year. We focused on *Asclepias viridis*, the most common milkweed species in this area. Milkweed densities were higher in the spring and declined by late summer and fall across all sites. Milkweed was present in late summer in previously burned rangeland sites, but nearly absent in unburned sites. *Asclepias viridis* also regrew after mowing, but in the absence of disturbance, it senesced and was no longer available as a host plant for monarchs. We discuss the potential for management practices to modify milkweed availability in the southern Great Plains, focusing on the implications for late breeding monarchs.

INTRODUCTION

Monarchs in eastern North America breed in the United States and Canada and overwinter in central Mexico. Monarchs recolonize North America over two generations, with overwintering individuals recolonizing the south-central and southeastern United States, and their offspring moving farther north and east to the Midwest, eastern United States, and southern Canada (Cockrell et al. 1993; Malcolm et al. 1993). Monarchs reproduce in the southern Great Plains (a region that encompasses Texas, Oklahoma, and New Mexico) during their spring migration northward but are uncommon or absent in this region during midsummer (Calvert 1999; Baum and Sharber 2012). The peak fall migration occurs in late September to early October in north-central Oklahoma (latitude approximately 36°N; Monarch Watch 2012). However, some monarchs return to the southern Great Plains in August and September, prior to peak fall migration (Calvert 1999; Prysby and Oberhauser 2004; Baum and Sharber 2012; Batalden and Oberhauser, this volume, Chapter 19). These monarchs, sometimes referred to as "premigrants," are a reproductively active subset of fall monarchs (the remainder of fall monarchs are in reproductive diapause); therefore, monarchs use milkweed host plants in this region during reproductive periods associated with spring and fall migration.

In North America, monarchs use at least 27 milkweed species as host plants (Malcolm and Brower 1989), with the dominant species varying by location. For example, *Asclepias syriaca* is the dominant host plant in the Midwest and Canada (Malcolm et al. 1993; Hartzler 2010; Pleasants and Oberhauser 2012), while *A. viridis* (Figure 17.1) is locally abundant and readily used by monarchs in the southern Great Plains (Malcolm and Brower 1989; Calvert 1999; Baum and Sharber 2012). In Oklahoma, *A. viridis* typically flowers in May and June, beginning to senesce by July or August in undisturbed (e.g., not burned or mowed) sites (Baum and Sharber 2012). Climatic conditions such as precipitation and temperature also influence milkweed phenology (Batalden and Oberhauser, this volume, Chapter 19), and plants may not follow the same pattern of senescence across space and time; however, activities that remove the aboveground portion of *A. viridis* plants, such as burning or mowing, cause the plants to regrow (Plate 13a), generating the tender new growth preferred by ovipositing monarchs (Zalucki and Kitching 1982a). The timing and frequency of disturbance will influence the phenology of milkweeds and their subsequent availability for monarchs during their spring and fall breeding periods.

Information about the abundance of milkweed species across much of the monarch's breeding range is scarce (but see Calvert 1999; Hartzler and Buhler 2000; Hartzler 2010; Baum and Sharber 2012; Pleasants and Oberhauser 2012). In the past, monarchs relied heavily on milkweed in agricultural fields in the Midwest for reproduction (Oberhauser et al. 2001), but the development of glyphosate-tolerant crops, especially corn and soybeans, and the resultant increase in herbicide applications have decreased milkweed availability in recent years (Pleasants and Oberhauser 2012). The effects of other management practices on the distribution and abundance of milkweed have not been evaluated, but this information may be especially important throughout the late summer and early fall breeding range; therefore, we evaluated milkweed availability in rangelands, managed prairies, and roadsides in north-central Oklahoma, focusing on the management practices of prescribed fire and mowing. We discuss the implications of our findings for milkweed habitat in the southern Great Plains, with an emphasis on the implications for late-breeding monarchs.

METHODS

We collected data on milkweed density at a total of nine sites, three each in rangelands, managed prairies, and roadsides. Monitoring took place from May through October during 2010–2012 in the vicinity of Stillwater in north-central Oklahoma, USA, containing primarily mixed grass and tallgrass prairie. Rangeland sites ranged in size from 45 to 65 ha and were located at Oklahoma State University's Stillwater Research Range, approximately 21 km southwest of Stillwater in north-central Oklahoma. These sites are grazed by cattle and managed with patch burning, which involves patchily applying prescribed fire to the landscape to generate a heterogeneous pasture consisting of patches with different seasons and times since burn (Fuhlendorf and Engle 2001, 2004). The three managed prairie sites ranged in size from 6.8 to 19 ha and were located within Stillwater. One site is owned by Oklahoma State University and mowed for recreational purposes, and the other two sites are privately owned and mowed for hay. These sites are typically mowed once per year in July and the hay usually baled and removed. The roadside sites are managed by the Oklahoma Department of Transportation and located along Highway 177 (a two-lane road), within 24.4 km of Stillwater. The roadside study sites varied in width, ranging from 11.8 to 53.1 m, and are typically mowed two to three times per year (in late May/early June if necessary, early July, and early September) with the majority of the hay removed; however, mowing regimes are altered in response to climatic conditions and their effect on plant growth, with reduced mowing during periods of drought (e.g., 2012 in Oklahoma).

Within each site in areas where milkweed was present, we randomly established three 5 × 50 m transects for estimating milkweed density. We also recorded basic morphological and phenological data for the rangeland and managed prairie sites, including plant height (centimeters above ground level), plant width (diameter of the smallest circle encompassing the plant), and number of stalks (based on aboveground separation; Figure 17.1). Because the data were collected at different times from different sites, direct statistical analyses comparing treatments and time periods were not performed. Therefore, we limit our results and discussion to a comparison of general trends, identification of additional research needs, and potential implications for conservation and management.

RESULTS

Asclepias viridis was more common during the spring and early summer than during the late summer and fall across all sites (Figures 17.2, 17.3).

Figure 17.1. A single *Asclepias viridis* plant with multiple stalks based on aboveground separation.

Milkweed plants regrew after mowing and were present during the late summer and fall. They were 2.8 times taller (Figure 17.4) and 2.1 times wider (June 2012 (mean ± SE): 32.77 cm ± 0.88, n = 338; September 2012: 15.67 cm ± 0.83, n = 123) earlier in the year, and the number of stalks per plant declined by approximately 34% (June 2012 (mean ± SE): 2.01 ± 0.08, n = 338; September 2012: 1.50 ± 0.10, n = 123). At a managed prairie site not mowed during 2012, milkweed was not present during August (Figure 17.3). In rangelands, *A. viridis* was abundant during August through October at sites burned during the summer, and mostly absent from areas that were not burned that summer (Figure 17.5). Milkweed densities at burned sites followed a pattern similar to those observed at mowed sites, with higher densities during the spring and early summer than in late summer and fall (Baum, pers. observ.).

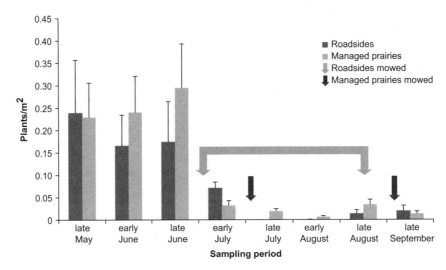

Figure 17.2. Density of *Asclepias viridis* plants (mean + SE) at managed prairie ($n = 3$) and roadside sites ($n = 3$) during 2011. Arrows denote when sites were mowed, with mowing of roadsides (twice at each site) indicated by dark shaded arrows and of prairie sites by light shaded arrows (each prairie was mowed once during the time interval shown).

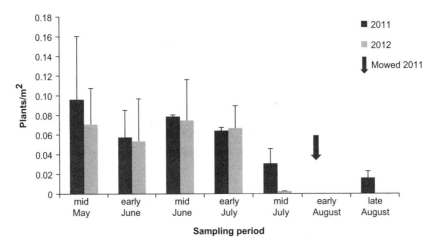

Figure 17.3. Density of *Asclepias viridis* plants (mean + SE) at a managed prairie site ($n = 3$ transects per sampling period) mowed once during early August 2011 (denoted by the dark shaded arrow), but not mowed in 2012.

DISCUSSION

Milkweed was more abundant during the spring and early summer than in late summer and fall, and this pattern held across all sites. Plants were also larger, in terms of both height and width, and possessed more stalks in the spring and early summer. These findings suggest that milkweed availability may be less limiting (or not limiting) for monarchs reproducing in north-central Oklahoma during their spring migration as they move toward their summer breeding grounds in the Midwest, eastern United States, and southern Canada. Milkweed availability could be limiting for monarchs reproducing in north-central Oklahoma in August and September, prior to the peak fall migration; however, it is important to note that all monarchs moving north through Oklahoma in the spring are breeding, whereas only a subset of the monarchs that move south through Oklahoma in the late summer and early fall are reproductively active.

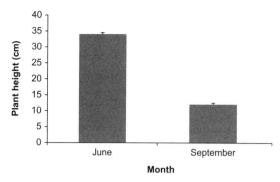

Figure 17.4. *Asclepias viridis* height (mean + SE) during June and September 2012 at a managed prairie site (*n* = 3 transects per sampling period) mowed in early July 2012.

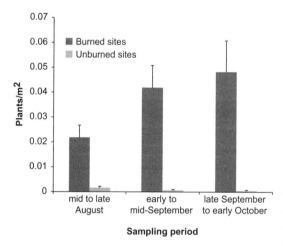

Figure 17.5. Density (mean + SE) of *Asclepias viridis* plants at rangeland sites (*n* = 3) burned and not burned during July 2010.

Disturbance regimes (mowing and prescribed fire) influenced *A. viridis* abundance. Plants regrew after mowing, both at managed prairies and roadsides, but in the absence of mowing, *A. viridis* senesced by August and was not available as a host plant for monarchs. In plots burned during the summer (July), *A. viridis* regrew and was available as a host plant for late breeding monarchs (see also Baum and Sharber 2012), whereas plants were no longer available at unburned sites. The timing of mowing and burning likely influences the availability of milkweed. For example, if plants senesced prior to the disturbance, then mowing or burning might have minimal effects. If this is the case, then shifting these management activities a few weeks earlier could increase milkweed availability in August and September. Later mowing or burning also might not allow adequate time for milkweed regrowth prior to the arrival of monarch butterflies. It is not clear whether regrowth following mowing or prescribed fire influences milkweed characteristics (including quality for monarch larvae) the following year, compared with plants that senesced prior to the prescribed fire and did not exhibit regrowth. Additional research is needed to evaluate the longer-term effects of disturbances on milkweeds and monarchs within and between years.

The arrival of reproductively active monarchs in August and September in the southern Great Plains corresponds to the time needed for newly laid eggs to hatch and develop into adult monarchs (Zalucki 1982) that can join the fall migration to their overwintering grounds in Mexico (Baum and Sharber 2012). Prior work showed most monarchs overwintering in Mexico originate from the Midwest, with relatively few from the southern Great Plains (Malcolm et al. 1993; Wassenaar and Hobson 1998). This pattern could be changing if breeding activity in late summer and early fall is increasing south of the main summer breeding grounds (Batalden and Oberhauser, this volume, Chapter 19).

It is not clear whether the incidence of reproductively active monarchs in the late summer and fall corresponds to the availability of milkweed in the southern Great Plains, or how it relates to the overall size of the eastern North American monarch population; both of these relationships warrant additional study. For example, we do not know whether monarchs break diapause in response to the availability of their host plants, or whether a proportion of the population is reproductively active as they migrate south regardless of host plant availability. Furthermore, milkweed availability under current disturbance regimes might differ from that expected based on historical disturbance regimes. In many areas, land use change has resulted in fewer fires (Pyne 2001), although in some parts of the southern Great Plains fire frequency has increased while fire intensity has decreased (e.g., Allen and Palmer 2011). The relative importance of changing anthropogenic disturbance regimes for milkweed abundance and characteristics, including leaf tissue quality, requires further investigation.

Disturbance regimes might also influence monarch interactions with predators and parasites. For example, *Ophryocystis elektroscirrha* is an obligate, spore-forming, protozoan parasite of monarch butterflies (McLaughlin and Myers 1970; Altizer and de Roode, this volume, Chapter 7). Infected adults deposit parasite spores on plants during oviposition that are then ingested by larvae. Disturbances could modify these interactions by reducing spore availability in the environment through the removal of the aboveground portion of plants, or they could increase spore availability by concentrating monarchs (including infected individuals) in areas with regrowing milkweed, especially if milkweed has senesced in other areas and availability is limited throughout the landscape.

Additionally, *Lespesia archippivora* is a fly parasitoid that lays its eggs in monarch caterpillars throughout their breeding range (Etchegaray and Nishida 1975b; Prysby 2004; Oberhauser et al. 2007; Oberhauser 2012). Fly maggots usually emerge prior to monarch pupation, killing the monarch caterpillar host. Parasitism rates vary among regions, years, and seasons, and may also vary with host plant species (Oberhauser et al. 2007; Oberhauser 2012). Disturbances could influence fly parasitism through a variety of mechanisms, such as concentrating monarch larvae in recently mowed or burned areas with new milkweed growth.

Based on the considerations discussed above, many questions remain about the potential role of disturbance in monarch ecology. The timing and frequency of disturbance will influence host plant availability for monarchs, and they could also determine milkweed distribution, morphology, phenology, and reproductive success. Standardized comparisons among different disturbance regimes and their roles as sources or sinks for monarchs need to be evaluated. Patches could serve as sources if they provide abundant milkweed, but they could serve as sinks if new growth concentrates monarchs in areas with high parasitism or if late mowing or prescribed fire, or both, reduce milkweed availability.

From a conservation perspective, it is possible that mowing and burning regimes could be better managed to increase host plant availability for monarchs during their late summer and early fall breeding period, while still meeting the needs of roadside, prairie, and rangeland managers. The potential contribution of these different land use types as habitat for monarchs depends on their distribution and abundance throughout the landscape, and the source-sink considerations discussed above. Other important factors that require consideration from the conservation perspective include the effects of mowing and prescribed fire on nectar resource availability, as well as effects on other wildlife species, including host plant availability for other butterflies.

ACKNOWLEDGMENTS

We thank K. Oberhauser, S. Altizer, D. Pérez-Salicrup, P. Guerra, and an anonymous reviewer for comments on the chapter. We thank A. Collier, R. Henry, G. Ingalls, A. Knoch, S. Livsey, S. McCoshum, K. Monroe, S. Parker, M. Parrish, J. Rojas, M. Thompson, J. Tidwell, E. Villert, M. T. Wade, H. Wehde, and G. R. Williams for assistance with data collection, and S. Fuhlendorf, C. Stansberry, and the Oklahoma Department of Transportation for access to and maintenance of field sites. Funding was provided by a USDA-NRI Managed Ecosystems Program grant (2009–35101–05170) to K.A.B. and a Payne County Audubon Society grant to E.K.M.

PART V

New Perspectives on Monarch Migration, Evolution, and Population Biology

An Overview

ANDREW K. DAVIS AND SONIA ALTIZER

The past decade has seen major advances in several areas of monarch biology, most notably in the mechanisms of monarch migration, global dispersal and evolution, and population trends and their underlying ecological drivers. It is an exciting time to be studying monarchs, with technological advances and newly available data that allow us to address questions at unprecedented scales. Additionally, the high visibility of this butterfly has given rise to monitoring efforts focused on every phase of its life cycle. The data from these programs open the door for powerful analyses of monarch population ecology at vast spatiotemporal scales.

MIGRATION

Arguably the best studied aspect of monarch behavior and ecology is their long-distance migration. While monarch migration has long captivated scientists and the public alike, new work continues to generate interesting and surprising findings. For example, pioneering work on the neurological basis of monarch migration (e.g., Reppert et al. 2010; Heinze and Reppert 2012) has shown that antennae house circadian clocks that function independently of other neural tissue (e.g., Merlin et al. 2009; Reppert et al. 2010; Guerra et al. 2012; Heinze and Reppert 2012). Guerra and Reppert (2013) solved the mystery of what triggers the reversal of flight direction from southward in the fall to northward in the spring, demonstrating that exposure to cold temperatures (similar to those experienced by monarchs at their high-altitude overwintering sites) for as few as 24 days was enough to cause this switch.

Recent studies using citizen science data are also improving our understanding of the spatial and temporal patterns of spring and fall monarch migration (Howard and Davis 2009, 2011; Davis et al. 2012a). In their chapter in this section, Howard and Davis use sightings of fall roosts submitted by Journey North volunteers to describe variation in the pace of southward migration throughout the fall, and its increase over the past seven years, perhaps as a result of expansion of the breeding range northward. Depending on milkweed availability, further expansion of the monarchs' breeding range into Canada could result from climate warming (Batalden et al. 2007). Climate warming could also alter monarch migration behavior in the southern United States, as demonstrated by Batalden and Oberhauser in their chapter. Warming temperatures and the availability of exotic (tropical) milkweeds may be causing migrating monarchs to break diapause, stop migrating south, and breed in Texas. This phenomenon may also occur when fall migrants along the east coast enter southern Florida and Cuba and mix with resident populations there (Knight and Brower 2009); thus, the future might see an expansion of winter breeding and a gradual replacement of the long-distance migratory/overwintering population with local patches where monarchs breed year-round and do not migrate.

Much can be learned about the migration of monarchs by studying their southern cousin, *Danaus erippus* (Plate 10). In Chapter 20, Malcolm and Slager present the first-ever detailed study of this species, with evidence of many similarities in migration characteristics, such as nectaring behavior, seasonal fat accumulation, clustering activity, and directed flight behavior. They also describe phenomena that have not been seen in monarchs, such as a tendency toward longer wings at higher altitudes. Given that longer wings are associated with longer-distance migratory behaviors in monarchs (Altizer and Davis 2010), these data suggest the presence of an altitudinal migration; however, because milkweed species also varied with altitude at their study area, the possibility of host plant effects on wing size warrants further study. Malcolm and Slager's study should pave the way for future investigations into the ecology and migration of this closely related species and, perhaps, better understanding of *D. plexippus*.

GLOBAL DISPERSAL AND EVOLUTION

Ancestral monarch populations occur throughout Central America, northern South America, and the Caribbean Islands (Ackery and Vane-Wright 1984; A. V. Z. Brower and Jeansonne 2004). After European colonization of the Americas post-1800, monarchs spread throughout the Pacific, establishing in Australia, New Zealand, Hawaii, and other small islands (Zalucki and Clarke 2004). Periodic historical reports exist of monarchs appearing in England following major weather events (described in Pyle 1999). In Chapter 22, Fernández-Haeger and colleagues summarize reports of persistent monarch colonies on islands across the Atlantic Ocean and in Spain, Portugal, and Morocco. Although many sightings of monarchs in Europe are isolated events, Fernández-Haeger et al. document successful breeding populations that have become established following the introduction and spread of host plants.

Until recently, a major gap in our understanding of monarchs concerned the extent of genetic differences between populations. On one hand, past work supports evolutionary divergence in monarch wing morphology and coloration across multiple migratory and nonmigratory populations (Davis et al. 2005; Altizer and Davis 2010), as might arise from selection pressures driven by local climatic differences and migratory behaviors. Nevertheless, as Pierce and colleagues discuss in their chapter, only very recently have modern molecular genetics tools been applied to determine the extent of genetic similarity across wild monarch populations. The recent development of microsatellite markers for monarchs (Lyons et al. 2012) and publication of the monarch genome (Zhan et al. 2011; Zhan and Reppert 2013) offer the potential to explore many aspects of monarch biology from a genetic perspective. For example, Pierce et al. show evidence for high rates of gene exchange among mainland monarchs from Central and North America, but evidence for genetic isolation of island populations that are surrounded by large water bodies. Further genetic analyses are helping to resolve the historical routes by which monarchs colonized new areas and have identified genes associated with the propensity for long-distance migration (Zhan et al. 2014, and Pierce et al. 2014). For example, of the more than 16,000 described protein-coding genes of monarchs (Zhan and Reppert 2013), several are already known to affect flight metabolism and navigation. More recent genome-wide analyses from approximately 20 locations globally showed evidence for genetic differences between migratory and non-migratory populations and will pave the way for exciting future research (Zhan et al. 2014).

Two chapters in this section (Pyle, Chapter 21, and Pierce et al., Chapter 23) shed light on the long-running debate regarding interpopulation mixing between monarchs from eastern and western North America. This debate is relevant for current USDA restrictions on the long-distance transfer of live monarchs for commercial sale and release (Brower et al. 1995; Bartel and Altizer 2012). The observation of distinct wintering sites in California (west) and central Mexico (east), combined with the Rocky Mountains as geographic barriers to movement, suggested that eastern and western monarchs were geographically and genetically distinct. However, Pyle summarizes recent evidence that some degree of population mixing occurs and that the previous model of distinct populations is outdated. These findings are especially true for monarchs originating in the southwestern United States, which have been reported to overwinter in both areas. Neutral molecular markers further support a lack of genetic differentiation between eastern and western monarchs (Lyons et al. 2012). On the other hand, Stevens and Frey (2010) showed a lack of correspondence between the numbers of overwintering monarchs in eastern and western North America, indicating that drivers of abundance could differ between east and west. Furthermore, the dynamics and virulence of a protozoan parasite, *Ophryocystis elektroscirrha*, also differ between eastern and western North America (reviewed by Altizer and de Roode, this volume, Chapter 7), reinforcing the idea that ecological and evolutionary differences exist across populations despite the absence of differentiation based on neutral molecular markers.

POPULATION ESTIMATES AND TRENDS

In the final chapter in this section, Ries and colleagues address issues of monarch abundance and population ecology that are paramount to monarch conservation. Only recently have rich data sets and analytical tools become available to allow careful analyses of the variation in abundance over space and time, and identification of key determinants of monarch abundance. Ries et al. compare estimates of population size from multiple citizen science data sets to determine the impact of monarch numbers at each phase on numbers during the following phase. Pinpointing where in the life cycle the numbers do not match up from one phase to the next allows us to identify stages where a high degree of mortality (or population augmentation) takes place. In fact, this work suggests that the two most perilous stages of the annual migratory cycle for eastern North American monarchs are the spring and fall migration.

Recent work shows sobering evidence of a long-term decline in the size of overwintering colonies (and this decline has been pointed out by Brower et al. 2012a; Vidal and

Rendón-Salinas 2014) and a strong relationship between the declining availability of breeding habitat in agricultural areas and overwintering colony size (Pleasants and Oberhauser 2012); however, standardized counts of breeding (Ries et al., this volume, Chapter 24) and fall migratory (Davis 2012) monarchs do not show this same decline. These discrepancies could indicate that mortality during migration is increasing over time, or that many monarchs are migrating to destinations other than the traditional colonies in central Mexico; clearly, more work is needed to tease out the factors that drive monarch population numbers from one year to the next.

On the positive side, the answers to these complex questions are now within reach, thanks to the large number of citizen science monitoring programs (Oberhauser et al., this volume, Chapter 2), the long-term data sets they generate, and the multitude of dedicated individuals who volunteer their time to advance scientific knowledge about this amazing insect. Only by understanding the factors that are causing declining numbers will we be able to address them.

18

Tracking the Fall Migration of Eastern Monarchs with Journey North Roost Sightings

New Findings about the Pace of Fall Migration

Elizabeth Howard and Andrew K. Davis

We used sightings of fall roosts submitted to Journey North to derive estimates of the pace of migration throughout the central flyway, as well as for discrete time periods within seasons. We regressed the date of all sightings (2005–2011, $n = 1284$) against their latitude to determine the change in latitude per day, which was converted to distance. The migration progresses southward at a rate of 32.2 km/d during fall migration, and the rate is slower in the first half (~13 km/d) than in the second half of the season (~42 km/d). The increased pace is not because the time spent at stopover sites becomes shorter later in the season. Our estimate of migration rate was slower than prior estimates of individual flight speed, because it includes time for both flight and stopover time. The migration rate increased in the 7 years examined, and the first 20 roost sightings from each year increased in latitude over this time. This pattern may be an indication of breeding range expansion, which is occurring now in other nonmigratory butterfly species, and may result in farther migration distances for monarchs. Citizen science data will be critical for identifying future changes in these patterns.

INTRODUCTION

To conserve the migration of monarchs in eastern North America, we need a thorough understanding of all aspects of this migration. In the past decade numerous advances have been made in this area, most notably with the use of citizen science observations. The Journey North program (Journey North 2013) has been especially effective at advancing scientific understanding of both spring and fall migration biology because of its continent-wide scope and wide range of observational data. In this program, participants submit sightings of monarchs online; their observations are used to generate maps that track the spring and fall migrations in real time.

The sightings of nocturnal roosts submitted to Journey North have already been used to track the southward migration flyways of the eastern population; they show one main "central" flyway in North America that points directly to the Mexican overwintering sites, and a second, smaller flyway along the Atlantic coast (Howard and Davis 2009). In addition, a recent study examined the online notes made by observers, including using GIS techniques to document both the actual trees used by roosting monarchs and the landscape characteristics around roost sites. This analysis showed that these "habitat preferences" changed throughout the flyway (Davis et al. 2012a). The primary trees used by roosting monarchs included pines and maples in the northern regions, and oaks, pecans, and willows in the southern United States. Few clear preferences were shown for particular landscape features in selection of roost sites except in the Texas area, where most roosts were in landscapes dominated by grasslands (Davis et al. 2012a).

Here, we use Journey North roost data to estimate the pace of the fall migration. This information will allow us to estimate transit times through regions where habitat conservation is of utmost importance, such as in Texas, where monarchs show a heightened degree of habitat preference (Davis et al. 2012a) and where a lack of nectar resources can have a negative impact on the whole population (Brower et al., this volume, Chapter 10). The ability to estimate changes in the pace of the migration will allow us to assess potential impacts of anthropogenic changes that may alter patterns of fall migration. For example, the removal of agricultural milkweeds from farms in the Midwest (by the use of genetically modified crops that allow for widespread herbicide use, Pleasants and Oberhauser 2012; Pleasants, this volume, Chapter 14), an area that historically produced a large portion of the migratory generation (Wassenaar and Hobson 1998), could alter the fall migration by shifting the breeding distribution away from regions with intensive agriculture. Human activities are also resulting in increasing temperatures, which may cause monarchs to shift their breeding range northward (Batalden et al. 2007). If breeding ranges change over time because of either of these phenomena, we might expect corresponding changes in fall roosting patterns.

Here, we report on the pace of the entire migration, and whether that pace changes throughout the flyway (i.e., does the migration speed up or slow down as the butterflies get closer to their destination?). In addition, we screened the notes associated with a subset of the observations to determine how long monarchs typically spend at roosts and tested whether this changes throughout the flyway. Finally, we put this information into context by summarizing prior estimates of monarch migration rates and flight speed.

METHODS

Journey North roost observations

Detailed descriptions of the Journey North program are provided elsewhere (Howard and Davis 2009, 2011). Briefly, every fall since 2005, Journey North participants have been encouraged to report observations of nocturnal roosts, which monarchs form during their fall migrations, and which can be of any size (often hundreds or even thousands of monarchs). All roost observations are archived online (Journey North 2013). Each observation is associated with a date (of the first night of observation), latitude and longitude (of the center of the town in which the roost was seen), as well as anecdotal notes about the roost itself. For the purposes of this study, we used all fall roost data from 2005 through 2011. Furthermore, we used roosts only in the primary, central flyway, as defined in a prior study (Howard and Davis 2009), since monarchs from Atlantic coastal locations make up only a small fraction of the overwintering cohort in Mexico (Wassenaar and Hobson 1998); thus, we did not include data from states or provinces on the Atlantic coast. These criteria resulted in a total of 1284 roost observations over seven years (Figure 18.1), although some locations were represented in multiple years, and even multiple times within years if two roosts were seen in the same town. The sample sizes for each year, from 2005 through 2011, were 178, 143, 231, 129, 153, 296, and 154, respectively.

Estimating migration pace

Since monarchs are moving primarily southward during the fall migration, the latitude of roost sightings becomes progressively lower over the course of the migration season (Howard and Davis 2009). With this in mind, we plotted the latitude of all sightings in a given year against the date of the sighting (in days since 1 January), resulting in a scatterplot with a downward-pointing pattern for each year (Figure 18.2). Then, we fitted a linear regression line to these data, with the slope of this line representing the *rate of reduction in latitude per day*, or in other words, the pace of the *southward* migration. Using this regression approach we obtained a single value for each year that reflects the average southward pace of the entire migration throughout the whole flyway, although this number does not take into account possible variation within seasons, which was one of our goals. Therefore, we obtained separate rates (using the procedure above) for four time periods within each season, which we arbitrarily defined as 20-day intervals starting at day 220 (10 August) and ending on day 300 (27 October). The number of roost sightings within each time interval (1–4) was 185, 412, 368 and 287. While the migration continues past 27 October, insufficient data points were reported beyond 27 October to analyze. In the end we had estimates of migration rate (i.e., slopes of the time-latitude plots) for 4 time intervals over 7 years ($n = 28$) for statistical analyses (see below).

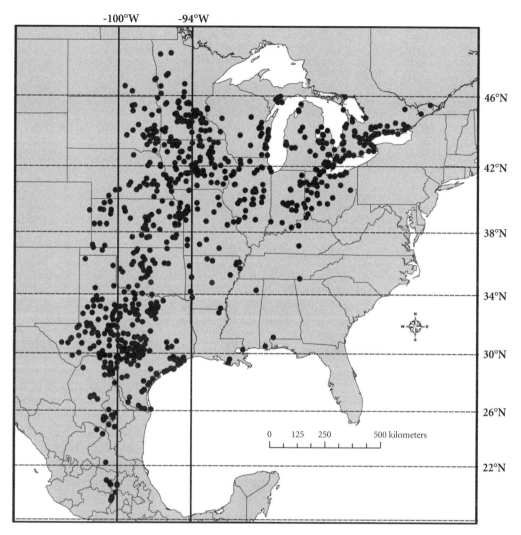

Figure 18.1. Locations of monarch roost observations submitted to Journey North between 2005 and 2011. Note that roosts from states bordering the Atlantic Ocean are not shown; this study used roost observations from only the main "central" flyway. Solid vertical lines indicate region used to examine the pace of "southward-only" migration (see methods for description).

Early in the migration, the primary flight direction for monarchs from the northeastern United States is to the southwest (Figure 18.1). This direction may result in slower rates of southward advancement in the early phase of the fall migration; therefore, we repeated the steps above (linear regression of date and latitude) for a narrow range of roost sightings running down the approximate center of the flyway (from −100°W to −94°W longitude, see Figure 18.1). Restricting the analysis to these points ensured that the estimates of migration pace reflected "southward-only" advancement. Too few roosts were sighted in certain years to obtain separate rate values for each time period, so for this subset we determined the annual migration rate for periods 1 and 2 combined, and for periods 3 and 4 combined.

Roost duration

For a subset of the roost data (2005–2008) we screened the written notes submitted along with the observations, recorded how many nights the roost was occupied, and categorized these data according

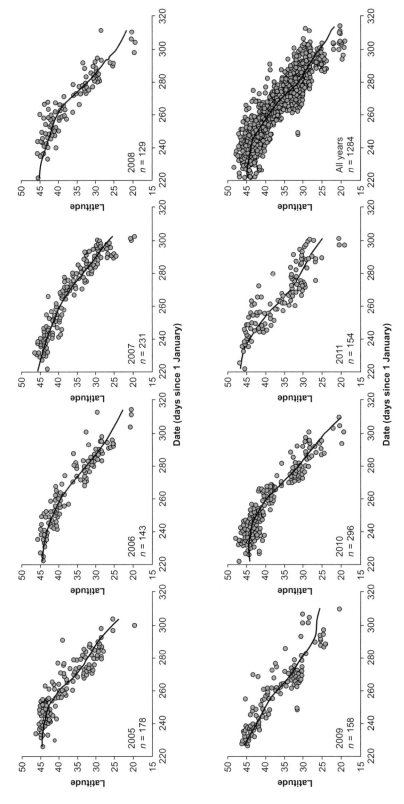

Figure 18.2. Plots of the latitudes of roost observations (y-axes) in relation to the date (in days since 1 January, x-axes) for all years separately and combined. A distance-weighted least-squares regression line is fitted to each plot.

to the four 20-day time intervals indicated above. Since not all observations contained this information, the final sample size here was $n = 158$.

Data analyses

We examined the migration rate data from the entire flyway (the slopes of the time-latitude scatterplots, $n = 28$) with a two-way ANOVA, where the time interval and year were predictor variables. Year was included as a categorical predictor variable, since we had no a priori expectation of an increase or decrease over the years of the study (but see results). This test therefore examined whether the rate of migration changed over time within each season, and whether it was different from year to year. We next examined the rates obtained using the subset of roosts from the center of the flyway (see Figure 18.1). Here, we compared the rates of the first half of the migration (periods 1 and 2, pooled, $n = 7$) to those from the second half (pooled, $n = 7$) using a t-test. The data set reflecting the roost durations ($n = 158$) was not normally distributed, so we log-transformed (+1) these values to approximate a normal distribution. We used a two-way ANOVA to examine roost durations across the 4 time intervals with year included as a categorical variable, although we had data spanning only 4 years in this subset.

RESULTS

With roost observations from all years considered together, including those after 27 October, the slope of a linear regression line fitted to the time-latitude graph (Figure 18.2) was −0.29. In other words, the migration progressed southward at a rate of 0.29° (or 32.2 km) per day. In the ANOVA examining migration rates within seasons, there was no effect of year ($F_{6,18} = 1.54$, $P = 0.221$) when it was included as a categorical predictor (but see below); however, the pace of the migration did vary with time interval ($F_{3,18} = 26.07$, $P < 0.001$). To depict this variation, the average rates across intervals are graphed in Figure 18.3. Based on Tukey's post-hoc tests, the rates for the first two time intervals (10 August–17 September) were not significantly different from each other, nor were the rates of the last two intervals (18 September–27 October); however, the pace of the migration (i.e., the rate of southward-only movement) approximately doubled in the second half of the fall season; in the first half it was between −0.06° and −0.17° latitude (7–19 km/d, or a combined average of 13 km/d), and in the second it is between −0.32° and −0.42° per day (36–47 km/d, or a combined average of 42 km/d). The relatively slower pace of the initial part of the migration can also be visualized by closely examining Figure 18.2. In most years there is little to no southward movement near the beginning of the season; this is especially evident if a distance-weighted regression line is fitted to the points. Examination of migration rate for the subset of roosts between −100°W and −94°W longitude showed a similar pattern; a significantly slower average pace during periods 1 and 2 (mean = −0.22°/d) compared with periods 3 and 4 (mean = −0.31°/d; t-test, df = 12, $t = 3.45$, $P = 0.005$). The faster early migration for this subset (0.22°/d vs. 0.06°–0.17°/d) reflects the fact that butterflies moving southwest were removed from the analysis.

On average, roost durations lasted about 2 nights considering all roost observations for which we had this information ($n = 158$). There was no significant change in durations across time intervals ($F_{3,151} = 2.07$, $P = 0.107$). We found an unexpected effect of year ($F_{3,151} = 3.53$, $P = 0.016$), but upon further examination it appears this effect was driven by one year,

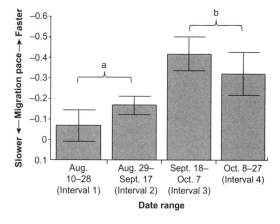

Figure 18.3. Average migration rates during the four date ranges (time intervals) used in this study. Rate was the slope of the regression line in a plot of latitude versus date, so values indicate the degrees of latitude covered per day (they are negative because latitudes become lower as the migration progresses southward). Values on the y-axis are shown in reverse order so that faster rates are at the top. Whiskers on bars represent 95% confidence intervals. Letters above bars indicate homogeneous groups based on Tukey's post-hoc tests.

2008, when roost durations averaged about 3 nights in length (significantly greater than other years, Tukey's post-hoc tests, $P = 0.020$).

Finally, when we inspected the migration rates for each year, we noticed that the annual rates appeared to be increasing over subsequent years. To test this observation, we examined the same data using ANCOVA, with time interval as a predictor as before, but with year included as a continuous covariate. This approach takes into account the ordinal nature of year (i.e., 2007 follows 2006, which follows 2005, etc.), which the ANOVA does not. Again there was an effect of time interval ($F_{3,23} = 29.29$, $P < 0.001$), but more importantly, in this model the effect of year was significant ($F_{1,23} = 7.62$, $P = 0.011$). The direction of this effect can be seen in Figure 18.4; when we averaged the 4 slope estimates (from each time interval) for each year and plotted them against year, there was a highly significant correlation ($r = -0.86$, $P = 0.014$). In other words, within the 7 years of data examined here, it seems that the migration progressed faster over time. To ensure that this pattern was not an artifact of increasing participation in Journey North observations, we compared the number of roost observations with year using Pearson correlation and found no significant relationship ($r = 0.20$, $P = 0.664$). To help interpret this pattern we extracted from the roost data the first 20 sightings from each year and compared their latitudes and dates across years. There was a small but positive correlation between year and latitude ($r = 0.21$, $p = 0.013$), but no relationship between date and year ($r = 0.02$, $P = 0.734$). This means that the fall migration did not change in terms of when it started, but in the 7 years we examined the first roosts sighted shifted northward (from about 44°N latitude to about 45°N, or about 100 km).

DISCUSSION

Journey North roost data indicate that the overall southward pace of the fall monarch migration is about 32 km/day. In other words, new roosts are formed about 32 km farther south than in the prior day (although this does not take into account variation within seasons; see below). This pace is not an estimate of the flight speed of individual butterflies; rather, this estimate includes both flight and stopover time (the latter reflecting time for feeding, resting, and waiting for appropriate wind and weather conditions). To demonstrate this point, we compiled a list of published and unpublished estimates of individual flight speed of monarch butterflies (Table 18.1). Assuming that migrating monarchs spend approximately 10 hours per day in flight, our daily estimate translates to about 3.2 km/hour, on the low end of flight speed estimates (Table 18.1). For example, Moskowitz et al. (2001) watched individual monarchs flying during one fall day (when exceptionally large numbers were flying) and estimated their flight speed at 7.2 km/h. Garland and Davis (2002) reported that a tagged monarch flew 226 km in a single day (although with a strong tailwind), a speed of approximately 14 km/h. On the other hand, in terms of the pace of the entire migratory cohort, the estimate we obtained here is consistent with dividing the total distance of the migration by the total time of the migration. If we consider the entire migration distance of approximately 3000 km (from northern Minnesota to Central Mexico), and the typical duration of the entire migration season, which is roughly 85 days (based on first roost reports and the arrival dates at the overwintering sites), the result is 35 km/d.

As we found with the spring migration (Davis and Howard 2005), the pace of the fall migration varies throughout the season; the migration appears to speed up in the second half of the season. The seemingly slow pace of the migration in the first half may be partly influenced by the fact that many of these

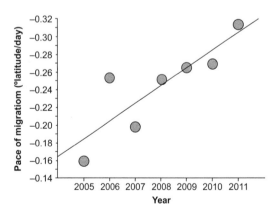

Figure 18.4. Plot of the annual pace of fall migration from 2005 to 2011. Each point on the graph represents the average rate (slope of the time-latitude graph, in degrees per day) of the 4 time intervals. $R = -0.86$, $P = 0.0139$. Values on the vertical axis are reversed so that faster rates are higher.

Table 18.1. Summary of published and unpublished estimates of monarch flight speed

Season	Estimate	Method of estimation	Source
Fall	18 km/h	Tracking "cruising" monarchs by car (and reading car speedometer).	Urquhart 1960
Fall	14 km/h	Calculated from the recapture of a monarch in eastern Virginia that was tagged the previous day at Cape May, NJ (226 km away). There was a strong tailwind that day.	Garland and Davis 2002
Fall	7.5 km/h	Derived from the time it takes the fall migration wave front to go from Minnesota to the Mexico border (~2250 km over 30 days, or 75 km/d).	MonarchWatch website
Fall	7.2 km/h	Viewing low-flying migrating monarchs as they passed by an open parking lot during an exceptionally large flight day in fall 1999.	Moskowitz et al. 2001
Fall	3.9 km/h	Average speed of 100+ healthy monarchs when attached to a flight mill and monitored remotely by computer. All were reared in captivity under late summer conditions.	Davis et al. 2012b
Spring	71.5 km/d or 7.2 km/h	Using GIS to measure rate of expansion of the spring migration wave front from sightings of adults submitted to Journey North. Average rate over 7 years used here.	Davis and Howard 2005
Spring	24 km/d or 2.4 km/h	Based on the slope of a regression of oviposition date and latitude.	Cockrell et al. 1993

Notes: Data in this table are for comparison with the estimate of migration pace from this study (which takes into account both flight time and stopover time). In cases where rates or speeds were reported in km/d, we report them here as km/h, assuming monarchs fly for 10 hours in a day.

monarchs are migrating southwest (Figure 18.1), reducing the overall rate of southward advancement; however, restricting our analyses to the subset of roosts observed in the center of the flyway (so that only southward movement is captured, Figure 18.1) also showed slower average rates in the first half of the migration. We also point out that the total number of roost sightings in the first half of the migration ($n = 597$) was roughly equivalent to the total number in the second half ($n = 655$), meaning these results should not have been influenced by uneven distribution of observers. Thus, we can be confident in concluding that the rate of southward advancement really is slower during the first half of the migration.

Given the difference in migration pace throughout the flyway, it seems surprising that roost *durations* did not vary over the course of the season; roosts tended to last about 2 nights on average, regardless how far along the migration was. If the overall migration pace quickens as the cohort moves south, one would expect roost durations to shorten as the migration advances (if monarchs spend less time at stopover sites). A possible explanation for this apparent discrepancy lies in our assumption that roost "durations" reflect actual stopover lengths of individual monarchs, which may or may not be the case, and can really be addressed only by using tagging data. In fact, two prior investigations using tagging data found that individual stopover lengths were actually longer in the southern site (in South Carolina) than in a more northern site (in Virginia) (Davis and Garland 2004; McCord and Davis 2012). Thus, if the migration advances more quickly as the season progresses, it does not appear to be because the time spent at stopover sites becomes shorter. If this is the case, then the only other explanation is that the monarchs must cover more ground during the day as the season progresses, either with faster individual flight speeds or by simply flying with fewer temporary daytime stops at ground sites.

Perhaps the most intriguing result of this study was the one we did not set out to examine; over the 7 years of roost observations we examined (2005–2011), the pace of the fall migration increased (Figure 18.4). This trend should be monitored closely in the future; indeed, there are good reasons to expect that certain aspects of the fall migration will change because of climate change (Batalden et al. 2007) or the loss of agricultural milkweeds (Pleasants and Oberhauser 2012), either of which could shift breeding ranges. In fact, this may already be happening. The northward shift in latitudes of the first roost sites each year suggests that the breeding areas of monarchs may be shifting northward, a phenomenon that was predicted to occur in response to climate change (Batalden et al. 2007). Climate-driven, northward range expansions are also being seen in other (nonmigratory) butterfly species (e.g., Crozier 2004b; Finkbeiner et al. 2011; Pateman et al. 2012). If this trend continues with eastern monarchs, their overall migration distance will also increase (assuming the ultimate destination will remain the overwintering sites in Central Mexico). Given the many risks associated with long-distance migration (McKenna et al. 2001; Howard and Davis 2012), anything that prolongs this sensitive period could ultimately result in fewer monarchs surviving the migration.

Finally, we point out that the information obtained in this study highlights how advances in monarch conservation and biology can be, and are continuing to be, made possible thanks to the dedication of the citizen scientists who participate in this and other monarch monitoring programs. Only by using large-scale data sets covering many years can we address important questions relating to migration pace, habitat selection, and climate impacts. With the answers to each new question, our ability to conserve this fascinating insect continues to improve.

ACKNOWLEDGMENTS

We thank the thousands of Journey North participants who have contributed observations of monarch roosts. Lincoln Brower has provided expert advice to the Journey North program over the years, and Kristen Baum, Steve Malcolm, Patrick Guerra, and Karen Oberhauser provided helpful comments on the manuscript. Funding for Journey North was provided by Annenberg Learner, a division of the Annenberg Foundation.

19

Potential Changes in Eastern North American Monarch Migration in Response to an Introduced Milkweed, *Asclepias curassavica*

Rebecca V. Batalden and Karen S. Oberhauser

We investigated fall and winter breeding behavior by eastern North American migratory monarchs. Using data from two citizen science projects (the Monarch Larva Monitoring Project and Journey North), we documented monarch egg and larva presence in Texas and other Gulf Coast states throughout the fall and winter. An experiment with caged migratory butterflies suggested that fall reproduction occurs in response to the presence of milkweed in good condition. Citizen science data and our own monitoring showed that female monarchs prefer the non-native *Asclepias curassavica* during the fall, and it is possible that the presence of this species triggers diapause termination. *A. curassavica* in Texas is more abundant in the fall and in better condition than the local native milkweeds. Our finding that monarch eggs and larvae are present throughout the fall and winter shows that some monarchs migrating southward in the fall reproduce, as do at least some of their offspring; however, our experiments with caged monarchs suggest that the vast majority of monarchs do not break diapause and reproduce, and that most of the offspring of those that do reproduce are in diapause.

INTRODUCTION

Each fall, monarchs in the eastern migratory population undertake the most well-known insect migration. As they fly from their breeding grounds to overwintering habitat in Mexico, most monarchs are thought to be in a state of reproductive diapause (Brower 1985, 1995). They remain in this state throughout the winter, until they travel north in the spring to lay eggs in the southern United States and northern Mexico. Diapause allows individuals to conserve energy and survive the migration and winter (Herman 1981; Herman and Tatar 2001), and to delay reproduction until conditions are favorable.

Recent monitoring efforts show adults, eggs, and larvae present throughout the fall and winter in Texas and along the U.S. Gulf Coast (Prysby and Oberhauser 2004; Howard et al. 2010). Monarchs are, for the most part, absent from this part of their breeding range during the summer (Prysby and Oberhauser 2004), presumably because of hot conditions (Batalden et al. 2007) and the dieback of most native milkweeds. Thus, we assume that monarchs breeding in the south in the fall have flown from the north. Note that we are not referring to continuously breeding resident, and likely nonmigratory, monarchs in southern Florida (well documented by Knight and Brower 2009). Rather, our references here to fall and winter monarchs are to those in the southern, temperate United States (Texas and the Gulf Coast states north of South Florida). We use the term "fall breeding" to refer to breeding in the southern United States by migrants from farther north, occurring primarily in September and October (although the timing of departure from the north suggests that late migrants may arrive in the southern United States as late as November). Fall breeders are often called "premigrants" if they precede the main migration. "Winter breeding" refers to breeding by the progeny and grandprogeny of fall migrants; we assume that it occurs from approximately 15 November to 15 March. Finally, "spring breeding" refers to breeding by remigrants from Mexico (or by the offspring of winter breeders), which occurs after 15 March. Note that these temporal divisions are approximate; there may be overlap across both the fall to winter and winter to spring divisions.

Two nonexclusive hypotheses could explain breeding during the fall and winter in southern U.S. areas other than those used by the permanent breeding populations in Florida. First, some monarchs may not enter diapause in their natal grounds and thus lay eggs as they migrate south, similar to the egg laying that occurs as they move north in the spring. The fact that migratory behavior and diapause are not always coupled (Pérez and Taylor 2004; Zhu et al. 2009) allows this hypothesis. Second, some monarchs could be breaking diapause (undergoing diapause termination) in response to environmental cues as they migrate. Many gardens and parks throughout the southern United States contain propagated tropical milkweed, *Asclepias curassavica*. While the native range of this species is conjectural, it is not native to the United States, but rather to South or Central America or Mexico (Woodson 1954). Woodson noted that tropical milkweed appeared only occasionally in Florida, Louisiana, Texas, and California, suggesting that it is more common now than it was 60 years ago. Because gardens are often watered throughout the summer, this species remains healthy and possibly attractive to migrating monarchs into the fall. It has also escaped into nongarden locations. In some places in southern Florida, large fields of *A. curassavica* grow wild in pastures (L. P. Brower, pers. comm.); this is also true in wet areas in southern Spain (Fernández-Haeger et al., this volume, Chapter 22).

Offspring of fall breeding individuals could either join the wintering population in Mexico, or stay in the United States. Monarchs continue to arrive in the Mexican overwintering colonies through December (Eduardo Rendón, pers. comm.), timing that is consistent with the first possibility. These monarchs could either be in diapause or not; some mating occurs in the overwintering sites throughout the winter (Oberhauser, pers. observ.), and it is possible that late arrivers are less likely to be in diapause. Any that stayed in the southern United States could either overwinter in reproductive diapause, or could be reproductively active. There is evidence that some diapause individuals remain in the United States; for example, Brower and Calvert (unpublished records)

observed overwintering clusters during the late 1980s on Honeymoon Island, Florida (near Tampa), and observers in Virginia Beach, Virginia, tagged a wild monarch on 25 September 2005 and recovered the same individual on 2 March 2006 (Journey North 2013). It is unlikely that a breeding individual could live for five months; additionally, Howard et al. (2010) documented overlap of monarch eggs, larvae, and adults along the Gulf Coast and up to the Georgia-South Carolina border, but only adults on the South Carolina, North Carolina, and southern Virginia coasts. This suggests that the more northern individuals were in diapause. However, the presence of monarch eggs and larvae in Texas and along the Gulf Coast throughout the winter (Howard et al. 2010 and this chapter) suggests that at least some of the offspring of fall breeding individuals remain in this region and themselves reproduce.

While we have no long-term records of monarchs in the eastern United States breeding throughout the winter except in southern Florida, there is evidence that monarchs are capable of changing their migration patterns and distributions over relatively short time scales (Vane-Wright 1993). Monarchs extended their range in the mid-nineteenth century to islands in the South Pacific and arrived on Australia's east coast around 1871 (Vane-Wright 1993; Clarke and Zalucki 2004; Zalucki and Clarke 2004). Until recently, many monarchs in Australia migrated to avoid hot and dry summers (Dingle et al. 2000). Now increased irrigation by suburban sprinklers allows milkweed to persist throughout the summers, and some monarchs in Perth and Adelaide no longer migrate (Zalucki and Rochester 1999).

Here, we investigate the two hypotheses for factors that lead to fall breeding by monarchs in the eastern migratory population, and we evaluate evidence of the degree to which the new fall generation migrates to Mexico or remains to breed or overwinter in the southern United States.

METHODS

Assessing monarch presence

We used 2000–2010 monarch sightings by volunteers in the Monarch Larva Monitoring Project (MLMP 2013) and Journey North (Journey North 2013). MLMP data came from regularly monitored sites, anecdotal sightings reported to the MLMP website, and e-mail reports to Oberhauser. Volunteers at regularly monitored sites survey a random selection of milkweed (or all milkweed plants on smaller sites), record the number of plants they observe and the numbers of eggs or larvae observed on these plants, and enter their data into an online database. These regular surveys provide presence and absence data. In addition to reporting data on monarch density, MLMP volunteers report several characteristics of their sites. Data relevant to this research include location (to geographical coordinates identified through Google Maps or GPS), milkweed species at the site, and practices such as planting milkweed, fertilizing, or weeding. Detailed methods are described elsewhere (Prysby and Oberhauser 2004; MLMP 2013). Anecdotal sighting reports allow volunteers to enter monarch sightings they observe off-season or in locations other than regularly monitored sites; the online form has fields for date, location, and monarch stage. Finally, archived e-mail reports were used if they included date, location, and monarch stage data.

Journey North data included reports of monarchs observed in January and February (winter sightings). These reports included a location (although only to zip code), and stages of monarchs and other data were gleaned from the comment fields. Because our analysis does not require detailed information about specific sites, the lack of precise location measurements does not affect our conclusions. If a sighting could not be identified to a specific date, location, and stage (egg, larva, pupa, or adult), we did not include it. For either project, if only eggs were reported at a site during a particular time period, we did not include them, since monarchs and queens (*D. gilippus*) co-occur, have eggs that are indistinguishable, and use the same host plants. Their larvae, however, are easily distinguishable.

To document the timing of fall breeding in more detail, we used MLMP data to identify peak weeks of egg density (weeks that are preceded and followed by multiple weeks of lower values) at sites through the northern, middle, and southern parts of monarchs' central flyway (defined by Howard and Davis 2009). There are broad regional patterns in egg density, with two peaks in each region, one when monarchs first arrive and again when a second generation is produced in that location (or a second and third generation combined in the U.S. Upper Midwest) (Batalden 2011). We calculated the week of the second peak in MLMP monitoring sites in MN, WI, and IA (northern region); IL, MO, and NE (middle region); and TX (southern region). We excluded data for which egg densities were clearly inaccurate (see Pleasants and Oberhauser 2012) and, with the exception of TX sites, site/year combinations that lacked continuous data from the monarchs' spring arrival until the second peak. The latter criterion ensured that we were analyzing a true peak and not an artifact of erratic monitoring; however, in TX, many volunteers stop monitoring during the hot summer months, so this criterion would exclude most sites. MLMP data from a typical TX monitoring location show a first peak in the spring that tapers off into nothing throughout the summer, similar to the pattern documented by Cockrell et al. (1993), then another peak in the fall (Prysby and Oberhauser 2004). We included TX sites that monitored at least one week before and after the fall peak, again to ensure that we were analyzing a true peak in monarch abundance.

Assessing milkweed availability and condition in Texas

We used MLMP data from 2001 to 2009 to estimate the change in relative availability of native milkweed and the introduced *A. curassavica* over the course of a season in Texas. We identified 11 Texas sites that were monitored at a minimum from April through November in a single year. Most Texas volunteers check all the milkweed plants on their site each time they monitor, so we assumed that the number of plants they observed represented the number available. We calculated the mean number of plants examined per monitoring session at each site for the entire year and the mean number examined per session in each month, then divided the monthly mean by the yearly mean to obtain relativized values for each month-site combination. These relativized values describe how the amount of milkweed at each site varies throughout the season in a consistent way, independent of the actual number of plants in each site. We compared the relative abundance of milkweed on sites that did and did not have *A. curassavica* (Ac) with a least-squares linear regression in JMP v. 5.1.2 (SAS Institute, Inc.); the model included precipitation during the previous month, month, presence of Ac, and all month*Ac interactions as potential predictors of the relativized number of milkweed plants.

In mid-October 2007, we surveyed 14 sites, all within 230 km of San Antonio, Texas. Two of these sites contained both native milkweed and *A. curassavica*, while the other 12 contained one or the other. The two sites with both native and non-native milkweed included a backyard and a nature center and were maintained by humans. The sites with only non-native milkweed were all gardens, also maintained by humans. The sites with only native milkweed included natural areas, roadsides, and restored prairies; while some of these sites may have been managed, management was minimal and did not include regular watering or weeding. We assessed monarch egg and larval densities on every milkweed plant at each site, recording the species, condition, herbivory damage, height, and "local density" of the plants, using methods described by Prysby and Oberhauser (2004). We assessed condition and herbivory in ordered categories, with *condition* referring to the percentage of the plant that had senesced (1 = <5%, 2 = 5–40%, 3 = 41–80%, 4 = >80% yellowed or dying) and *herbivory* to the amount of damage from herbivores (1 = 0%, 2 = <5%, 3 = 5–25%, 4 = >25% eaten). Our measurement of local plant density is not overall site density, but rather patch density, measured by counting the number of milkweed plants within the square meter with each plant at its center.

Assessing diapause

We used monarchs reared in the late summer and fall in Minnesota and collected from roosts in Texas to assess the conditions that might lead to diapause termination in fall migrants. We raised monarchs from eggs laid by wild-caught individuals captured in late summer 2007 in and near St. Paul, MN. They were reared outdoors from mid-September to early October in 0.3 × 0.4 × 0.7m screen cages with about 50 individuals per cage. Larvae received fresh cuttings of naturally growing common milkweed, and their cages were cleaned daily. These rearing conditions are known to induce diapause (Goehring and Oberhauser 2002). They emerged as adults 8–15 October, and after being checked for the protozoan parasite *Ophryocystis elektroscirrha*, 167 males and 164 females were either carried by plane or shipped overnight to Texas. On 15 October, we collected 253 roosting adults (119 males, 134 females) from a stand of trees about 240 km west of San Antonio.

Butterflies were individually marked and randomly assigned to one of four 2 × 2 × 2 m mesh cages (with males and females from each location assigned separately to ensure approximately equal sex and origin ratios in each cage). One cage contained no milkweed; one, wild-growing native milkweed *A. asperula*; one, wild-growing *A. oenotheroides*; and one, potted non-native milkweed *A. curassavica* purchased from a local organic nursery. All cages contained established native Texas grasses and were arranged in a square with about 7 m between cages. Butterflies had ad libitum access to sponges filled with 20% honey-water. They were kept in the cages for 11 days and observed twice a day (monarchs remain in copula for several hours, so this observation frequency allowed us to observe all matings). If males mated, they were considered reproductive. Males that did not mate were considered to be in diapause. Our observations allowed us to document female mating, but because females in diapause sometimes mate (Goehring and Oberhauser 2004) and we did not dissect females to ascertain whether they contained oocytes, we are not sure of the reproductive status of females. All butterflies were released after the 11-day period, except the mated females, which were kept in a cage with milkweed for an additional 7 days to determine whether they laid eggs. Our use of only behavioral assessments of both male and female reproductive status introduces potential error into our conclusions; however, our research group has a policy of minimizing destructive sampling of wild individuals, and we discuss the possibility of error below.

To assess the reproductive status of the offspring of fall migrants, we collected 57 wild Texas monarch eggs and larvae in late October 2007, then reared them outdoors in a shady spot in a residential yard near San Antonio. Larvae were grouped by collection site and placed in rearing cages ranging in size from 1 liter deli containers with ventilated lids to mesh cages of 1 m^2, depending on the number of individuals per group. They were fed fresh *A. curassavica* leaves daily. When the adults (29 males, 28 females) emerged, they were shipped overnight to Minnesota, where they were immediately placed in 1 m^2 mesh cages in a greenhouse with 16 hours of light per day. If males had mated, they would have been considered reproductive, though none mated before they died. Males that did not mate were considered to be in diapause. They had ad libitum access to sponges soaked with 20% honey-water while they were in the mating cages.

RESULTS

Fall and winter breeding

The timing of the second peak of eggs across MN, WI, IA, MO, NE, IL, and TX is correlated with latitude (Figure 19.1), with later peaks farther south. Around 45° latitude (the northern part of our range), the second peak of eggs occurs, on average, at week 29, the last week in July. At 38° (the middle of our range), the second peak occurs in week 32, the second week in August. Finally, at 30° (the southern part of our range), the second peak of eggs occurs in week 39, the second week in September. The fact that the eggs in the middle and southern latitudes are usually observed after a period of absence during the summer (Prysby and Oberhauser 2004) suggests that monarchs from the north are laying eggs as they move southward in the late summer and early fall.

Table 19.1 illustrates the number of sites with immature monarchs reported by MLMP and Journey North observers in the southern United States between 15 November and 15 March each year (evidence of winter breeding), omitting observations in southern Florida (lat <30°N, long <83°W). If an immature monarch was observed at a site during a month in a given year, the site was given a value of "present" for that month. Sampling effort varies across years and months, so we have not tested for an effect of time, nor are we including absence data; we are simply showing that observers documented winter breeding every year and in most months. Immature monarchs were observed in all but 15 of the 50 year-month combinations (2001–2010 November–March). Because neither project explicitly requested winter observations, this summary underrepresents the degree to which winter breeding occurred; we have received many reports from observers who tell us that they see monarch larvae throughout every winter, but we have not used these reports because they do not specify dates or monarch stages.

There was a small, but significant association between date of immature sighting and latitude. Earlier winter sightings occurred farther north (least squares linear regression of date [where 15 Nov = 1] on latitude: coeff for lat = −1.61, T = −2.38, R^2 = 0.013, P = 0.018, N = 365). This suggests that monarchs are moving farther south to breed over the course of the winter.

Milkweed status

A. curassavica is more likely to be in managed gardens—sites that are planted, fertilized, and weeded—at least in the sites that we studied (Table 19.2). Conversations with MLMP volunteers suggest that native milkweed senesces during the summer and returns in the fall only during years with sufficient rainfall, while *A. curassavica* in maintained areas is

Table 19.1. Number of sites that reported immature monarchs in Gulf Coast states (excluding sightings from the Florida peninsula) by winter month (15 November–15 March)

Winter start year	November	December	January	February	March
2001	0	0	2	9	0
2002	1	0	5	9	2
2003	4	7	11	10	3
2004	4	11	3	4	0
2005	0	0	4	4	2
2006	2	1	0	3	0
2007	0	0	4	4	0
2008	0	2	6	10	0
2009	2	1	1	0	0
2010	3	5	8	11	10

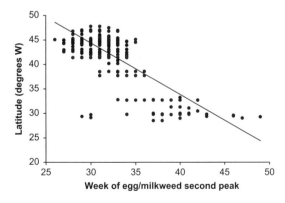

Figure 19.1. Weeks of peak egg abundance (after the first spring peak). The second peak in abundance occurs later with decreasing latitude. MLMP data from MN, IA, WI, IL, MO, NE, and TX. Linear regression equation: week = 53.19 − 0.53*latitude (R^2 = 0.568, $F_{1,420}$ = 552, P < 0.0001). Each point represents a site-year combination from 2000 to 2010. Week 27 = 8 July, Week 37 = 16 September, Week 47 = 25 November.

green throughout the year. The relative abundance of milkweed in sites with and without *A. curassavica* across the season shows strikingly different patterns. Milkweed abundance in sites with only native species peaks in the spring and is low in the summer and fall, while milkweed abundance *increases* through May and then remains stable throughout the fall in sites with *A. curassavica* (Figure 19.2, see month*Ac interactions in Table 19.3). Milkweed abundance is also predicted by the previous month's precipitation (Table 19.3). Fall 2008 and 2009 were extremely wet; the effect of these wet years is illustrated in Figure 19.2, where sites without *A. curassavica* in 2008 and 2009 are shown separately from those in 2001–2007. In August 2008, San Antonio, the nearest large city to sites analyzed for milkweed quantity, received 12.6 cm of rain. The following month, a spike in counts of native milkweed blurred the distinction between native and non-native milkweed abundance. In October 2009, San Antonio received 30.2 cm of rain and November 2009 saw a similar increase in native milkweed.

MLMP volunteers record both the milkweed species present at their sites and monarch absence data, so we could use their data to determine whether the presence of the non-native *A. curassavica* affected the likelihood of monarch presence. Of 25 MLMP sites that were monitored during the winters from 1997 to 2010, 17 contained *A. curassavica*. Because volunteers sampled unevenly over the winter, we used the entire winter period (15 November–15 March) as the unit of analysis, assigning a value of "present" if the volunteer observed any immature monarchs during this period. Sites with *A. curassavica* had monarchs present during a higher proportion of years: on average, 64% versus 31% of years on sites without *A. curassavica* (least squares linear regression of arcsin-transformed proportions weighted by the number of years of observations: coeff for Ac = 0.039, T = 2.2, R^2 = 0.17, P = 0.039, N = 25 sites). Only 3 of the 8 sites without *A. curassavica* ever observed immature monarchs during the winter, while 13 of the 17 sites with *A. curassavica* observed monarchs during at least one winter.

Across the 14 sites that we surveyed near San Antonio in fall 2007, we found averages of 1.1 eggs per *A. curassavica* and 0.016 per native milkweed plant (Figure 19.3). *A. curassavica* was in significantly better condition than the native milkweeds, but there was no difference in the amount of herbivory damage, suggesting that neither milkweed was preferred earlier in the season or by other insects (Table 19.4). *A. curassavica* plants were taller than the native species and thus probably contained more leaf area, which could explain the higher number of eggs found on *A. curassavica* plants; however, the significant difference in egg counts remained when weighted by plant height (P < 0.0001). *A. curassavica* also occurs in higher density (Table 19.4), which is expected because it was planted in gardens. With the introduced milkweed being denser and in better condition, its presence could simply attract more monarchs to these sites; however, in the two sites in which both introduced and native milkweeds occurred, there were more eggs on *A. curassavica* than on native milkweed (compare the two black bars in Figure 19.3), indicating that females are choosing not just the sites, but the plants themselves.

Table 19.2. Management practices in Texas MLMP sites with and without *A. curassavica* (Ac)

	N	Planted (%)	Fertilized (%)	Weeded (%)
Sites with Ac	24	88	38	67
Sites without Ac	26	15	4	19

Table 19.3. Model results for the analysis of relativized milkweed counts for Texas sites, 2001–2009

Term	Coefficient	Std error	tratio	P
Intercept	0.79	0.06	12.80	<0.0001
Prev. Month Precip.	0.05	0.01	4.13	<0.0001
March	−0.47	0.25	−1.86	0.06
May	0.66	0.12	5.39	<0.0001
June	0.45	0.12	3.68	0.0003
August	−0.24	0.14	−1.77	0.08
November	−0.27	0.14	−1.93	0.06
December	−0.32	0.17	−1.83	0.07
Ac * April	−0.38	0.16	2.39	0.02
Ac * May	−0.48	0.12	3.95	<0.0001
Ac * June	−0.34	0.12	2.78	0.01
Ac * August	0.33	0.14	−2.40	0.02
Ac * September	0.44	0.12	−3.59	<0.0001

Notes: Results correspond to Figure 19.2. Data were insufficient to include February, and July is the dummy variable. All terms where the coefficient estimate ± 1 SE does not include zero and with $P < 0.09$ are included. Overall $R^2 = 0.633$, $F_{20,159} = 12.0$, $P < 0.001$.

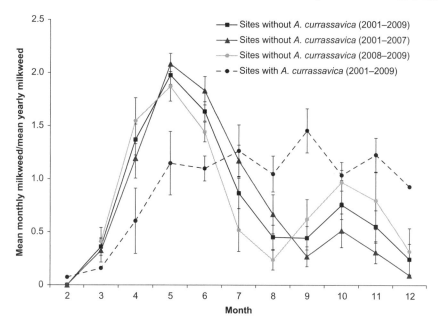

Figure 19.2. Relativized milkweed abundance in Texas. Ratio of mean monthly milkweed counts to mean yearly counts in sites, across months and in the presence and absence of *A. curassavica*. Sites without *A. curassavica* are shown across all years, and also separately for 2001–2007 and 2008–2009, as precipitation was markedly higher in 2008 and 2009. Sites with *A. curassavica* have more milkweed during the summer and fall, except when fall rains promote regrowth of native milkweed. Error bars are standard error.

Monarch reproductive condition

The probability of mating did not differ between the Minnesota-raised monarchs and monarchs collected while roosting in Texas; 2 out of 119 wild and 9 out of 167 Minnesota male monarchs mated (Fisher's Exact Test, $P = 0.13$). Butterfly origin also had no effect on the number of days that elapsed between being put in a cage and mating; the two wild-caught butterflies mated 4 and 5 days, and the Minnesota butterflies 3–8 days after being put into the cage. Given these data, we assumed no difference in reproductive condition at the start of this experiment, with the majority of both groups being in diapause and migratory. In addition, butterflies congregated in the southwest corner of the mating cages, not tracking the sun throughout the day, as nonmigratory summer butterflies do (pers. observ.).

No monarchs mated in the cage without milkweed or in the cage with *A. asperula*, but 7.8% and 5.6% of the males mated in the cages with *A. curassavica* and *A. oenotheroides*, respectively. Examining all four cages separately, we found significant differences in the probability of mating between the four cages (Pearson $\chi^2 = 17.06$, df = 3, $P = 0.0007$), but no significant differences between the probability of

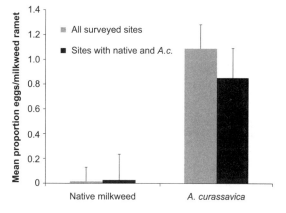

Figure 19.3. Eggs per milkweed ramet. Gray bars indicate mean number of eggs per ramet across all sites surveyed ($F_{1,172} = 22.99$, $P < 0.0001$). Black bars indicate mean number of eggs per ramet only at surveyed sites that contained both native and *A. curassavica* milkweeds ($F_{1,61} = 6.77$, $P = 0.01$). There were fewer eggs on native milkweed plants, even when they were growing in sites that also contained *A. curassavica*. Error bars indicate standard error.

mating in the cages with *A. curassavica* or *A. oenotheroides* (Fisher's Exact Test, $P = 0.76$). If we assume that something in these two cages provided a cue for diapause termination, 6.6% of them responded to this cue (95% binomial CI = 3.3–11%). If we assume

Table 19.4. Summary of milkweed quality from field study

	Condition	Herbivory	Height (cm)	Density (plants/m²)	N
Native milkweed	2.09 ± 0.09	2.34 ± 0.10	27.2 ± 1.59	2.55 ± 0.27	129
A. curassavica	1.24 ± 0.15	2.27 ± 0.17	84.7 ± 2.69	4.16 ± 0.45	45

Notes: Mean ± SE. Average condition rating (scale of 1–4) of *A. curassavica* is significantly better ($F_{1,172} = 24.1$, $P < 0.0001$). Average herbivory rating (scale of 1–4) does not differ between *A. curassavica* and native milkweed ($F_{1,172} = 0.15$, $P = 0.70$). *A. curassavica* is taller ($F_{1,172} = 339$, $P < 0.0001$), and grows in higher average densities than native milkweed ($F_{1,172} = 9.47$, $P = 0.0024$).

that all males were exposed to a common environmental cue and include all of them in the calculation, 3.8% of them responded to the cue (95% binomial CI = 1.9–6.8%). Eggs were produced in the cage in which we left the mated females, so we know that at least one of them was not in diapause after mating. Because they were unfortunately not kept in individual cages, we do not know how many of them were reproductive.

None of the monarchs (29 males, 28 females) collected as eggs or larvae and reared in outdoor conditions in Texas mated in our Minnesota greenhouse cage with extended daylight conditions (95% binomial CI = 0–12%). We did not assess female reproductive status, but this result suggests that at least a large majority of these individuals were in diapause.

DISCUSSION

We have long known that some monarchs reproduce after moving into Texas and other southern U.S. states in the fall (Calvert 1999; Borland et al. 2004; Prysby and Oberhauser 2004; Howard et al. 2010), but this analysis provides some quantification of the degree to which this occurs and documents its correlation with the presence of non-native milkweed.

We outlined two hypotheses that could explain breeding during the fall migration: (1) a subset of the migratory monarch population migrates south in a reproductive, nondiapause state, and (2) a subset terminates diapause in response to environmental cues in the south. The timing of monarch arrival in the Mexican overwintering sites, and the fact that the vast majority of monarchs arriving in Mexico are in reproductive diapause (Herman 1981; Herman and Tatar 2001) demonstrate that most monarchs are in diapause as they migrate south. However, migratory behavior is not necessarily linked to diapause (Pérez and Taylor 2004; Zhu et al. 2009), and Figure 19.1 shows that some monarchs lay eggs all along the fall migratory route. Thus, at least some monarchs are reproductive as they migrate (Hypothesis 1). However, data from the mating cages suggest that monarchs are capable of breaking diapause during the fall migration (Hypothesis 2), since the monarchs were collected at roosting sites or raised in conditions that have produced 100% diapause individuals in past studies (Goehring and Oberhauser 2002). Hypothesis 2 was also supported by Borland et al. (2004), who found that a higher proportion of fall migratory females collected in Texas than in Minnesota had mated. In our cage study, only 6.6% of the males mated in the two cages in which mating occurred, but because so many monarchs migrate through Texas, this proportion could result in significant egg production by fall migrants. Borland et al. (2004) reported that 17.5% of 645 females collected during the falls of 1998–2001 had mated, with this proportion decreasing as the fall migration progressed, also suggesting significant fall reproduction.

Our study provides insights into the cues that trigger diapause termination and the degree to which this occurs. It is very likely that the butterflies in the mating cages in Texas were in diapause and migratory at the start of the experiment, given the time that elapsed until they mated and their clustering behavior in the southwest corner of the cages. The mating was likely to have been in response to milkweed presence, as none was observed in the cage without milkweed. Goehring and Oberhauser (2004) showed that the presence of milkweed can hasten reproductive development in the spring, and our findings suggest that this cue may be effective in the fall as well. While we are confident in our interpretation that environmental conditions triggered diapause termination in some of the monarchs in the cages, it is important to note that our behavioral assessment of mating status could under- or overestimate the number of reproductive individuals. While we have used this technique often in the summer, reproductive males may have been rejected by females in diapause, or reproductive females may not have found a mate. We did not see mating attempts by the males that did not mate, suggesting that there were not more reproductive males in the cages; however, it is possible that there were reproductive females that did not mate. It is also possible that our cage studies overestimated the number of reproductive females, since diapause females may have been forced to mate (Oberhauser and Frey 1999).

Our field studies indicate a preference for *A. curassavica* as a substrate for fall oviposition, and the presence of this species could trigger diapause termination. The preference for *A. curassavica* is not apparent throughout the year; in the spring, MLMP sites with and without *A. curassavica* are used equally by monarchs (Batalden 2011). *A. curassavica* is more likely to be available and in good condition in the fall than native species, and more likely to be in sites that are planted, fertilized, and weeded; however, in our cage study, mating occurred in the cage with *A. oenotheroides*, complicating this interpretation. The conditions in our mating cages deviate from those a monarch would normally encounter; the butterflies were held in close proximity to host plants for an extended period of time. Typically, a monarch in diapause would be likely to contact host plants during brief encounters, and in most years it is less probable that a migratory monarch would encounter a native host plant in good condition. Thus, simply by being available and healthy enough to be included in this experiment, the native plants were in better condition than typical native plants in the fall. More work is needed to further quantify the role played by *A. curassavica* in diapause termination; at minimum, these data suggest that healthy milkweed presence is correlated with mating behavior.

The offspring of fall breeding individuals could join the wintering population in México, in diapause or as reproductive individuals, or they could remain in the southern United States in diapause or as reproductive individuals. Our data suggest that multiple outcomes occur. The fact that none of the individuals reared outdoors in Texas in the fall mated after being transferred to summerlike conditions in a greenhouse suggests that they were in diapause; while this does not guarantee their migratory status, it seems likely that at least some of them would continue the journey to Mexico.

Indeed, a second influx of individuals often occurs in December (Eduardo Rendón, pers. comm.). We cannot assess whether these late-arriving monarchs are more likely than early arrivers to be reproductive and thus responsible for the small amount of mating that occurs in Mexico even early in the winter. The presence of monarch roosts throughout the winter (Howard et al. 2010), very long-lived monarchs, and adult monarchs in areas where no immature monarchs are reported (Howard et al. 2010) suggest that some individuals in diapause remain in the United States instead of migrating to Mexico. The presence of immature monarchs in Texas and along the Gulf Coast throughout the winter (Howard et al. 2010 and Table 19.1) demonstrates that at least some of the offspring of fall breeding individuals remain in this region and reproduce. The fact that earlier winter sightings tended to occur further north suggests that those that do stay in the United States could move southward to avoid freezing temperatures (see Nail et al., this volume, Chapter 8). We have no solid estimates of the degree to which any of these outcomes occur, but the fact that we observed no mating by any of the 29 males and 28 females reared from larvae collected in the fall suggests that continued breeding for another generation is rare.

Winter reproduction, and thus possibly the presence of *A. curassavica*, could have both negative and positive fitness consequences. Negative effects could stem from both biotic and abiotic interactions. First, Bartel et al. (2011) showed that monarch migration allows individuals to escape from environments that are contaminated with the protozoan parasite *O. elektroscirrha* (OE). Because OE can build up on host plants that do not die back in the winter, continuous breeding in one location could lead to higher levels of infection by this potentially debilitating parasite (Altizer et al. 2000). A second, population-level impact could stem from loss of the "migratory culling effect" also suggested by Bartel et al. (2011), who showed that migration tends to result in the removal of individuals infected with OE from the population. Third, immature monarchs and their host plants may be exposed to lethally cold temperatures or suffer sublethal effects from cool temperatures that lead to slower development and thus longer exposure to larval predators and parasitoids (Nail et al., this volume, Chapter 8). *A. curassavica* cannot tolerate freezing temperatures; the plant may remain viable underground and grow again under favorable conditions, but its foliage will die. Thus, even if a monarch larva survives a freeze itself, it will not have milkweed to eat. The consequences of a lack of milkweed is likely to vary with monarch age (Nail et al., this volume, Chapter 8); eggs and first instars may have enough to eat if the plant regenerates leaves quickly, and fifth instars may be able to pupate. The average low temperature in January in Dallas is 0.4 °C; San Antonio, 4.3 °C; and Houston, 6.1 °C (Office of the Texas State Climatologist). With these average lows, a freeze could occur in a significant portion of winters, making monarch survival unlikely.

On the other hand, breeding in the south could have positive fitness effects for some monarchs. The payoff of breeding versus migrating to Mexico might be higher for individuals in poor condition, if they are less likely to survive the rigors of migration and overwintering. Additionally, individuals that encounter high-quality milkweed in the fall might, on average, experience higher fitness if they take advantage of the milkweed by breeding, even if this behavioral decision could have catastrophic consequences in some years. Monarchs sometimes experience lethal weather conditions in their Mexican overwintering sites that can lead to mass mortality events (Brower et al. 2004), and global climate change could make the sites less suitable for monarchs and the oyamel fir trees on which they roost (Oberhauser and Peterson 2003; Ramírez et al., this volume, Chapter 13). Thus, from a population perspective, behavioral plasticity that results in a portion of the population breeding in the southern United States may have positive consequences.

It is informative to compare the situation documented in this study with those in southern Florida (Knight and Brower 2009) and southern Spain (Fernández-Haeger et al., this volume, Chapter 22), locations with introduced *A. curassavica* that is present all year. In these locations, *A. curassavica* appears to have naturalized to a greater extent than in Texas, growing in pastures (Florida) and frost-free areas with sufficient moisture (springs, creeks, irrigation channels) in Spain. In these areas, monarchs breed year-round in areas not tended by humans, an occurrence that is rare in Texas, at least in most years.

Our findings demonstrate that some monarchs migrating southward in the fall and some of their offspring reproduce, providing support for the hypothesis that they are responding to environmental cues, most likely the presence of milkweed in good condition. The findings also demonstrate variability in monarch responses to these cues, and we recommend further research on the degree to which this variability is condition-dependent. While we have made some estimates of the degree to which fall breeding occurs, more work is needed to refine these estimates. Additionally, population modeling could help us understand the relative costs and benefits of fall and winter breeding versus migration to the Mexico winter sites, and, as a corollary, the costs and benefits of the availability of *A. curassavica*.

ACKNOWLEDGMENTS

We thank MLMP and Journey North volunteer monitors and trainers for their work collecting monarch and milkweed density and occurrence data, especially the MLMP volunteers in Texas who spent many hours showing Batalden their milkweed. Elizabeth Howard provided the Journey North data and, along with other monarch working group NCEAS participants, helped read through the comment fields for monarch sightings. Mary Kennedy hosted the mating cage experiments and reared larvae, and Jolene Lushine and Sarah Kempke provided field assistance. Lincoln Brower and Juan Fernández-Haeger provide useful comments on an earlier version of the manuscript. This work was supported by a grant from the National Science Foundation (DBS-0710343), the Dayton and Wilkie Natural History Fund at the University of Minnesota, and the National Center for Ecological Analysis and Synthesis.

20

Migration and Host Plant Use by the Southern Monarch, *Danaus erippus*

Stephen B. Malcolm and Benjamin H. Slager

Migration is a life history strategy employed by mobile animals to move predictably between spatially separated resources. The monarch butterfly, *Danaus plexippus*, is the preeminent insect example of migration, and much is known about its migration in North America. In contrast, little is known about its sister species, *Danaus erippus*, the southern monarch, in South America. We describe mixed life histories in *D. erippus* with nonmigratory butterflies in the Bolivian lowlands, possible altitudinal migrants in the Bolivian Andes, and latitudinal migrants in Argentina that exploit both larval and adult food resources across a "milkweed landscape" that is radically different from North America. The latitudinal migrants in Argentina show nectaring behaviors, seasonal fat accumulation, clustering activity, and directed flight behaviors that are very similar to these characteristics shown by the monarch butterfly in North America. We argue that *D. erippus* fly south in autumn in Argentina to locate sufficiently cool conditions in the eastern cordillera of the Andes at the southern end of the Yungas ecosystem, thus allowing them to conserve stored lipids through the winter.

INTRODUCTION

Understanding the processes that generate patterns of distribution and abundance in animals and plants is a primary goal of ecology. A prime determinant of these patterns is variation in resource availability in both space and time, and the life histories of species are shaped by natural selection to maximize use of variable resource availability (Holland et al. 2006; Dingle and Drake 2007). In some cases this adaptation involves migration between resources separated by space and time. The monarch butterfly, *Danaus plexippus* (L.), is the preeminent insect example of such migration, in which adults move between overwintering resources in Mexico and coastal California, and breeding resources distributed across much of North America (Brower 1985, 1995). Monarchs show complex patterns of movement to exploit their breeding resources in North America (Cockrell et al. 1993; Malcolm et al. 1993), which include more than 100 species of milkweeds in the genus *Asclepias* (Woodson 1954; Fishbein et al. 2011).

In contrast to the monarch, almost nothing is known about its sister species in South America, *Danaus erippus* (Cramer), the southern monarch (Ackery and Vane-Wright 1984) (Plate 10). We know from an early series of publications by Hayward (1928, 1953, 1955, 1962a, 1962b, 1963, 1964, 1967, 1972) that *D. erippus* moves on a seasonal basis in Argentina, and it has been suggested that this common South American butterfly is a seasonal migrant like *D. plexippus* (Williams 1958; Johnson 1969; Ackery and Vane-Wright 1984); however, unlike the situation in North America, the diversity of available host plants is very low, with only nine species in the genus *Asclepias* throughout South America (Bollwinkel 1969; Goyder 2007). Such differences provide an opportunity to compare the life histories of *D. plexippus* and *D. erippus* to determine the significance of host plant diversity in the evolution of migration.

Here, we examine *D. erippus* in Bolivia and northwestern Argentina to determine whether butterfly wing size, flight orientation, and fat content at different times of the year are consistent with migratory or nonmigratory life history predictions. We predict that migrant *D. erippus* should show seasonal evidence of directional flight orientation similar to that shown by *D. plexippus* in North America (Schmidt-Koenig 1993; Perez et al. 1997; Mouritsen and Frost 2002; Reppert et al. 2010), longer wings (Altizer and Davis 2010), and seasonal increases in fat content to fuel migratory behavior (Masters et al. 1988; Alonso-Mejía et al. 1997; Brower et al. 2006). We also examine host plant use in Bolivia and Argentina and speculate on the role of these plants in selection for both migratory and nonmigratory life histories in *D. erippus*. Since so little is known about *D. erippus*, we began our research by visiting museum collections to establish the distribution of the species in South America. We then used herbarium records of milkweed distributions to determine locations of field trips to Bolivia, northwestern Argentina, and southern Brazil to search for milkweeds and both immature and adult *D. erippus*.

METHODS

Distribution of *D. erippus* from museum records

We gathered information from pinned *D. erippus* in the insect collections of the Natural History Museum (London), Museo Noel Kempff Mercado (Santa Cruz, Bolivia), Instituto Miguel Lillo (Tucumán, Argentina), University of Campinas (Campinas, Brazil), McGuire Center (Gainesville, Florida, USA), and the Natural History Museum of the Universidad Nacional Mayor de San Marcos (Lima, Peru). Locations were then identified using Google Earth and plotted as coordinates (Figure 20.1) for comparison with published estimates of the distribution of *D. erippus* by Urquhart (1960) and Ackery and Vane-Wright (1984).

Field observations of host plant use and adult behavior

We visited Bolivia in July 2002, December 2005, May 2006, October 2009, January 2010, and April 2010, and Argentina in January/February 2010 and April 2010. During each field trip we visited locations where *Asclepias* species had been recorded in herbaria. The most significant *Asclepias* species we observed are described and pictured in Plate 9. To find *D. erippus* we also visited sites mentioned by Hayward (1962a, 1962b, 1964, 1967), at Horco Molle, west of the city of Tucumán (Figure 20.1), and the villages of El Siambón and Raco in the mountains northwest of Tucumán.

During all field trips we attempted to take vanishing bearing readings of *D. erippus* adults in flight by recording the compass bearing of adults observed through 8×42 binoculars as they disappeared from view. This required adults to show constant, directed flight behaviors, with no stops, rather than random movements among and within milkweed patches or nectar resources.

At high densities of mobile adults we also estimated the numbers of *D. erippus* by counting the number observed per minute in the 42 m field of view of 8×42 Nikon Monarch binoculars (linear field of view M = 17.45 × A° = 139.6 mm/m, at 300 m = 41.88 m) from 2:56 to 4:50 p.m. on 28 April 2010 at Horco Molle (26°46′32.75″S, 65°19′44.77″W). We estimated the width of the flight stream of butterflies at 5× the linear field of view of the binoculars, 5×42 = 210 m. Flight speed of *D. erippus* was estimated by observing the time it took individuals to fly a marked distance of 70 m.

We measured the wing lengths of 413 adult butterflies caught in the field during all field trips. The right forewing length was measured in mm from the point of wing attachment at the base to the apex at the wing tip. These butterflies were then frozen, freeze dried, ground finely in 6 ml diethyl ether with a motorized tissue homogenizer, and then centrifuged. The ether supernatant was decanted, dried, and weighed in a glass tube to measure the mass of ether-extractable fat (Brower et al. 2006).

Statistical analyses

Flight orientation data were analyzed with Oriana version 4 to generate a circular frequency distribution with a mean vector and tests of uniformity using both the nonparametric Rayleigh test and Rao's spacing test. Wing length and fat data were analyzed using JMP version 9 (SAS Institute) by examining variation against day, latitude, and altitude with linear, polynomial, and stepwise regressions. Stepwise regression models of butterfly fat content were selected on the basis of the lowest corrected Akaike's

Figure 20.1. Distribution of 524 *D. erippus* throughout South America. Triangles indicate the lat-long coordinates of individuals recorded from museum collections in Tucumán (Miguel Lillo Institute), Santa Cruz (Noel Kempff Mercado Museum), Florida (McGuire Center), Lima (Universidad Nacional Mayor de San Marcos), and London (BMNH). (a) Location of field sites (closed circles) with *Asclepias candida* and the invasive milkweed *Calotropis procera* in the vicinity of Santiago de Chiquitos (open circles) in eastern Bolivia, (b) field sites in Bolivia in the lowlands near Santa Cruz and Huaytú near Buena Vista on the edge of Amboró National Park, and the highlands near Sucre, (c) field sites near Salta in northern Argentina, and (d) field sites near the city of Tucumán and the suburb of Horco Molle to the west of Tucumán where we observed migrating *D. erippus*, and the village of El Siambón near where we observed roosting and nectaring *D. erippus* on the banks of the Río Grande.

RESULTS

Distribution of *D. erippus* from museum records

Plotted latitude and longitude data from 524 museum specimens show a distribution of *D. erippus* in South America extending south from the Amazon River in Brazil and Peru, to approximately 40°S in Argentina, with a curious record of the butterfly in the Falkland Islands (Islas Malvinas) (Figure 20.1) (Note: Gerardo Lamas pointed out to us that this butterfly was reported by Weir [1894], who stated that it was the only *D. erippus* recorded over a 20-year period in the Falklands, and that the three records of *D. erippus* from Ecuador and central Peru are dubious). It appears to be largely absent from the Amazon basin and is uncommon in Chile. Most records occur in Bolivia, Paraguay, Uruguay, the northern half of Argentina, and southern Brazil, and extend from sea level to just above 4000 m altitude. This distribution is also largely in agreement with the niche modeling predictions of Zalucki et al. (this volume, Chapter 11) for the closely related *D. plexippus*, which includes *D. p. plexippus* in North America, *D. p. megalippe* in the Caribbean and Central America, and *D. p. nigrippus*, the South American subspecies that extends as far south as Peru and the region north of the Amazon River and west of the Ucayali River (Hay-Roe et al. 2007; Gerardo Lamas, pers. comm.).

Field observations of host plants

According to Bollwinkel (1969), the South American milkweed flora is limited to 12 taxa in the genus *Asclepias* (although it is likely that three of these are actually part of other species, so it is likely that there are only nine species). Of these, only seven—*A. barjoniifolia, A. boliviensis, A. bracteolata, A. candida, A. curassavica, A. mellodora* and *A. pilgeriana* (= *A. flava*)—occur within the range of *D. erippus* (Bollwinkel 1969; Goyder 2007). We found all these species except *A. bracteolata* (Plate 9), although we searched for *A. bracteolata* near Campinas in southern Brazil without success. The most common species that occurred in our field samples in Bolivia and Argentina were *A. barjoniifolia* (at high altitudes), *A. boliviensis* (at mid altitudes), and *A. curassavica* (abundant at low to mid altitudes). We think that *A. mellodora* may be a significant spring ephemeral host plant in lowland grasslands, based on our observations in grasslands at Warnes, and near the Viru-Viru airport north of Santa Cruz, Bolivia, in October, and extensive herbarium specimens including those listed in Bollwinkel (1969). We found *A. candida* only in eastern Bolivia near Santiago de Chiquitos and, based on our field work and comments by John Wood (pers. comm.), this species also appears to be a spring ephemeral like *A. mellodora* in the lowlands and *A. pilgeriana* in the highlands. In October 2009 (spring) we observed *D. erippus* ovipositing on *A. pilgeriana* in the mountains above the Río Grande valley near Villa Serrano, Bolivia (19°01′31.09″S, 64°20′23.07″W) at an elevation of 2600 m. *A. pilgeriana* also occurs in the eastern Andes immediately to the west of Tucumán (Alfredo Grau, pers. comm.).

Adult butterfly flight orientation

On most occasions, *D. erippus* flights were short, random movements, with patrolling flights by males guarding milkweed patches and mating attempts by males that responded to almost any intruder to the airspace over a milkweed patch. These movements precluded measurements of vanishing bearings because the butterflies remained in the immediate vicinity unless they were alarmed and showed escape behaviors.

In contrast, the butterflies we observed in April 2010 at Horco Molle at the edge of the Andes Mountains and the Yungas forest ecosystem (26°46′32.75″S, 65°19′44.77″W) were clearly showing directed and migratory flight behavior. In the morning they landed on dew-covered grass in fields and drank water while basking in the morning sun and then took flight. For the rest of the day they flew with constant directional flight, and it was easy to determine their vanishing bearings. On 27 April (10:20 a.m. to 4:25 p.m.) and 28 April 2010 (1:45–4:50 p.m.) we measured 280 independent vanishing bearings at an observer altitude of 746 m and with air temperatures ranging from 16.8°C at 10:23 a.m. to 22.3°C at 4:00 p.m. These flight directions had a mean vector of 215.2° (±30.1° SD) (Figure 20.2a). Both the Rayleigh test for mean circular direction

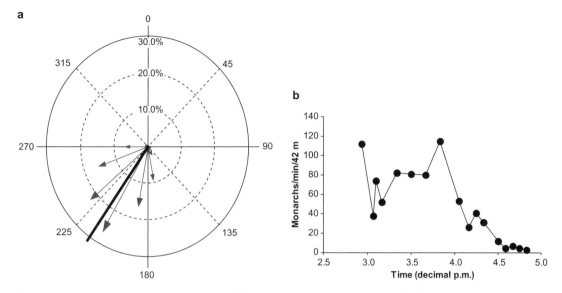

Figure 20.2. (a) *D. erippus* vanishing bearings (*n* = 280) and mean vector (thick black line = 215.2°). Data collected during 25–29 April 2010 (autumn) from Horco Molle, on the eastern edge of the Andes Mountains in Tucumán, Argentina. (Rayleigh's test for mean circular direction, Z = 212.30, $P < 0.001$) (Rao's spacing test for circular uniformity, U = 244.14, $P < 0.01$). (b) Change in the numbers of *D. erippus* observed per minute in the 42 m field of view of 8×42 binoculars (linear field of view M = 17.45 × $A°$ = 139.6 mm/m at 300 m = 41.88 m) from 2:56 to 4:50 p.m. on 28 April 2010 at Horco Molle. The peak numbers at 2:56 of 112/min and at 3:50 of 115/min represent, across an estimated flight front of 210 m, a total of 33,600 and 34,500 *D. erippus* per hour, respectively, or 168,000 and 172,500 *D. erippus* per day (for 5 hours of flight a day), respectively.

and Rao's spacing test for circular uniformity were significant.

On 28 April we observed *D. erippus* flying overhead at Horco Molle through a range of heights from 1 to 50 m (with most approximately 15–20 m) and a flight front that was approximately 210 m wide (5× the linear field of view of our binoculars at 42 m) (Figure 20.2b). Using the number of butterflies observed in our binoculars' fields of vision, at the maximum recorded observation of 115 butterflies/minute we estimated that 172,500/day flew over this area. At the average of 76 butterflies/minute from 2:56 to 4:03 p.m., this is 22,900 butterflies/hour or 114,500/day. We estimated the flight speed of low-flying *D. erippus* as varying between 14 and 19 seconds to fly 70 m (18.0 and 13.25 km/h, respectively).

While observing these directed flight orientations of *D. erippus*, we observed large numbers of blue-and-white swallows (*Notiochelidon cyanoleuca*) flying among the *D. erippus*. The butterflies paid no attention to the birds unless large numbers resulted in an aerial collision, when the struck *D. erippus* would perform a roll, drop, and then resume directed flight.

Wing length and fat content

The mean wing length of all females sampled was 46.34 mm ± 0.24 SE (*N* = 137) and for males, 47.46 mm ± 0.17 SE (*N* = 267). Linear regressions against *altitude* showed that female, but not male, wing length increased significantly with altitude (Figure 20.3, left and right, respectively). *D. erippus* adults appear to have longer wings above 1000 m altitude, although the weak relationship is probably driven by large numbers of butterflies collected further north in Bolivia at low altitude. Regressions against *day* indicated significant decreases in both female and male wing lengths through the year, with larger individuals up to autumn in April and smaller individuals in winter and spring after day 125 (5 May, Figure 20.3). Regressions against *latitude* also suggest that *D. erippus* had significantly smaller wings north of the Tropic of Capricorn (currently at −23°26′16″), although the data show considerable scatter and all R^2 values are low (Figure 20.3).

We investigated interactions among the independent variable effects on wing length and found that latitude had a significant negative impact on wing

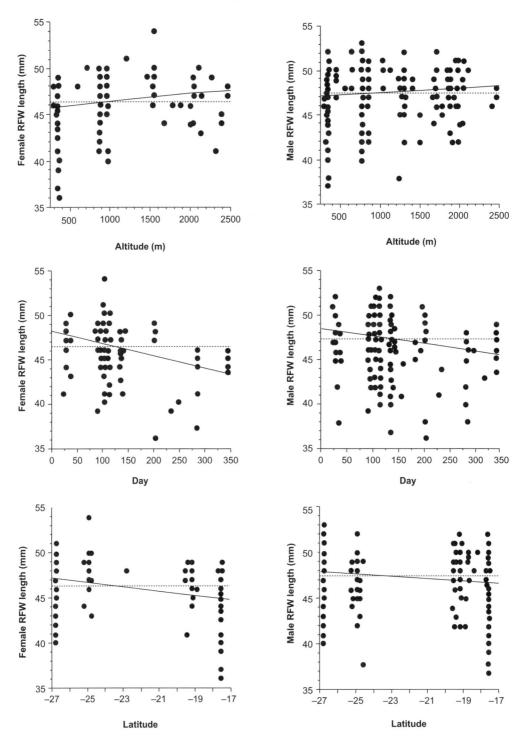

Figure 20.3. Wing lengths of *D. erippus* adults plotted against altitude (m), date (day of year), and latitude (°S) for females (left) and males (right). Linear regressions against *altitude* for female wing length = 45.49 + 0.00087*altitude, $R^2 = 0.03$, $F_{1,136} = 4.37$, $P = 0.038$; male wing length = 47.00 + 0.00045*altitude, $R^2 = 0.01$, $F_{1,266} = 2.84$, $P = $ NS. Against *date* for female wing length = 48.08 − 0.014*day, $R^2 = 0.09$, $F_{1,136} = 13.34$, $P = 0.0004$; male wing length = 48.48 − 0.008*day, $R^2 = 0.03$, $F_{1,266} = 8.80$, $P = 0.003$. Against *latitude* for female wing length = 41.05 − 0.23*latitude, $R^2 = 0.11$, $F_{1,136} = 16.87$, $P < 0.0001$; male wing length = 44.76 − 0.12*latitude, $R^2 = 0.03$, $F_{1,266} = 8.65$, $P = 0.004$.

length ($t = -3.12$, $P = 0.002$), and the only significant interaction term was between day and latitude ($t = -2.13$, $P = 0.034$); however, we stress that these data are not random samples and are strongly biased by our ability to sample butterflies at different locations and at different times, thus our analyses are tentative indications of trends.

The fat content of adult butterflies was represented as a percent of dry body weight; there were significant polynomial relationships with *altitude*, *day*, and *latitude* (Figure 20.4). Most fat variance was accounted for by latitude ($R^2 = 0.58$), followed by altitude ($R^2 = 0.49$) and day ($R^2 = 0.24$) (Figure 20.4). Butterflies at intermediate altitudes had more fat than butterflies at both low and high altitudes. There was a distinct, and marked increase in fat content in April (days 110–120 = 19–29 April) consistent with extensive nectaring behavior in autumn. *D. erippus* south of the Tropic of Capricorn in Argentina were markedly fatter than those north of the Tropic in Bolivia.

We also examined interactions among altitude, day, and latitude for possible influences on fat content of butterflies, using stepwise regression and the minimum AICc to select the best models. In this analysis the influence on fat content was ranked with latitude as the strongest influence, followed by altitude and day. The minimum AICc scores occurred for the influence of latitude on % fat and interactions between latitude*day and latitude*day*altitude.

Nectaring and roosting behavior

During our field observations in Argentina in April we noted extensive nectaring by *D. erippus* adults at *Eupatorium arnottianum* (Asteraceae) (= *Chromolaena arnottiana*) in the valleys of the mountains immediately west of Tucumán (Figures 20.1 and 20.5). Nectaring was especially evident in the region of the Río Grande near El Siambón and La Sala where thousands of *D. erippus* cascaded out of tree roosts in the morning to nectar at *E. arnottianum* with other lepidopteran migrants, including *Episcada philoclea* (Ithomiinae), *Doxocopa cyane burmeisteri* (Apaturinae), *Tegosa claudina* (Nymphalinae), *Eunica tatila bellaria* (Biblidinae), *Libytheana carinenta* (Libytheinae), and the aposematic ctenuchiine moth *Cyanopepla hurama* (Arctiidae). In a preliminary analysis of freeze-dried *E. arnottianum* flowers we collected by the Río Grande at El

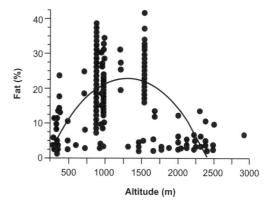

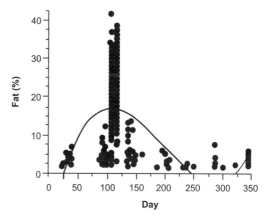

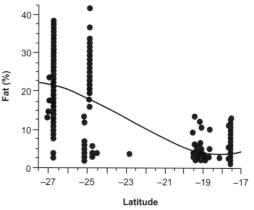

Figure 20.4. Fat content (% dry weight) of adult *D. erippus* plotted against *altitude (m)* (% fat = 10.49 + 0.01*altitude − 0.000019*(altitude − 1019.3)^2, $R^2 = 0.49$, $F_{2,370} = 177$, $P < 0.0001$); *day* (% fat = 24.76 − 0.07*day − 0.001*(day − 129.5)^2 + 0.0000069*(day − 129.5)^3, $R^2 = 0.24$, $F_{3,392} = 41.6$, $P < 0.0001$); and *latitude (°S)* (% fat = −52.12 − 2.83*latitude − 0.05*(latitude − 23.4)^2 + 0.04*(Latitude − 23.39)^3, $R^2 = 0.58$, $F_{3,370} = 170.55$, $P < 0.0001$) for males and females combined.

Siambón, Dr. José Trigo at UNICAMP in Campinas, Brazil identified a lycopsamine-like pyrrolizidine alkaloid (PA), which suggests that the butterflies are nectaring for both sugar and PA resources at these flowers and may explain the dense aggregations of aposematic butterflies and moths. *D. erippus* was by far the most abundant species, and individuals were so intent on nectaring that they could easily be caught by hand. We estimated at least 5000 *D. erippus* in a hectare patch of flowering *E. arnottianum* along the eastern bank of the Río Grande, north of the road bridge near El Siambón. We also observed them late in the afternoon in roosting clusters in trees above the flowering *Eupatorium* and, in the morning, dispersing from the clusters to move onto adjacent flowers.

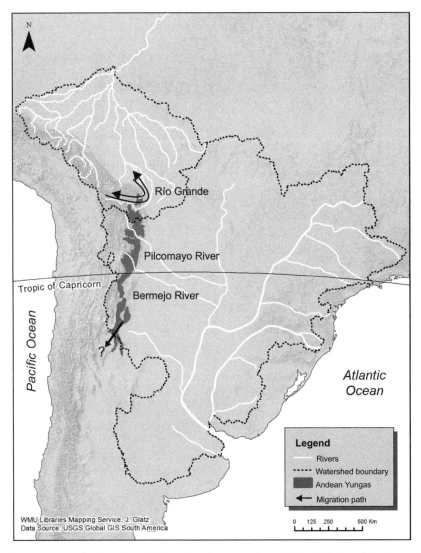

Figure 20.5. Principal watersheds of southern Bolivia and northern Argentina and the distribution of the Andean Yungas forest ecosystem. The arrows in Bolivia show suggested movements of *D. erippus* following rivers such as the Río Grande between Andean uplands with seasonally available milkweeds and tropical lowlands with milkweeds available throughout the year. The arrow in Argentina, south of the Tropic of Capricorn, shows the mean orientation of *D. erippus* that we observed (215.2°; see Figure 20.3) at Horco Molle, immediately west of Tucumán. The trajectory starts at Horco Molle and ends at the end of the Yungas ecosystem in the valleys of the eastern cordillera of the Andes in the province of Catamarca (Navarro et al. 2009).

Our April observations in Argentina were the only occasions on which we saw roosting *D. erippus*. Along the banks of the Río Grande near El Siambón in Tucumán province we found clusters of *D. erippus* roosting in both native and non-native tree species. Near La Sala in an adjacent valley we saw them forming a roost at 5:00 p.m. on 29 April 2010 in a large *Eucalyptus globulus* tree, much as *D. plexippus* cluster in California in the same tree species. When the roosts broke up in the morning sunshine, adult butterflies gently cascaded to the vegetation below the trees where they basked in the same manner as described by Masters et al. (1988). Both males and females perched on vegetation with their abdomens raised until body temperatures were sufficiently warm for flight. They then flew short distances to their nectar resources on *E. arnottianum*.

DISCUSSION

When we began this research we found very little information about *D. erippus*. In his volume on insect migration, Williams (1958) states (p. 15) that *D. erippus* "is believed to migrate there [South America] in a way similar to *plexippus* in the north—that is to say away from the equator [south] in their spring and towards the equator [north] in their autumn." He describes records of northward flights in March and April (autumn) in Argentina and writes that the ornithologist F. Alexander Wetmore reported overwintering in the Chaco of Northern Argentina. Unfortunately no citations are given. The only information we found was the series of papers published by Hayward (1928, 1953, 1955, 1962a, 1962b, 1963, 1964, 1967, 1972) that focus on *D. erippus* movement around Tucumán in northwestern Argentina. Although Hayward's records include spring, summer, and autumn observations, *D. erippus* appears only in the April and May lists. In a total of approximately 30 days of observations, *D. erippus* was observed moving south in autumn except for two observations (13 April 1928 and 30 April 1970) when it was seen flying north. Hayward (1972) states that the northward flight was unusual and that *D. erippus* is always seen flying south in the autumn parallel to the Andes. These published observations are thus at odds with the somewhat inductive speculations of Williams (1958), Johnson (1969), and Ackery and Vane-Wright (1984) based on the behavior of *D. plexippus* in North America.

Our data, especially those from Horco Molle on the edge of the Yungas forest ecosystem, are in clear agreement with Hayward's and colleagues' observations. Many of Hayward's observations were actually made by José Haedo Rossi from the field in front of his house. With the help of Dr. Fernando Navarro, we found this house and the location of the observations on property owned by the National University of Tucumán and its field station at Horco Molle. Our observations support and add to those of Hayward and strongly support southward migration of *D. erippus* in autumn (Figure 20.2).

In addition to these observations, we found that *D. erippus* tended to have longer wings further south in Argentina than in Bolivia and some tendency toward longer wings at higher altitudes (Figure 20.3). Butterflies earlier in the year also tended to have longer wings than those later in the year. Based on parallel observations of *D. plexippus* in North America (Altizer and Davis 2010), these observations are consistent with seasonal migration in *D. erippus*. Additionally, butterflies further south show stronger evidence of migratory life histories than those further north.

Our fat content data support this interpretation, with fatter butterflies at mid altitude, mid to late April, and south of the Tropic of Capricorn (Figure 20.4). The mid altitude butterflies with high fat content were almost all caught in April nectaring at *Eupatorium arnottianum* in the mid altitude valleys of the Andes Mountains immediately west of Tucumán, especially in the valley of the Río Grande near El Siambón (Figure 20.1). Nectaring at these plants was striking because thousands of *D. erippus* in pristine condition were aggregated at the flowers from which they were ingesting (and perhaps sequestering) both sugars and pyrrolizidine alkaloids. These behaviors, plus our observations of roosting behaviors, support a migratory interpretation for these autumn butterflies in Argentina.

While we were unable to determine where the autumn migrants are headed, we suspect that they are distributed during the winter in the moist valleys along the extreme eastern edge of the Andes from Salta south to La Rioja, or even as far south as Mendoza. The large numbers of migrating butterflies at Horco Molle suggest that most butterflies are heading further south than Tucumán, and it is interesting to speculate that the southern end of the Yungas ecosystem may be as far south as they migrate (Figure 20.5). The presence of both moisture and cool temperatures in these mountain valleys during the winter is likely to reduce stored lipid expenditure and maintain survivorship during the winter, while milkweed resources are scarce. The autumn abundance of nectar resources in the Asteraceae, such as *Eupatorium arnottianum*, suggests that *D. erippus* are behaving like northern monarchs to accrue sugar resources in the autumn to fuel overwintering behaviors as described by Masters et al. (1988) and Brower et al. (2007, this volume, Chapter 10). They are also clearly showing overnight roosting behaviors like northern monarchs.

In conclusion, we consider that southern monarchs, *D. erippus*, in Argentina show a seasonal migration with behaviors strikingly similar to those exhibited by *D. plexippus* in North America. The southward flight in autumn is an interesting difference, but it is probably explained by the need for cool temperatures and moisture provided by the abiotic conditions of the eastern edge of the Andes in Argentina between Tucumán and La Rioja. In contrast to *D. erippus* in Argentina, we believe that the butterflies we observed in Bolivia are nonmigratory in low-altitude, tropical regions, such as those we observed at Huaytú, near Buena Vista on the northern edge of Amboró National Park, northwest of Santa Cruz (Figure 20.1). At this location, both *A. curassavica* and *A. boliviensis* are available as larval food resources throughout the year. At higher altitude in Bolivia we found *D. erippus* using both *A. barjoniifolia* and *A. boliviensis* distributed along roadsides and riverbanks, or dry riverbeds. While these plants were also observed at different times of the year, they showed more evidence of seasonality than lower altitude plants. Thus, we suggest that *D. erippus* in Bolivia are able to use seasonal migration to exploit milkweeds at higher altitudes by using watercourses as flyways (Figure 20.5), much as suggested by Dingle et al. (2005) for *D. plexippus* monarchs in western North America.

Many of our observations were made in the region of Sucre, Potosí, and Tarija, a mountainous region that forms the separation between watersheds draining to the north and south (Figure 20.5). The Río Grande begins near Cochabamba in Bolivia and flows south and east through the mountains, eventually turning north past Santa Cruz and Buena Vista to drain into the Mamoré and Madeira Rivers, which flow to the Amazon River near Manaus in Brazil (Figure 20.5). Thus the Río Grande provides a navigable and predictable route for *D. erippus* between the tropical lowlands and the Andean highlands, where a different suite of milkweed resources is available. Immediately south of this watershed the drainage flows south from southern Bolivia to northern Argentina and Paraguay (Figure 20.5). The Pilcomayo River flows from near Sucre southeast to Asunción in Paraguay and the Bermejo River runs parallel further to the south and both contribute to the Paraná River which flows to Buenos Aires (Figure 20.5). These two separate drainages thus give *D. erippus* a choice in the Andes of southern Bolivia. They can behave either as altitudinal migrants and follow the Río Grande, or as latitudinal migrants and keep to the moist eastern cordillera of the mountains, flying south in autumn until they reach the end of the Yungas ecosystem (Figure 20.5). Latitudinal migrants can benefit from seasonal diapause (like northern monarchs) and use the cool, moist climate of the eastern Andes valleys with nectar resources and trees for roosting during storms and cold temperatures. Altitudinal migrants in Bolivia may not need to undergo seasonal deposition of fat or reproductive diapause and may simply move to lower altitudes in winter and back to higher altitudes in spring, following river valleys, as suggested by Dingle et al. (2005) for monarchs in western North America.

While we do not know where *D. erippus* in Argentina spend the winter, our data are consistent with a pattern of autumn migration, overwintering, spring migration, and host plant use similar to that exhibited by monarchs in North America, albeit with a much smaller diversity of *Asclepias* species to exploit. In Bolivia, they may be a mix of altitudinal migrants and nonmigrants, consistent with current thinking about "partial migration" (Chapman et al. 2011). *D. erippus* in Brazil and Uruguay may be nonmigratory, with a life history that may be ancestral for the subgenus *Danaus* (*Danaus*) and the three sister species, *plexippus*, *erippus*, and *cleophile* (Young 1982; Ackery and Vane-Wright 1984; Brower et al. 2007).

ACKNOWLEDGMENTS

We thank Robin Clarke of Buena Vista, Bolivia, and Thomas Malcolm, Allister Malcolm, and Dr.

Barbara Cockrell (Western Michigan University) for field assistance; Julieta Ledezma (Museo Noel Kempff Mercado, Santa Cruz, Bolivia) for access to collections and assistance with permit applications; John R. I. Wood (University of Oxford, England) and Dr. Michael Nee (New York Botanical Garden) for milkweed locations in Bolivia; Dr. David Goyder (Kew Gardens, London) for access to the herbarium; Dr. Gerardo Lamas (Universidad Nacional Mayor de San Marcos, Lima, Peru) and Blanca Huertas (Natural History Museum, London) for access to collections; and Dr. Mirian Hay-Roe for record data from the McGuire Collection in Gainesville, Florida. In Argentina, Dr. Fernando Navarro, Dr. Adriana Chalup, Eugenia Drewniak, Noelia Villafañe, Hernán Beccacece, and Germán San Blas (Instituto Miguel Lillo, University of Tucumán) took us on field trips, gave us access to their field station at Horco Molle, and arranged for access permissions. Dr. Alfredo Grau (Instituto Miguel Lillo/University of Tucumán) identified nectar resources. Dr. José Trigo (UNICAMP, Campinas, Brazil) identified pyrrolizidine alkaloids and hosted a visit to Campinas. Ezra Piller, Derrick Hilton, Christine Carpenter, Andrew Graham, Kerry Steinke, Jennifer Bailey, Mike Demapan, Josh Armagost, Andrew Johnson, Logan Rowe, and Joel Stevens of Western Michigan University provided laboratory assistance. Jason Glatz (WMU Map Library) provided Figure 20.5, and Dr. Karen Oberhauser, Dr. Gerardo Lamas, Patrick Guerra, and two anonymous reviewers commented on earlier versions of our chapter. Our research was funded by the National Geographic Society Committee for Research and Exploration; and the College of Arts and Sciences, Office of Research and Sponsored Programs, and the Haenicke Institute for Global Education at Western Michigan University.

ns on Monarch Distribution in the Pacific Northwest

21

Monarchs in the Mist

New Perspectives on Monarch Distribution in the Pacific Northwest

Robert Michael Pyle

New and reviewed information gives the first cumulative view of monarchs and their annual movements in the Pacific Northwest. This summer subpopulation is at best marginal to the overall western population; still, a persistent occurrence in southern British Columbia, Washington, Oregon, northern California, and Idaho may achieve modest importance in good years and grow in years to come, given predicted climatic change. This chapter parses historic and recent records of Cascadian monarchs to discern patterns in their distribution and movement, and to suggest how the Northwest component contributes to the overall migratory phenomenon. The blurring effects of monarch releases are discussed, along with historical misconceptions based on transfer exercises. The relationship of Northwest breeding monarchs to both Californian and Mexican winter populations is considered in view of recent field research. Recommendations are made for future work to illuminate and conserve monarchs and their habitat in the Pacific Northwest.

INTRODUCTION

Our knowledge of the monarch butterfly and its movements, far poorer in the West than in the East and Midwest, has been even murkier in the Pacific Northwest (PNW). The general public knows the insect in iconic terms, though the citizens of this region actually see living monarchs much less frequently than people do in most other states. When Northwesterners report that they have seen monarchs, interviews reveal that they are often referring to swallowtails or another species. The actual incidence of monarchs throughout most of the region is sparse, and west of the Cascade Mountains, north of Salem, Oregon, it drops to almost nil. Even so, monarchs are a cherished part of the Northwest's fauna. Here I consider the facts and future of this species' occurrence in the PNW, the region defined as British Columbia, Washington, Oregon, and Idaho. "Cascadia," a term also used here, refers to the greater bioregion, which extends into northern California.

Published descriptions of monarchs' range often describe the species as occurring throughout the United States and southern Canada. While this may describe the grosser picture, it is inaccurate at a finer scale. As a remarkably vagile insect ("the wanderer" is one of its alternative names), the monarch may appear almost anywhere in the United States, and occasionally well north of the Canadian border. The early, warm spring of 2012, for example, saw numerous monarchs reaching Edmonton, Alberta, and even farther north (Alberta Press 2012; John Acorn, pers. comm.). However, as a breeding resident, the monarch is restricted to the range of its host plants, species of milkweeds (*Asclepias* spp.). Certain subregions of the continent are without native or naturalized milkweeds, including all of western British Columbia and western Washington, and northwestern Oregon. Even the portions of these states and province that do support milkweeds, abundantly in places, possess them at a much lower density and frequency than most of the continent to the east. The region is also subject to the fluctuation of remigrants, not only from Mexico, but also (and probably chiefly) from California. Therefore, monarch occurrence in the Pacific Northwest is patchier, rarer, and less consistent than in much of the United States and southern Canada. The actual breeding incidence in this region is severely restricted, as shown in Figure 21.1, which illustrates the limits of milkweed in the PNW.

Actual records of Northwest monarchs can be seen in the state atlases of Hinchliff (1994, 1996, derived from the database of the Northwest Lepidoptera Survey), Guppy and Shepard (2001), and Dingle et al. (2005). Together, these reveal a pattern of widespread occurrence over much of Oregon, thinning into certain parts of Washington (with outliers), running out into a narrow zone of south-central and southeastern British Columbia. Dingle et al (2005) usefully illustrate the seasonal expansion and contraction of these PNW records. These extensive observations over the past century can be summarized as follows:

1. *A random scatter of vagrant records throughout the PNW*, including the cities and western inlets of the interior basin (Puget Sound Trough and adjacent lowlands) from Portland, Oregon, to and beyond Vancouver, British Columbia. The references cited above contain records far from milkweed; it is likely that these reflect not only vagility on the return migration (= "overshot"),

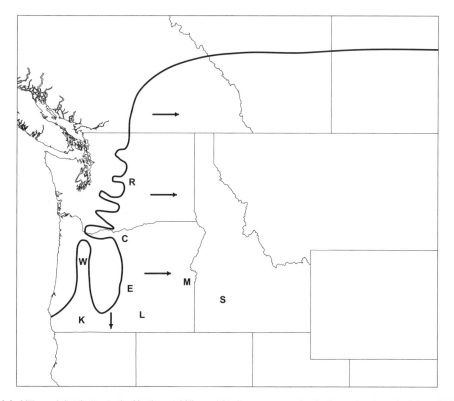

Figure 21.1. Milkweed distribution in the Northwest. Milkweeds' indigenous range lies to the east and south of the solid line, noted by arrows on the map. This approximate border is based on the author's field experience and on regional floras and herbaria. Milkweed distribution is highly patchy within this area. Limiting factors include precipitation, latitude, altitude, and human development and management. Records of introduced A. syriaca are not included here, nor are outliers such as ephemeral stands along rail lines south of Seattle. Areas of denser occurrence include the following: **K** = Klamath-Siskiyou, SW Oregon: *A. fascicularis, A. speciosa,* and *A. cordifolia*; **L** = "Oregon Lake District" of SE OR and NE CA: mostly *A. speciosa*; **M** = Malheur-Owyhee aridlands of SE OR: *A. cryptoceras* and *A. fascicularis*; **E** = Eastern Cascades, Warners, Wallowas, Blues, and related ranges: *A. speciosa* and *A. fascicularis*; **W** = Willamette Valley of western OR, dispersed stands as far north as Salem: *A. speciosa*; **C** = Columbia Gorge, from the Cascade Crest eastward: mostly *A. fascicularis, A. speciosa*; **S** = Snake River Plain, ID: *A. speciosa*; **R** = Rivers, canyons, and adjacent uplands in eastern WA and southernmost BC (as far north as Kamloops), chiefly the Columbia, Snake, Yakima, Cle Elum, Wenatchee, Okanogon, Methow, Spokane, and Similkameen drainages: mostly *A. speciosa*, some *A. fascicularis* toward the south.

but also collecting bias owing to concentration of entomologists and other observers, such as a cluster of records in Wahkiakum County, southwest Washington, my home. Furthermore, no recent records in greater Seattle, Portland, and other heavily populated areas are entirely trustworthy, because of releases from weddings and other events (see below, and J. P. Pelham, pers. comm., 2013).

2. *Breeding records, concentrated where* Asclepias *spp. occur in sufficient density to attract and nourish incoming migrants.* The locations and particular *Asclepias* species of these breeding locations are illustrated in Figure 21.1. In most cases, milkweed in these areas is very patchy, although dense stands, especially of *A. speciosa*, sometimes occur. In some cases, milkweed is much less common than it once was. For example, the Columbia Basin in eastern Washington and Oregon, and the Snake River Plain in Idaho (areas C and S on Figure 21.1), have historically held much *A. speciosa* associated with robust monarch populations, but much of this habitat has been lost to intensive chemical agriculture (Pyle 1999, 2000, and unpublished data). Vagrant monarchs have occasionally found and bred on adventitious stems of milkweed along railroad lines south of Seattle, but these are ephemeral and biologically insignificant.

3. *Migratory records, also related to rivers and their watershed uplands.* Many of these come from my observations while following the southward migration (Pyle 1999) and from other lepidopterists' reports from migration routes (D. James and others, pers. comm.; NW Leps Yahoo list-serve; Lepidopterists' Society Season's Summaries; others). Important nectar plants used by monarchs in autumn migration from the PNW include common sunflower, rabbitbrush, goldenrods, coreopsis, purple loosestrife, and milkweeds. Willows, locusts, junipers, and Russian olives are frequently employed for roosting. The fidelity of migratory and breeding monarchs to watercourses reported by Pyle (1999) was further demonstrated by Dingle et al. (2005).

Both breeding and migrating monarchs face significant threats around Cascadia. These threats include cutting and spraying of milkweed by farmers and roadside management crews. For example, James (2013) reported showy milkweed being targeted by herbicide spray crews in the area around Weed, Mt Shasta, and Yreka in northern California in May 2013. Additional dangers involve fire, development, and other vegetation removal. In one dramatic example, both larvae-bearing milkweed and important roost trees were lost at the same location, from two different causes. On 17 August 2008, Sue Anderson and colleagues (pers. comm.) counted more than 50 monarch larvae on *A. fascicularis* at Lava Beds National Monument in northeastern Siskiyou County, California, near the Oregon border. That night, lightning started a wildfire that consumed all those larvae. Such wildfires are common and increasing in Cascadia, as warming and drying intensify. The previous winter, National Park Service workers had removed juniper trees near Gillems Camp on the Monument, in an effort to restore the historic appearance of former army encampments during the final wars with the Modoc Indians. In doing so, they inadvertently cut down all the monarch roost trees in that area. In earlier years, at this same monument, roadside management crews were aggressive in removing milkweed from the margins of the roadways. So, in a single management unit, we can see a panoply of the threats that can befall monarchs even on protected public lands. On private lands the risks only intensify, since the landowner has no obligation to listen to citizens such as public officials do. Pesticide drift from orchards, for example, can render nearby milkweeds unsuitable or lethal for monarchs, as I witnessed in the Okanogan Valley, Washington (Pyle 1999).

Historical reports of large numbers of monarchs moving through the region (and some such anecdotal accounts do exist) almost certainly refer to influxes of painted ladies (*Vanessa cardui*) or mass movements of irrupting California tortoiseshells (*Nymphalis californica*). Almost every year I receive reports of "monarchs flying in huge numbers" on Cascadian passes and peaks. These always apply to the latter species, which often does exactly that (Pyle 2000). Pyle (1974, p. 80), apparently based on a personal communication, reported that "Don Frechin once noted a northerly migration over Bremerton at 7:30 on a cloudy day, the only record I have for a mass flight in the state." Bremerton is on the west side of Puget Sound, distant from any milkweed stands. No further details are available and this report must be considered dubious. However, Frechin was an experienced lepidopterist, and it is difficult to imagine what else he might have seen. This could indeed have been a mass of monarchs blown westerly, at a time when numbers were much higher in California than today; or pushing northward fruitlessly from the breeding habitat in the Willamette Valley of western Oregon.

MOVEMENT PATTERNS

As everyone who has spent much time with migrating monarchs knows, the spring and autumn generations differ in their mode of travel. Spring migrants tend to fly near the ground, foraging for milkweed, nectar, and mates, and are widely dispersed across the countryside. Autumn fliers soar and glide at altitude, dropping periodically to feed and roost, and grow increasingly channeled as their numbers increase toward the south. Spring monarchs fan out, and may appear almost anywhere, with incidence diminishing away from rivers and milkweed-bearing areas. Fall monarchs bunch, and to some extent, follow flyways, especially farther south.

In *Chasing Monarchs* (Pyle 1999), I describe monarchs as being wedded, but not welded, to rivers. In my experience, they tend to follow rivers (sources of moisture, nectar, host plants, and roost sites) as long as the rivers hew essentially to the general direction of the migration. When a river diverges from that direction, monarchs are able to abandon the watercourse and strike off cross-country, presumably shifting from visual cues to another system for orientation. I observed them moving along the Snake River southeasterly through three states, until it diverged north toward its origin in the Grand Tetons at Burley, Idaho, then detouring southward over Adobe Pass into the Great Basin. This behavior also occurs with monarchs following the Green and Colorado rivers (M. Monroe and G. Nabhan, pers. comm., in Pyle 1999), turning south when the Colorado takes a right to the west in the Grand Canyon. My observations suggest that monarchs have great flexibility of movement, and impressive powers of geographic perception and facultative response.

It has long been assumed that all northwestern monarchs come from California, and that all successful PNW autumn migrants fly to Californian wintering sites (e.g., Urquhart 1960; Pyle 1974; Urquhart and Urquhart 1977; Brower 1995). To some extent this assumption was a product of common sense, reinforced by weak quasi-experimental data (see below) and by copious repetition in popular and educational media as well as in the scientific literature. As such, it became a classic *factoid*, "an item of unreliable information that is repeated so often that it becomes accepted as fact" (Simpson and Weiner 2008). The default version in virtually every mention of monarch migration, that western monarchs (those originating west of the Rocky Mountains) all fly to and from the California coast, was repeated until it became canon that few questioned.

The first published exception was Zahl (1963), who, having considered all sources to date, created a hypothetical map for *National Geographic* that suggested movement to and from Mexico both west and east. Almost every other pre-2000 representation I have seen shows a clear east-west break, some of them depicting it with such robust graphics as to suggest the existence of a massive body of proof: broad arrows emerging from the East, converging on Mexico; another quiver-full of fat arrows emanating from the West toward California. Anyone viewing such a map would naturally assume that it was based on real, and substantial, justification. However, when I examined the available data, I found that this all-but-universal assumption was grounded in almost no evidence, none of it sufficient to support the reigning model. In fact, some of the available historical evidence suggests quite a different conclusion (Figure 21.2), one that has been confirmed by recent fieldwork.

Some data exist showing movement of wild monarchs between the PNW and California, but these are few. First, the tagging-recovery records of two teachers in Boise, Idaho (Pyle 1999): between them, Faye Sutherland and Mary Henshall and their students tagged and released hundreds of local, wild-caught monarchs in the 1970s and 1980s. Both women were later induced by monarch enthusiast Paul Cherubini to release transferred California monarchs as well as the wild ones they had always tagged. Twenty or so of their monarchs were recovered along the California coast, at least some of them wild in origin (Pyle 1999; F. Sutherland, M. Henshall, and P. Cherubini, pers. comm.). Second, I tagged a female monarch netted by David Branch at Roosevelt, Washington, on the Columbia River, on 26 September 1997, which was recovered in Aptos, California, by Jeremy Lovenfosse

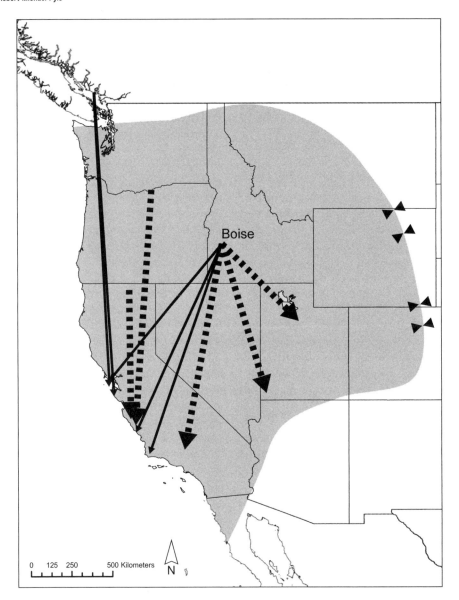

Figure 21.2. Long-distance recovery records of tagged monarchs from the Pacific Northwest, wild and transferred, as of 1998. Note the discrepancy between the paltry data actually available, and the massive extrapolation described above; also note the wild (as opposed to California transfer) flights indicating SSE movement. Narrow solid lines = Urquhart's release-recaptures (all transfers, except for some of those originating from Boise) (Urquhart 1977); bold broken lines = subsequent release-recaptures (all wild butterflies). Shaded area indicates Urquhart's concept of the range of the "western population"; note the exclusion of Arizona (Urquhart 1977). Diamonds indicate his idea of "areas where eastern migrants enter the field of the western populations."

on 26 October (Pyle 1997, 1999). Third, a male tagged by Sue Anderson on 30 August 1997 at Lava Beds National Monument in northeastern California was recovered at the Verde Road autumnal cluster south of Half Moon Bay south of San Francisco on 2 November. This is the entire body of data demonstrating wild, nontransferred Cascadian monarchs migrating to California. (See Note Added in Proof, p. 246.)

Many recoveries have been made in California of monarchs shipped from southern or central California and released in the PNW, or of the offspring of imported farm stock (Urquhart 1960; Urquhart and Urquhart 1977; informal newsletters; P. Cherubini, pers. comm.). These are not considered relevant here, since distant transfers do not provide acceptable surrogates for wild monarchs (see below). David James (pers. comm.), in a program for rearing monarchs in a Washington state prison, has had some recoveries from 2381 tagged monarchs released in 2012. Seven captures were reported within 10 miles of the release point, four recoveries from 50 miles or more, two having flown long distances. One was released at Yakima, Washington, on 17 September 2012 and found at Bolinas, California, on 30 October (600 miles). This individual came from stock reared through two generations in Idaho, from material originating on a mid-California farm, so it cannot contribute to the short list of wild monarchs tagged in the PNW and recovered elsewhere; however, half of James's releases were offspring of three females he collected in northern California, at the confluence of Eagle Creek with the Trinity River in the Shasta-Trinity National Forest, on 2 June 2012. This site, only 56 miles from Oregon, is of Cascadian affinity. His other long-distance recovery was a daughter of one of these females, released in Walla Walla, Washington, on 11 July and recovered at an I-84 rest stop just north of Brigham City, Utah, on 3 August, 450 miles southeast of the release point. This record, though involving a transfer, is of real interest, as it represents a southeastern vector by a monarch from the greater Northwest.

This slender body of results demonstrates that some southward migration does indeed occur from southwest Idaho, southeast Washington, and northeast California to the northern California coast. The data do not support the conclusion that the entire regional population migrates to the California coast; nor do they show where the summer monarchs of the PNW originated, Remarkably, this scanty handful of records is all that actually stands behind a monumental tenet of American natural history.

So how did the bicameral model of North American monarch distribution become so well established? The notion that all western monarchs fly to and from California received a major boost from a set of transfer-and-release exercises conducted by F. A. Urquhart and associates in the 1970s (Urquhart and Urquhart 1977). This involved many hundreds of California monarchs (Urquhart and Urquhart 1977; P. Cherubini, pers. comm.) that were transported to various locations around the West, tagged, and released. Californian monarchs released in Idaho, Oregon, and Nevada resulted in some recoveries in California, up to 10% of some releases (*fide* Cherubini).

Some of the monarchs used in Urquhart's transfer experiments came from much farther away. In August 1972, Don Davis of Toronto, Canada, shipped 900 pretagged Ontario monarchs to Gibsons, north of Vancouver, British Columbia, for release there as part of one of Urquhart's transfer experiments (Davis, pers. comm.). These monarchs were released in Gibsons by Maryanne West. Ten were recovered: five in southwestern British Columbia, one in Bellingham, Washington, one in Seattle, Washington, one in Central Point, Oregon, and two in California (Bolinas, 12 October, and Stinson Beach, 21 November, both found in clusters by Paul Cherubini).

These trajectories were then plotted and published by the Urquharts (1977; they are shown in Figure 21.2. While a small number of lines are shown for transfers flying east or southeast for modest distances before recovery, all the longer vectors indicate coastal or interior migration to the Californian sites. Urquhart concluded, "Recapture of [transferred Californian] migrants tagged in the Rocky Mountains and areas west of the mountains has established the origin of colonies overwintering in California" (Urquhart and Urquhart 1977, p. 1585). Examination of Figure 21.2 suggests that the "origin" he means is the entire catchment west of the Rockies, even though most recaptures came from elsewhere in California; Gibsons, British Columbia (a place outside the range of milkweed, hence monarch breeding); Boise, Idaho; Salem, Oregon; and Reno, Nevada, all locations to which the monarchs had been transferred from California or Ontario. The only reliable migration data (that is, recoveries based on wild, untransferred butterflies), were separate monarchs from Boise; therefore, the terms of Urquhart's conclusion greatly outstripped his support for it, based on his belief that California and Ontario transfers introduced no bias into the exercise.

Furthermore, recoveries of wild Idaho monarchs flying southeastward into Utah were never taken into account in Urquhart's model (at least one of them took place well after his paper was written). But even when he learned of a Utah recovery, Urquhart oddly said of it in a letter to Faye Sutherland on 2 June 1989, "Also your tagged butterfly 78237 was recaptured at Orem, Utah on 21/10/88. This is a unique record as it is the first record we have of a butterfly going through the mountains in a southeasterly direction showing what we have maintained for many years that the migrants do go through the mountain passes in order to reach the coast. So we must assume that the reverse is true when the butterflies leave the coast of California." Actually, there is no significant mountain barrier between Boise and Orem. That aside, even when presented with a western monarch on a clear southeastern trajectory toward Mexico, Urquhart adduced a coastal destination for it, so wedded was he to the Californian conclusion.

The Urquharts' 1977 paper was the primary statement on the subject at its time, and his map, the chief visual image for years to come. The overwhelming impression it conveyed (frequently repeated and extrapolated by others) was that western monarchs spring from and return to California. It was to test this hypothesis that I followed the western migration in 1996 (Pyle 1999) and found the facts on the ground to be otherwise. Yet in spite of his false assumptions and conclusions, Urquhart and his associates were the pioneers of this work. We owe them the debt that all who come later always owe the breakers of messy ground.

Another, larger body of information has been gathering that suggests the old model must be rejected. First, some of the Boise school releases pointed toward Mexico rather than California. The same year (1989) that Mary Henshall had a recovery in Ventura (an area with many observers), one of her monarchs was also recaptured southeast of Salt Lake City in Orem, Utah (where very few people were looking for them), and one of Faye Sutherland's Boise releases turned up near St. George, Utah, also far southeast of Idaho (Pyle 1999; Figure 21.2). Also see the Utah recovery of a Cascadian monarch reared and released by James (above).

Furthermore, most of the monarchs I observed during the autumn migration of 1996, throughout the inland Northwest and Great Basin along a transect roughly from Cawston, British Columbia, to Douglas, Arizona, gave vanishing bearings toward the south and southeast (Figure 21.3; Brower and Pyle 2004); only 9 of 62 bearings oriented toward California. This was a surprising and very marked result, but entirely in line with the Idaho women's Orem and St. George recoveries.

Finally, in October 1996, in southeast New Mexico well west of the Continental Divide, I observed monarchs flying south toward the Mexican border, less than 10 miles distant (Pyle 2001). And then on 5 October 1998, Eve and Rob Gill observed monarchs actually crossing into Mexico west of the Continental Divide, in Sonoita Basin, Organ Pipe National Monument, Arizona (Brower and Pyle 2004). The link between western monarchs and Mexico was established.

In 2003, extensive work tagging Arizona monarchs was begun by more than 300 volunteers for Southwest Monarch Study (SWMS). Their early results were confused by mixing farmed and transferred California monarchs with wild Arizona monarchs, but those data are now being separated. According to SWMS Coordinator Gail Morris (pers. comm.), the trend is toward tagging wild monarchs only. When only wild-monarch data are considered, they reveal Arizonan recoveries in both Mexico and California winter sites (Southwest Monarch Study 2013b). This is the first evidence of wild monarchs from west of the Continental Divide, tagged and released on their home ground (= not farmed or transferred) ending up in the Mexican overwintering sites.

Taking all these data into account, the conclusion is inescapable that the old bicameral model of the North American monarch migration is wrong. A new, more complex and supple model is necessary, an initial rendition of which was essayed by Brower and Pyle (2004). This model posits some unknown degree of U.S.-Mexico interchange in the West, and the possible refreshment of the Californian population with monarchs of Mexican origin in some years. The interior West, and probably the PNW, exchanges monarchs with both California and Mexico. The nearer the coast that monarchs are observed migrating or breeding in the PNW, the more likely they are to have come from California. The proportion of the western population coming from or going to Mexico is unknown. It may vary depending on latitude, as well as stochastic factors such as wind, other weather, and survivorship in both overwintering regions.

Dingle et al. (2005) concurred with the need for a new model, stating that "the model of monarchs from west of the Rocky Mountains wintering exclusively along the California Coast is no longer valid." They further concluded that monarchs that overwintered in more northerly California sites migrate into the broader interior Northwest, "the same area which Frey and Schaeffner (2004) postulated would

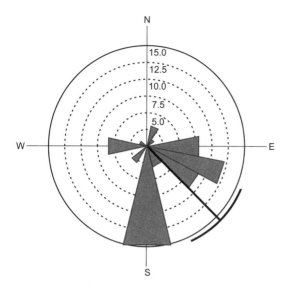

Figure 21.3. Vanishing bearings for autumn emigrants observed by author in 1996 in the interior West, along a rough transect from Cawston, British Columbia, to Douglas, Arizona. The greater part of the measured movement was oriented SSE. Data from "Chasing Monarchs" exercise (Pyle 1999). Dates and capture localities appear in Brower and Pyle (2004). Radii represent vanishing bearings of monarchs tagged and released in the West during the autumn of 1996, with the total number of individuals flying in each direction indicated by the length of each radius: total $N = 62$. Many more individuals were sighted; these data include only individuals with a discrete vanishing bearing as opposed to nondirectional flight, surface foraging, or roosting. Note: the one WNW datum was for a monarch in central Arizona caught up in a mass movement of snout butterflies (*Libythea carinenta*).

be the recruitment range for Californian overwintering aggregations."

One still sees the old bicameral model put forth, but the revised pattern is beginning to be broadcast by more informed sources. For example, migration maps that appear on the websites, posters, and brochures of Monarch Watch (2013a) and the Monarch Joint Venture (2013) show the western-Mexican link.

HOW TRANSFERS AND RELEASES IMPAIR UNDERSTANDING OF MONARCH MIGRATION

If one wishes to understand the natural movements and range of an organism, one must first observe those movements and study its distribution over the landscape with as little interference as possible. Just as clearly, one cannot logically move an animal from point A to point B, recover it at point C, and thereby conclude that animals originating at point B would necessarily travel to point C. The Urquhart transfers described above are a signal example of how incorporating recoveries of transferred monarchs into a data set subverts the conclusions that may reliably be drawn from those data. When the provenance of any given animal is made suspect, the reliability of every individual datum in the data set becomes suspect; and any distributional picture drawn from those data is damaged as a consequence. This is actually the case with monarch releases, particularly untagged ones, because there is no way for the observer to discriminate between wild monarchs and farmed, transferred ones; thus there is no logical way to establish criteria for eliminating suspect data: all data become suspect. For this reason, Jonathan Pelham, co-coordinator of the Northwest Butterfly Survey, no longer considers any new monarch records for Washington or Oregon to be reliable. As he stated in an e-mail to monarch researchers in the region: "Actually, all Monarch records are void of scientific information because some of them are, and we don't know which. This is a very unfortunate situation but is unavoidable [as long as releases persist]" (J. Pelham, pers. comm., 31 August 2013).

I can imagine that certain release exercises could be instructive in some ways, if they were well marked, based on an experimental protocol that included controls, and involved equivalent recovery efforts for both wild and released monarchs. But just scattering monarchs across the countryside, either for poorly founded "experiments" or for weddings and other ceremonies, creates a serious impediment to the scientific study of monarch biogeography. This is why I consider the transfer and release of western monarchs to be generally unhelpful to distributional studies, and to the conservation efforts that depend on a solid understanding of monarch whereabouts and movements. This was not appreciated in Urquhart's time; he was simply trying to extend his methods to regions where monarchs were not readily available. Now that the negative side of releases has been fully exposed, it is important that we stop allowing them, and stop treating all recovery data as if they were equal. This topic is more fully discussed by Brower and Pyle (2004), and Pyle (1998, 2010).

Washington and Oregon have recently tightened their regulatory controls on these activities (A. Potter and D. Hilburn, pers. comm.), and many monarch biologists hope that USDA will follow their example by ceasing to permit interstate transfers around the West, as is already the case with east-west transfers for reasons argued by Brower et al. (1995). It is important to note that the fall of the bicameral model of monarch migration in no way weakens the case for the east-west transfer ban by USDA, since it is unclear how many western monarchs fly south and mingle with eastern monarchs in Mexico. Western monarchs follow rivers and cross passes between watersheds, and some individuals undoubtedly cross the Continental Divide in this fashion, as Urquhart (1960, 1987) supposed and I have observed (Pyle 2001). It is likely that the integrity of the mostly eastern and mostly western populations remains largely intact, even though there is no doubt that some gene flow occurs via the Arizona border. The new model does not suggest or support a condition of cross-continental panmixis among monarchs. Recent genetic data suggest that the amount of mixing that does occur maintains strong genetic similarity between the populations (Pierce et al., this volume, Chapter 23), but this conclusion is based so far on a limited number of individuals, locations, and genes sequenced.

MONARCH RESEARCH AND CONSERVATION ACTIVITY IN THE PACIFIC NORTHWEST

Compared with tagging activity in California, the Midwest, and the East, few monarch taggings have been conducted in the Pacific Northwest, apart from those cited above. Dan Hilburn organized an Oregon-based tagging program in the 1990s, reporting the results in the newsletter *1000: 1*. Those years produced good numbers of monarchs by Northwest standards, and some hundreds of tags were used. The two recoveries that resulted exceeded expectations inherent in the program's name by a factor of two, but 1000:1 did not continue into a subsequent series of poor monarch years.

A program to restore sparse wild stands of showy milkweed (*A. speciosa*) in the Willamette Valley of western Oregon, and to monitor monarch response, has been underway by the group Cascadia Monarchs, founded and conducted primarily by Jim Kiser (Portland Tribune 2006). Cooperators have included wineries, public agencies, and specialty farmers who are including ecological and endangered species restoration among their goals. While this effort has expanded milkweed growth, there have been setbacks from road maintenance crews mowing milkweed when occupied by monarch larvae (G. Pearson, pers. comm.). Steve Northway and colleagues (pers. comm.) are working to conserve milkweed farther south in the valley, around Eugene and beyond. The task of gauging monarch response to plantings has been frustrated by the extremely low numbers of monarchs in recent years of drought in the south and wet, cool weather in the north. Wedding releases of monarchs in the cities of the Willamette Valley (Portland, Salem, Corvallis, and Eugene) hold the potential to falsify impressions of wild monarch recruitment through these host plant restoration efforts, making it difficult to evaluate the program's success.

In the summer of 2012, David James (who has published extensively on monarch migration in Australia) began a monarch rearing and tagging program in eastern Washington (see above). Based on a scheme in western Washington whereby female inmates of a state correctional facility have been employed in rearing state-endangered butterflies for restoration purposes (Seattle Times 2012), James has involved male inmates of the state penitentiary in Walla Walla to rear monarchs for tagging and release (pers. comm.). The objective, apart from the social aims (which have been successful), is to study migratory movements of the tagged, released monarchs. Frustrated by the virtual absence of wild monarchs in Washington in recent summers, James began his program with wild monarchs captured in Northern California, reinforced by Idaho-bred descendants of mid-California farm stock. The latter cannot satisfy the stated goals, as explained above, and were discontinued in any case because of rampant disease in the Idaho lab from these California transfers. And while the first breeders were at least of Cascadian origin, the fact that they were transferred nonetheless compromises results from them to some extent. In 2013, this exercise was able to rely on rearing stock obtained chiefly from wild Washington monarchs. All the late-season releases emanated from females caught at Lower Crab Creek, in east-central Washington. James hoped for a total of some 2000 tagged individuals from this wild-parent rearing, giving a good chance of some recoveries.

In order that the numbers, fortunes, migration, and adaptation of monarchs in the Northwest might better be understood and the population enhanced and protected, I recommend the following actions:

1. Lepidopterists, butterfly watchers, and photographers, including members of the Washington and Eugene-Springfield Butterfly Associations, should be recruited to intensify search and observation efforts for monarchs, and to carefully record their occurrence, numbers, movements, and behavior throughout the Northwest. Butterflies and Moths of North America (BAMONA 2012b) and eButterfly (eButterfly 2013) are complementary citizen science projects that collect, store, and display digital data on species occurrence. Both programs could prove helpful with monarchs here, if enough observers use them; however, as is the case for most observations, there is no way of telling whether any given record in these systems represents a wild monarch or one that has been transferred and released.
2. A new Cascadian tagging program should be instituted, promoted, and curated, to advance knowledge and management of the bioregion's monarchs and their sources. Of course the success of any such effort will depend on the availability of returning monarchs. David James (pers. comm.) succeeded at tagging of 27 wild monarchs at Lower Crab Creek in 2013, giving hope that a concentrated effort of tagging wild monarchs in Washington might not be as fruitless as feared in light of the poor return in recent years. He has also launched a regional tagging program, through the Facebook page "Monarchs in the Pacific Northwest," which also gives news and a great deal of other valuable information pertaining to this chapter (James 2013). The Washington Butterfly Association has indicated enthusiasm for taking part in monarch tagging (A. Wagar, pers. comm.).
3. For the results of such tagging efforts to be meaningful, commercial and private interstate transfer and release of monarchs should be disallowed through local, state, and federal regulation.
4. Programs to conserve, enhance, and restore native milkweed stands and nectar corridors, such as those of The Xerces Society (Xerces 2012a) and Monarch Watch (2013a), should be supported in every way possible. I suggest that an especially fruitful area of action will be milkweed culture for monarchs on the lands of organic and IPM-managed vineyards, already pioneered in Oregon and planned in Washington (J. Kizer, D. James, pers. comm.). David James (2012) has demonstrated that showy milkweed has significant benefits to pollinators generally, and to biological control agents of arthropod pests of grapes and other crops in eastern Washington.
5. To prevent losses of milkweed, nectar sources, or roost trees through ignorance of their occurrence (as in the Fossil Beds National Monument instance cited above), everyone engaged in monarch studies in the PNW should notify public lands managers of the existence of important resources for monarchs, wherever they occur in the region. Ongoing monitoring will be necessary to safeguard landscape features essential to monarch success in the Pacific Northwest.

SIGNIFICANCE OF PACIFIC NORTHWEST MONARCHS

While to my knowledge no estimates have been made of the numbers of monarchs produced in the Northwest, it is likely that their overall contribution to the total number of western monarchs is small. In the most favorable years, when there is both a strong influx of monarchs from the south and favorable conditions for breeding and migration, such as 1992 and 2002 (pers. observ.), their numbers may achieve modest significance (Swengel 1994). In poorer years, such as 2009–2011 (pers. observ.), almost no records are reported; nonetheless, Cascadian monarchs hold theoretical interest as the outermost immigrants and breeders of the entire western phenomenon.

The region may increase in importance for monarchs and their migratory behavior if both milkweeds and overwintering behavior expand to the north with climate change, a reasonable conjecture in light of a drying and warming California. It is possible that monarchs will eventually be found overwintering in the Redwoods, even into Oregon, and breeding along the Columbia River to its mouth, over the next several decades. (A fictional account of similar displacement of the eastern population owing to global warming [Kingsolver 2012] may have once seemed fantastical; but today, or tomorrow, it might be quite plausible.) On the other hand, if changing climate brings more La Niña years, with wetter winters and springs along the north coast and maritime valleys, a northward shift of overwintering habitat and westward expansion of breeding habitat might fail regardless of warmer temperatures. If drought grows more acute in the interior PNW, territory with conditions amenable to monarchs might contract even more.

CONCLUSIONS

Much remains to be learned about the monarchs in the upper left corner of the United States. For a number of fundamental questions, we have no answer. For example, as Dingle et al (2005) ask, how many spring immigrants to the PNW are themselves last year's overwinterers, versus the offspring of overwinterers that bred farther south? For another, what proportion of our autumn émigrés succeed in reaching one overwintering colony or another?

Even if survivorship of the northern generations is high, the contribution of the Pacific Northwest to the entire western monarch population is probably minor, at least for now. Its relevance to the Mexican population must be still more minute, if only numbers of butterflies are considered. Yet, no matter how sparse, monarchs matter to the human residents of this region. And as climate change proceeds, the Cascadian component could become increasingly important to the fluctuating West Coast wintering masses. Still vague in its specifics, the recently discovered Mexican connection may remain largely a matter of mere intellectual interest. But to the extent that the outer bounds of a major phenomenon can help to illuminate its reach, amplitude, and ultimate mechanism, this too will matter.

ACKNOWLEDGMENTS

I would like to thank all the monarch workers of the Pacific Northwest, past, present, and future, whether cited in this chapter or not. Through their ongoing efforts and impact, we may stand a better chance to sustain a remarkable resource that seems at once marginal and major. Special thanks are also owed to Thea Pyle, David Branch, Lincoln Brower, David James, Myron Zalucki, Stephen Malcolm, Dan Hilburn, Sue Anderson, Sarina Jepsen, Mía Monroe, Gail Morris, Jonathan Pelham, Don Davis, and Paul Cherubini. Their data, special assistance, critiques, and challenging questions have made this a sharper study than I could have conducted alone. I am also grateful to two anonymous reviewers, to David James and Patrick Guerra for additional review, to Kelly Nail for helping greatly in preparation of the maps and figures, to Katherine Hue-Tsung Liu for her care at Cornell University Press, and especially to Karen Oberhauser for her extraordinary attention to the betterment of this paper and the entire volume.

NOTE ADDED IN PROOF

As this paper goes to press, David James and colleagues (pers. comm.) have made four additional California recoveries of Washington-origin monarchs tagged in autumn 2014: Glen Ellen from Yakima (570 miles); Santa Cruz from Yakima (685 miles); Goleta from Walla Walla (825 miles); and Santa Cruz from Yakima (685 miles). In addition, Linda Kappen (pers. comm.) had a recovery of a wild monarch from Applegate, OR, in San Mateo, CA (330 miles). These recoveries significantly increase the data summarized above and reinforce the conclusions based on them. There will likely be more recoveries, and it will be of great interest to see whether any of the WA-tagged monarchs will be found in Arizona or Mexico. Total wild- or wild-offspring WA monarchs tagged and released by the James team in 2014: N = 2027 individuals.

22

Monarchs across the Atlantic Ocean

What's Happening on the Other Shore?

Juan Fernández-Haeger, Diego Jordano, and Myron P. Zalucki

We document the introduction, establishment, distribution, and abundance of monarchs in Europe and North Africa based on an analysis of the historic literature, estimated climatic suitability, and our own surveys. Monarchs spread across the Atlantic in the middle and late 1800s. Although monarch sightings on the Azores, Madeira, Canary Islands, British Isles, Gibraltar, and Portugal were first reported during these times, and milkweed (*Gomphocarpus* spp. and *Asclepias curassavica*) had been naturalized in many of these localities much earlier, breeding populations did not seem to establish until much later. Populations are now well established in the southern Iberian Peninsula and North Africa in areas that have been predicted to be climatically suitable. In southern Spain, monarchs breed year-round in a metapopulation of milkweed patches restricted to areas where the plants can survive the dry summer. We predict that monarchs may continue to spread throughout the Mediterranean, colonizing areas that are climatically suitable and have milkweed established. The greatest threat to the species' persistence are campaigns to eradicate milkweeds, listed in Spain as invasive species.

INTRODUCTION

During the mid- to second half of the nineteenth century, monarchs spread from North America throughout the Pacific, colonizing both small islands and larger land masses, such as Australia and New Zealand. This rapid spread invoked the interest of entomologists at the time (see summary in Zalucki and Clarke 2004). At about the same time, monarchs also spread eastward across the Atlantic, establishing on the few available island groups and making forays into Europe (see below).

Here we examine the history and status of the monarch butterfly in Europe and North Africa, addressing the issue from different perspectives and at different spatial scales. First, we review and synthesize the scattered information available on the presence of monarchs in this part of the world, relating their establishment to the history and ecology of their food plants and the local climate. Successful invasion by herbivores requires both a suitable climate and the presence of host plants. With the help of bioclimatic models, we have identified potentially suitable regions for monarchs and predict an expansion of their range in Europe and around the Mediterranean basin. The model encompasses the areas where monarchs have established well. We have recently investigated the occurrence of established populations in southern Spain, southern Portugal, northern Morocco, and the Azores and summarize those investigations here. At a more detailed level, for three consecutive years, we have investigated the spatial structure and dynamics of the population of monarchs in an area around the Gibraltar strait (southern Spain), closely linked to that of its food plants in the region. These surveys add to those published in Fernández-Haeger and Jordano (2009).

METHODS

We conducted an extensive literature search to reconstruct the history of monarch arrival and establishment in Europe, seeking data on both sightings of migratory vagrants and specimen citations, as well as reported work on established local breeding populations (Web of Science, Scopus, Science Direct, Wiley Online Library, Google Scholar, and Google).

To study the current status of monarchs, we conducted surveys utilizing all available access roads in the Algarve region of southern Portugal (September 2012), in northern Morocco (October 2012), in the Azores (September 2012), and, especially, in southern Spain (1997 to date). During these surveys, we paid particular attention to areas potentially suitable for larval food plants (predominantly *Asclepias curassavica* and *Gomphocarpus fruticosus*) and located all stands of these plants to determine whether monarchs were present in any stage of their life cycle. Once located, we inspected a sample of milkweeds by walking around the patch and looking closely at each entire plant sample, recording immatures observed as well as sightings and activity of adults.

Since 1997, we have regularly recorded monarchs along the southernmost point of continental Europe (36°N) close to the Gibraltar strait (Fernández-Haeger and Jordano 2009). Starting in June 2008, we systematically surveyed a coastal area of 900 km^2 in this region. We located 62 milkweed patches that were classified as monospecific stands of *G. fruticosus* or *A. curassavica* or as mixed stands with both species present, and given a score for plant abundance. In July 2009 we actually counted the number of plants by species in each patch, a task that was more complicated and necessarily less accurate in larger stands at Carrizales (3000 plants) and Cortijo Zambrana (4500 plants). In addition, we measured patch size by walking along the edge and gathering the track coordinates with a Garmin 60Cx GPS device. The tracks were imported to ArcGIS 9.3 (ESRI 2008) to map locations and to determine their area and perimeter. Patches were visited monthly (with some exceptions) from June 2008 to December 2010, for a total of 1803 patch visits (about 30 visits per patch).

During each visit we counted the total number of adult monarchs in the patch. In addition, during 10–30 minutes per patch (time roughly proportional to patch size) we also counted the eggs, larvae, and pupae in a sample of plants haphazardly chosen. A patch was considered occupied if at least one adult or one immature was counted; otherwise it was considered an empty patch. We also recorded data on host plant phenology and gathered information on any disturbance that occurred at the site (Fernández-Haeger et al. 2011a). From 2008 to 2010, patch occupancy was computed monthly and quarterly as the frequency of visits with positive sightings or records. From January 2011 onward, monthly monarch counts were carried out only in nine patches where the monarchs had been more abundant to date. We visited the other 53 patches once in September–October 2011, at the peak of monarch abundance, to reassess patch occupancy.

We use the CLIMEX model to show which areas in Europe and North Africa would be climatically suitable for monarchs. The methodology used to make the geographic predictions can be found in Zalucki and Rochester (1999, 2004), while that needed to make predictions on temporal dynamics due to climate can be found in Zalucki et al. (this volume, Chapter 11).

RESULTS

Monarch distribution on the eastern side of the Atlantic Ocean

There are regular records of "migrants" or vagrants that have strayed on winds from North America and arrived in the various islands of the eastern Atlantic and continental Europe (e.g., Cruz and Gonçalves 1973, 1977; Bivar de Sousa 1984–85; Owen and Smith 1993; Vieira 1997, 1999; Asher et al. 2001; Neves et al. 2001); however, local breeding populations appear to have established much later than the initial sightings (summarized below).

Monarchs that had crossed the Atlantic Ocean were first reported in several European countries after the mid-1800s and, more recently, in northern Africa (Plate 16). The earliest record for Madeira was in 1860 (Leestmans 1975), 1864 in the Azores (Godman 1870), 1876 in the British Isles (Llewelyn 1876), 1880 in the Canary Islands (Fernández-Rubio 1991; Báez 1998) and continental northern Portugal (Cruz and Gonçalves 1973), and 1886 in Gibraltar (Walker 1886). The first record for northern Africa (Morocco) was not made until 1988 (Steiniger and Eitschberger 1989), probably because of lack of interest and reporting from this geographical area.

In Great Britain and Ireland, most sightings have occurred in September and October, with high numbers of individuals in particular years (Asher et al. 2001). The dates of most records correlate well with the southern migrations of monarchs in eastern North America and with favorable meteorological conditions (easterly flow of warm air) (Asher et al. 2001). In any case, the absence of food plants in the British Isles precludes any realistic prospect of breeding or colonization, and further sightings depend on immigration from America or, perhaps, southern Spain (see below). This limitation also appears to be true for the northern Iberian Peninsula (Fernández Vidal 2002) and points further north.

Several sightings have been made in Madeira since the first observations in 1860 (Meyer 1993). A resident breeding population appears to be more recent (Bivar de Sousa 1984–85, 1991), as the species was not recorded by Baker (1891) or Cockerell (1923). In the Azores, monarchs were first detected in 1864 (Godman 1870). A permanent breeding population appears not to have established there until the early 1990s (Neves et al. 2001). The latter is interesting, as one of the major hosts on the Azores, *G. fruticosus*, was naturalized there in the nineteenth century (Seubert 1844; Watson 1844, 1870). Currently, monarchs seem to be present on the nine main islands of this archipelago (Borges et al. 2010) but apparently at very low densities. Food plants (*A. curassavica*, *G. physocarpus*, and *G. fruticosus*) are restricted mainly to gardens of big private properties (*quintas*) and even during the best season (late summer and early autumn) adult monarchs are difficult to find. In September 2012, we visited the islands of Terceira, Pico, San Miguel, and Faial searching for monarchs and their host plants, traversing the entire islands, including localities where monarchs had been sighted previously, but we failed to find a single monarch or host plant.

According to Wiemers (1995), monarch colonies were established in the Canary Islands in 1887 (citing Rebel and Rogenhafer 1894). In any case, they seem to be at low densities in subtropical coastal areas of the islands (Wiemers 1995) and have recently colonized the eastern and driest islands of Fuerteventura and Lanzarote (Strecker and Wilkens 2000). As a consequence, monarchs can now be found on all seven main islands of the archipelago (Izquierdo et al. 2004) (Figure 22.1).

In continental Europe, monarchs seem established only in the Iberian Peninsula (Spain and Portugal) (Figure 22.1 and Plate 16). In 2008, a few vagrant individuals were sighted in the Balearic Islands and, despite *G. fruticosus* being locally abundant, no breeding colonies have been reported to date (Encinas and Vicens 2008). Monarchs were not included in the checklist of butterflies of Portugal (Cruz and Gonçalves 1977), and Maravalhas (2003) considered that migrant individuals could not survive the winter in continental Portugal. However, Palma and Bivar de Sousa (2003) described 15 colonies in southern Portugal feeding on *G. fruticosus* and *A. curassavica* growing on riverbanks. We visited four of these localities in September 2012 and found butterflies in just two of them, where we counted around 60 adults as well as 23 larvae and one pupa. Most of the butterflies were males actively flying along creeks in abandoned orange groves, where only *G. fruticosus* was present.

In southern Spain, an established breeding population of monarchs has existed since at least the 1980s (Bretherton 1984; Fernández-Haeger and Jordano 2009 and references therein), although the first observation of adults in the region was in 1886 (Walker 1886). Breeding colonies have been repeatedly sighted in different localities of the southernmost provinces of Huelva, Cádiz, Málaga, Granada, and Almería, with some additional sightings further north along the Mediterranean coast (Figure 22.1). Most of these breeding colonies were ephemeral and some disappeared as a result of urban development in recent years (Fernández-Haeger and Jordano 2009). The key host plants here, as on the Atlantic Isles, are *A. curassavica* and *G. fruticosus* (Fernández-Haeger and Jordano 2009); these have been present for centuries (Quer 1762). In Granada and Almeria, monarchs also use *Cynanchum acutum* (Asclepiadoidea, Apocinaceae) as a host plant (Gil-T. 2006). This native host plant is extremely rare in our study area near the Gibraltar strait (province of Cádiz).

The occasional occurrence of migrant monarchs has been reported in the Ebro Delta, in the province of Tarragona in northeast Spain (Pérez De-Gregorio and Rondós Casas 2005), as well as in the Balearic Islands (Encinas and Vicens 2008) (Figure 22.2). Although host plants are available, breeding has not been observed yet. Other sightings in northern Spain (Pontevedra and La Coruña) are not likely to lead to permanent colonies because of the lack of host plants (Fernández Vidal 2002).

In Morocco, a single monarch was reported in 1988 in the locality of Taroudannt (30°N) (Steiniger and Eitschberger 1989). Ten years later, Tarrier

Figure 22.1. Records of monarchs in Spain, continental Portugal, and northern Morocco. Closed circles: colonies where breeding was confirmed; open circles: isolated, migrant butterflies (reproduction not yet observed). Notice that several breeding colonies were ephemeral and no longer persist.

(2000) confirmed the presence of monarchs in Larache and Rabat-Agdal (both on the Atlantic coast of Morocco). After these sightings, monarchs were detected in gardens in Casablanca and near Ceuta, using *A. curassavica* as a food plant (Tarrier and Delacre 2008). During October 2012 we located monarchs at high densities in Ben Younech, a coastal valley close to the Gibraltar strait in northern Morocco. In this valley, monarchs were very abundant and larvae were found concurrent with *Danaus chrysippus* (the plain tiger, or African monarch), often feeding on the same individual of *G. fruticosus* and even on the same stem. Other localities explored further away from the coast yielded negative results, both for monarchs and their host plants.

The records cited above suggest that in both southern Europe and northern Africa, monarchs are well established but limited to coastal areas, probably because of the restricted distribution of their host plants, which are not found in very dry or cold locations that are typical of inland areas. These coastal areas are climatically suitable for monarchs year-round. The Ecoclimatic Index, a measure of overall suitability of a location (for details see Zalucki et al., this volume, Chapter 11), is low, ranging from 6 to 20, but indicates areas where populations could persist. The Growth Index varies seasonally from 6 to 28 and suggests that monarchs could extend their range seasonally if milkweeds were present (for details on the CLIMEX model for monarchs see Zalucki et al., this volume, Chapter 11). Many areas predicted to be climatically suitable year-round and in which milkweed is present now have established monarch populations (blue rings in Plate 16). Note that our climate model predictions are on a coarse scale and that locally, microclimates may be much more suitable.

The monarch in southern Spain

In our surveys in southern Spain, we found that *G. fruticosus*, *A. curassavica*, and to a lesser extent *G. physocarpus* are patchily distributed, always growing near localized seeps where groundwater naturally

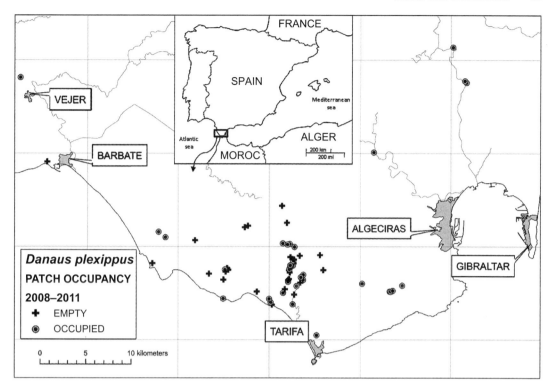

Figure 22.2. Patches of milkweeds (*n* = 62) occupied or unoccupied by monarchs during 2008–2011 in study areas in southern Spain. The crosshatch indicates urban areas.

discharges, near springs and creeks with plenty of water, or in places where soil moisture is available during the hot, dry Mediterranean summer. Our data suggest that the three species benefit from perturbations like clearing, mowing, flooding, and heavy grazing. In the absence of perturbations, the patches tend to be overgrown by native species like brambles and reeds, which clearly are superior competitors. These disturbances promote these Asclepiadaceae, which regrow quickly, keeping native species that otherwise might outcompete them in check.

Asclepias curassavica is less tolerant than *G. fruticosus* of both summer water shortages and low winter temperatures, and *G. physocarpus* is scarce and restricted to just a few patches. We found 62 host plant patches in the whole study area (Fernández-Haeger and Jordano 2009), most of them of *G. fruticosus* (45%) or mixed with both *G. fruticosus* and *A. curassavica* (32%), while monospecific patches of *A. curassavica* were less frequent (23%). *Gomphocarpus physocarpus* was present in just two patches where *A. curassavica* also occurred, but we observed fine-scale spatial segregation between species in both patches.

Overall, *G. fruticosus* was more abundant than *A. curassavica* (22,923 vs. 3778 plants in total). Patch size ranged from 10 m^2 to 44,743 m^2 (mean ± SE, 1848 ± 769 m^2, median = 286 m^2), and patches were smaller when they included only *A. curassavica* (Table 22.1).

Host plant patches are vital for monarch persistence, as they provide both food for the larvae and nectar for adults. The role of milkweeds as nectar sources is very important, since during the hot and dry summer *A. curassavica*, *G. fruticosus*, and *G. physocarpus* flowers are virtually the only nectar sources available. Their flowering phenology is quite distinct from that of native species. In fact, during the dry season, native herbaceous plants are withered (Polunin and Huxley 1972) while milkweeds are in full bloom; thus, the complex matrix of habitats surrounding the host plant patches are unsuitable for monarchs, because of lack of both food plants and nectar sources.

Table 22.1. Patch size statistics for patches with *G. fruticosus*, with *A. curassavica*, and with both species

Patch area (m²)	*G. fruticosus*	*A. curassavica*	Both spp.
Mean (± SE)	1383 (± 610)	612 (± 193)	3365 (± 2223)
Median	283	239	409
Range	10–16,771	10–2445	15–44,743
Total	38,736	8569	67,305
N	28	14	20

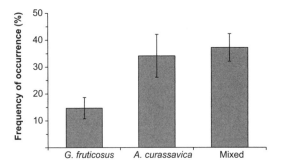

Figure 22.3. Frequency of monarch occurrence (average ± SE) in patches of *G. fruticosus* (n = 28), *A. curassavica* (n = 14), and mixed with both species present (n = 20), 2008–2010. Kruskal-Wallis test, H_c (tie corrected) = 11.62, P = 0.003. Mann-Whitney pairwise comparisons, Bonferroni corrected: *Gf* vs. *Ac*, P = 0.1005; *Gf* vs. mixed patches, P = 0.00306; *Ac* vs. mixed patches, P = 1.

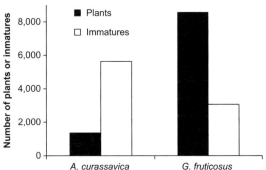

Figure 22.4. Recorded number of plants by species in all the mixed patches of *A. curassavica* and *G. fruticosus*, and total number of immature stages (combined number of eggs and larvae) counted on each host plant species.

In this spatially structured system, 47 patches were occupied at least temporarily by monarchs during the period 2008–2011, while 15 never contained monarchs (Figure 22.2). The total number of occupied patches varied slightly between years, being 29, 31, 41, and 34 for 2008–2011, respectively. Both colonization and local extinction seem to have occurred, suggesting metapopulation dynamics (Hanski 1999).

In the 1803 visits to our 62 patches, we detected a highly variable frequency of monarch occurrence. In some patches they were almost always present (>80% of our visits) but in others they were only sporadically detected (<10% of visits). Up to 15 patches remained unoccupied all the time, of which 11 were patches of *G. fruticosus*, 2 of *A. curassavica*, and another 2 with both species. In general, the frequency of occurrence of monarchs was lower in patches of *G. fruticosus* than in mixed and *A. curassavica* patches (Figure 22.3).

In addition, patch occupancy, measured as the percentage of visits to any given patch with positive results (presence of any monarch stages), was higher in patches of *A. curassavica* (41%) and mixed patches (43%) than in patches of *G. fruticosus* (23%). Considering only mixed patches, immature stages were more likely found on *A. curassavica* than on *G. fruticosus* (Figure 22.4), even though *G. fruticosus* was more abundant.

We also detected a seasonal change in patch occupancy, which consistently increased from winter to the summer and autumn (Figure 22.5). This means that during the winter monarchs are restricted to fewer favorable patches and expand to other patches during the summer and early autumn.

Patch size was not correlated with abundance of adults (Spearman r = 0.066, P = 0.62), egg (Spearman r = 0.069, P = 0.60), or larval (Spearman r = 0.024, P = 0.85). Just 7 patches made a large contribution to the whole system, with 83% of the adults and 84% of the immature stages counted in these patches; however, even some of these patches were occasionally poor habitat, becoming sinks for butterflies during the winters of 2009 and 2010, when persistent rain and flooding caused severe damage to the host plants growing on riverbanks and consequently to the immatures.

In the study area, monarchs are multivoltine with no diapause, breeding throughout the year. We have consistently observed egg-laying females, as well as many eggs and larvae during the winter since 1983. We recorded immature stages on 66% of our fieldwork days ($N = 313$), even during the winter; however, immature abundance tended to be lower during the winter and increased during the summer, showing large fluctuations (Figure 22.6).

In the last five years, we have recorded major population outbreaks in several patches, leading to severe patch defoliation (Plate 16, Figure 22.6). After plants were defoliated, adults dispersed to other patches. Severe defoliation in two patches was subsequently followed by flooding in 2009 and 2010, and the plants did not recover until the following late spring.

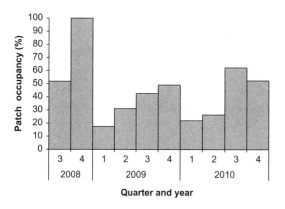

Figure 22.5. Seasonal variation of patch occupancy during three years (quarter: 1 = winter, 2 = spring, 3 = summer, and 4 = autumn). Number of patches occupied by butterflies increased yearly during summer and autumn.

DISCUSSION

The native distribution of *Gomphocarpus* centers mostly on semiarid parts of eastern and southern Africa, extending into the Arabian Peninsula, Sinai, and north to the Dead Sea (Goyder and Nicholas

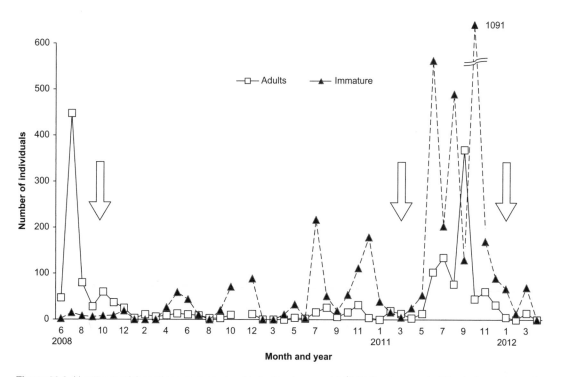

Figure 22.6. Variation in adults and immature stages of butterflies in one patch (#49, Guadarranque) of the study area by month, from 2008 to 2012. Arrows indicate dates when defoliation was recorded. River flood happened during January 2009, December 2009, and February 2010.

2001). These short-lived perennial shrubs are pioneers of open, disturbed habitats. Both *G. fruticosus* and *G. physocarpus* are native in the former Portuguese colonies in southern Africa (Mozambique and Angola). This distribution suggests that they were probably introduced by the Portuguese from sailing vessels along the northwestern coast of Africa, and in the Azores and Madeira possibly as early as the 1600s. It is likely that the Portuguese introduced them to the Iberian Peninsula as well. In Morocco, *Gomphocarpus* was planted and used by Arabs with herbal infusions (Vidal y López 1921).

Gomphocarpus fruticosus was first recorded in Spain during the eighteenth century (Quer 1762) as common in coastal areas of the southeastern part of the country. In his inventory of the Spanish flora, Willkomm (1893) notes this species around Gibraltar and Algeciras, the area in which our research project was focused.

On the other hand, *A. curassavica* was introduced to Spain from Central America (Cuba or Mexico, or both) during the eighteenth century or even earlier, after the arrival of Spaniards to the North American continent (Puerto 2002). It was cultivated as an ornamental and is relatively frequent in gardens of coastal frost-free areas both in the Iberian Peninsula and northern Africa (Tarrier and Delacre 2008; pers. observ.).

Therefore, both *G. fruticosus* and *A. curassavica* were widely distributed in the Iberian Peninsula and probably in northern Africa before the nineteenth century, when the first monarchs were reported east of the Atlantic Ocean. *G. physocarpus* seems a more recent introduction, at least in southern Spain, and appears to be expanding (Fernández-Haeger et al. 2010).

Both *Gomphocarpus* and *Asclepias* are in full bloom with fruit ripening during the summer (Fernández-Haeger et al. 2011b). In their native range in Africa, high summer temperatures coincide with the rainy season, but in Mediterranean areas, there is almost no rain during the summer and most herbaceous species die after seed ripening at the beginning of the summer. Therefore, under Mediterranean conditions these milkweeds are restricted to growing in naturally (such as in riverbeds, or near springs and ponds) or artificially wet locations (such as irrigation ditches) with enough moisture available for flowering and fruit ripening during the summer. Heavily grazed areas are also favorable, since livestock (sheep and goats) browse almost any other species except milkweeds. As a consequence, these plants show a strong patchy distribution around wet places heavily grazed by domestic ungulates, while the matrix surrounding the patches is effectively empty of host plants. Therefore monarchs concentrate in the patches, seeking both food plants for oviposition and nectar sources. Plants belonging to the subfamily Asclepiadoidea (Apocinaceae) produce high amounts of nectar (Wyatt and Broyles 1994) and are very attractive to butterflies during the summer drought, when almost no other nectar sources are available. During the summer only a few native butterfly species are present as adults in the study area, and all of them in quite low densities. *Cynthia cardui*, *Lycaena phlaeas*, and *Celastrina argiolus* have also been occasionally observed nectaring on *A. curassavica* or *G. fruticosus*.

Once their host plants were established by anthropogenic introduction from southern Africa or Central America, vagrant monarch individuals blown off course by winds during their migration from North America to Mexico could successfully colonize areas where food plants were available. While it is possible that monarchs were also introduced inadvertently or on purpose by humans, the fact that monarchs are still blown to Europe on the wind suggests that humans did not need to be involved. There is also some evidence of vagrant individuals arriving in Europe during the spring. Baynes (1966) reported the presence of a single individual in May 1965 in Ireland and there are records for the spring of 1896, 1947, and 1948 in England; however, reports of arrivals of monarchs during the spring have been far less numerous than those reported in autumn (Asher et al. 2001).

Our summary and analysis of the literature suggests that successful establishment in the Atlantic archipelagos, southern Iberian Peninsula, and northern Africa occurred many years after the first sighting records. Some of these records (1876 for the United Kingdom and 1886 for Portugal and Gibraltar) coincided with reports of large migrations along the Atlantic coast of North America (Brower 1995 and references therein). Repeated observations of butterflies on the rigging of ships at different distances from the coast in the Atlantic ocean (Harker 1883; Barret 1893) and successful arrivals at coastal areas in mainland Europe suggest that wind assisted dispersal and hitchhiking are complementary explanations of long-distance movement over the Atlantic ocean (Brower 1995). Since monarchs have a large exploratory capacity and easily find isolated milkweed patches (Shapiro 1981; Zalucki and Lammers 2010), they are likely to establish in areas where one or more milkweed species are naturalized (Brower 1995). Mild temperatures in southern Spain in autumn (average temperature in Tarifa for September: 20.9 °C; October, 18.6 °C) may favor egg maturation of arriving butterflies (Johnson 1963). These temperatures combined with the availability of milkweeds throughout the year favor the colonization process.

These areas are climatically suitable (Figure 22.1) and have milkweed, suggesting, as has been found in many biological control campaigns, that the size of the original arriving propagules was too small for successful establishment. It is also fairly common for introduced biological control agents of weeds to "disappear," and the program initially deemed a failure, only for the agent to become common many years later (Myers 1987). It would be very difficult without direct genetic evidence (Shephard et al. 2002), cardenolide fingerprints (Malcolm et al. 1989), or other markers (Hobson et al. 1999) to distinguish occasional vagrants from North America from local breeding butterflies, especially at such very low populations post establishment. Recent use of microsatellite markers may enable us to determine the phylogeographic history of monarch colonization of Europe (Pierce et al., this volume, Chapter 23).

In latitudes further north, monarch establishment fails in spite of their repeated arrivals (Williams et al. 1942; Cruz and Gonçalves 1977; Bretherton 1984; Asher et al. 2001; Fernández Vidal 2002) because of either the absence of food plants or the marginal climate. As a result, monarchs' spatial distribution is closely related to that of their larval food plants, which also act as one of the main nectar sources for adults, especially during the summer.

Monarchs are multivoltine in southern Spain and Portugal, and most likely on the Atlantic Islands, where they fly and breed throughout the year. Basically, this reproductive strategy is the same as that we observed during February 2013 in local monarch populations at Tuxpan (Michoacán, Mexico; 19°33.968′N, 100°27.379′W, elevation 1758 m) and Tiripitiu (near Tuzantla, Mexico: 19°15.115′N, 100°35.229′W, elevation 640 m). These two sites are located just 27.5 and 50 km, respectively from the Mexican overwintering colony of El Rosario (Sierra el Campanario, Michoacán, Mexico).

In our study area, about two thirds of the 62 patches studied were occupied at least in some years, and several of these patches were occupied during more than 80% of our visits, while 15 remained empty throughout the four years of our study. These data suggest that some patches could act as sources and some as sinks in a metapopulation system. But patch quality also varies seasonally. Some attractive patches along riverbeds that may function as source habitats during the summer become sinks during the winter, when the river floods and devastates plant patches, as in the very rainy winters of 2009 and 2010. Both frequency of adult occurrence and abundance of immatures were in general consistently higher in patches of *A. curassavica* and mixed patches than in patches of *G. fruticosus*. These results suggest that patches of *G. fruticosus* are less suitable habitat in southern Spain. In Australia, where monarchs utilize a similar mixture of milkweeds, Zalucki et al. (1989) also found that *A. curassavica* was more likely to be used.

As far as we can determine, monarchs seem to have a similar patchy distribution in coastal areas along the eastern margin of the Atlantic Ocean with a Mediterranean climate (southern Portugal, southern Spain, and northern Morocco). In northern Spain, France, and the United Kingdom, the climate is too cold to allow the milkweed species introduced so far to establish, and southern Morocco is probably too dry. Further expansion of monarchs through coastal areas of the Mediterranean basin is probable, because food plants are widely distributed and the area is also climatically suitable to monarchs (Figure 22.1). In the central and eastern Mediterranean (Sicily, southern Italy, Croatia, Greece, Turkey) monarchs have, to our knowledge, not been detected, despite the wide distribution of their food plant (*G. fruticosus*), which is exploited by their congener, *D. chrysippus* (Pamperis 1997; Grillo 1999; Perkovic 2006; Van der Heyden 2009). However, as in the successful colonization of the Iberian Peninsula, we believe it is just a matter of time until monarchs move into the central and eastern Mediterranean region.

The patchy distribution of monarchs is determined by plant distribution, which depends on a source of water during the summer drought. Since milkweeds are widely distributed along coastal areas in the Mediterranean basin, monarch persistence seems not in danger; however, both *G. fruticosus* and *A. curassavica* are included in the checklist of invasive plants of southern Spain (Dana et al. 2005), and attempts at eradication have occurred in some protected areas with varying success. A control action conducted several years ago in a stand in our study area, in the Natural Park los Alcornocales, was followed by strong regeneration of *G. fruticosus* shortly afterward. In the National Park of Doñana, situated between our study area and southern Portugal, systematic eradication tasks, including manually uprooting and burning all milkweeds, have been undertaken from 1990 to date (L. Cobo, pers. comm.). While *A. curassavica* was always scarce and completely eradicated in 2005, *G. fruticosus* is still present each year in 10–15 patches growing inside nesting territories of imperial eagles, where protection measures during the eagle breeding season preclude any control action until September. A *G. fruticosus* patch occupied by a flourishing monarch population was eradicated in autumn 2004 (Fernández-Haeger and Jordano 2009). Yet another breeding colony recolonized the site some years later during the summer, until it was destroyed again the following autumn (Paz et al. 2010). In Portugal, decree no. 565/99 concerning "Non-native species of flora and fauna" includes *G. fruticosus* in the list of invasive plant species, and as such this species can be affected by a national plan of control and eradication. Surprisingly, *A. curassavica* is not cited in this decree. Notwithstanding, the eradication of a widely distributed weed, with a high dispersal capacity, growing in managed soils with great anthropic influence, seems to be rather difficult. Occasional efforts made by conservation authorities to destroy plant patches have proven ineffective, since both species have great dispersal capacity and the ability to regenerate after human disturbances. Our results in southern Spain show that monarchs are able to find and recolonize newly regenerated patches after both human management and catastrophic events, like river floods.

ACKNOWLEDGMENTS

We wish to thank the Fundación Migres for providing financial support for this research. Charo Rivas, Carlos Camacho, Mateo León, Rafael Obregón, and Álvaro helped with our fieldwork, especially searching for patches and counting butterflies and their immature stages.

23

Unraveling the Mysteries of Monarch Migration and Global Dispersal through Molecular Genetic Techniques

Amanda A. Pierce, Sonia Altizer, Nicola L. Chamberlain, Marcus R. Kronforst, and Jacobus C. de Roode

Monarchs are found across the world, but little is known about their movement between sites or its effect on evolutionary processes. Work in population genetics is elucidating routes by which monarchs have colonized and the levels of gene flow between migratory and nonmigratory populations. This work has shown a lack of neutral genetic differentiation between populations with divergent migration strategies. We summarize published work on monarch population genetics, beginning in the 1970s with allozyme markers, transitioning to mitochondrial DNA markers in the 1990s, and leading to microsatellites and genomic sequencing. We present new results from a genetic analysis of monarchs from Central and North America, Bermuda, and Puerto Rico. Island monarchs are genetically differentiated from mainland populations, but we find a lack of differentiation at neutral loci for monarchs across the mainland. These results show that differences in monarch migration can be maintained despite high gene flow. We describe techniques that can help determine whether migration differences are maintained by natural selection on specific genes or by differential gene expression in migratory and nonmigratory populations.

INTRODUCTION

Monarchs are iconic insects best known for their spectacular annual fall migration from Canada and the United States to Mexico (Urquhart and Urquhart 1978). Despite decades of study, many aspects of this migration remain clouded in mystery (Brower 1995, 1996b). Unsolved questions regarding the monarchs' southward fall migration focus on the mechanisms by which monarchs orient towards and locate their overwintering sites (Reppert et al. 2004; Zhu et al. 2008a, 2008b, 2009; Merlin et al. 2009; Zhan et al. 2011; Guerra et al. 2012) and the relative role of active navigation and passive wind-based movement (Wenner and Harris 1993). Importantly, monarchs occupy locations ranging from the New World tropics to more recently colonized Pacific islands (Ackery and Vane-Wright 1984) to Europe (Fernández-Haeger et al., this volume, Chapter 22), and most of these populations are nonmigratory (James 1993; Altizer et al. 2000). A population is deemed nonmigratory if its habitat range does not significantly differ throughout the year. The wide distribution of monarchs raises important questions regarding the genetic differences between and interconnectedness among existing populations (Lyons et al. 2012). For example, is genetic separation a prerequisite for monarch populations to maintain different migration strategies and destinations, and have nonmigratory populations repeatedly arisen from migratory ancestors? The occurrence of monarchs around the world raises the additional question of where all these monarchs came from. Which ancestral populations served as sources for more recent colonization events, and in what patterns did monarchs spread across the Atlantic and Pacific Oceans from the New World (Vane-Wright 1993; Zalucki and Clarke 2004)?

Given the popularity of monarchs and the long history of scientific study focused on them, it is perhaps surprising that many questions regarding their migration and evolutionary history remain unanswered; however, modern molecular genetic approaches are required to investigate many of these issues, and these techniques have not been widely available for monarchs until recently. The recent development of microsatellite markers (Lyons et al. 2012) (for definitions of this and other genetic terms, see the glossary at end of the chapter), the publication of the monarch's genome sequence (Zhan et al. 2011), and the genomic work that preceded it (Zhu et al. 2008a, 2008b) now offer the potential to explore many aspects of monarch biology from a genetic perspective. In this chapter, we address some of these questions and discuss answers offered by recent analyses. We begin by summarizing early genetic work on monarchs based on studies of allozyme variation and mitochondrial DNA. We then describe how the use of microsatellite markers provides insights into the genetic connectedness between monarch populations and worldwide monarch dispersal, and how these markers may change our thinking on monarch migration. Finally, we briefly describe insights into monarch navigation revealed by genomic work, and how it is likely to improve our understanding of monarch migration.

ALLOZYME MARKERS SHOW SEASONAL MIXING AND SHED LIGHT ON THE ORIGINS OF PACIFIC MONARCHS

Migration can affect levels of gene flow within and genetic differentiation among populations, especially when species occupy multiple breeding grounds and migration destinations (Haig et al. 1997). Divergent migratory pathways and destinations could lower opportunities for genetic mixing, and hence result in genetic divergence; for example, the Old World noctule bats, *Nyctalus noctula*, which migrate between hibernating and summer nursing sites, are genetically differentiated with respect to overwintering sites and migration flyways (Petit and Mayer 2000). Similarly, beluga whales, *Delphinapterus leucas*, migrate between wintering sites in arctic pack ice and summering grounds in arctic and subarctic offshore waters, and there are considerable levels of differentiation between belugas using different summering grounds (O'Corry-Crowe et al. 1997). In contrast, the use of common migratory flyways, breeding grounds, or overwintering areas can lead to high levels of genetic mixing, even when populations experience different selection pressures or population substructuring at other points in their migratory cycle. For example, red-billed quelea birds, *Quelea quelea*, in southern Africa undergo long-distance migrations in response to seasonal patterns of rainfall and grass seed production. Although different groups of birds move in northwesterly or southeasterly directions (Dallimer and Jones 2002), genetic analysis shows high levels of mixing between these groups, which probably occurs when birds recolonize the same areas in the following season (Dallimer et al. 2003).

As with other species, the use of widely dispersed breeding grounds, distinct wintering sites, and different migratory flyways could cause local genetic differences in migratory monarch populations. The earliest published population genetic study on monarchs examined this issue using allozymes, which are different forms of enzymes (proteins) that result from different amino acid sequences; these enzyme variants can be detected via differential electrophoretic mobility. In a seminal paper, Eanes and Koehn (1978) examined the population structure of eastern North American monarchs by collecting 30 geographic samples throughout the monarchs' summer breeding grounds and along their fall migration routes in the eastern United States. Using six allozyme loci, Eanes and Koehn found differentiation between monarch groups during the summer breeding season. They hypothesized that this differentiation might result from genetic drift, involving random processes that alter allelic frequencies such as founder effects. Because monarchs are more regionally contained during the summer than during the migration, such random effects could result in differences in allele frequencies between subpopulations. It is also possible that differential selection could cause allele frequencies to vary among sampling locations. Such differential selection may act on allozyme markers, some of which have been linked with flight metabolism (Hughes and Zalucki 1993; Zalucki et al. 1993; Solensky and Oberhauser 2009b). Eanes and Koehn also found that the annual migration erased the genetic differentiation detected across summer breeding sites, by mixing monarchs from different breeding regions; thus, as with red-billed quelea birds (Dallimer et al. 2003), monarchs originating from a range of breeding sites appear to mix randomly during the migration season.

Allozymes have also helped elucidate the genetic origin of monarchs in Australia and shown that the lack of migration of Australian monarchs affects their genetic structure (Shephard et al. 2002). Monarchs likely spread beyond the New World within the last 200 years, across the Pacific and Atlantic oceans to destinations as distant as Australia and Spain (Figure 23.1). But the exact routes by which they did so remain unclear (Vane-Wright 1993; Zalucki and Clarke 2004); in particular, did monarchs spread in a stepwise fashion from North America to far-flung locations across the Pacific and Atlantic oceans (Vane-Wright 1993), or did multiple independent dispersal events occur (Zalucki and Clarke 2004)? To address this question, Shephard and colleagues (2002) collected 1194 butterflies from 15 sites in Australia, North America, and Hawaii. They found that the North American monarchs had more allelic diversity than monarchs from Hawaii and Australia, and that Australia and Hawaii had different subsets of alleles. These results suggest that both Hawaiian and Australian monarchs are derived from North America, but that the colonization of each location resulted from an independent dispersal event. Had the Australian population derived from the Hawaiian population, the alleles found in Australia would have been a subset of those found in Hawaii. Shephard and colleagues also found that monarchs obtained from different regions in Australia were more similar genetically than they were to monarchs from either Hawaii or North America, suggesting that monarchs colonized Australia in a single event. The lack of genetic differentiation among Australian sites also suggests seasonal mixing; this supports the hypothesis that, despite their lack of a two-way migration, Australian monarchs undergo alternating bouts of seasonal dispersal and range contraction through which monarchs from different regions mix (James 1993).

EAST MEETS WEST: ON THE ORIGINS OF AND MIXING BETWEEN NORTH AMERICAN MONARCHS

One long-standing question in monarch biology has been whether monarchs in the eastern and western regions of the United States and Canada are genetically similar or distinct. Although monarchs are best known for their migration from eastern Canada and the United States to the oyamel fir forests in central Mexico, monarchs in the western states embark on a shorter-distance migration to overwinter in eucalyptus and Monterey pine groves along coastal California (Urquhart and Urquhart

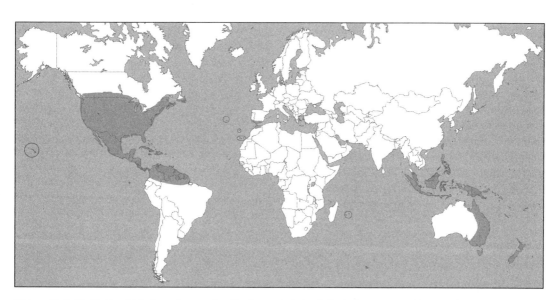

Figure 23.1. Worldwide distribution of monarchs. Shading and circles indicate known monarch range, following Ackery and Vane-Wright (1984) and updated following Neves et al. (2001) and personal communications.

1977; Tuskes and Brower 1978; Brower 1995; Dingle et al. 2005). Conventional wisdom suggested that these monarchs were geographically separated by the Rocky Mountains and continental divide (Figure 23.2a), although recent evidence, summarized by Pyle (this volume, Chapter 21), makes it clear that there is some interchange between these populations. Moreover, because of their different overwintering sites, eastern monarchs can travel up to 3500 km during the fall migration (Urquhart and Urquhart 1979), while western monarchs generally fly shorter distances, usually less than 500 km. Until recently, it was unknown whether the different migratory pathways and destinations of eastern and western monarchs depend on, or have resulted in, genetic divergence of these butterflies, and disagreement on the amount of gene flow is ongoing (Urquhart and Urquhart 1977; Shephard et al. 2002; Brower and Pyle 2004; Monarch Watch 2011).

A. V. Z. Brower and Boyce (1991) used mitochondrial DNA markers to determine whether eastern and western migratory monarchs were genetically differentiated, and to compare these monarchs with those from ancestral populations in the Neotropics. Mitochondrial DNA (mtDNA) is a useful genetic marker for several reasons. First, it is relatively easy to replicate and thus generates high numbers of copies to study in the laboratory. Second, it contains highly conserved regions (found across a wide range of species) that make it possible to use the same primers for replication across species and populations. These conserved regions surround regions with an elevated mutation rate that can cause differences between isolated populations of the same species. Brower and Boyce used 12 butterflies from each of the eastern and western migratory groups, and also included monarchs from the West Indies islands of Trinidad and Tobago. Surprisingly, in the fragments of mtDNA they examined, they found virtually identical patterns in all populations. The only unique variants discovered were in one individual from the eastern population and one individual from the western population. This indicates that based on mtDNA fragments alone, eastern and western migratory monarchs cannot be distinguished from each other, nor can they be distinguished from monarchs in the Neotropics, despite the large distances and geographic barriers separating them (Brower and Boyce 1991).

Brower and Boyce hypothesized three nonexclusive explanations for such low polymorphism and variability: low mutation rates, stabilizing natural selection, and random processes; however, they believed the most plausible of these explanations lay in random processes driven by a recent genetic bottleneck. A bottleneck would reduce overall levels of genetic diversity, especially because mtDNA is maternally inherited and genetic recombination does not occur. Thus, their data indicate that monarchs may have experienced a significant population reduction sometime in the recent past, followed by a rapid radiation into the temperate zone from the tropics. This was later confirmed by A. V. Z. Brower and Jeansonne (2004), when mtDNA markers indicated a lack of genetic divergence between North American and South American monarchs, despite clear differences in morphology and behavior.

To better resolve genetic variation within and among contemporary monarch populations, Lyons et al. (2012) recently developed microsatellite markers. Microsatellites are selectively neutral markers comprising repeats of nucleotide sequences that are scattered throughout the genome and tend to show extreme variability, which makes them ideal for studying genetic variation within and between populations. To compare eastern and western migratory monarchs, Lyons and colleagues collected 100 butterflies from St. Marks, Florida, a stopover location along the eastern autumn migration to Mexico (Urquhart and Urquhart 1978), and another 100 butterflies from Pismo Beach and Ellwood, California, two western North American overwintering sites. To ensure their microsatellite markers were able to detect subtle genetic differentiation, they also included monarchs from Hawaii and New Zealand in their analysis (Figure 23.2a).

Using a set of 11 polymorphic microsatellite markers and a series of population genetic analysis tools, Lyons et al. (2012) found that, despite differences in migration destination and the Rocky Mountains serving as a potential barrier, eastern and western North American monarchs are genetically indistinguishable on the basis of their microsatellite genetic make-up. Using a genetic clustering analysis to determine the most likely number of genetic populations from which these 262 butterflies were derived (Figure 23.2b), this work offers support for three, rather than four, genetically distinct populations. Briefly, this clustering analysis

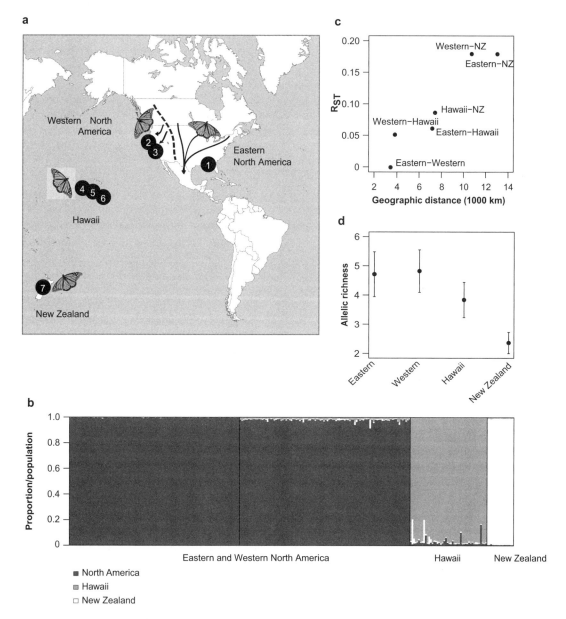

Figure 23.2. East meets west (adapted from Lyons et al. 2012). (a) Location and migratory patterns of sampled populations (Lyons et al. 2012): 1. St. Marks, FL ($n = 100$) 2. Pismo Beach, CA ($n = 84$) 3. Santa Barbara, CA ($n = 16$) 4. Kauai, HI ($n = 15$) 5. Oahu, HI ($n = 14$) 6. Maui, HI ($n = 17$) 7. Christchurch, New Zealand ($n = 16$). Eastern and western North America populations are at least somewhat separated by the Rocky Mountains (dashed line), but see discussion in text and Pyle (this volume, Chapter 21). Hawaii and New Zealand populations are nonmigratory. (b) Inferred genetic proportion of individual butterflies from each of three populations. Genetic clustering analysis was used to determine the likely proportion of alleles of each butterfly that originates from each of three genetic populations: eastern and western North America, Hawaii, and New Zealand. Individual monarchs are indicated by vertical bars; bars for butterflies with alleles from different populations are divided into different portions accordingly. Genetic assignments were determined on the basis of 11 microsatellite loci using the software STRUCTURE (Pritchard et al. 2000). (c) A strong pattern of isolation by distance is demonstrated by a correlation between geographic and genetic distance ($r = 0.95$, $P = 0.037$). R_{ST} values closer to zero indicate a lack of genetic differentiation. (d) Allelic richness was highest in North America and significantly lower in New Zealand. Error bars show ±1 SE across 11 loci. From Lyons et al. 2012. © 2012 Blackwell Publishing. Printed by permission.

assigned individual butterflies to a source population based on the alleles present in each butterfly, and whether those alleles appear to be shared across all populations, or restricted to a single population. In some cases, an individual butterfly will show up as a single-color bar in Figure 23.2b; if a butterfly has alleles representative of two or more populations, it will show up as a bar with multiple colors. The results of this clustering analysis show that eastern and western monarchs belong to a single genetic population (all North American monarchs are represented by dark gray bars in Figure 23.2b). Thus, sufficient gene flow exists between eastern and western migratory monarchs to homogenize the selectively neutral molecular variation examined in this analysis.

The results of Lyons et al. (2012) indicate that eastern and western migratory monarchs regularly exchange genes, despite the Rocky Mountains separating their breeding ranges. Such a conclusion was suggested by Shepard and colleagues, who found high levels of gene flow between monarchs from California and Michigan based on allozyme analysis (Shephard et al. 2002). How these genetic exchanges occur is not clear, although it has been suggested that monarchs dispersing from Mexico in the spring can populate areas in the western United States in high numbers (Brower and Pyle 2004; Vandenbosch 2007), and some monarchs tagged in the west are retrieved at overwintering sites in Mexico (Pyle, this volume, Chapter 21, Southwest Monarch Study 2013a).

Results of Lyons et al. (2012) further suggest that large-scale genetic differentiation is neither a prerequisite for, nor a result of, differential migration of eastern and western monarchs. This does not necessarily mean that differential selection is not operating on eastern and western monarchs; despite a lack of genetic differentiation of neutral genetic markers, eastern and western monarchs could still show divergence of particular genes that are involved in migration and that are under strong selection. Alternatively, divergent migration pathways might arise from differential gene expression (based on varying and seasonally changing environmental conditions) rather than genetic differences per se (Liedvogel et al. 2011). Such a scenario has been suggested for North American populations of Mexican free-tailed bats, *Tadarida brasiliensis*, which are not genetically differentiated despite their varying migration routes and overwintering sites (Russell et al. 2005).

Next-generation sequencing of the transcriptome (which provides information on the genes that are being transcribed at a given time) of eastern and western migratory butterflies might reveal differential expression of genes resulting in divergent migrations; such an approach has already revealed differential expression of genes in breeding versus migratory monarchs in the eastern United States and Canada (Zhu et al. 2008a, 2008b, 2009).

Although Lyons and colleagues did not find genetic differentiation between eastern and western migratory butterflies, they did find that Hawaii and New Zealand monarchs are differentiated from North American monarchs as well as from each other (Figure 23.2b). They also found that monarchs were more differentiated from one another when they were farther apart geographically (Figure 23.2c), suggesting that greater geographic distances reduce levels of gene flow. Levels of genetic diversity, as measured by allelic richness, appeared similar in the eastern and western United States, decreased in Hawaii and decreased further in New Zealand (Figure 23.2d). This trend of decreasing genetic diversity with increasing distance from North America is consistent with the hypothesis that monarchs dispersed across the Pacific Ocean from an origin in North America (Vane-Wright 1993; Clarke and Zalucki 2004; Zalucki and Clarke 2004). Moreover, the lower genetic diversity in New Zealand than Hawaii is consistent with serial dispersal events, each leading to an additional loss in genetic diversity. However, an alternative hypothesis for this pattern is that monarchs dispersed to Hawaii on more occasions than to New Zealand. Further studies that include a greater number of Pacific islands are necessary to distinguish between these hypotheses.

DOES OPEN WATER IMPEDE GENE FLOW?

The long-distance migration and dispersal of monarchs across the globe demonstrate that monarchs have strong flight ability and therefore the potential to exchange genes between distant geographic regions. At the same time, the study by Lyons et al. (2012) suggests that although monarchs can traverse seas and oceans to colonize distant islands, such large water bodies also present barriers (albeit imperfect) to high and recurrent gene flow. We tested this idea in a separate analysis by estimating the amounts of genetic exchange between several North American monarch populations that are varying distances from each other and either connected by land or separated by sea.

Between 2007 and 2012, we obtained 144 monarchs from Mexico overwintering sites, Bermuda, Belize, Costa Rica, Puerto Rico, and Ecuador (see Figure 23.3 for locations and sample sizes), and compared them with the 200 butterflies collected from St. Marks, FL and coastal California for the study by Lyons et al. (2012). We refer to the original 200 butterflies as the United States sample from here on. Monarchs collected from Ecuador represent the subspecies *Danaus plexippus megalippe*, while all other butterflies represent *Danaus plexippus plexippus*. Like Lyons et al. (2012), we used 11 polymorphic microsatellite markers to determine whether the butterflies are genetically differentiated by location, or whether extensive gene flow occurs. DNA extractions and PCR protocols followed those published by Lyons et al. (2012). To investigate population structure, we used the same genetic clustering analysis and software settings in STRUCTURE version 2.3.2.1 (Pritchard et al. 2000) as described in Lyons et al. (2012). The parameters were sensitive enough to detect subtle or newly formed population structure, and thus to determine the most likely number of genetic populations present.

Our analysis showed that monarchs from the United States, Mexico, Belize, and Costa Rica are genetically indistinguishable (i.e., they are all derived from a single genetic population, indicated in dark gray in Figure 23.4), suggesting significant genetic mixing between the monarchs from these locations. In addition, although butterflies from Bermuda and Puerto Rico are genetically similar to mainland North American monarchs, we found moderate genetic differentiation between island and mainland populations (island butterflies carry high proportions of alleles from both the dark and light gray populations in Figure 23.4). Furthermore, monarchs from Ecuador form a very distinct population (indicated in black in Figure 23.4), different from all other mainland and island groups, a finding that is not surprising given that these monarchs have been previously characterized as a different subspecies.

To confirm our STRUCTURE results, we used F_{ST} and R_{ST} statistics (Holsinger and Weir 2009) to measure genetic population substructuring among the six *D. plexippus plexippus* populations. These

Figure 23.3. Location and hypothesized historical dispersal of sampled monarch populations. Letters represent the sample sites as follows: A. Mexico ($n = 27$) B. United States ($n = 200$) C. Puerto Rico ($n = 29$) D. Bermuda ($n = 13$) E. Belize ($n = 31$) F. Costa Rica ($n = 30$) G. Ecuador ($n = 14$).

statistics are frequently used to measure genetic differentiation, with levels of 0 indicating that individuals belong to the same panmictic population, and values higher than 0 indicating genetic differentiation. Typically, population geneticists consider a value of 0–0.05 to indicate little differentiation, 0.05–0.15 as moderate, 0.15–0.25 as great, and >0.25 as very great; permutation tests are often used to determine whether a value is considered significantly different than 0. R_{ST} was developed as a more suitable statistic for microsatellite markers, based on its dependence on a stepwise mutation model (Slatkin 1995) instead of the infinite alleles model that underlies F_{ST} statistics (Balloux and Lugon-Moulin 2002); however, because neither of these mutation models

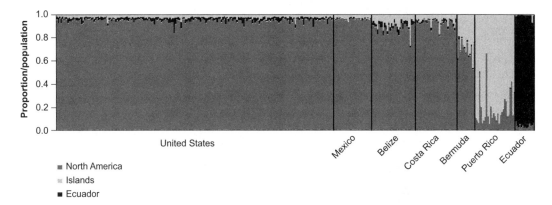

Figure 23.4. Inferred genetic proportion of individual butterflies to each of seven populations. Genetic clustering analysis suggests three genetic populations: North America (United States, Mexico, Belize, Costa Rica), island populations (including Bermuda and Puerto Rico), and Ecuador. Individual monarchs are indicated by vertical bars. Genetic assignments were determined on the basis of 11 microsatellite loci using the software STRUCTURE (Pritchard et al. 2000).

perfectly reflect natural mutation rates of microsatellites, studies on microsatellites often report both measures (Balloux and Lugon-Moulin 2002), and we followed this practice. We calculated pairwise F_{ST} and R_{ST} values between the six *D. plexippus plexippus* populations using Genepop version 4.1.0 software (Rousset 2008). Permutation tests (using 10,000 permutations), as implemented in the "Population comparisons" calculations in Arlequin 3.5.1.2 (Excoffier and Lischer 2010) were used to determine whether pairwise F_{ST} and R_{ST} values differed significantly from 0 (i.e., to determine whether populations are differentiated from each other).

Our analyses of F_{ST} and R_{ST} confirm the results from our STRUCTURE analysis; we found no significant genetic differentiation between monarchs obtained from the United States, Mexico, Belize, and Costa Rica, but Bermuda and Puerto Rico were differentiated from each other and from the mainland populations (Table 23.1). Thus, populations separated by land masses did not show genetic differentiation, whereas populations separated by water bodies did. This suggests that monarchs are able to travel across land freely enough for gene flow to occur, even across large geographic distances; however, large expanses of open water appear to limit the amount of genetic mixing, resulting in population differentiation of island monarchs.

Apart from generating insights into monarch dispersal ability, our results might also change the way scientists view monarch migration. The migration of monarchs from Canada and the United States to Mexico and back again has generally been viewed as a directed two-way migration. A major reason for this view is that monarch migration has been best studied by United States and Canadian citizens and scientists who observe monarchs flying south/southwest toward Mexico in the fall and returning in the spring; however, an alternative scenario is that North American monarchs aggregate in Mexico during the winter, and then disperse in all directions, not just northward to the United States and Canada, during the spring. Such undirected dispersal has been suggested by Wenner and Harris (1993) with regard to spring migrants in California, although the number of monarchs returning to the south-central United States in the spring is evidence that at least a large number of monarchs from Mexico do return to the breeding range of the eastern migratory population (Miller et al. 2012). Importantly, either of these scenarios would result in the genetic mixing of monarch populations north and south of the Mexican overwintering sites shown by our analyses. Further study into dispersal from the overwintering sites is required to understand the full implications of these results as well as their relevance to the navigational and homing mechanisms involved in monarch migration.

A NEW VIEW OF MONARCH MIGRATION AND EVOLUTION IN THE GENOMIC ERA

Past studies of gene flow within and among monarch populations were aided by genetic markers

Table 23.1. Pairwise R_{ST} and F_{ST} values between monarchs from 7 geographic regions.

	United States	Mexico	Belize	Costa Rica	Bermuda	Puerto Rico
Mexico	R_{ST} −0.01046 F_{ST} −0.00044					
Belize	R_{ST} −0.02144 F_{ST} −0.00062	R_{ST} −0.01301 F_{ST} 0.00044				
Costa Rica	R_{ST} −0.00964 F_{ST} −0.00153	R_{ST} 0.00658 F_{ST} 0.00565	R_{ST} 0.00074 F_{ST} −0.00042			
Bermuda	R_{ST} 0.00458 F_{ST} 0.07429*	R_{ST} 0.01936 F_{ST} 0.06768*	R_{ST} 0.04018* F_{ST} 0.08379*	R_{ST} 0.00521 F_{ST} 0.08776*		
Puerto Rico	R_{ST} 0.01794* F_{ST} 0.08264*	R_{ST} 0.09878* F_{ST} 0.11018*	R_{ST} 0.08454* F_{ST} 0.08456*	R_{ST} 0.03750* F_{ST} 0.09715*	R_{ST} 0.06209* F_{ST} 0.16965*	
Ecuador	R_{ST} 0.13180* F_{ST} 0.18721*	R_{ST} 0.25540* F_{ST} 0.24336*	R_{ST} 0.20782* F_{ST} 0.21644*	R_{ST} 0.20677* F_{ST} 0.20134*	R_{ST} 0.28394* F_{ST} 0.30676*	R_{ST} 0.27552* F_{ST} 0.34122*

Note: Values near zero indicate that populations are not genetically distinct, whereas higher values indicate differentiation. Asterisks and shading denote values that are significantly different from zero (indicating that those populations are significantly differentiated from each other).

that are selectively neutral, including allozymes, mitochondrial DNA markers, and microsatellites. But different genetic approaches are required to examine evolutionary changes in monarchs that might have arisen from selection pressures, including those driven by long-distance migration. For example, even though selectively neutral markers have suggested a lack of genetic differentiation between monarchs with different migration destinations in eastern and western North America (Lyons et al. 2012), monarchs from these different locations could still be differentiated at loci that are under differential selection in their respective geographic areas. Thus, it is possible that locally adapted variants of specific genes are selected for during the breeding and migration seasons despite a high influx of neutral genes via gene flow. Likewise, despite the lack of differentiation of neutral markers across migratory and nonmigratory monarch populations in the New World, these same populations could be differentiated at loci that influence traits such as flight ability or metabolism, being favored in populations that migrate annually, but not in populations that do not migrate.

Work on phenotypic variation across wild monarch populations already supports the idea that populations experiencing gene flow can continue to diverge at traits under selection. For example, a recent study focused on wing morphology across multiple wild monarch populations and showed that forewings were larger and more angular in shape (higher aspect ratio) in North American migratory monarchs relative to nonmigratory monarchs in Hawaii, South Florida, and Costa Rica (Altizer and Davis 2010); this work further showed that wing traits were heritable and among-population differences maintained even when monarchs were reared in common garden experiments. A similar study on monarchs in Cuba showed that resident individuals (identified based on stable isotopes and cardenolide fingerprints) had shorter wings than migrants from eastern North America (Dockx 2007, 2012).

Molecular genetics approaches allow us to examine variation across multiple traits, including those that might be difficult to measure phenotypically, such as flight metabolism or navigational systems. One way to identify such "migration genes" is to carry out a candidate gene approach, in which specific genes are sequenced in migratory and non-migratory monarchs. One likely candidate is the phosphoglucose isomerase (*Pgi*) gene; *Pgi* is a central enzyme in glycolysis and affects the flight and dispersal ability of monarchs and other butterflies (Hughes and Zalucki 1993; Niitepõld et al. 2009). It is likely that butterflies that migrate seasonally will carry different alleles of this gene than nonmigratory butterflies; indeed, preliminary analyses have found that migratory and nonmigratory monarch populations from North and South America are nearly fixed for different *Pgi* haplotypes (N. Chamberlain and M. Kronforst, unpublished data).

The downside of a candidate gene approach is that most candidate sites that could be responsible for differences in migration are currently unknown. With this in mind, a genome-wide analysis of genetic differences between migratory and nonmigratory monarchs might offer more promise, and the recently published sequence of the full monarch genome (Zhan et al. 2011) will make this approach feasible. In particular, by sequencing the genomes of monarchs from migratory and nonmigratory populations, as well as related *Danaus* species that do not migrate, we can investigate associations between particular genes and alleles and the extent to which populations undergo long-distance migration. Such sequencing, currently under way, will provide a powerful way to uncover novel migration genes, and could confirm whether migratory monarchs have variations in genes that are involved in flight ability, such as *Pgi*.

Recent work in developing an expressed sequence tag library and the sequencing of the monarch genome have opened the door to almost unlimited possibilities in monarch genetics (Zhu et al. 2008b; Zhan et al. 2011). This work has identified multiple genes that may be involved in circadian rhythm and other aspects involved in migration (Zhu et al. 2008b). For example, by comparing expressed sequences in the brains of summer breeding versus fall migratory monarchs, Zhu and colleagues showed that genes like *turtle*, which affects fruit fly locomotion, are upregulated during migration and could be involved in migratory locomotor behavior. Another gene, *rosy*, related to increased longevity in fruit flies, was also found to be upregulated in migratory monarch adults, which in North America live for up to eight months, compared with only about a month for their summer counterparts. Thus, along with 70 other genes identified in the genome, *rosy* could account for the increased life span of migratory monarchs (Zhu et al. 2008b). Genes relating to eye development and neural processing have also been identified, which may provide insights into the monarchs' use of a sun compass (Zhan et al. 2011). These genomic tools have just begun to allow researchers to discover genes related to the monarch's fascinating migration. They will surely prove themselves invaluable in the future understanding of monarch biology in general.

OUTLOOK

The field of monarch genetics has rapidly expanded and this growth will continue with improving technologies. Recent years have seen the transition from traditional population genetics approaches to the use of genomics and the investigation of genome-wide gene expression, and the recently sequenced monarch genome has much to offer for determining the underlying mechanisms of monarch migration. Efforts to resequence, in which part of an individual's genome is sequenced and compared with the standard genome to detect differences, will help identify genes that underlie migration, and transcriptomics will continue to offer insights into the role of differential gene expression in monarch migration. Moreover, the monarch genome can now be mined for thousands of single nucleotide polymorphisms, which will help scientists more accurately quantify the evolutionary history of monarch populations worldwide. The enticing aspect of genetic studies is that they can show us what is currently happening with monarchs, and they can also give insights into their past. Differing allele frequencies and genetic diversity levels provide glimpses into monarchs' historical colonization pathways and evolutionary history. We can use the monarch genome to better understand how and when monarchs colonized the world, and how monarchs' exposure to novel habitats and environments has affected their genetic adaptation.

Although most previous genetic studies on monarchs have focused on some aspect of migration, scientists have now reached an era when genetics can be used to investigate a wide variety of monarch features and behaviors. For example, recent genomic studies have identified the genetics underlying the chemical defense mechanism of monarchs, which includes a variant of the sodium/potassium pump that makes them more resistant to the toxic effects of milkweed chemicals (Zhu et al. 2008b; Zhan et al. 2011; Dobler et al. 2012; Zhen et al. 2012). We expect that the ongoing development of genetic tools will help scientists better understand the evolution of monarch metamorphosis, host plant specialization, warning coloration, and resistance to disease, to name just a few topics. Undoubtedly, future work in genetics, coupled with traditional studies and observation, will lead to exciting new breakthroughs in the field of monarch biology.

GLOSSARY OF TERMS

allele: Variant of a gene

allelic richness: The number of alleles per locus

allozyme: Variant of a protein that represents underlying genetic variation

bottleneck effect: Changes in allele frequency (and often reductions in genetic diversity) that result when population size is drastically reduced

fixation index (F_{ST}): A measure of population differentiation with a value closer to 0 indicating complete mixing and a value closer to 1 indicating functional isolation

founder effect: The loss of genetic diversity that results when a limited number of individuals found a new population

gene: DNA "blueprint" that codes for a heritable trait

gene expression: Activity of a gene; the process by which the instructions in genes are converted to messenger RNA before protein manufacture

gene flow: The transfer of alleles from one population to another, typically through dispersal and mating

genetic drift: Random changes in allele frequencies due to chance rather than natural selection

genome: The sum of all genetic material in a cell

locus: A specific site in the genome (plural: loci); often used synonymously with 'gene'

microsatellite: short sequence of DNA that is repeated in tandem

mitochondrial DNA: DNA located in the mitochondria and inherited only from the mother

panmixia: Random mating. In a panmictic population, all individuals are potential mates.

phenotype: Expressed traits such as morphology or behavior

polymorphism: The existence of two or more forms of a trait or more than one allele for a given gene

R_{ST}: A measure of population differentiation similar to F_{ST}, but designed for microsatellites by incorporating a stepwise mutation process

single nucleotide polymorphism (SNP): A site on the DNA strand at which the base sequence differs among individuals by one nucleotide (for example, one individual might have sequence ATTC and another might have ATCC)

transcriptome: The total set of activated and expressed genes in a particular cell, tissue, or organism at a given time

24

Connecting Eastern Monarch Population Dynamics across Their Migratory Cycle

Leslie Ries, Douglas J. Taron, Eduardo Rendón-Salinas, and Karen S. Oberhauser

The eastern North American monarch population has a complex annual cycle with four phases: (1) aggregation of most individuals within a small overwintering zone in Mexico; (2) northward spring migration and breeding through the southern United States; (3) summer expansion and breeding throughout the eastern United States and southeastern Canada; and (4) autumn migration to the same overwintering sites. We followed monarch population dynamics throughout this annual cycle, using data from seven large-scale monitoring programs, most relying on citizen scientists. We looked for evidence that dynamics at one step of the migratory cycle carry over to subsequent steps using linear regression. Our results confirm earlier findings that dynamics during the spring recruitment phase have a critical influence on the ultimate size of the breeding population each year. We also found a disconnect between summer and winter numbers that deserves further study. We highlight the need to reexamine these results as new data continue to become available, to develop models that can tease apart multiple interacting factors, and to bolster monitoring programs where data are currently lacking, especially during the spring migration.

INTRODUCTION

The eastern migratory population of North American monarchs (hereafter, "monarchs" refers to the eastern population unless otherwise specified) follows a fairly consistent annual pattern (Figure 24.1). Individuals that have spent the winter at overwintering sites in central Mexico fly north to lay eggs in northern Mexico and the southern United States. Those offspring then travel to the summer breeding grounds in the north-central and northeastern United States and southern Canada that produced their parents' generation the year before. There they breed and produce two to three additional generations. In late summer and early fall, the last generation undergoes a southward migration and travels to the overwintering colony sites, where it remains until the following spring, when the cycle repeats. Thus, the population can be characterized by fairly consistent spatial and numerical expansion each year, followed by contraction into wintering sites, a period during which little reproduction occurs. We refer to this pattern of migrating north, expansion and breeding, migrating south, and overwintering as the monarch annual life cycle (or annual cycle) to distinguish it from the individual life cycle of egg, larva, pupa, and adult.

There are deviations from this "normal" annual cycle. During the winter some monarchs remain in southern regions of the United States and reproduce (Prysby and Oberhauser 2004; Howard et al. 2010; Batalden and Oberhauser, this volume, Chapter 19) and some winter breeding occurs in Mexico (Oberhauser, pers. observ.). Even the existence of an eastern population completely distinct from the West is probably a myth (Pyle, this volume, Chapter 21); however, the evidence is overwhelming that the cycle described above characterizes the vast majority of a monarch population that is largely spatially separated from the population in the West. Even

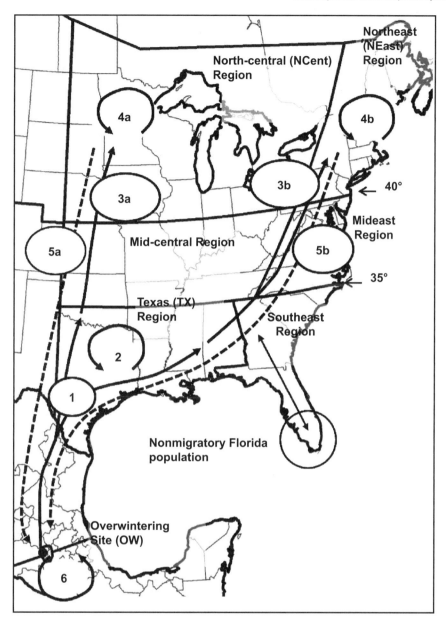

Figure 24.1. Range map of the eastern migratory population of monarchs illustrating major steps in their annual cycle. Arrows and numbers represent transitions from one stage of the cycle to the next, which often involve movement from one region to another. Steps include: (1) Spring migration from overwintering sites in Mexico. (2) Reproduction by the migratory generation to produce first new generation of the year in the southern United States (Texas region). (3) Migration to the North-central (3a) and Northeast (3b) regions. (4) Reproduction and population build-up in the North-central (4a) and Northeast (4b) regions. (5) Fall migration from the North-central (5a) and Northeast (5b) regions. (6) Overwintering in Mexico from late fall through early the following spring.

though this population is widely distributed during the summer, it nonetheless constitutes a single, cohesive population, as documented by Eanes and Koehn (1978) and several subsequent studies (summarized by Pierce et al., this volume, Chapter 23). Thus, data collected throughout eastern North America can inform our understanding of what is happening to the population as a whole.

Recent evidence suggests that the eastern monarch population is declining (Brower et al. 2011; Rendón-Salinas and Tavera-Alonso 2013), and the advent of herbicide-tolerant crops and resultant loss of milkweed in crop fields in the U.S. Upper Midwest has been identified as a possible primary threat contributing to this decline (Pleasants and Oberhauser 2012; Pleasants, this volume, Chapter 14); however, population surveys of migrating individuals in some locations do not show a similar downward trend (Davis 2011). To reconcile the conflicting patterns documented during different phases of their annual life cycle, and to identify critical phases in this cycle, we need to understand monarch population dynamics throughout the year.

To follow annual cycle dynamics, we divided the eastern monarch range into seven biologically relevant regions (Figure 24.1): overwintering sites, Texas Region, Southeast, Mid-central, Mideast, North-central, Northeast. These regions were developed to reflect when and where activity is concentrated in each stage of the annual cycle, but we note that boundaries and seasonal definitions are approximations that can shift from year to year. For the analyses presented here, we focus on four regions and use the following abbreviations in figures and tables: OW (overwintering sites), TX (Texas), NCent (North-central), and NEast (Northeast).

We present the following simplified series of events (steps) throughout the monarch annual cycle to inform specific questions about monarch population dynamics (Figure 24.1). In Step 1, surviving monarch adults that arrived the previous fall leave their Mexican wintering sites in late February and early March, fly northward, and begin arriving in the Texas region in mid-March. Few individuals are observed north of 35°N latitude until several weeks later (Howard and Davis 2004), so we defined this latitude as the northern limit for spring breeding (Figure 24.1). We are less sure about the southern boundary of the spring breeding region because there are few spring observers in Mexico, but we assume that some egg laying occurs south of the U.S.-Mexico border. We also separated Florida, Georgia, and South Carolina into a separate region because individuals from the nonmigratory Florida population may fly north in the spring (Dockx 2012) and our work focuses only on the migratory population. For that reason, we tracked spring dynamics only in the Texas region. In Step 2, monarchs that migrated from Mexico lay eggs in the Texas region. This first pulse of egg laying and development continues until about early May (Prysby and Oberhauser 2004) and produces that year's first generation of adults.

In Step 3, this first generation flies northward, with some individuals flying toward the North-central region and others toward the Northeast region (Steps 3a,b in Figure 24.1), in a split that is roughly around the Appalachian Mountains (Journey North 2013). Individuals usually arrive north of 40°N latitude by mid-May, but slightly later in the Northeast (Howard and Davis 2004). The northern boundary in Figure 24.1 is based on the northern limit of most observation records. In Step 4, monarchs produce two to three additional generations in the North-central and Northeast regions (Steps 4a,b in Figure 24.1) over a period of about three months, with first generation adults continuing to lay eggs until late June (MLMP 2013). Although some recruitment continues in all regions, we focus on the northern regions because there is little summer breeding south of our 40°N latitude cutoff (Prysby and Oberhauser 2004, Baum and Mueller, this volume, Chapter 17) and because most individuals that migrate to the overwintering sites originate from the northern regions (Malcolm et al. 1993; Wassenaar and Hobson 1998).

In Step 5, most individuals enter reproductive diapause and migrate south. Although the timing varies by year and latitude, breeding generally winds down by mid- to late August in the northern regions (Prysby and Oberhauser 2004; MLMP 2013). There appear to be fairly separate migrations from the Central and East regions (Calvert and Wagner 1999), with more sightings commonly reported in the Central region (Howard and Davis, this volume, Chapter 18). We know that egg laying occurs along the fall migratory pathway (Batalden and Oberhauser, this volume, Chapter 19; Baum and Mueller, this volume, Chapter 17), and it is unlikely that reproductive individuals fly all the way to the overwintering sites; however, because the vast majority of fall migrants are nonreproductive (Batalden and Oberhauser, this volume, Chapter 19), for the sake of simplicity we focus on the individuals that fly directly to the Mexican overwintering sites and ignore fall or winter reproduction in the United States. Individuals begin to arrive at the Mexican overwintering sites in early November, with stragglers arriving throughout November and probably into December (Rendón-Salinas, pers. observ.).

Step 6 represents winter survival. The individuals that arrive in the overwintering sites in late November remain there until they begin migrating north again in February. To survive the winter, they must arrive with or obtain enough energy reserves to last the winter, survive any extreme weather conditions that occur, and avoid predation. Those that survive are part of the group that moves northward in the spring and begins the cycle again.

Here, we track year-to-year dynamics to determine the extent to which yearly variation in population size in one step explains yearly variation in the subsequent step. Specifically, a strong, positive relationship between steps suggests the number of individuals feeding into the next step has a strong influence on subsequent population size and indicates "carry-over" effects (Harrison et al. 2011). We do not include other potential explanatory factors in our models (e.g., climate, predator numbers, or resource availability); instead, we are simply attempting to determine the degree to which population dynamics at one stage are predictive of the next. The lack of a predictive link between steps could indicate that environmental factors (e.g., local climate) are swamping any carry-over effects; however, we are cautious about concluding that there is no link, especially when we observe nonsignificant trends or sample sizes are low. We have two reasons for our caution. First, there are steps for which we still have few data; second, citizen science data tend to be particularly variable for several reasons, including different skill levels, nonrandom placement of surveys, and irregular survey intervals. Low sample sizes and increased variability reduce statistical power; however, as data continue to accumulate, we will revisit these patterns to see whether suggested trends are supported or not. Further, as the data resources grow, power will increase to perform more sophisticated modeling that accounts for survey design and multiple interactions. For now, we offer a broad overview of the links between each step in the cycle by exploring the transitions (steps) illustrated in Figure 24.1 (note that we had insufficient data to explore Step 2):

Step 1: Do the numbers of adults at the end of the winter in Mexico predict the number of adults recorded during the spring season in the Texas region?

Step 3: Do numbers of adults (or their eggs) during the spring breeding season predict the numbers of first-generation adults (or their eggs) that arrive in the North-central or Northeast region?

Step 4: Do the numbers of first-generation adults arriving in the north (or their eggs) predict how large the population grows during the summer in the North-central or Northeast region?

Step 5: Does the size of the summer breeding population predict the number of fall migrants in the North-central or the Northeast regions or the size of the winter colonies soon after their arrival?

Step 6: Does the size of the overwintering colonies in Mexico at the beginning of the winter season predict the size of the colonies at the end of the season?

METHODS

Continental-scale data from volunteer monitoring networks

A vast network of citizen science monitoring programs (Figure 24.2) covers the range of the eastern monarch, making this continental-scale, multi-year examination of monarch population dynamics possible. For these analyses, we used data from seven monitoring programs, briefly described below; full descriptions are given by Oberhauser et al. (this volume, Chapter 2).

Data on overwintering colony size in the Mexican Reserve (star in Figure 24.2) were provided by the World Wildlife Fund-Mexico (WWF) and Monarch Butterfly Biosphere Reserve (MBBR) personnel (Rendón-Salinas and Tavera-Alonso 2013). The area (combined across all sites) supporting roosting monarchs is calculated during 10 two-week periods throughout the season, starting in early November and going through late February; we used this area as a proxy for monarch abundance. For 1993–2003, only one estimate of the size of the arriving colony was made each year, but colony size estimates were made every two weeks throughout the winters of 2004–2011.

Data on relative adult population sizes in the spring and summer breeding grounds are provided by three general butterfly survey programs, the continental-scale North American Butterfly Association (NABA) Count Program (gray dots in Figure 24.2) and butterfly monitoring programs in Illinois (IL) and Ohio (OH) (black dots in Figure 24.2). The NABA program uses count circles of 25

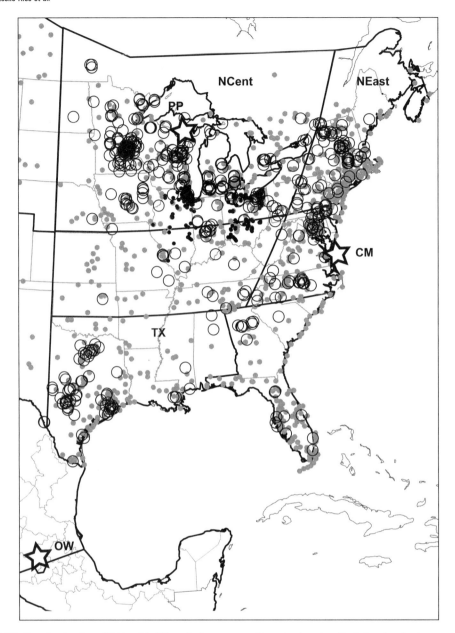

Figure 24.2. Sampling locations (2011 only) of 7 monitoring programs throughout eastern North America. Analyses are restricted to four focal regions: the North-central (NCent), Northeast (NEast), and Texas (TX) regions and the overwintering sites in Mexico (OW). During the spring and summer breeding seasons, adult monarchs are sampled by the North American Butterfly Association (gray dots), and by butterfly monitoring networks in Illinois and Ohio (black dots). Egg density is sampled by the Monarch Larva Monitoring Project (open, gray circles). Surveys of fall migrants occur in Peninsula Point MI (PP) and Cape May NJ (CM), locations shown with stars. A star also shows the location of the overwintering (OW) sites in Mexico.

km diameter in which groups of volunteers count all butterflies they see in a single day. We extracted monarch data and calculated the numbers of monarchs observed adjusted by the total number of hours spent by each group in the field for 1984–2011. In the IL and OH programs, single volunteers walk set transects multiple times per year and record all butterflies observed. We extracted monarch records and calculated monarchs observed adjusted by the total number of hours spent on each transect. Data from IL cover 1987–2011, and OH, 1996–2008. Values used for analysis were averaged within regions during each time period corresponding to the transitions explored for each step (Table 24.1).

We used egg densities recorded by the Monarch Larva Monitoring Project (MLMP) as a proxy for adult abundance. MLMP volunteers establish sites at milkweed patches, which they monitor weekly (open circles in Figure 24.2). They record the number of eggs observed each week, and the number of milkweed plants they search. We used per-plant egg densities that met certain screening criteria (see Pleasants and Oberhauser 2012). Because we do not account for milkweed abundance, egg densities will not correspond exactly to adult densities, especially when milkweed abundance varies from year to year. Data span 1997–2011. We averaged values used for analysis within regions during the time periods corresponding to each step (Table 24.1).

Indices of fall migration population abundance are taken from two locations that support consistent stopover sites from year to year (stars in Figure 24.2). Peninsula Point is on the southern shore of Michigan's Upper Peninsula. Cape May is also located on a peninsula at the southernmost point in New Jersey. Peninsulas often serve as funnel points on the southward journey and regularly host large populations (Meitner et al. 2004). These stopover sites are surveyed multiple times during the migration season, and we used the mean value across surveys for our analyses. We used data from 1996–2010 and 1992–2010 from Peninsula Point and Cape May, respectively.

Consistency between data sets

Although citizen science data often have large spatial and temporal coverage, they frequently do not use strict protocols, survey locations are not established randomly, and surveys may not be performed at the same time each year; nevertheless, they can provide robust data that allow many types of ecological and evolutionary questions at scales not possible from traditional academic surveys (Dickinson et al. 2012). Further, several studies have shown that overlapping bird monitoring programs tend to recover similar data patterns (e.g., Greenberg and Droege 1999; Lepage and Francis 2002; Link et al. 2006). Less comparison work has been done for butterflies (but see van Strien et al. 1997), but citizen science monitoring data underlies much of what we know about butterfly responses to land-use and climate change in Europe (Settele et al. 2009) and examinations from two U.S. programs show that both are effective at capturing local community patterns (Matteson et al. 2012).

To test for consistency between data sets, we compared data from within the same region and season to determine whether we saw similar year-to-year trends. Here, the lack of a strong relationship among programs would indicate that the programs are not providing robust metrics of year-to-year patterns. This conclusion is distinct from conclusions we draw when testing for patterns across the steps in the annual cycle (the main focus of this chapter), where the presence or absence of a strong pattern is used to determine whether there are carry-over effects from one step to the next. To test for consistency, we used three data sets that measure adult densities (NABA, IL, OH) and one that measures egg densities (MLMP). Although variability in milkweed density may erode the relationship between adult and egg densities, we tested the utility of MLMP data as a proxy for adult numbers. We compared year-to-year densities for these four data sets during the Northcentral summer breeding season, calculating correlation coefficients for each comparison.

Abundances of adults as measured through the three monitoring programs were remarkably consistent (Figure 24.3), with high correlation coefficients that were statistically significant for all three comparisons, which is consistent with earlier results (Oberhauser 2007). As anticipated, egg densities correlated less closely with adult abundances. Still, the year-to-year trends showed similar patterns and all correlation coefficients were positive, although none statistically significant. Based on these results, we are confident that indices developed from adult data sets reflect relative abundances between years and seasons; however, because of lack of strong congruence between adult and egg data, we do not compare adult and egg data sets directly in any of our analyses.

Relationship between steps in the annual cycle

To explore population dynamics between transitions, we drew data from the monitoring data sets displayed in Figure 24.2. Although we have a tremendous amount of data, some region/season combinations have greater data availability than others.

Table 24.1. Analyses used to explore five of the six steps illustrated in Figure 24.1

	Explanatory Variable			Response Variable			N	Results						
St	Region	Date Range	Data (no. sites)	Region	Date Range	Data (no. sites)	(yrs)	Int	Slope	Sqr	LRT	adj P	R^2	Fig
1	OW	15 Feb–1 Mar	WWF (1)	TX	15 Mar–16 May	NABA (6)	7	0.70	0.14		0.42	0.32	0.20	–
3	TX	15 Mar–16 May	NABA (6)	NC	24 May–13 Jun	NABA (2.4)	7	0.27	0.55		0.27	1	0.09	–
3	TX	15 Mar–16 May	NABA (6)	NC	24 May–13 Jun	IL (55.9)	7	0.75	0.12		0.08	1	0.03	–
3	TX	15 Mar–16 May	NABA (6)	NE	24 May–4 Jul	NABA (6)	7	0.54	0.38		0.12	1	0.17	–
3	TX	15 Mar–16 May	MLMP (9.7)	NC	24 May–13 Jun	MLMP (44.5)	10	0.13	0.13		0.37	0.16	0.45	–
3	TX	15 Mar–16 May	MLMP (9.9)	NC	24 May–13 Jun	MLMP (42.8)	9	0.16	0.04		0.03	1	0.10	–
3	TX	15 Mar–16 May	MLMP (10.2)	NE	24 May–4 Jul	MLMP (11.4)	9	0.26	–0.09		0.85	1	0.05	–
4	NC	24 May–13 Jun	NABA (1.9)	NC	28 Jun–15 Aug	NABA (83.1)	16	1.73	2.68	–0.53	<0.01	0.001	0.68	24.4a
4	NE	24 May–4 Jul	NABA (15.1)	NE	5 Jul–15 Aug	NABA (41.6)	22	0.45	1.62		0.86	<0.001	0.53	24.4a
4	NC	24 May–13 Jun	IL (30.4)	NC	28 Jun–15 Aug	IL (65.8)	22	1.83	2.74		0.09	<0.001	0.58	–
4	NC	24 May–13 Jun	IL (31.2)	NC	28 Jun–15 Aug	IL (66.2)	21	1.42	3.46		0.65	<0.0001	0.76	24.4b
4	NC	24 May–13 Jun	OH (25.9)	NC	28 Jun–15 Aug	OH (27.4)	14	1.05	3.84		0.26	0.056	0.34	–
4	NC	24 May–13 Jun	OH (25.9)	NC	28 Jun–15 Aug	OH (26.8)	13	1.49	0.61		0.24	0.82	<0.01	24.4b
4	NC	24 May–13 Jun	MLMP (35.5)	NC	28 Jun–15 Aug	MLMP (51.1)	14	0.21	0.44		<0.05	0.1	0.48	–
4	NC	24 May–13 Jun	MLMP (33.6)	NC	28 Jun–15 Aug	MLMP (54.7)	13	0.13	0.95		0.95	0.02	0.52	24.4c
4	NE	24 May–4 Jul	MLMP (10.4)	NE	5 Jul–15 Aug	MLMP (16.3)	13	0.12	0.63		0.03	<0.01	0.81	–
4	NE	24 May–4 Jul	MLMP (8.6)	NE	5 Jul–15 Aug	MLMP (13.6)	12	0.07	1.86		0.35	0.0233	0.49	24.4c
5	NC	28 Jun–15 Aug	NABA (86.1)	NC	16 Aug–17 Oct	PP (1)	15	–32.18	46.30	–6.76	<0.01	0.03	0.56	24.5a
5	NE	5 Jul–15 Aug	NABA (44.2)	NE	16 Aug–17 Oct	CM (1)	19	5.01	14.79		0.90	0.04	0.41	24.5b
5	NC	28 Jun–15 Aug	IL (58.76)	NC	16 Aug–17 Oct	IL (18.6)	25	7.75	0.13		0.50	0.79	0.00	24.5c
5	NC	28 Jun–15 Aug	OH (29.5)	NC	16 Aug–17 Oct	OH (29.2)	13	1.87	1.27		0.41	0.32	0.17	24.5c
5	NC	28 Jun–15 Aug	IL (73.3)	OW	16–30 Nov	WWF (1)	19	13.60	–4.27	0.51	0.02	0.55	0.26	24.6a
5	NC	28 Jun–15 Aug	OH (29.5)	OW	16–30 Nov	WWF (1)	13	7.07	0.13		0.40	1	0.00	24.6b
5	NC	28 Jun–15 Aug	NABA (80.9)	OW	16–30 Nov	WWF (1)	19	7.59	–0.29		0.87	1	0.01	24.6c
5	NE	5 Jul–15 Aug	NABA (44.5)	OW	16–30 Nov	WWF (1)	19	7.00	–0.13		0.88	1	0.00	–
6	OW	16–30 Nov	WWF (1)	OW	16–31 Jan	WWF (1)	7	–0.81	0.87		0.55	0.01	0.74	24.7

Notes: Step number is shown in the left column. For each comparison, we performed ordinary linear regressions. The explanatory and response variables are shown separately in each row, including region (OW = overwinter sites, TX = Texas region, NC = North-central region, NE = Northeast region), date range, data source (WWF = World Wildlife Fund, NABA = North American Butterfly Association, MLMP = Monarch Larvae Monitoring Project, IL = Illinois, OH = Ohio) and the average number of sites (no. sites) over all years used in the analysis. The number of sites for the response variables were used as weighting factors. For each comparison, we show the intercept, slope, and, if appropriate, the square term from the model chosen (either linear or quadratic) by a likelihood ratio test. We present R^2 and P-values (adjusted using Holm-Bonferroni method to account for multiple comparisons within each question) to illustrate the fit of models. If a significant result was achieved, but a single point had excessive leverage (Cook's D score > 1), that point was dropped from the analysis and a new analysis conducted, shown shaded in gray. For results illustrated in figures, the figure number is shown; points identified as outliers are circled within those figures.

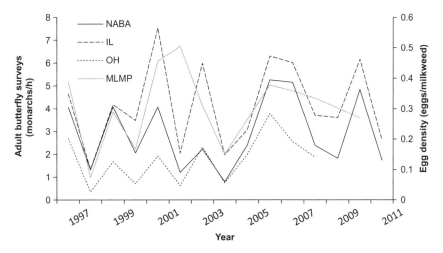

Figure 24.3. Similar year-to-year trends recorded by programs estimating monarch abundance during the summer breeding season in the North-central region. Pearson's correlation coefficients between programs: NABA and OH ($R^2 = 0.69^{**}$), NABA and IL ($R^2 = 0.68^{**}$), OH and IL ($R^2 = 0.68^{*}$). $^{*}P < 0.05$, $^{**}P < 0.01$, all P-values calculated with a sequential Bonferroni correction (Holm 1979). None of the comparisons with MLMP data were significant, but all had positive correlation coefficients (MLMP and NABA, IL, and OH $R^2 = 0.40, 0.49, 0.50,$ and Bonferroni-corrected P-values $= 0.20, 0.20,$ and 0.10, respectively).

NABA and MLMP, the only programs that cover the entire breeding range, have more summer than spring and fall data, limiting our power to test spring or fall linkages with data from those programs. We have more data throughout the full season from the IL and OH butterfly monitoring networks, although these programs allow explorations only in the North-central region. We parsed the region and date ranges for exploring each question based largely on the timeline of events described above, but also adjusted to decrease overlap between generations and for general data availability. We performed ordinary linear regressions where the predictor variable was always the abundance index of the earlier step and the response variable the abundance measured in the subsequent step. Multiple comparisons were possible for most steps (Table 24.1).

We developed a monarch abundance index for each monitoring program (see program descriptions above) and calculated its value for each year/region/season combination. Although for the purposes of analysis, each year/region/season index contributes one data point to the analysis, each point is actually based on multiple surveys within a region. For example, if we compare trends over 10 years, the sample size for our regression test will be 10, but data for each point will be drawn from multiple survey sites, sometimes numbering in the dozens (Figure 24.2).

Table 24.1 shows the average number of sites, over all years, available for each abundance index we calculated. Further, except for the NABA counts, all sites have multiple surveys performed during the year, and those numbers are pooled for each site and specified data ranges within years. We used the number of sites underlying the response data as a weighting factor in each analysis, and we performed analyses only when we had at least five years of data.

Earlier exploratory analysis suggested that some relationships might be curvilinear, so we ran each model both including and excluding a squared term. We used a likelihood ratio test to choose between the linear and quadratic models (Table 24.1). We also present the R^2 and P-values for the best model to illustrate the strength of fit. P-values were adjusted using a sequential Bonferroni correction (Holm 1979) when there were multiple comparisons within a step. Whenever we had multiple tests examining the same step, we always considered whether there was a dominating pattern in the multiple tests, especially when there were conflicting results or non-significant trends. As noted earlier, observed trends can easily be retested as more data become available. To consider the possible influence of outliers, we calculated a measure of Cook's Distance and identified any data points with a value greater than 1 as outliers (Cook and Weisberg 1982) and we present results with and without these outliers. Analyses from which outliers were removed are highlighted in gray in Table 24.1 and are always shown beneath the model with all data included. All analyses were done in R 2.15.2 (R Core Team 2012) using the lm procedure. Likelihood ratio test results were generated using the lmtest package (Horthorn et al. 2013).

The transition between the arrival of migratory individuals from Mexico and the first generation of the year is Step 2 in Figure 24. 1. We had insufficient early and late spring data to analyze this step. Further, to explore patterns within Steps 1 (relationship between winter numbers and Texas arrivals) and 3 (relationship between first-generation adults and arrivals in the north), it would be ideal to separate adult data for the spring breeding season into an early (arrivals) and late (first-generation recruits) period; however, we lacked enough NABA spring data in TX to allow this separation. Therefore, we used an index of the entire spring generation time for both questions, assuming that numbers across the entire season reflect both arrivals and the amount of recruitment. For Step 3, only 4 years of overlap occurred between WWF and OH data, which did not meet our 5-year minimum criterion. For Step 4, we used NABA and MLMP data to compare within-region abundance indices for the spring and summer in the North-central and Northeast regions. We used IL and OH butterfly monitoring network data to perform additional analyses on the same question in the North-central region only.

We examined two points along the pathway of Step 5 (migration from summer breeding to the overwintering sites, Figure 24.1). First, we compared late summer numbers to fall migration abundances using Cape May and Peninsula Point data (first two rows for Step 5 in Table 24.1). Note that only Cape May occurs south of the region that is the source of migrating adults, while Peninsula Point is closer to the northern limit of adult activity. Since monarchs are migrating south, we have an a priori expectation that comparisons using Cape May data (Northeast region) will provide a better test than Peninsula Point (North-central region). Nevertheless, we present comparisons from both sites. We also present comparisons between IL and OH summer and fall indices to explore this segment of Step 5 (third and fourth rows for Step 5 in Table 24.1), but cannot use NABA data since there are too few fall counts. However, the key comparison for Step 5 is determining the success of the end-of-summer population at making it to the overwintering grounds in Mexico; therefore, in a second set of comparisons for Step 5, we compared end-of-summer adult numbers (NABA, IL, OH) with colony size in Mexico during the two-week arrival period at the end of November (fifth through eighth rows for Step 5 in Table 24.1).

To explore Step 6, overwintering survival in Mexico (Figure 24.1), we compared WWF monitoring data from the beginning and end of the season. In addition to quantifying the relationship, by comparing final overwinter size to a 1:1 line indicating no mortality, we are able to estimate overwintering mortality.

RESULTS

Results are shown for all comparisons in Table 24.1 with the focal Step indicated in the first column. For each comparison, the data sets and date ranges used for each test are shown along with sample sizes. Here, sample size is the number of years of data available for comparison (n), but we also note the average number of sites that contributed to each year's single datum. Recall that for each site, there are also multiple surveys done within the season (except for NABA), but these numbers are not shown. Parameter estimates from either the linear or quadratic models are shown depending on the likelihood ratio test, and the R^2 and adjusted P-values are also shown for each comparison. Slope and intercept estimates are shown for all models and the square term only if the quadratic was the chosen model.

There are few spring surveys for any of the monitoring programs, so comparisons that include the spring stage (Steps 1 and 3) have data only from 7–10 years and those data points are estimated from only 6–10 individual sites on average (Table 24.1). There was only one comparison possible to test the relationship between late winter colony size and the spring population in the Texas region (Step 1). A positive slope suggests there may be a relationship (figure not shown), but at this point there are too few years of data based on too few sites for a robust test. For Step 3, we were able to make multiple comparisons to test the relationship between spring migrants in TX and spring arrivals in the North-central and Northeast regions; none were significant (but 5 of 6 had positive slopes). The only comparison that even approached significance was that between egg densities in Texas and North-central regions, but when an identified outlier was removed, the relationship disappeared entirely (Table 24.1).

Although we have very few survey data tracking dynamics during the spring growing season in the South, both the IL and OH data sets provide substantial data on spring arrivals in the North. Surveys for these programs begin in spring, and any monarchs observed in early May are almost exclusively adults that were recruited further south in the spring. Therefore, we had the most data available to compare spring arrivals in the north with the size of the summer breeding population (Step 4). In the North-central region, we compared four data sets (NABA, IL, OH, MLMP) at the beginning and end of the breeding season, and in the North-east we compared NABA and MLMP. All comparisons were highly or nearly significant, even with the Bonferroni correction (Table 24.1), and all showed a positive relationship (Figure 24.4). Removing outliers caused only one relationship to lose significance (OH).

We found mixed results when examining the first segment of Step 5, population size at the end of the summer and abundance of fall migrants. In the Northeast region, there was a positive relationship between summer numbers and fall counts at Cape May (Table 24.1 and Figure 24.5a); however, there was an unexpected parabolic relationship between summer adult populations in the North-central region and stopover sizes at Peninsula Point (Figure 24.5b). We again note that the subregion funneling into Peninsula Point is largely unmeasured. Interestingly, there was no significant relationship between summer and fall population sizes within the IL or OH data sets (Table 24.1, Figure 24.5c).

For the full journey illustrated in Step 5 (Northeast and North-central to overwinter sites in Mexico, Figure 24.1), we found no significant relationships between summer (NABA, IL, OH) population indices and the size of the arriving population in Mexico, suggesting a disconnect between the summer and overwinter numbers; however, although no data points were identified as outliers using the Cook's D test, visual examination suggests two outliers (circled in Figure 24.6). A relationship is suggested if those outliers are removed, and this relationship should continue to be examined as more data become available. Finally, for the seven years of data we had available, there was a strong relationship between the colony size at the beginning and end of winter (Table 24.1, Figure 24.7). The points lie close to the 1:1 line, suggesting that for the seven years for which we have data, little overwinter mortality occurred; we note, however, that this interpretation of the data assumes that monarch densities within the measured colony areas are the same at the beginning and end of the winter, which may not be the case.

DISCUSSION

Our step-by-step analysis of the annual cycle of eastern monarchs suggests that the most critical factor impacting the size of each year's breeding population appears to be the size of the first generation of migrants that arrives each year in the north. In both the North-central and Northeast regions, the number of individuals arriving was highly predictive of the eventual size of that summer's population (Step 4), whether we looked at adult or egg density data (Figure 24.4). This suggests that previous transitions that determine the number of monarchs arriving in the north are crucial, and that factors that occur over the course of the summer are less important in driving population, at least during the years for which we had data. However, because we lack sufficient data on dynamics during the spring recruitment and subsequent migration period, it is impossible to say if the crucial step driving that result is the number of spring arrivals from Mexico (Step 1), spring recruitment (Step 2), or successful northward migration of the first generation of adults (Step 3). Recent modeling showed that temperature and precipitation in Texas during the spring had a stronger impact on summer monarch population growth in Ohio than did temperature and precipitation during the summer (Zipkin et al. 2012); however, it is important to note that while the Zipkin et al. model suggests that spring recruitment (our Step 2) is one important driver, it does not address the potential contributions of Steps 1 or 3. A focus of future research should be teasing apart the contributions of these early stages on the number of arrivals in the Northeast and North-central regions.

We were able to provide preliminary explorations of patterns for Steps 1 and 3 (no data for Step 2 were available). No significant relationship was found between the area occupied by adult monarchs at the end of the overwintering period in Mexico and

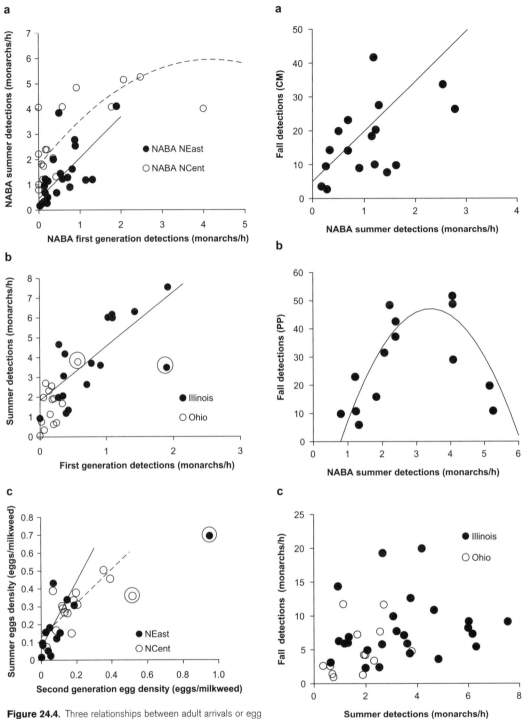

Figure 24.4. Three relationships between adult arrivals or egg density in the north at beginning of season compared with population sizes at end of season: (a) NABA data in the NEast and NCent regions, (b) Illinois and Ohio data in the NCent region, and (c) MLMP egg data in the NEast and NCent regions. In all panels, dashed lines indicate models fit to white dots and solid lines indicate models fit to black dots. Model parameters and significance levels in Table 24.1 are presented with and without the outliers that are circled here. Outliers are excluded from model fits presented in the figure.

Figure 24.5. Relationships between summer adult detections and fall migration numbers. The relationship between NABA counts and the two fixed stopover site surveys, (a) Cape May and (b) Peninsula Point, showed significant relationships, but no relationship was found between late summer and fall detections in (c) IL or OH. Model parameters and significance levels shown in Table 24.1.

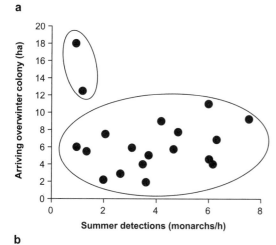

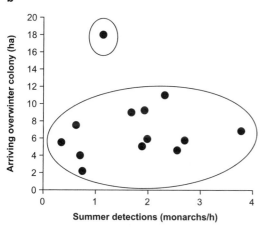

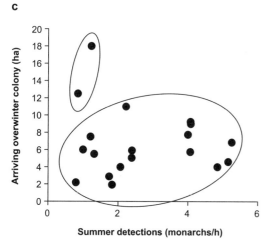

Figure 24.6. Relationships between end-of-summer adult detections in the NCent region and colony size at the beginning of the overwinter period from (a) Illinois, (b) Ohio, and (c) NABA. None of these relationships are significant, even if the outliers are removed, but still suggest a relationship that should be reexamined as more data become available.

the numbers of adults arriving in the Texas region (Step 1); however, it is important to note that we had only one data set (NABA) to compare with colony size data from Mexico, data were available from both sources during only seven years, and NABA data were based on very few surveys (Table 24.1). A positive relationship is suggested, and it will be valuable to continue to track this transition. If the lack of a relationship holds, it suggests that environmental conditions during the northward migration through Mexico swamp the relationship between numbers leaving the wintering sites and entering the Texas region, and thus that this is a crucial transition for monarchs. Increased monitoring throughout the spring migration is essential to do a better job of quantifying this relationship and determining the important drivers during this critical phase.

There was also no significant relationship between the number of monarchs observed in the Texas region in the spring and the number of first-generation individuals in the north (Step 3). In fact, the only comparison showing a trend (MLMP in the Texas region compared with MLMP in the North-central region) was lost when an outlier was removed. It is possible that these results are simply a detection problem resulting from the relatively small number of spring surveys on which the yearly indices are based; it is important to note that five of six comparisons had positive trends (Table 24.1). Further, as noted earlier, we had insufficient data to divide the spring population into early arrivals

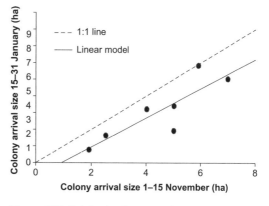

Figure 24.7. Relationship between colony size at the beginning and end of winter at the overwinter sites (solid line, $y = 6.1901x - 0.8091$). Because there is no recruitment during the winter but always mortality, the numbers at the end of the winter should always be lower, so points are a priori expected to fall below the 1:1 line (dashed line).s

and the population size after spring recruitment. Using population numbers from late in the spring would be the most rigorous approach to exploring this question and enable us to capture the impact of spring weather on spring recruitment, which has been shown to be important (Zipkin et al. 2012). Ultimately, until we have more spring data, it will be difficult to have enough power to test this question rigorously. Further, if data were sufficient to split between early and late spring numbers, we could address population growth during the spring (Step 2 in Figure 24.1). Increased spring monitoring in the south would greatly enhance our ability to explore dynamics during this phase; more structured surveys of adults (repeated surveys at the same sites accounting for effort, such as those done by NABA, IL, and OH) would allow us to assess both the size of population arriving from Mexico and numbers of their adult offspring, which migrate northward.

Numbers at the end of the breeding season appear to have variable relationships with numbers observed during the fall (first segment of Step 5). Ideally, late summer population observations should be made north of fall migratory stopover sites with which they will be compared. We were able to do this only for the comparison between late summer observations in the Northeast and those in Cape May, and found a positive relationship (Figure 24.5a). We do not see that relationship when making the same comparison at Peninsula Point, possibly because this count measures butterflies that have flown from more northerly regions. We also saw no pattern when comparing late summer OH and IL counts with fall OH and IL counts; however, fall counts in OH and IL probably include migrants from further north, and thus might not reflect local summer abundances. Detection dynamics may also be a factor here since migrating monarchs tend to come in spurts and cluster in overnight roosting locations, meaning they are both temporally and spatially clumped during the fall and thus more difficult to detect. We may need to use more sophisticated modeling to explore this relationship more rigorously. We have no explanation for the parabolic shape of the relationship between North-central and Peninsula Point (Figure 24.5b) and suspect this may be spurious, but as we continue to accumulate monitoring data we will be able to determine whether that unexpected pattern holds.

There were also no significant relationships between late summer (NABA, IL, OH) population indices and the size of the wintering population (the full pathway of Step 5) despite the fact that we have a great deal of data available to develop end-of-summer abundance indices. This lack of relationship suggests that variable migration success may be a key driver of observed patterns, or that other factors, such as breeding habitat loss (Pleasants and Oberhauser 2012; Pleasants, this volume, Chapter 14), are swamping our ability to detect the relationships. Brower et al. (this volume, Chapter 10) suggest that fall nectar availability is important to migratory success, and it is likely that this availability varies from year to year. Another large unknown is the proportion of individuals that remain in the southern United States (Batalden and Oberhauser, this volume, Chapter 19) and whether this varies from year to year. Finally, our late summer and fall abundance indices measure local densities, while the measurements in Mexico measure the entire population. Overwintering colony size has declined over the past several years (Brower et al. 2011; Rendón-Salinas and Tavera-Alonso 2013), yet summer counts do not reflect that, nor do patterns at fall stopover sites (Davis 2011). Trying to pin down the causes of this mismatch, if it holds with more data, should be a focus of future research because it suggests that fall migration is a key stage in the annual cycle.

The strong relationship between early and late winter population size suggests that overwintering mortality generally is not variable enough to erode the relationship between how many monarchs arrive and how many are alive at the end of each winter. Further, for the seven years we examined, overall mortality appeared to be low since observed values fell close to the 1:1 line (Figure 24.7). It is important to note, however, that none of the years included in this analysis had a catastrophic mortality event, which happens periodically (Brower et al. 2004).

Our results highlight (1) steps that appear to be critical and thus important for conservation focus, but that are also understudied, and (2) critical steps that should be the focus of more sophisticated modeling. These points are well illustrated by our results focused on dynamics in the spring. Numbers arriving in the northern United States are critical to the size of the breeding population for the rest of the summer (Figure 24.4), and previous models (Zipkin et al. 2012) show that Texas climate in the spring has the strongest influence on summer growth, yet we do not know the extent to which each of the earlier steps influences those numbers. Another critical step highlighted by this analysis appears to be the fall migration (Step 5), illustrated by the disconnect between population trends in the summer and winter stages. Understanding the lack of a significant relationship between summer numbers and winter colony size, and the role of events during the migration, should be a focus of future research. This relationship will be easier to understand as more data accumulate, but more sophisticated modeling will be required to tease apart potential interacting factors.

Our analyses also highlight the value of the citizen science programs that provided the data. Not only do the results show that valuable and consistent information is contained within the data (Figure 24.3), but they also highlight the spatial and temporal scales over which we can now ask (and answer) questions. The fact that these data continually stream in each year means that we will be able to reexamine our results, especially those for which we had too little data to rigorously explore in this analysis (Steps 1, 2, and 3). While we were fortunate to have anywhere from 7 to 22 years of data with which to explore our questions (Table 24.1), those years are still a sample of the different combination of factors that could arise to impact population dynamics. For example, this work, as well as that of previous modeling efforts (Zipkin et al. 2012), suggests that conditions during the summer are not as critical as the spring for growth; however, in 2012, which was not included in any of our analyses, spring conditions may have been conducive for growth, yet record summer temperatures and a severe drought in the Upper Midwest likely had a negative impact on the year's population, and the arriving population in Mexico was the lowest ever at that time (Rendón-Salinas and Tavera-Alonso 2013). As the climate changes, we need to constantly examine and reexamine the critical transitions between each region and stage; these continuing investigations can only be done thanks to the hard work and dedication of thousands of citizen science volunteers.

ACKNOWLEDGMENTS

This paper was the result of a working group sponsored by the National Center for Ecological Analysis and Synthesis. Additional funding for LR was provided by NSF (Award DBI-1147049 and DBI-1052875 to SESYNC). We thank Andrew Davis, John Pleasants, and one anonymous reviewer for comments on our manuscript. We thank the program and data managers from all the monitoring programs who provided their data, including Jerry Wiedmann and Rick Ruggles from the Ohio Lepidopterist Society, Jeff Glassberg and Jim Springer from the North American Butterfly Association, Gina Badgett from Peninsula Point, and Dick Walton from Cape May. Most importantly, we thank the thousands of volunteers in the Ohio Lepidopterists' Society, the North American Butterfly Association, the Illinois Butterfly Monitoring Network, the Monarch Larva Monitoring Project, the Monarch Monitoring Project of the U.S. Forest Service in Peninsula Point, Michigan, the Monarch Monitoring Project in Cape May, New Jersey, and the Monarch Butterfly Biosphere Reserve and WWF-Mexico, without whose participation this analysis would not have been possible.

References

Abrams, P.A. 1995. Implications of dynamically variable traits for identifying, classifying, and measuring direct and indirect effects in ecological communities. Am. Nat. 146:112–134. (Ch. 7)

Ackery, P.R., and R.I. Vane-Wright. 1984. Milkweed butterflies: Their cladistics and biology. Ithaca, NY: Cornell University Press. (Ch. 7, 9, 20, 23, Parts 2, 5)

Adamo, S. 1998. The specificity of behavioral fever in the cricket *Acheta domesticus*. J. Parasitol. 84:529–533. (Ch. 7)

Agee, J.K. 1993. Fire ecology of the Pacific Northwest forests. Washington, DC: Island Press. (Ch. 15)

Agrawal, A.A. 2005. Natural selection on common milkweed (*Asclepias syriaca*) by a community of specialized insect herbivores. Evol. Ecol. Res. 7:651–667. (Part 2)

Agrawal, A.A. 2007. Macroevolution of plant defense strategies. Trends Ecol. Evol. 22:103–109. (Ch. 4)

Agrawal, A.A., and M. Fishbein. 2006. Plant defense syndromes. Ecology 87:S132–S149. (Ch. 4)

Agrawal, A.A., and M. Fishbein. 2008. Phylogenetic escalation and decline of plant defense strategies. Proc. Natl. Acad. Sci. 105:10,057–10,060. (Ch. 4, Part 2)

Agrawal, A.A., and K. Konno. 2009. Latex: A model for understanding mechanisms, ecology, and evolution of plant defense against herbivory. Annu. Rev. Ecol. Evol. Syst. 40:311–331. (Ch. 4, Part 2)

Agrawal, A.A., M.J. Lajeunesse, and M. Fishbein. 2008. Evolution of latex and its constituent defensive chemistry in milkweeds (*Asclepias*): A phylogenetic test of plant defense escalation. Entomol. Exp. Appl. 128:126–138. (Ch. 4, 6)

Agrawal, A.A., M. Fishbein, R. Halitschke, A.P. Hastings, D.L. Rabosky, and S. Rasmann. 2009a. Evidence for adaptive radiation from a phylogenetic study of plant defenses. Proc. Natl. Acad. Sci. 106:18,067–18,072. (Ch. 4)

Agrawal, A.A., M. Fishbein, R. Jetter, J.P. Salminen, J.B. Goldstein, A.E. Freitag, and J.P. Sparks. 2009b. Phylogenetic ecology of leaf surface traits in the milkweeds (*Asclepias* spp.): Chemistry, ecophysiology, and insect behavior. New Phytol. 183:848–867. (Ch. 4)

Agrawal, A.A., J.P. Salminen, and M. Fishbein. 2009c. Phylogenetic trends in phenolic metabolism of milkweeds (*Asclepias*): Evidence for escalation. Evolution 63:663–673. (Ch. 4, Part 2)

Agrawal, A.A., E.E. Kearney, A.P. Hastings, and T.E. Ramsey. 2012a. Attenuation of the jasmonate burst, plant defensive traits, and resistance to specialist monarch caterpillars on shaded common milkweed (*Asclepias syriaca*). J. Chem. Ecol. 38:893–901. (Ch. 14)

Agrawal, A.A., G. Petschenka, R.A. Bingham, M.G. Weber, and S. Rasmann. 2012b. Toxic cardenolides: Chemical ecology and coevolution of specialized plant-herbivore interactions. New Phytol. 194:28–45. (Ch. 4, 6, Part 2)

Alatorre, A. 2007. Gana a medias Monarca: Reporta WWF reducción de 48% en pérdida anual por tala. Reforma. Diario publicado el 22 de julio de 2007. (Ch. 13)

Alaux, C., F. Ducloz, D. Crauser, and Y. Le Conte. 2010. Diet effects on honeybee immunocompetence. Biol. Lett. 6:562–565. (Ch. 7)

Alberta Press. 2012. CBC News. Monarch butterflies make surprise appearance in central Alberta. http://www.cbc.ca/news/canada/calgary/story/2012/06/20/monarch-butterfly-migration-alberta.html. Accessed January 2013. (Ch. 21)

Alizon, S., A. Hurford, N. Mideo, and M. Van Baalen. 2009. Virulence evolution and the trade-off hypothesis: History, current state of affairs and the future. J. Evol. Biol. 22:245–259. (Ch. 7)

Allen, C.D., A.K. Macalady, H. Chenchouni, D. Bachelet, N. McDowell, M. Vennetier, T. Kizberger, A. Rigling, D.D. Breshears, E.H. Hogg, P. Gonzalez, R. Fensham, Z. Zhang, J. Castro, N. Demidova, J.H. Lim, G. Allard, S.W. Running, A. Semerci, and N. Cobb. 2010. A global overview of drought and heat-induced tree mortality reveals emerging climate change risks for forests. For. Ecol. Manage. 259:660–684. (Ch. 9, 13)

Allen, M.M., and K.B. Snow. 1993. The monarch project: A program of practical conservation in California. *In* S.B. Malcolm and M.P. Zalucki, eds., Biology and conservation of the monarch butterfly, pp. 393–394. Los Angeles: Natural History Museum of Los Angeles County. (Ch. 12)

Allen, M.S., and M.W. Palmer. 2011. Fire history of a prairie/forest boundary: More than 250 years of frequent fire in a North American tallgrass prairie. J. Veg. Sci. 22:436–444. (Ch. 17)

Alonso-Mejía, A., and A. Arellano-Guillermo. 1992. Influence of temperature, surface body moisture and height aboveground on survival of monarch butterflies overwintering in Mexico. Biotropica 24:415–419. (Ch. 9)

Alonso-Mejía, A., J.I. Glendinning, and L.P. Brower. 1993. The influence of temperature on crawling, shivering, and flying in overwintering monarch butterflies in Mexico. *In* S.B. Malcolm and M.P. Zalucki, eds., Biology and conservation of the monarch butterfly, pp. 309–314. Los Angeles: Natural History Museum of Los Angeles County. (Ch. 9)

Alonso-Mejía, A., E. Rendón-Salinas, E. Montesinos-Patino, and L.P. Brower. 1997. Use of lipid reserves by monarch butterflies overwintering in Mexico: Implications for conservation. Ecol. Appl. 7:934–947. (Ch. 8, 9, 10, 20)

Altizer, S.M. 2001. Migratory behaviour and host-parasite co-evolution in natural populations of monarch butterflies infected with a protozoan parasite. Evol. Ecol. Res. 3:1–22. (Ch. 7)

Altizer, S., and A.K. Davis. 2010. Populations of monarch butterflies with different migratory behaviors show divergence in wing morphology. Evolution 64(4):1018–1028. (Ch. 20, 23, Part 5)

Altizer, S.M., and K. Oberhauser. 1999. Effects of the protozoan parasite *Ophryocystis elektroscirrha* on the fitness of monarch butterflies (*Danaus plexippus*). J. Invertebr. Pathol. 74:76–88. (Ch. 7)

Altizer, S.M., K.S. Oberhauser, and L.P. Brower. 2000. Associations between host migration and the prevalence of a protozoan parasite in natural populations of adult monarch butterflies. Ecol. Entomol. 25:125–139. (Ch. 7, 19, 23)

Altizer, S.M., K. Oberhauser, and K.A. Geurts. 2004. Transmission of the protozoan parasite, *Ophryocystis elektroscirrha*, in monarch butterfly populations. *In* K.S. Oberhauser and M.J. Solensky, eds., The monarch butterfly: Biology and conservation, pp. 203–218. Ithaca, NY: Cornell University Press. (Ch. 7)

Altizer, S., R. Bartel, and B.A. Han. 2011. Animal migration and infectious disease risk. Science 331:296–302. (Ch. 7, Part 2)

Anderson, J.B., and L.P. Brower. 1996. Freeze-protection of overwintering monarch butterflies in Mexico: Critical role of the forest as a blanket and an umbrella. Ecol. Entomol. 21:107–116. (Ch. 8, 9, 11, 12, 13, Parts 3, 4)

Anderson, R.M., and R.M. May. 1982. Coevolution of hosts and parasites. Parasitology 85:411–426. (Ch. 7)

Anderson, R.P., D. Lew, and A.T. Peterson. 2003. Evaluating predictive models of species' distributions: Criteria for selecting optimal models. Ecol. Model. 162:211–232. (Ch. 11)

Ángeles-Cervantes, E., and L. López-Mata. 2009. Supervivencia de una cohorte de plántulas de *Abies religiosa* bajo diferentes condiciones postincendio. Bol. Soc. Bot. Méx. 84:25–33. (Ch. 15)

Anon. 2009. Talarán 10 mil árboles en reserva mariposa monarca por plaga. La Jornada. 20 October 2009. (Ch. 13)

Anon. 2011. Se incrementa en mas de 100% superficie de bosque ocupapda por la mariposa Monarca en la temporada 2010–2011. Comision Nacional de Areas Naturales Protegidas (CONANP) and Secretaría de Medio Ambiente y Recursos Naturales (SEMARNAT) Press Release, Mexico City. 14 February 2011:1–2. (Ch. 9)

Arango, N.V. 1996. Stabilizing selection in migratory butterflies: A comparative study of queen and monarch butterflies. Master's thesis, University of Florida (Gainesville). (Ch. 8)

Aridjis, H. 2004. Twilight of the monarchs. Reforma (1 February 2004). (Ch. 9)

Arnaud, P.H. Jr. 1978. A host-parasite catalog of North American Tachinidae (Diptera). Misc. Pub. 1319. Washington, DC: U.S. Department of Agriculture. (Ch. 6)

Arribere, M.C., A.A. Cortadi, M.A. Gattuso, M.P. Bettiol, N.S. Priolo, and N.O. Caffini. 1998. Comparison of Asclepiadaceae latex proteases and characterization of *Morrenia brachystephana* Griseb. cysteine peptidases. Phytochem. Anal. 9:267–273. (Ch. 4)

Asher, J., M. Warren, R. Fox, P. Harding, G. Jeffcoate, and S. Jeffcoate. 2001. The millennium atlas of butterflies in Britain and Ireland. Oxford University Press. 433 pp. (Ch. 22)

Askew, R.R. 1971. Parasitic insects. London: Heinemann Educational Books. (Ch. 6)

Audubon 2012. Christmas Bird Count. birds.audubon.org/christmas-bird-count. Accessed November 2012. (Ch. 2)

Báez, M. 1998. Mariposas de Canarias. Editorial Rueda, S.L., Madrid. 216 pp. (Ch. 22)

Bailey, R.G. 1995. Description of the ecoregions of the United States. Misc. Pub. 1391, 2nd ed. Washington, DC: U.S. Department of Agriculture. (Ch. 3)

Baker, G.T. 1891. Notes on the Lepidoptera collected in Madeira by the late T. Vernon Wollaston. Trans. Entomol. Soc. Lond. 1891:197–221. (Ch. 22)

Baker, H.G., and I. Baker. 1982. Chemical constituents of nectar in relation to pollination mechanisms and phylogeny. *In* H.M. Nitecki, ed., Biochemical aspects of evolutionary biology, pp. 131–171. Chicago: University of Chicago Press. (Ch. 10)

Balloux, F., and N. Lugon-Moulin. 2002. The estimation of population differentiation with microsatellite markers. Mol. Ecol. 11:155–165. (Ch. 23)

[BAMONA] Butterflies and Moths of North America. 2012a. Butterflies and Moths of North America: Collecting and sharing data about Lepidoptera. www.butterfliesandmoths.org/. Accessed January 2013. (Ch. 2)

BAMONA. 2012b. http://www.butterfliesandmoths.org/species/Danaus-plexippus. Accessed January 2013. (Ch. 21)

Barker, J.F., and W.S. Herman. 1976. Effect of photoperiod and temperature on reproduction of the monarch butterfly, *Danaus plexippus*. J. Insect Physiol. 22:1565–1568. (Ch. 9)

Barret, C.G. 1893. *Anoxia plexippus* (*Danais archippus*) in the Atlantic. Entomol. Mon. Mag. 29:163. (Ch. 22)

Barron, M.C. 2007. Retrospective modeling indicates minimal impact of non-target parasitism by *Pteromalus puparum* on red admiral butterfly (*Bassaris gonerilla*) abundance. Biol. Control 41:53–63. (Ch. 6)

Barron, M.C., N.D. Barlow, and S.D. Wratten. 2003. Non-target parasitism of the endemic New Zealand red admiral butterfly (*Bassaris gonerilla*) by the introduced biological control agent *Pteromalus puparum*. Biol. Control 27:329–335. (Ch. 5, 6)

Bartel, R.A., and S. Altizer. 2012. From protozoan infection in monarch butterflies to colony collapse disorder in bees: Are emerging infectious diseases proliferating in the insect world? *In* A. Aguirre, P. Daszak, and R. Ostfeld, eds., Conservation medicine: Applied cases of ecosystem health, pp. 284–300. New York: Oxford University Press. (Part 5)

Bartel, R.A., K.S. Oberhauser, J.C. de Roode, and S.M. Altizer. 2011. Monarch butterfly migration and parasite transmission in eastern North America. Ecology 92:342–351. (Ch. 2, 7, 19, Part 2)

Batalden, R.V. 2006. Possible changes in monarch fall migration detected in Texas. Monarch Larva Monitoring Project Newsletter 7:3. mlmp.org. Accessed November 2012. (Ch. 2)

Batalden, R.V. 2011. Potential impacts of climate change on monarch butterflies, *Danaus plexippus*. Ph.D. thesis, Univ. of Minnesota (Twin Cities). (Ch. 19)

Batalden, R.V., K. Oberhauser, and A.T. Peterson. 2007. Ecological niches in sequential generations of eastern North American monarch butterflies (Lepidoptera: Danaidae): The ecology of migration and likely climate change implications. Environ. Entomol. 36:1365–1373. (Ch. 2, 3, 7, 8, 11, 18, 19, Parts 3, 5)

Baucom, R.S., and J.C. de Roode. 2011. Ecological immunology and tolerance in plants and animals. Funct. Ecol. 25:18–28. (Ch. 7)

Baum, K.A., and W.V. Sharber. 2012. Fire creates host plant patches for monarch butterflies. Biol. Lett. doi: 10.1098/rsbl.2012.0550. (Ch. 16)

Baum, K.A., and W.V. Sharber. 2012. Fire creates host plant patches for monarch butterflies. Biol. Lett. Published online:1 August 2012. (Ch. 17)

Baynes, E.S.A. 1966. Report on migrant insects in Ireland for 1965. Irish Nat. Jour. 15 (6):157–159. (Ch. 22)

Becerra, J.X. 1997. Insects on plants: Macroevolutionary chemical trends in host use. Science 276:253–256. (Ch. 4)

Bell, E., L.P. Brower, W.H. Calvert, J. Dayton, D. Frey, K. Leong, D. Murphy, R.M. Pyle, W. Sakai, K.B. Snow, and S. Weiss. 1993. Conservation and management guidelines for preserving the monarch butterfly migration and monarch overwintering habitat in California. The Xerces Society. (Ch. 3, 12)

Bellows, T.S. Jr., R.G. Van Driesche, and J.S. Elkinton. 1992. Life-table construction and analysis in the evaluation of natural enemies. Annu. Rev. Entomol. 37: 587–614. (Ch. 6)

Berman, M.G, J. Jonides, and S. Kaplan. 2008. The cognitive benefits of interacting with nature. Psychol. Sci. 19:1207–1212. (Ch. 1)

Beutelspacher, C.R. 1988. Las mariposas entre los antiguos Mexicanos. Fondo de Cultura Econonica, Avenida de la Univeridad, 975. 03100 Mexico. (Part 1)

Bhowmik, P.C. 1994. Biology and control of common milkweed (Asclepias syriaca). Rev. Weed Sci. 6:227–250. (Ch. 14)

Bivar de Sousa, A. 1984–85. Lepidoptera Papilionoidea dos Arquipélagos da Madeira e Selvagens. Bol. Soc. Port. Cienc. Nat. 22:47–53. (Ch. 22)

Bivar de Sousa, A. 1991. Novas citações de Lepidópteros para os Açores. Bol. Soc. Port. Entomol. 133 (V-1):1–15. (Ch. 22)

Blackman, C.J., T.J. Brodribb, and G.J. Jordan. 2009. Leaf hydraulics and drought stress: Response, recovery and survivorship in four woody temperate plant species. Plant Cell Environ. 32:1584–1595. (Part 3)

Boettner, G.H., J.S. Elkinton, and C.J. Boettner. 2000. Effects of a biological control introduction on three nontarget native species of saturniid moths. Conserv. Biol. 14:1798–1806. (Ch. 6)

Bojórquez-Tapia, L.A., L.P. Brower, G. Castilleja, S. Sánchez-Colón, M. Hernández, W. Calvert, S. Díaz, P. Gómez-Priego, G. Alcantar, E.D. Melgarejo, M.J. Solares, L. Gutiérrez, and M. del L. Juárez. 2003. Mapping expert knowledge: Redesigning the Monarch Butterfly Biosphere Reserve. Conserv. Biol. 17:367–379. (Ch. 9, 13, Part 4)

Bollwinkel, C.W. 1969. A revision of South American species of Asclepias L. Ph.D. dissertation, Southern Illinois University, Carbondale. (Ch. 20)

Boots, M., and M. Begon. 1993. Trade-offs with resistance to a granulosis virus in the Indian meal moth, examined by a laboratory evolution experiment. Funct. Ecol. 7:528–534. (Ch. 7)

Boots, M., and R.G. Bowers. 1999. Three mechanisms of host resistance to microparasites—avoidance, recovery and tolerance—show different evolutionary dynamics. J. Theor. Biol. 201:13–23. (Ch. 7)

Boppré, M., and R.I. Vane-Wright. 2012. The butterfly house industry: Conservation risks and education opportunities. Conserv. Soc. 10:285–303. (Ch. 1)

Borges, P.A.V., A. Costa, R. Cunha, R. Gabriel, V. Goncalves, A. Frias Martins, A. Melo, M. Parente, R. Raposeiro, P. Rodríguez, R. Serrao Santos, L. Silva, P. Vieira, and V. Vieira. 2010. Listagem dos organismos terrestres e marinhos dos Azores. Principia Editora Ltda. 429 pp. (Ch. 22)

Borkin, S.S. 1982. Notes on shifting distribution patterns and survival of immature Danaus plexippus (Lepidoptera: Danaidae) on the food plant Asclepias syriaca. Great Lakes Entomol. 15:199–206. (Ch. 5, 6)

Borland, J., C. Johnson, T. Crumpton, T. Thomas, S. Altizer, and K. Oberhauser. 2004. Characteristics of fall migratory monarch butterflies, Danaus plexippus, in Minnesota and Texas. In K.S. Oberhauser and M.J. Solensky, eds., The monarch butterfly: Biology and conservation, pp. 97–104. Ithaca, NY: Cornell University Press. (Ch. 19)

Bossard, C.C., J.M. Randall, and M.C. Hoshovsky. 2000. Invasive plants of California's wildlands. Berkeley: University of California Press. (Ch. 12)

Boyer, C.R. 2005. Contested terrain: Forestry regimes and community responses in Northeastern Michoacán, 1940–2000. In D.B. Bray, L. Merino, and D. Barry, eds., The community forests of Mexico: Managing for sustainable landscapes, pp. 27–48. Austin: University of Texas Press. (Ch. 13)

Bradley, C.A., and S. Altizer. 2005. Parasites hinder monarch butterfly flight ability: Implications for disease spread in migratory hosts. Ecol. Lett. 8:290–300. (Ch. 7, Part 2)

Breed, G.A., S. Stichter, and E.E. Crone. 2012. Climate-driven changes in northeastern US butterfly communities. Nat. Clim. Change 3:142–145. (Ch. 2, 8)

Bremermann, H.J., and H.R. Thieme. 1989. A competitive exclusion principle for pathogen virulence. J. Math. Biol. 27:179–190. (Ch. 7)

Brenner, L. 2006. Áreas naturales protegidas y ecoturismo: El caso de la reserva de la Biosfera Mariposa Monarca, México. Relaciones 27(105):236–265. (Ch. 15)

Brenner, L. 2009. Aceptación de políticas de conservación ambiental: El caso de la Reserva de la Biosfera Mariposa Monarca. Econ. Soc. Territ. 9(30):259–295. (Ch. 15)

Breshears, D.D., N.S. Cobb, P.M. Rich, K.P. Price, C.D. Allen, R.G. Balice, W.H. Romme, J.H. Kastens, M.L. Floyd, J. Belnap, J.J. Anderson, O.B. Myers, and C.W. Meyer. 2005. Regional vegetation die-off in response to global-change-type drought. Proc. Natl. Acad. Sci. 102:15,144–15,148. (Ch. 13)

Bretherton, R.F. 1984. Monarchs on the move—Danaus plexippus (L.) and D. chrysippus (L.). Proc. Trans. Br. Entomol. Nat. Hist. Soc. 17:65–66. (Ch. 22)

Brower, A.V.Z., and T.M. Boyce. 1991. Mitochondrial DNA variation in monarch butterflies. Evolution 45:1281–1286. (Ch. 23)

Brower, A.V.Z., and M.M. Jeansonne. 2004. Geographical populations and "subspecies" of New World monarch butterflies (Nymphalidae) share a recent origin and are not phylogenetically distinct. Ann. Entomol. Soc. Am. 97:519–523. (Ch. 4, 23, Part 5)

Brower, L.P. 1977. Monarch migration. Nat. Hist. 86:40–53. (Ch. 10)

Brower, L.P. 1984. Chemical defence in butterflies. *In* R.I. Vane-Wright and P.R. Ackery, eds., The biology of butterflies, pp. 109–134. London: Academic. (Ch. 6)

Brower, L.P. 1985a. New perspectives on the migration biology of the monarch butterfly, *Danaus plexippus* L. *in* Rankin, M.A., ed., Migration: Mechanisms and adaptive significance, pp. 748–785. Austin: University of Texas Press. (Ch. 9, 10, 19, 20, Part 4)

Brower, L.P. 1985b. The yearly flight of the monarch butterfly. Pac. Discov. 38:4–12.

Brower, L.P. 1995. Understanding and misunderstanding the migration of the monarch butterfly (Nymphalidae) in North America: 1857–1995. J. Lepid. Soc. 49:304–385. (Ch. 7, 9, 11, 12, 14, 19, 20, 21, 22, 23)

Brower, L.P. 1996a. Forest thinning increases monarch butterfly mortality by altering the microclimate of the overwintering sites in Mexico. Decline and Conservation of Butterflies in Japan 3:33–44. (Ch. 9)

Brower, L.P. 1996b. Monarch butterfly orientation: Missing pieces of a magnificent puzzle. J. Exp. Biol. 199:93–103. (Ch. 23)

Brower, L.P. 1999. Biological necessities for monarch butterfly overwintering in relation to the oyamel forest ecosystem in Mexico. *In* J. Hoth, L. Merino, K. Oberhauser, I. Pisenty, S. Price, and T. Wilkinson, eds., 1997 North American conference on the monarch butterfly: Paper presentations, pp. 11–28. Montreal: Commission for Environmental Cooperation. (Ch. 9, 10)

Brower, L.P. 2001. Canary in the cornfield: The monarch and the Bt corn controversy. Orion 20:32–41. (Ch. 11)

Brower, L.P., and W.H. Calvert. 1985. Foraging dynamics of bird predators on overwintering monarch butterflies in Mexico. Evolution 39:852–868. (Ch. 6)

Brower, L.P., and L.S. Fink. 1985. A natural toxic defense system—cardenolides in butterflies versus birds. Ann. N. Y. Acad. Sci. 443:171–188. (Ch. 7)

Brower, L.P., and S.B. Malcolm. 1991. Animal migrations: Endangered phenomena. Am. Zool. 31:265–276. (Ch. 7, 11)

Brower, L.P., and M. Missrie. 1998. Fires in the monarch butterfly sanctuaries in Mexico, spring 1998. Que Pasa (Toronto, Canada) 3:9–11. (Ch. 9)

Brower, L.P., and R.M. Pyle. 2004. The interchange of migratory monarchs between Mexico and the western United States, and the importance of floral corridors to the fall and spring migrations. *In* G.P. Nabhan, ed., Conserving migratory pollinators and nectar corridors in western North America, pp. 144–166. Tucson: University of Arizona Press. (Ch. 2, 12, 21, 23)

Brower, L.P., J. Van Zandt Brower, and J.M. Corvino. 1967. Plant poisons in a terrestrial food chain. Proc. Natl. Acad. Sci. 57:893–898. (Ch. 4)

Brower, L.P., W.N. Ryerson, L. Coppinger, and S.C. Glazier. 1968. Ecological chemistry and the palatability spectrum. Science 161:1349–1351. (Ch. 7, Part 2)

Brower, L.P., P.B. McEvoy, K.L. Williamson, and M.A. Flannery. 1972. Variation in cardiac glycoside content of monarch butterflies from natural populations in eastern North America. Science 177:426–429. (Ch. 4)

Brower, L.P., W.H. Calvert, L.E. Hedrick, and J. Christian. 1977. Biological observations on an overwintering colony of monarch butterflies (*Danaus plexippus*, Danaidae) in Mexico. J. Lepid. Soc. 31:232–242. (Ch. 9)

Brower, L.P., J.N. Seiber, C.J. Nelson, S.P. Lynch, M.P. Hoggard, and J.A. Cohen. 1984. Plant-determined variation in cardenolide content and thin-layer chromatography profiles of monarch butterflies, *Danaus plexippus* reared on milkweed plants in California. 3. *Asclepias californica*. J. Chem. Ecol. 10:1823–1857. (Ch. 4)

Brower, L.P., B.E. Horner, M.M. Marty, C.M. Moffitt, and B. Villa-R. 1985. Mice (*Peromyscus maniculatus labecula, P. spicelegus* and *Microtus mexicanus*) as predators of monarch butterflies (*Danaus plexippus*) in Mexico. Biotropica 17:89–99. (Ch. 6, 9)

Brower, L.P., L.S. Fink, A.V.Z. Brower, K. Leong, K. Oberhauser, S. Altizer, O.R. Taylor, D. Vickerman, W.H. Calvert, T. Van Hook, A. Alonso-M., S.B. Malcolm, D.F. Owen, and M.P. Zalucki. 1995. On the dangers of interpopulational transfers of monarch butterflies. Bioscience 45:540–45. (Ch. 1, 21, Part 5)

Brower, L.P., L.S. Fink, A.V.Z. Brower, K. Leong, K. Oberhauser, S. Altizer, O.R. Taylor, D. Vickerman, W.H. Calvert, T. van Hook, A. Alonso-M, S.B. Malcolm, D.F. Owen, and M.P. Zalucki. 1996. Response to Keiper: Monarch transfer: A real concern? Bioscience 46:563–564. (Ch. 1, 2)

Brower, L.P., G. Castilleja, A. Peralta, J. López-García, L. Bojórquez-Tapia, S. Díaz, D. Melgarejo, and M. Missrie. 2002. Quantitative changes in forest quality in a principal overwintering area of the monarch butterfly in Mexico, 1971–1999. Conserv. Biol. 16(2):346–359. (Ch. 3, 7, 9, 11, 13, 15)

Brower, L.P., D.R. Kust, E. Rendon-Salinas, E. García-Serrano, K.R. Kust, J. Miller, C. Fernández del Rey, and K. Pape. 2004. Catastrophic winter storm mortality of monarch butterflies in Mexico during January 2002. *In* K.S. Oberhauser and M.J. Solensky, eds., The monarch butterfly: Biology and conservation, pp. 151–166. Ithaca, NY: Cornell University Press. (Ch. 9, 11, 19, 24, Parts 3, 4)

Brower, L.P., L.S. Fink, and P. Walford. 2006. Fueling the fall migration of the monarch butterfly. Integr. Comp. Biol. 46:1123–1142. (Ch. 2, 4, 9, 10, 20, 21, Part 4)

Brower, L.P., K.S. Oberhauser, M. Boppré, A.V.Z. Brower, and R.I. Vane-Wright. 2007. Monarch sex: Ancient rites, or recent wrongs? Antenna 31(1):12–18. (Ch. 20, Part 1)

Brower, L.P., E.H. Williams, L.S. Fink, R.R. Zubieta-Hernández, and M.I. Ramírez. 2008. Monarch butterfly clusters provide microclimatic advantages during the overwintering season in Mexico. J. Lepid. Soc. 62:177–188. (Ch. 9, 11)

Brower, L.P., E.H. Williams, D.A. Slayback, L.S. Fink, M.I. Ramírez, R.R. Zubieta, M.I. Limón-García, P. Gier, J.A. Lear, and T. Van Hook. 2009. Oyamel fir forest trunks provide thermal advantages for overwintering monarch butterflies in Mexico. Insect Conserv. Divers. 2(3):163–175. (Ch. 9, 11, 13, Part 4)

Brower, L.P., L.S. Fink, I. Ramírez, R. Zubieta, and D. Slayback. 2010. Updated report on the impact of storms in the monarch overwintering area. Accessed 2 Nov 2012. http://www.monarchbutterflyfund.org/es/node/242 (Ch. 9)

Brower, L.P., E.H. Williams, L.S. Fink, D.A. Slayback, M.I. Ramírez, M.I. Limon García, R.R. Zubieta, S.B. Weiss, W.H. Calvert, and W. Zuchowski. 2011. Overwintering clusters of the monarch butterfly coincide with the least hazardous vertical temperatures in the oyamel forest. J. Lepid. Soc. 65:27–46. (Ch. 9)

Brower, L.P., O.R. Taylor, E.H. Williams, D.A. Slayback, R.R. Zubieta, and M.I. Ramírez. 2012a. Decline of monarch butterflies overwintering in Mexico: Is the migratory phenomenon at risk? Insect Conserv. Divers. 5:95–100. (Ch. 2, 9, 10, Part 4)

Brower, L.P., O.R. Taylor, and E.H. Williams. 2012b. Response to Davis: Choosing relevant evidence to assess monarch population trends. Insect Conserv. Divers. 5:327–329. (Ch. 2, 9)

Brown, J.J., and G.M. Chippendale. 1974. Migration of the monarch butterfly, *Danaus plexippus*: Energy sources. J. Insect Physiol. 20:1117–1130. (Ch. 10)

Brown, M.J.F., R. Loosli, and P. Schmid-Hempel. 2000. Condition-dependent expression of virulence in a trypanosome infecting bumblebees. Oikos 91:421–427. (Ch. 7)

Buhler, D., D.E. Stoltenberg, R.L. Becker, and J.L. Gunsolus. 1994. Perennial weed populations after 14 years of variable tillage and cropping practices. Weed Sci. 42:205–209. (Ch. 14)

Bullock, A.A. 1952. Notes on African Asclepiadaceae. I. Kew Bull. 7:405–426.

Bureau of Land Management. 2012. Technical protocol for the collection, study, and conservation of seeds from native plant species for seeds of success. http://www.nps.gov/plants/sos/protocol/protocol.pdf. Accessed 09/27/2013. (Ch. 16)

Burnet, M., and D.O. White. 1972. The natural history of infectious disease. Cambridge: Cambridge University Press. (Ch. 7)

Busby, J.R. 1991. BIOCLIM—A bioclimatic analysis and prediction tool. Plant Prot. Q. 6:8–9. (Ch. 11)

Butterfield, H.M. 1935. The introduction of *Eucalyptus* into California. Madroño 3:149–154. (Ch. 12)

Byers, B. 2004. La Reserva de la Biosfera Mariposa Monarca y el papel del comportamiento de la gente en su conservación. Resumen provisional preparado para WWF-México, Programa Mariposa Monarca. (Ch. 15)

Calvert, W.H. 1998. Do monarchs rest at the same roost sites every fall? Journey North 22 September 1998:3. http://www.learner.org/jnorth/tm/monarch/FallRoostsQ.html. Accessed 3 November 2012. (Ch. 10)

Calvert, W.H. 1999. Patterns in the spatial and temporal use of Texas milkweeds (Asclepiadaceae) by the monarch butterfly (*Danaus plexippus* L.) during fall, 1996. J. Lepid. Soc. 53:37–44. (Ch. 16, 17, 19)

Calvert, W.H. 2001. Monarch butterfly (*Danaus plexippus* L., Nymphalidae) fall migration: Flight behavior and direction in relation to celestial and physiographic cues. J. Lepid. Soc. 55:162–168. (Ch. 10)

Calvert, W.H. 2004a. The effects of fire ants on monarch breeding in Texas. *In* K.S. Oberhauser and M.J. Solensky, eds., The monarch butterfly: Biology and conservation, pp. 47–53. Ithaca, NY: Cornell University Press. (Ch. 5, 6)

Calvert, W.H. 2004b. Two methods of estimating overwintering monarch population size in Mexico. *In* K.S. Oberhauser and M.J. Solensky, eds., The monarch butterfly: Biology and conservation, pp. 121–128. Ithaca, NY: Cornell University Press. (Ch. 9, 11)

Calvert, W.H., and L.P. Brower. 1981. The importance of forest cover for the survival of overwintering monarch butterflies (*Danaus plexippus*, Danaidae). J. Lepid. Soc. 35:216–225. (Ch. 9)

Calvert, W.H., and L.P. Brower. 1986. The location of monarch butterfly (*Danaus plexippus* L.) overwintering colonies in Mexico in relation to topography and microclimate. J. Lepid. Soc. 40:164–187. (Ch. 3, 9, 10, 11, 13 Part 3)

Calvert, W.H., and J.A. Cohen. 1983. The adaptive significance of crawling up onto foliage for the survival of grounded overwintering monarch butterflies (*Danaus plexippus*) in Mexico. Ecol. Entomol. 8:471–474. (Ch. 9, 12)

Calvert, W.H., and M. Wagner. 1999. Patterns in the monarch butterfly migration through Texas—1993 to 1995. *In* J. Hoth, L. Merino, K. Oberhauser, I. Pisanty, S. Price, and T. Wilkinson, eds., 1997 North American conference on the monarch butterfly: Paper presentations, pp. 119–126. Montreal: Commission for Environmental Cooperation. (Ch. 10, 24)

Calvert, W.H., L.H. Hedrick, and L.P. Brower. 1979. Mortality of the monarch butterfly (*Danaus plexippus* L.): Avian predation at five overwintering sites in Mexico. Science 204:847–851. (Ch. 6)

Calvert, W.H., W. Zuchowski, and L.P. Brower. 1982. The impact of forest thinning on microclimate in monarch butterfly (*Danaus plexippus* L.) overwintering areas of Mexico. Bol. Soc. Bot. Méx. 42:11–18. (Ch. 9)

Calvert, W.H., W. Zuchowski, and L.P. Brower. 1983. The effect of rain, snow, and freezing temperatures on overwintering monarch butterflies in Mexico. Biotropica 15:42–47. (Ch. 9, 11, Part 4)

Calvert, W.H., W. Zuchowski, and L.P. Brower. 1984. Monarch butterfly conservation: Interactions of cold weather, forest thinning and storms on the survival of overwintering monarch butterflies (*Danaus plexippus* L.) in Mexico. Atala (Portland) 9:2–6. (Ch. 9)

Calvert, W.H., M.B. Hyatt, and N.P. Mendoza-Villasenor. 1986. The effects of understory vegetation on the survival of overwintering monarch butterflies, (*Danaus plexippus* L.) in Mexico. Acta Zool. Mex. 18:1–17. (Ch. 9)

Calvert, W.H., S.B. Malcolm, J.I. Glendinning, L.P. Brower, M.P. Zalucki, T. Van Hook, J.B. Anderson, and L.C. Snook. 1989. Conservation biology of monarch butterfly overwintering sites in Mexico. Vida Silv. Neotrop. 2:38–48. (Ch. 9)

Carius, H.J., T.J. Little, and D. Ebert. 2001. Genetic variation in a host-parasite association: Potential for coevolution and frequency-dependent selection. Evolution 55:1136–1145. (Ch. 7)

Carpenter, G., A.N. Gillison, and J. Winter. 1993. DOMAIN: A flexible modelling procedure for mapping potential distributions of plants and animals. Biodivers. Conserv. 2:667–680. (Ch. 11)

Carranza, J., I. Paniagua, K.A. Oceguera, and L. Ruiz. 2010. Análisis del impacto por la 5ª tormenta invernal del 2010 en la Reserva de la Biosfera Mariposa Monarca. Comisión Nacional de Áreas Naturales Protegidas. Unpublished report. 35 pp. (Ch 13)

Carreno, R.A., D.S. Martin, and J.R. Barta. 1999. *Cryptosporidium* is more closely related to the gregarines than to coccidia as shown by phylogenetic analysis of apicomplexan parasites inferred using small-subunit ribosomal RNA gene sequences. Parasitol. Res. 85:899–904. (Ch. 7)

Carrier, S., 2009. Environmental education in the schoolyard: Learning syles and gender. J. Environ. Educ. 40:2–12. (Ch. 1)

Carrillo, M.A., N. Kaliyan, C.A. Cannon, R.V. Morey, and W.F. Wilcke. 2004. A simple method to adjust cooling rates for supercooling point determination. CryoLetters 25:155–60. (Ch. 8)

Casagrande, R.A., and J.E. Dacey. 2007. Monarch butterfly oviposition on swallow-worts (*Vincetoxicum* spp.). Environ. Entomol. 36:631–636. (Ch. 3)

Casey, T.M. 1992. Biophysical ecology and heat exchange in insects. Am. Zool. 32:225–237. (Ch. 9)

Castella, G., M. Chapuisat, and P. Christe. 2008. Prophylaxis with resin in wood ants. Anim. Behav. 75:1591–1596. (Ch. 7)

Caughley, G. 1994. Directions in conservation biology. J. Anim. Ecol. 63:215–244. (Ch. 11)

[CEC] Commission for Environmental Cooperation. 2008. North American monarch conservation plan. Montreal: CEC Office of the Secretariat. (Preface, Ch. 3, 11, 16, Part 4)

CEC. 2009. Monarch monitoring in North America: Overview of initiatives and protocols. Montreal: CEC. (Ch. 2)

Cervo, R., F. Zacchi, and S. Turillazzi S. 2000. *Polistes dominulus* (Hymenoptera, Vespidae) invading North America: Some hypotheses for its rapid spread. Insectes Soc. 47:155–157. (Ch. 6)

Cevallos, D. 2002. Monarcas sin trono (monarchs with no throne). Mexico City: Tierramerica, p. 3. http://www/tierramerica.net.2002/0120/articulo.shtml/. Accessed July 2012. (Ch. 9)

Chalwa, L., 1998. Significant life experiences revisited: A review of research on sources of environmental sensitivity. J. Environ. Educ. 29:11–21. (Ch. 1)

Chang, Y.F., M.J. Tauber, and C.A. Tauber. 1995. Storage of the mass-produced predator *Chrysoperla carnea* (Neuroptera: Chrysopidae): Influence of photoperiod, temperature, and diet. Environ. Entomol. 24:1365–1374. (Ch. 6)

Chaplin, S., and P. Wells. 1982. Energy reserves and metabolic expenditures of monarch butterflies overwintering in southern California. Ecol. Entomol. 7:249–256. (Ch. 12)

Chapman, B.B., C. Brönmark, J. Nilsson and L. Hansson. 2011. The ecology and evolution of partial migration. Oikos 120:1764–1775. (Ch. 20)

Chapman, R.F. 1998. The insects: Structure and function, 4th ed. Cambridge: Cambridge University Press. (Ch. 10)

Chapuisat, M., A. Oppliger, P. Magliano, and P. Christe. 2007. Wood ants use resin to protect themselves against pathogens. Proc. R. Soc. Lond. B 274:2013–2017. (Ch. 7)

Chawkins, S. 2010. Anger flutters over "butterfly town, USA." L.A. Times. http://articles.latimes.com/2010/aug/29/local/la-me-butterflies-20100829. Accessed February 2010. (Ch. 12)

Cherry, C. 2006. The nectaring biology of the overwintering monarch butterfly (*Danaus plexippus*). Unpublished undergraduate research paper. Sweet Briar College, Department of Biology, Sweet Briar, VA. (Ch. 10)

Chokkalingam, U., and W. DeJong. 2001. Secondary forest: A working definition and typology. Int. For. Rev. 3:19–26. (Ch. 13)

Chown, S.L., K.J. Gaston, M. van Kleunen, and S. Clusella-Trullas. 2010. Population response within a landscape matrix: A macrophysiological approach to understanding climate change impacts. Evol. Ecol. 24:601–616. (Ch. 11)

Christe, P., A. Oppliger, F. Bancala, G. Castella, and M. Chapuisat. 2003. Evidence for collective medication in ants. Ecol. Lett. 6:19–22. (Ch. 7)

Christensen, B.M., J. Li, C.C. Chen, and A.J. Nappi. 2005. Melanization immune responses in mosquito vectors. Trends Parasitol. 21:192–199. (Ch. 7)

Clark, T.L., and F.J. Messina. 1998. Foraging behavior of lacewing larvae (Neuroptera: Chrysopidae) on plants with divergent architectures. J. Insect Behav. 11:303–317. (Ch. 6)

Clarke, A.R., and M.P. Zalucki. 2004. Monarchs in Australia: On the winds of a storm? Biol. Invasions 6:123–127. (Ch. 19, 23)

[CNDDB] California Natural Diversity Database. California Department of Fish and Game. http://www.dfg.ca.gov/biogeodata/cnddb/. Accessed October 2012. (Ch. 12)

Cockerell, T.D.A. 1923. The Lepidoptera of the Madeira Islands. Entomologist 56:244–247. (Ch. 22)

Cockrell, B.J., S.B. Malcolm, and L.P. Brower. 1993. Time, temperature, and latitudinal constraints on the annual recolonization of eastern North America by the monarch butterfly. *In* S.B. Malcolm and M.P. Zalucki, eds., Biology and conservation of the monarch butterfly, pp. 233–251. Los Angeles: Natural History Museum of Los Angeles County. (Ch. 11, 16, 17, 18, 19, 20)

Cohen, J.E., S.L. Pimm, P. Yodzis, and J. Saldana. 1993. Body sizes of animal predators and animal prey in food webs. J. Anim. Ecol. 62:67–78. (Ch. 5)

Combes, C. 2001. Parasitism: The ecology and evolution of intimate interactions. Chicago: Chicago University Press. (Ch. 7)

CONAFOR. 2012. Reporte semanal de resultados de incendios forestales 2012. Datos acumulados del 01 de enero al 27 de septiembre. http://www.conafor.gob.mx/portal/. Accessed February 2013. (Ch. 15)

CONANP. 2001. Programa de manejo Reserva de la Biosfera Mariposa Monarca. Mexico City: CONANP-SEMARNAT. 138 pp. (Ch. 15)

CONANP. 2011. Estrategia y lineamientos de manejo del fuego en Áreas Naturales Protegidas. CONANP. México, D.F. 36 pp. (Ch. 15)

Cook, R., and S. Weisberg. 1982. Residuals and influence in regression. New York: Chapman & Hall. (Ch. 24)

Cornejo-Tenorio, G., and G. Ibarra-Manríquez. 2008. Flora Illustrada de la Reserva de la Biosfera Mariposa Monarca. Comisión Nacional Para el Conocimiento y uso de la Biodiversidad, Mexico City. (Ch. 9)

Cornell, H.V., and B.A. Hawkins. 1995. Survival patterns and mortality sources of herbivorous insects: Some demographic trends. Am. Nat. 145:563–593. (Ch. 5)

Cory, J.S., and K. Hoover. 2006. Plant-mediated effects in insect-pathogen interactions. Trends Ecol. Evol. 21:278–286. (Ch. 7)

Cory, J.S., and J.H. Myers. 2004. Adaptation in an insect host-plant pathogen interaction. Ecol. Lett. 7:632–639. (Ch. 7)

COSEWIC. 2010. COSEWIC Assessment and status report on the monarch, *Danaus plexippus*, in Canada. Ottawa: Committee on the Status of Endangered Wildlife in Canada. vii + 43 pp. (www.sararegistry.gc.ca/status/status_e.cfm). (Ch. 3)

CRC. 1992. CRC handbook of chemistry and physics, 73rd ed. Boca Raton, FL: CRC Press. (Ch. 9)

Crewe, T.L., J.D. McCracken, and D. Lepage. 2007. Population trend analysis of monarch butterflies using daily counts during fall migration at Long Point, Ontario, Canada (1995–2006). U.S. Fish and Wildlife Service. (Ch. 2)

Crookston, N.L. 2012. Research on forest climate change: Potential effects of global warming on forests and plant climate relationships in western North America and Mexico. Custom climate data request. http://forest.moscowfsl.wsu.edu/climate/customData/. Accessed 13 September 2012. (Ch. 13)

Crozier, L.G. 2004a. Field transplants reveal summer constraints on a butterfly range expansion. Oecologia 141:148–157. (Ch. 8)

Crozier, L.G. 2004b. Warmer winters drive butterfly range expansion by increasing survivorship. Ecology 85:231–241. (Ch. 18)

Crozier, L.G., and G. Dwyer. 2006. Combining population-dynamic and ecophysiological models to predict climate induced insect range shifts. Am. Nat. 167:853–866. (Ch. 11)

Cruz, M.A.S., and T. Gonçalves. 1973. Imigração ou sobrevivência—Lepidópteros aclimatados ou naturalizados em Portugal. An. Fac. Cienc. Port. 56(1–2):9–25. (Ch. 22)

Cruz, M.A.S., and T. Gonçalves. 1977.Catálogo sistemático dos macrolepidópteros de Portugal. An. Fac. Cienc. Port. 60 (1–4):1–48. (Ch. 22)

Daane, K.M., and G.Y. Yokota. 1997. Release strategies affect survival and distribution of green lacewings (Neuroptera: Chrysopidae) in augmentation programs. Environ. Entomol. 26:455–464. (Ch. 6)

Daane, K.M., G.Y. Yokota, Y. Zheng, and K.S. Hagen. 1996. Inundative release of common green lacewings (Neuroptera: Chrysopidae) to suppress *Erythroneura variabilis* and *E-elegantula* (Homoptera: Cicadellidae) in vineyards. Environ. Entomol. 25:1224–1234. (Ch. 6)

Dallimer, M., and P.J. Jones. 2002. Migration orientation behaviour of the red-billed quelea *Quelea quelea*. J. Avian Biol. 33:89–94. (Ch. 23)

Dallimer, M., P.J. Jones, J.M. Pemberton, and R.A. Cheke. 2003. Lack of genetic and plumage differentiation in the red-billed quelea *Quelea quelea* across a migratory divide in southern Africa. Mol. Ecol. 12:345–353. (Ch. 23)

Dana, E.D., M. Sanz, S. Vivas, and E. Sobrino. 2005. Especies vegetales invasoras de Andalucía. Dirección General de la RENPA. C.M.A. Junta de Andalucía. 233 pp. (Ch. 22)

Davis, A.K. 2012. Are migratory monarchs really declining in eastern North America? Examining evidence from two fall census programs. Insect Conserv. Divers. 5:101–105. doi: 10.1111/j.1752–4598.2011.00158.x. (Ch. 24, Part 5)

Davis, A.K., and M.S. Garland. 2004. Stopover ecology of monarchs in coastal Virginia: Using ornithological methods to study monarch migration. *In* K.S. Oberhauser and M.J. Solensky, eds., The monarch butterfly: Biology and conservation, pp. 89–96. Ithaca, NY: Cornell University Press. (Ch. 18)

Davis, A.K., and E. Howard. 2005. Spring recolonization rate of monarch butterflies in eastern North America: New estimates from citizen science data. J. Lepid. Soc. 59:1–5. (Ch. 18)

Davis, A.K., B. Farrey, and S. Altizer. 2004. Quantifying monarch butterfly larval pigmentation using digital image analysis. Entomol. Exp. Appl. 113:145—147. (Part 3)

Davis, A.K., B. Farrey, and S. Altizer. 2005. Variation in thermally induced melanism in monarch butterflies (Lepidoptera: Nymphalidae) from three North American populations. J. Therm. Biol. 30:410–421. (Parts 3, 5)

Davis, A.K., N.P. Nibbelink, and E. Howard. 2012a. Identifying large- and small-scale characteristics of migratory stopover sites of monarch butterflies with citizen-science observations. Int. J. Zool. Article ID 149026, 9 pp. (Ch. 2, 18, Part 5)

Davis, A.K., J. Chi, C.A. Bradley, and S. Altizer. 2012b. The redder the better: Wing color predicts flight performance in monarch butterflies. PLoS ONE 7:e41323. doi:10.1371/journal.pone.0041323. (Ch. 18)

Dawson, B. 2011. A drought for the centuries: It hasn't been this dry in Texas since 1789. Texas Climate News, 23 October 2011. http://texasclimatenews.org/wp/?m=201110&cat=4. Accessed 9 August 2012. (Ch. 10)

Day, K.R., M. Docherty, S.R. Leather, and N.A.C. Kidd. 2006. The role of generalist insect predators and pathogens in suppressing green spruce aphid populations through direct mortality and mediation of aphid dropping behavior. Biol. Control 38:233–246. (Ch. 6)

De Moraes, C.M., W.J. Lewis, P.W. Pare, H.T. Alborn, and J.H. Tumlinson. 1998. Herbivore-infested plants selectively attract parasitoids. Nature 393:570–573. (Ch. 6)

de Roode, J.C., and S. Altizer. 2010. Host-parasite genetic interactions and virulence-transmission relationships in natural populations of monarch butterflies. Evolution 64:502–514. (Ch. 7)

de Roode, J.C., and T. Lefèvre. 2012. Behavioral immunity in insects. Insects 3:789–820. (Ch. 7)

de Roode, J.C., L.R. Gold, and S. Altizer. 2007. Virulence determinants in a natural butterfly-parasite system. Parasitology 134:657–668. (Ch. 7)

de Roode, J.C., A.B. Pedersen, M.D. Hunter, and S. Altizer. 2008a. Host plant species affects virulence in monarch butterfly parasites. J. Anim. Ecol. 77:120–126. (Ch. 7, Part 2)

de Roode, J.C., A.J. Yates, and S. Altizer. 2008b. Virulence-transmission trade-offs and population divergence in virulence in a naturally occurring butterfly parasite. Proc. Natl. Acad. Sci. 105:7489–7494. (Ch. 7)

de Roode, J.C., J. Chi, R.M. Rarick, and S. Altizer. 2009. Strength in numbers: High parasite burdens increase transmission of a protozoan parasite of monarch butterflies (*Danaus plexippus*). Oecologia 161:67–75. (Ch. 7)

de Roode, J.C., C. Lopez Fernandez de Castillejo, T. Faits, and S. Alizon. 2011a. Virulence evolution in response to anti-infection resistance: Toxic food plants can select for virulent parasites of monarch butterflies. J. Evol. Biol. 24:712–722. (Ch. 7) de Roode, J.C., R.M. Rarick, A.J. Mongue, N.M. Gerardo, and M.D. Hunter. 2011b. Aphids indirectly increase virulence and transmission potential of a monarch butterfly parasite by reducing defensive chemistry of a shared food plant. Ecol. Lett. 14:453–461. (Ch. 7, Part 2)

de Roode, J.C., T. Lefèvre, and M.D. Hunter. 2013. Self-medication in animals. Science 340:150–151. (Ch. 7)

Dethier, V.G. 1941. Chemical factors determining the choice of food plants by *Papilio* larvae. Am. Nat. 75:61–73. (Ch. 4)

De Villiers, M., V. Hattingh, and D.J. Kriticos. 2013. Combining field phenological observations with distribution data to model the potential range distribution of the fruit fly *Ceratitis rosa* Karsch (Diptera: Tephritidae). Bull. Entomol. Res. 103:60–73. (Ch. 11)

Dickinson, J.L., and R. Bonney. 2012. Citizen science: Public participation in environmental research. Ithaca, N.Y.: Cornell University Press. (Ch. 2)

Dickinson, J.L., J. Shirk, D. Bonter, R. Bonney, R.L. Crain, J. Martin, T. Phillips, and K. Purcell. 2012. The current state of citizen science as a tool for ecological research and public engagement. Front. Ecol. Environ. 10:291–297. (Ch. 2, 24)

Diffendorfer, J.E., J.B. Loomis, L. Ries, K. Oberhauser, L. Lopez-Hoffman, D. Semmens, B. Semmens, B. Butterfield, K. Bagstad, J. Goldstein, R. Widerholt, B. Mattsson, and W.E. Thogmartin. 2013. National valuation of monarch butterflies indicates an untapped potential for incentive-based conservation. Conserv. Lett. doi: 10.1111/conl.12065. (Part 4)

Diggs, G.M. Jr., B.L. Lipscomb, and R.J. O'Kennon. 1999. Illustrated flora of North Central Texas. Austin: Austin College Center for Environmental Studies and the Botanical Institute of Texas. (Ch. 10)

Dingle, H., and A. Drake. 2007. What is migration? BioScience 57:113–121. (Ch. 20)

Dingle, H., W.A. Rochester, and M.P. Zalucki. 2000. Relationships among climate, latitude and migration: Australian butterflies are not temperate-zone birds. Oecologia 124:196–207. (Ch. 19)

Dingle, H., M.P. Zalucki, W.A. Rochester, and T. Armijo-Prewitt. 2005. Distribution of the monarch butterfly, *Danaus plexippus* (L.) (Lepidoptera : Nymphalidae), in western North America. Biol. J. Linn. Soc. 85:491–500. (Ch. 16, 20, 21, 23)

Dively, G.P., R. Rose, M.K. Sears, R.L. Hellmich, D.E. Stanley-Horn, D.D. Calvin, J.M. Russo, and P.L. Anderson. 2004. Effects on monarch butterfly larvae (Lepidoptera: Danaidae) after continuous exposure to Cry1Ab-expressing corn during anthesis. Environ. Entomol. 33:1116–1125. (Ch. 14)

Dobler, S., S. Dalla, V. Wagschal, and A.A. Agrawal. 2012. Community-wide convergent evolution in insect adaptation to toxic cardenolides by substitutions in the Na,K-ATPase. Proc. Natl. Acad. Sci. 109:13,040–13,045. (Ch. 4, 7, 23, Part 2)

Dockx, C. 2007. Directional and stabilizing selection on wing size and shape in migrant and resident monarch butterflies, *Danaus plexippus* (L.), in Cuba. Biol. J. Linn. Soc. 92:605–616. (Ch. 23).

Dockx, C. 2012. Differences in phenotypic traits and migratory strategies between eastern North American monarch butterflies, *Danaus plexippus* (L.). Biol. J. Linn. Soc. 106:717–736. (Ch. 23, 24)

Dockx, C., L.P. Brower, L.I. Wassenaar, and K.A. Hobson. 2004. Do North American monarch butterflies travel to Cuba? Stable isotope and chemical tracer techniques. Ecol. Appl. 14:1106–1114. (Ch. 23)

[DOF] Diario Oficial de la Federación. 1980. Decreto que declara zonas de reserva y refugio silvestre, lugares donde la mariposa inverna y se reproduce. Órgano del Gobierno Constitucional de los Estados Unidos Mexicanos, Mexico City, 9 April. (Ch. 3, 15)

DOF. 1986. Decreto que declara áreas naturales protegidas para fines de migración, invernación, y reproducción de la Mariposa Monarca. Órgano del Gobierno Constitucional de los Estados Unidos Mexicanos, Mexico City, 9 October: 33–41. (Ch. 3, 13, 15)

DOF. 2000. Decreto de la Reserva de la Biosfera Mariposa Monarca (RBMM). Órgano del Gobierno Constitucional de los Estados Unidos Mexicanos, Mexico City, 10 November: 5–29. (Ch. 3, 13,15)

DOF. 2009. Decreto que modifica el artículo primero del diverso que se declaró como Área Natural Protegida, con la Categoría de a Reserva de la Biosfera, la región denominada Mariposa Monarca. Órgano del Gobierno Constitucional de los Estados Unidos Mexicanos, Mexico City. (Ch. 15)

Dunn, P. 1990. Humoral immunity in insects. BioScience 40:738–744. (Ch. 7)

Dussourd, D.E., and T. Eisner. 1987. Vein-cutting behavior: Insect counterploy to the latex defense of plants. Science 237:898–900. (Ch. 4)

Dyer, L.A., C.D. Dodson, and J. Richards. 2004. Isolation, synthesis, and evolutionary ecology of Piper amides. *In* L.A. Dyer and A.N. Palmer, eds., *Piper*: A model genus for studies of evolution, chemical ecology, and trophic interactions, pp. 117–139. Boston: Kluwer Academic. (Ch. 6)

Eanes, W.F., and R.K. Koehn. 1978. Analysis of genetic structure in the monarch butterfly, *Danaus plexippus* L. Evolution 32:784–797. (Ch. 23, 24)

Earle, C.J., ed. 2011. The gymnosperm database. Accessed 7 June 2012. http://www.conifers.org/. (Ch. 9)

eButterfly 2013. www.ebutterfly.ca. Accessed May 2013. (Ch. 2, 21)

Edgar, C., and B. Carraway. 2011. Preliminary estimates show hundreds of millions of trees killed by 2011 drought. Texas Forest Service news release, 19 December 2011:pp1–2. http://txforestservice.tamu.edu/main/popup.aspx?id=14954. Accessed 29 May 2012. (Ch. 10)

Egan, T. 2009. The big burn: Teddy Roosevelt and the fire that saved America. Houghton Mifflin Harcourt. 324 pp. (Ch. 15)

Ehrlich, P.R., and P.H. Raven. 1964. Butterflies and plants: A study in coevolution. Evolution 18:586–608. (Ch. 4)

Elith, J., S.J. Phillips, T. Hastie, M. Dudik, Y.E. Chee, and C.J. Yates. 2011. A statistical explanation of MaxEnt for ecologists. Divers. Distrib. 17:43–57. (Ch. 11)

Elkinton, J.S., J.P. Buonaccorsi, T.S. Bellows Jr., and R.G. Van Driesche. 1992. Marginal attack rate, k-values and density dependence in the analysis of contemporaneous mortality factors. Res. Popul. Ecol. 34:29–44. (Ch. 6)

Elliot, S.L., S. Blanford, and M.B. Thomas. 2002. Host–pathogen interactions in a varying environment: Temperature, behavioural fever and fitness. Proc. R. Soc. Lond. B. 269:1599–1607. (Ch. 7)

Encinas, C., and P. Vicens. 2008. Primeres observacions de la papallona monarca, *Danaus plexippus* (Linnaeus, 1758), a les Illes Balears. Bol. Soc. Hist. Nat. Balears 51:225–228. (Ch. 22)

English-Loeb, G.M., A.K. Brody, and R. Karban. 1993. Host-plant-mediated interactions between a generalist folivore and its tachinid parasitoid. J. Anim. Ecol. 62:465–471. (Ch. 6)

Enquist, M. 1987. Wildflowers of the Texas Hill Country. Austin: Lone Star Botanical. (Ch. 10)

ERMEX. 2009. Images SPOT 585/310 and 585/311, January 04, 2009, Pancromatic 2.5 m resolution, Multiespectral 10 m resolution. Estación de Recepción México—Subdirección de Geomática, Comisión Nacional de Áreas Naturales Protegidas. Morelia, Mexico. (Ch. 13)

ERMEX. 2012. Images SPOT 585/310, 585/311, 586/311, January 17, 2012, Pancromatic 2.5 m resulution, Multiespectral 10 m resolution. Estación de Recepción México—Centro de Investigaciones en Geografía Ambiental, UNAM. Morelia, Mexico. (Ch. 13)

Eslin, P., and G. Prevost. 1996. Variation in *Drosophila* concentration of haemocytes associated with different ability to encapsulate *Asobara tabida* larval parasitoid. J. Insect Physiol. 42:549–555. (Ch. 7)

Esquivel R., S., G. Cruz J., L. Zizumbo V., C. Cadena I., and R. del C. Serrano B. 2011. Turismo rural, política ambiental y redes de política pública en la Reserva de la Biosfera Mariposa Monarca. Rosa dos ventos Revista do Programa do Pós-Graduação em Turismo. 3(2):290–300. (Ch. 15)

ESRI. 2004. ArcGIS 9. ArcGIS Desktop. Environmental Systems Research Institute, Redlands, CA. (Ch. 13)

ESRI. 2008. ArcGIS 9.3 Resource Center. Environmental Systems Research Institute, Redlands, CA. http://www.resources.esri.com. (Ch. 22)

Etchegaray, J.B., and T. Nishida. 1975a. Biology of *Lespesia archippivora* (Diptera: Tachinidae). Proc. Hawaii. Entomol. Soc. 12:41–49. (Ch. 6)

Etchegaray, J.B., and T. Nishida. 1975b. Reproductive activity, seasonal abundance and parasitism of monarch butterfly, *Danaus plexippus* (Lepidoptera, Danaidae) in Hawaii. Proc. Hawaii. Entomol. Soc. 22:33–39. (Ch. 17)

Evans, H.F. 1976. The searching behavior of *Anthocoris confuses* (Reuter) in relation to prey density and plant surface topography. Ecol. Entomol. 1:163–169. (Ch. 5)

Excoffier, L., and H.E.L. Lischer. 2010. Arlequin suite ver 3.5: A new series of programs to perform population genetics analyses under Linux and Windows. Mol. Ecol. Resour. 10:564–567. (Ch. 23)

[FAO] Food and Agriculture Organization. 1996. Forest resources assessment 1990. Survey of tropical forest cover and study of change processes. Forestry Paper 130, Rome. (Ch. 13)

FAO. 2001. Global forest resources assessment 2000. Forestry Paper 140, Rome. (Ch. 13)

Farrell, B.D., D.E. Dussourd, and C. Mitter. 1991. Escalation of plant defense: Do latex and resin canals spur plant diversification. Am. Nat. 138:881–900. (Ch. 4)

Feddeman, J.J., J. Shields, O.R. Taylor, and D. Bennett. 2004. Simulating the development and migration of the monarch butterfly. *In* K.S. Oberhauser and M.J. Solensky, eds., The monarch butterfly: Biology and conservation, pp. 229–240. Ithaca, NY: Cornell University Press. (Ch. 11)

Feener, D.H. Jr., and B.V. Brown. 1997. Diptera as parasitoids. Annu. Rev. Entomol. 42:73–97. (Ch. 6)

Feeny, P.P. 1976. Plant apparency and chemical defense. *In* J.W. Wallace and R.L. Mansell, eds., Biochemical interaction between plants and insects, pp. 1–40. New York: Plenum. (Ch. 4)

Fellowes, M., A. Kraaijeveld, and H. Godfray. 1999. Cross-resistance following artificial selection for increased defense against parasitoids in *Drosophila melanogaster*. Evolution 53:966–972. (Ch. 7)

Fernández-Haeger, J. 1999. *Danaus chrysippus* (Linnaeus, 1758) en la Península Ibérica: ¿migraciones o dinámica de metapoblaciones? (Lepidoptera: Nymphalidae, Danainae). SHILAP-Rev. Lepid. 27(107):423–430. (Ch. 22)

Fernández-Haeger, J., and D. Jordano-Barbudo. 2009. La mariposa monarca *Danaus plexippus* (L. 1758) en el estrecho de Gibraltar (Lepidoptera: Danaidae). SHILAP-Rev. Lepid. 37(148):421–438. (Ch. 22)

Fernández-Haeger, J., D. Jordano Barbudo, M. León Meléndez, and J. Devesa. 2010. *Gomphocarpus* R.Br. (Apocynaceae Subfam. Asclepiadoideae) en Andalucía occidental. Lagascalia 30:39–46. (Ch. 22)

Fernández-Haeger, J., D. Jordano, and M. León Meléndez. 2011a. Status and conservation of Asclepiadaceae and *Danaus* in southern Spain. J. Insect Conserv. 15:361–365. (Ch. 22)

Fernández-Haeger, J., D. Jordano Barbudo, and M. León Meléndez. 2011b. Ocupación de fragmentos, persistencia y movimientos de la mariposa monarca (*Danaus plexippus*) en la zona del Estrecho de Gibraltar. Migres 2:35–51. (Ch. 22)

Fernández-Rubio, F. 1991. Guía de mariposas diurnas de la Península Ibérica, Baleares, Canarias, Azores y Madeira. Madrid: Ediciones Pirámide. (Ch. 22)

Fernández Vidal, E.H. 2002. La *Danaus plexippus* (Linnaeus, 1758) en Galicia (España). Nuevos datos y noticias sobre la "operación monarca" (Lepidoptera; Danaidae). Bol. Soc. Entomol. Argent. 31:243–246. (Ch. 22)

Fink, L.S., and L.P. Brower. 1981. Birds can overcome the cardenolide defense of monarch butterflies in Mexico. Nature 291:67–70. (Ch. 6)

Fink, L.S., L.P. Brower, R.B. Waide, and P.R. Spitzer. 1983. Overwintering monarch butterflies as food for insectivorous birds in Mexico. Biotropica 15:151–153. (Ch. 6)

Finkbeiner, S.D., R.D. Reed, R. Dirig, and J.E. Losey. 2011. The role of environmental factors in the northeastern range expansion of *Papilio cresphontes* Cramer (Papilionidae). J. Lepid. Soc. 65:119–125. (Ch. 18)

Fischer, S.J., E.H. Williams, L.P. Brower, and P.A. Palmiotto. Forthcoming. Enhancing monarch butterfly reproduction by mowing fields of common milkweed. American Midland Naturalist. (Part 4)

Fishbein, M., D. Chuba, C. Ellison, R.J. Mason-Gamer, and S.P. Lynch. 2011. Phylogenetic relationships of *Asclepias*

(Apocynaceae) inferred from non-coding chloroplast DNA sequences. Syst. Bot. 36:1008–1023. (Ch. 4, 20)

Fittinghoff, C.M., and L.M. Riddiford. 1990. Heat sensitivity and protein synthesis during heat-shock in the tobacco hornworm, *Manduca sexta*. J. Comp. Physiol. B. 160:349–356. (Ch. 8)

Floater, G.J., and M.P. Zalucki. 2000. Habitat structure and egg distributions in the processionary caterpillar *Ochrogaster lunifer*- lessons for conservation and pest management. J. Appl. Ecol. 37:87–99. (Ch. 14)

Flockhart, D.T.T., T.G. Martin, and D.R. Norris. 2012. Experimental examination of intraspecific density-dependent competition during the breeding period in monarch butterflies (Danaus plexippus). PLoS ONE 7(9): e45080. doi:10.1371/journal.pone.0045080. (Ch. 16)

Flores-Nieves, P., M.A. López-López, G. Ángeles–Pérez, M.L. de la Isla-Serrano, and G. Calva-Vásquez. 2011. Biomass estimation and distribution models of *Abies religiosa* (Kunth Schltdl. et Cham.) in decline. Rev. Mex. Cienc. For. 2(8):9–20. (Ch. 9, 13)

Folstad, I., F.I. Nilssen, A.C. Halvorsen, and O. Andersen. 1991. Parasite avoidance: The cause of post-calving migrations in Rangifer? Can. J. Zool. 69:2423–2429. (Ch. 7)

Fondren, K.M., D.G. McCullough, and A.J. Walter. 2004. Insect predators and augmentative biological control of balsam twig aphid (*Mindarus abietinus* Koeh) (Homoptera: Aphididae) on Christmas tree plantations. Environ. Entomol. 3:1652–1661. (Ch. 6)

Franco, S., H.H. Regil, C. González, and G. Nava. 2006. Cambio de uso del suelo y vegetación en el Parque Nacional Nevado de Toluca, México, en el periodo 1972–2000. Investigaciones Geográficas 61:38–57. (Ch. 13)

Frank, S.A. 1996. Models of parasite virulence. Q. Rev. Biol. 71:37–78. (Ch. 7)

Freckleton, R.P., A.B. Phillimore, and M. Pagel. 2008. Relating traits to diversification: A simple test. Am. Nat. 172:102–115. (Ch. 4)

Frey, D., and A. Schaffner. 2004. Spatial and temporal pattern of monarch overwintering abundance in western North America. *In* K.S. Oberhauser and M.J. Solensky, eds., The monarch butterfly: Biology and conservation, pp. 167–176. Ithaca, NY: Cornell University Press. (Ch. 2, 21)

Frey, D.F., S.L. Hamilton and J.W. Scott. 2003a. Andrew Molera State Park Cooper Grove Management Plan. Prepared for California Department of Parks and Recreation, Monterey, CA. (Ch. 2)

Frey, D., S.L. Stock, S. Stevens, J.W. Scott, and J.L. Griffiths. 2003. Monarch butterfly population dynamics in western North America—Emphasis on Monterey and San Luis Obispo Counties: Winter 2002–2003 report. Ventana Wilderness Society and California Polytechnic State University. (Ch. 12)

Frick, C., and M. Wink. 1995. Uptake and sequestration of ouabain and other cardiac-glycosides in *Danaus plexippus* (Lepidoptera: Danaidae)—evidence for a carrier-mediated process. J. Chem. Ecol. 21:557–575. (Ch. 4)

Fuhlendorf, S.D., and D.M. Engle. 2001. Restoring heterogeneity on rangelands: Ecosystem management based on evolutionary grazing patterns. Bioscience. 51:625–632. (Ch. 17)

Fuhlendorf, S.D., and D.M. Engle. 2004. Application of the fire-grazing interaction to restore a shifting mosaic on tallgrass prairie. J. Appl. Ecol. 41:604 614. (Ch. 17)

Fulé, P., and W.W. Covington. 1998. Spatial patterns of Mexican pine-oak forest under different recent fire regimes. Plant Ecol. 134(2):197–209. (Ch. 15)

Funk, R.S. 1968. Overwintering of monarch butterflies as a breeding colony in southwestern Arizona. J. Lepid. Soc. 22:63–64. (Ch. 16)

Futuyma, D.J. 1987. On the role of species in anagenesis. Am. Nat. 130:465–473.

Futuyma, D.J., and A.A. Agrawal. 2009. Macroevolution and the biological diversity of plants and herbivores. Proc. Natl. Acad. Sci. 106:18,054–18,061. (Ch. 4)

Galicia, L., and A. García-Romero. 2007. Land use and land cover change in highland temperate forests in the Izta-Popo National Park, Central Mexico. Mt. Res. Dev. 27:48–57. (Ch. 13)

García-Serrano, E. 2012. The Monarch Fund. Presentation delivered at the Monarch Biology and Conservation Meeting, June 2012, University of Minnesota. http://www.monarchlab.org/mn2012/Admin/upload/104.pdf. (Ch. 3)

Garcia-Serrano, E., J.L. Reyes, and B.X.M. Alvarez. 2004. Locations and area occupied by monarch butterflies overwintering in Mexico from 1993 to 2002. *In* K.S. Oberhauser and M.J. Solensky, eds., The monarch butterfly: Biology and conservation, pp. 129–133. Ithaca, NY: Cornell University Press. (Ch. 2, 9, 14)

Garland, M.S., and A.K. Davis. 2002. An examination of monarch butterfly (*Danaus plexippus*) autumn migration in coastal Virginia. Am. Midl. Nat. 147:170–174. (Ch. 18)

Geiger, R., R.H. Aron, and P. Todhunter. 2009. The climate near the ground, 7th ed. Lanham, MD: Rowman & Littlefield. (Ch. 9)

Gibbs, D., R. Walton, L.P. Brower, and A.K. Davis. 2006. Monarch butterfly (Lepidoptera: Nymphalidae) migration monitoring at Chincoteague, Virginia and Cape May, New Jersey: A comparison of long-term trends. J. Kans. Entomol. Soc. 79:156–164. (Ch. 2)

Gibo, D.L. 1986. Flight strategies of migrating monarch butterflies (*Danaus plexippus* L.) in southern Ontario. *In* W. Danthanarayan, ed., Insect flight: Dispersal and migration, pp. 172–184. Springer-Verlag, Berlin. (Ch. 11)

Gibo, D.L., and J.A. McCurdy. 1993. Evidence for use of water ballast by monarch butterflies, *Danaus plexippus* (Nymphalidae). J. Lepid. Soc. 47:154–160. (Ch. 10)

Gibo, D.L., and M.J. Pallett. 1979. Soaring flight of monarch butterflies, *Danaus plexippus* (Lepidoptera: Danaidae), during the late summer migration in southern Ontario. Canadian J. Zool. 57:1393–1401. (Ch. 10)

Gil-T., F. 2006. A new hostplant for *Danaus plexippus* (Linnaeus, 1758) in Europe. A study of cryptic preimaginal polymorphism within *Danaus chrysippus* (Linnaeus, 1758) in southern Spain (Andalusia) (Lepidoptera, Nymphalidae, Danainae). Atalanta 37(1/2):143–149. (Ch. 22)

Gilbert, O., T.B. Reynoldson, and J. Hobart. 1952. Gause's hypothesis: An examination. J. Anim. Ecol. 21:310–312. (Ch. 5)

Gillette, C. P. 1888. Parasites on *Danais archippus* and *Anthomyia raphani*. Can. Entomol. 20:133–134. (Ch. 6)

Giménez-Azcárate, J., M.I. Ramírez, and M. Pinto. 2003. Las comunidades vegetales de la Sierra de Angangueo (Estados de Michoacán y México, México): Clasificación, composición y distribución. Lazaroa. 24:87–111. (Ch. 15)

Glendinning, J.I., and L.P. Brower. 1990. Feeding and breeding responses of five mice species to overwintering aggregations of the monarch butterfly. J. Anim. Ecol. 59:1091–1112. (Ch. 6, 9)

Glendinning, J.I., A. Alonso-Mejía, and L.P. Brower. 1988. Behavioral and ecological interactions of foraging mice (*Peromyscus melanotis*) with overwintering monarch butterflies (*Danaus plexippus*) in Mexico. Oecologia 75:222–227. (Ch. 9)

Godfray, H.C.J. 1994. Parasitoids: Behavioral and evolutionary ecology, Princeton, NJ: Princeton University Press. (Ch. 6)

Godman, F.D-C., ed. 1870. Natural history of the Azores or Western Islands. London: John Van Voorst & Paternoster Row. 358 pp. (Ch. 22)

Goehring, L., and K.S. Oberhauser. 2002. Effects of photoperiod, temperature, and host plant age on induction of reproductive diapause and development time in *Danaus plexippus*. Ecol. Entomol. 27:674–685. (Ch. 19, Part 3)

Goehring, L., and K.S. Oberhauser. 2004. Environmental factors influencing postdiapause reproductive development in monarch butterflies. In K.S. Oberhauser and M.J. Solensky, eds., The monarch butterfly: Biology and conservation, pp 187–196. Ithaca, NY: Cornell University Press. (Ch. 19)

Goolsby, J.A., M. Rose, R.K. Morrison, and J.B. Woolley. 2000. Augmentative biological control of longtailed mealybug by *Chrysoperla rufilabris* (Burmeister) in the interior plantscape. Southwest. Entomol. 25:15–19. (Ch. 6)

Gordon R.D. 1985. The Coleoptera (Coccinellidae) of America north of Mexico. J. N. Y. Entomol. Soc. 93:1–912. (Ch. 6)

Goulson, D., and J.S. Cory. 1995. Responses of *Mamestra brassicae* (Lepidoptera: Noctuidae) to crowding: Interactions with disease resistance, colour phase and growth. Oecologia 104:416–423. (Ch. 7)

Goyder, D.J. 2001a. *Gomphocarpus* (Apocynaceae: Asclepiadeae) in an African and a global context—an outline of the problem. Biol. Skrif. 54:55–62. (Ch. 4)

Goyder, D.J. 2001b. A revision of the tropical African genus *Trachycalymma* (K. Schum.) Bullock (Apocynaceae: Asclepiadoideae). Kew Bull. 56:129–161. (Ch. 4)

Goyder, D. 2007. 585. *Asclepias barjoniifolia*. Apocynaceae: Asclepiadoideae. Curtis's Bot. 24(2):93–100. (Ch. 20)

Goyder, D.J. 2009. A synopsis of *Asclepias* (Apocynaceae: Asclepiadoideae) in tropical Africa. Kew Bull. 64:369–399. (Ch. 4)

Goyder, D.J., and A. Nicholas. 2001. A revision of *Gomphocarpus* R.Br. (Apocynaceae, Asclepiadaceae). Kew Bull. 56:769–836. (Ch. 22)

Granet, A., and H. Fonfrède. 2005. Desarrollo turístico en la región de la Mariposa Monarca: Situación actual y propuestas. SEMARNAT. 71 pp. (Ch. 15)

Greenberg, R., and S. Droege. 1999. On the decline of the rusty blackbird and the use of ornithological literature to document long term population trends. Conserv. Biol. 13:553–559. (Ch. 24)

Greischar, M.A., and B. Koskella. 2007. A synthesis of experimental work on parasite local adaptation. Ecol. Lett. 10:418–434. (Ch. 7)

Griffiths, J. 2012. The tree is dead, long live the tree: Do monarchs prefer or simply use Eucalyptus for overwintering roosts? Presentation at Monarch Biology and Conservation Meeting, Chanhassen MN. http://www.monarchlab.org/mn2012/Admin/upload/68.pdf. Accessed March 2013. (Ch. 12)

Griffiths, J.L. 2006. Micro-climate parameters associated with three overwintering monarch butterfly habitats in central California: A four year study. Ventana Wildlife Society Technical Report #037 to the California Department of Parksand Recreation, Big Sur, CA. (Ch. 2)

Grillo, N. 1999. Record of *Danaus chrysippus* (Linnaeus, 1758) in the province of Palermo (Sicily, Italy) (Lepidoptera: Danaidae). Linneana Belgica. 17(1):3–4. (Ch. 22)

Guerra, P.A., and S.M. Reppert. 2013. Coldness triggers northward flight in remigrant monarch butterflies. Curr. Biol. 23:419–423. (Parts 3, 5)

Guerra, P.A., C. Merlin, R.J. Gegear, and S.M. Reppert. 2012. Discordant timing between antennae disrupts sun compass orientation in migratory monarch butterflies. Nat. Commun. 3. (Ch. 23, Part 5)

Guppy, C.S., and J.H. Shepard. 2001. Butterflies of British Columbia. Vancouver: UBC Press and Royal British Columbia Museum. (Ch. 21)

Haig, S.M., C.L. Gratto-Trevor, T.D. Mullins, and M.A. Colwell. 1997. Population identification of western hemisphere shorebirds throughout the annual cycle. Mol. Ecol. 6:413–427. (Ch. 23)

Hainsworth, F.R., E. Precup, and T. Hamill. 1991. Feeding, energy processing rates and egg production in painted lady butterflies. J. Exp. Biol. 156:249–265. (Ch. 10)

Hanski, I. 1999. Metapopulation ecology. Oxford University Press. 309 pp. (Ch. 22)

Harker, G.A. 1883. *Anoxia plexippus* off the coast of Portugal. Entomol. Mon. Mag. 29:86. (Ch. 22)

Harrison X.A., J.D. Blount, R. Inger, D.R. Norris, and S. Bearhop. 2011. Carry-over effects as drivers of fitness differences in animals. J. Anim. Ecol. 80:4–18. (Ch. 24)

Hart, B.L. 2005. The evolution of herbal medicine: Behavioural perspectives. Anim. Behav. 70:975–989. (Ch. 7)

Hartzer, K.L., K.Y. Zhu, and J.E. Baker. 2005. Phenoloxidase in larvae of *Plodia interpunctella* (Lepidoptera: Pyralidae): Molecular cloning of the proenzyme cDNA and enzyme activity in larvae paralyzed and parasitized by *Habrobracon hebetor* (Hymenoptera: Braconidae). Arch. Insect Biochem. Physiol. 59:67–79. (Ch. 6)

Hartzler, R.G. 2010. Reduction in common milkweed (*Asclepias syriaca*) occurrence in Iowa cropland from 1999 to 2009. Crop Prot. 29:1542–1544. (Ch. 14, 16, 17)

Hartzler, R.G., and D.D. Buhler. 2000. Occurrence of common milkweed (*Asclepias syriaca*) in cropland and adjacent areas. Crop Prot. 19:363–366. (Ch. 14, 17)

Harvey, J.A., Gols, R., Wagenaar, R. & Bezemer, T.M. 2007. Development of an insect herbivore and its pupal parasitoid reflect differences in direct plant defense. J. Chem. Ecol. 33:1556–1569. (Ch. 6)

Hassell, M.P. 1978. The dynamics of arthropod predator-prey systems. Princeton University Press, Princeton, New Jersey. (Ch. 6)

Hawkins B.A., H.V. Cornell, and M.E. Hochberg. 1997. Predators, parasitoids, and pathogens as mortality agents in phytophagous insect populations. Ecology 78:2145–2152. (Ch. 5, 6)

Hay-Roe, M.M., G. Lamas, and J.L. Nation. 2007. Pre-and postzygotic isolation and Haldane rule effects in reciprocal crosses of *Danaus erippus* and *Danaus plexippus* (Lepidoptera: Danainae), supported by differentiation of cuticular hydrocarbons, establish their status as separate species. Biol. J. Linn. Soc. 91:445–453. (Ch. 20)

Hayward, K.J. 1928. Migration of insects in north-eastern Argentina, 1928. Entomologist 61:210–212. (Ch. 20)

Hayward, K.J. 1953. Migration of butterflies in Argentina during the spring and summer of 1951–52. Proc. R. Entomol. Soc. (A) 28(4–6):63–73. (Ch. 20)

Hayward, K.J. 1955. Migration of butterflies in Argentina, 1953–54. Proc. R. Entomol. Soc. (A) 30(4–6):59–62. (Ch. 20)

Hayward, K.J. 1962a. Migration of butterflies and a moth in Argentina, spring and summer 1960–61. Entomologist. 95:8–12. (Ch. 20)

Hayward, K.J. 1962b. Migration of butterflies in north western Argentina, spring and summer 1961–62. Entomologist. 95:237–239. (Ch. 20)

Hayward, K.J. 1963. Migration of butterflies and moths in north-western Argentina, late spring and summer, 1962–63. Entomologist. 96:258–264. (Ch. 20)

Hayward, K.J. 1964. Migration of butterflies in north-western Argentina during the summer of 1964. Entomologist. 97:272–273. (Ch. 20)

Hayward, K.J. 1967. Notes on Argentine butterfly migration, 1965–66. Entomologist. 100:29–34. (Ch. 20)

Hayward, K.J. 1972. Observations on migration of Lepidoptera in north-west Argentina, November 1968-June 1970. Entomologist. 105:206–208. (Ch. 20)

Heikkinen, R.K., M. Luoto, M.B. Araújo, R. Virkkala, W. Thuiller, and M.T. Sykes. 2006. Methods and uncertainties in bioclimatic envelope modelling under climate change. Prog. Phys. Geogr. 30:751–777. (Part 4)

Heinrich, B. 1983. Insect foraging energetics. *In* C.E. Jones and R.J. Little, eds., Handbook of experimental pollination biology, pp. 187–214. New York: Van Nostrand Reinhold. (Ch. 10)

Heinze, S., and S.M. Reppert. 2012. Anatomical basis of sun compass navigation I: The general layout of the monarch butterfly brain. J. Comp. Neurol. 520:1599–1628. (Part 5)

Hellmich, R.L., B.D. Siegfried, M.K. Sears, D.E. Stanley-Horn, M.J. Daniels, H.R. Mattila, T. Spencer, K.G. Bidne, and L. Lewis. 2001. Monarch larvae sensitivity to *Bacillus thuringiensis*-purified proteins and pollen. Proc. Natl. Acad. Sci. 98:11925–11930. (Ch. 14)

Herman, W.S. 1981. Studies on the adult reproductive diapause of the monarch butterfly, *Danaus plexippus*. Biol. Bull. 160:89–106. (Ch. 19)

Herman, W.S., and M. Tatar. 2001. Juvenile hormone regulation of longevity in the migratory monarch butterfly. Proc. R. Soc. Lond. B 268:2509–2514. (Ch. 19)

Hetru, C., D. Hoffman, and P. Bulet. 1998. Antimicrobial peptides from insects. *In* T. Brey and D. Hultmark, eds., Molecular mechanisms of immune responses in insects, pp. 40–66. London: Chapman & Hall. (Ch. 7)

Hill, J.K., H.M. Griffiths, and C.D. Thomas. 2011. Climate Change and Evolutionary Adaptations at Species' Range Margins. Annu. Rev. Entomol. 56:143 159 (Ch. 11)

Hinchliff, J. 1994. An atlas of Oregon butterflies. Corvallis: Oregon State University Bookstore. (Ch. 21)

Hinchliff, J. 1996. An atlas of Washington butterflies. Corvallis: Oregon State University Bookstore. (Ch. 21)

Hobson, K.A., L.I. Wassenaar, and O.R. Taylor. 1999. Stable isotopes (δD and $\delta^{13}C$) are geographic indicators of natal origins of monarch butterflies in eastern North America. Oecologia 120:397–404. (Ch. 11, 22)

Hoeksema, J.D., and S.E. Forde. 2008. A meta-analysis of factors affecting local adaptation between interacting species. Am. Nat. 171:275–290. (Ch. 7)

Hogg, E.H., J.P. Brandt, and B. Kochtubajda. 2002. Growth and dieback of aspen forests in northwestern Alberta, Canada, in relation to climate and insects. Can. J. For. Res. 32:823–832. (Ch. 13)

Holland, R.A., M. Wikelski, and D.S. Wilcove. 2006. How and why do insects migrate? Science 313:794–796. (Ch. 20)

Holm, S. 1979. A simple sequentially rejective multiple test procedure. Scand. J. Stat. 6:65–70. (Ch. 24)

Holsinger, K.E., and B.S. Weir. 2009. Genetics in geographically structured populations: Defining, estimating and interpreting F_{ST}. Nat. Rev. Genet. 10:639–650. (Ch. 23)

Holt, R.D. 1990. The microevolutionary consequences of climate change. Trends Ecol. Evol. 5:311–315. (Ch. 8)

Holtrop, J. 2010. Internal Directive to U.S. Forest Service Personnel. (Ch. 3)

Holzinger, F., and M. Wink. 1996. Mediation of cardiac glycoside insensitivity in the Monarch butterfly (*Danaus plexippus*): Role of an amino acid substitution in the ouabain binding site of Na+,K+-ATPase. J. Chem. Ecol. 22:1921–1937. (Ch. 4)

Honey-Rosés, J. 2009a. Disentangling the proximate factors of deforestation: The case of the monarch butterfly biosphere reserve in Mexico. Land Degrad. Dev. 20:22–32. (Ch. 13)

Honey-Rosés, J. 2009b. Illegal logging in common property forests. Soc. Nat. Resour. 22:916–930. (Ch. 13, 15)

Honey-Rosés, J., J. López-Garcia, E. Rendón-Salinas, A. Peralta-Higuera, and C. Galindo-Leal. 2009. To pay or not to pay? Monitoring performance and enforcing conditionality when paying for forest conservation in Mexico. Environ. Conserv. 36(2):120–128. (Ch. 3, 13, 15)

Honey-Rosés, J., K. Baylis, and M.I. Ramírez. 2011. A spatially explicit estimate of avoided forest loss. Conserv. Biol. 25(5):1032–1043. (Ch. 3, 15)

Horthorn, T., A. Zeileis, G. Millo, and D. Mitchell. 2013. R Package 'lmtest': Testing linear regression models. http://cran.r-project.org/web/packages/lmtest/lmtest.pdf. (Ch. 24)

Hoth, J. 1995. Mariposa monarca, mitos y otras realidades aladas. Ciencias 37:19–28 (Ch. 13, 15)

Hoth, J., L. Merino, K.S. Oberhauser, I. Pisantry, S. Price, and T. Wilkinson, eds. 1999. The 1997 North American conference on the monarch butterfly: Paper presentations. Montreal: Commission for Environmental Cooperation. (Preface)

Howard, E. 2011. Animated map of overnight roosts monarch butterfly fall migration 2011. Journey North. http://www.learner.org/jnorth/maps/galleries/2011/monarch_anima

tion_fall2011_roosts.html. Accessed 26 October 2012. (Ch. 10)

Howard, E. 2012. Fall monarch migration map archives. Location of overnight roosts, 2004 to present. Journey North. http://www.learner.org/jnorth/tm/monarch/Migration MapsFallRoost.html. Accessed 9 Nov 2012. (Ch. 10)

Howard, E., and A.K. Davis. 2004. Documenting the spring movements of monarch butterflies with Journey North, a citizen science program. In K.S. Oberhauser and M.J. Solensky, eds., The monarch butterfly: Biology and conservation, pp. 105–116. Ithaca, NY: Cornell University Press. (Ch. 2, 16, 24)

Howard, E., and A.K. Davis. 2009. The fall migration flyways of monarch butterflies in eastern North America revealed by citizen scientists. J. Insect Conserv. 13:279–286. (Ch. 2, 10. 18, 19, Part 5)

Howard, E., and A.K. Davis. 2011. A simple numerical index for assessing the spring migration of monarch butterflies using data from Journey North, a citizen-science program. J. Lepid. Soc. 65:267–270. (Ch. 2, 18, Part 5)

Howard, E., and A.K. Davis. 2012. Mortality of migrating monarch butterflies from a wind storm on the shore of Lake Michigan, USA. J. Res. Lepid. 45:49–54. (Ch. 2, 18)

Howard, E., H. Aschen, and A.K. Davis. 2010. Citizen science observations of monarch butterfly overwintering in the southern United States. Psyche 10:1–6. (Ch. 2, 7, 8, 19, 24)

Howard, L.O. 1889. The hymenopterous parasites of North American butterflies. Cambridge: Samuel H. Scudder. (Ch. 6)

Huffman, M.A., and M. Seifu. 1989. Observations on the illness and consumption of a possibly medicinal plant *Vernonia amygdalina* (Del.), by a wild chimpanzee in the Mahale Mountains National Park, Tanzania. Primates 30:51–63. (Ch. 7)

Hughes, J.M., and M.P. Zalucki. 1993. The relationship between the *Pgi* locus and the ability to fly at low temperatures in the monarch butterfly *Danaus plexippus*. Biochem. Genet. 31:521–532. (Ch. 23)

Hughes, L., and F.A. Bazzaz. 1997. Effect of elevated CO_2 on interactions between the western flower thrips, *Frankliniella occidentalis* (Thysanoptera: Thripidae) and the common milkweed, *Asclepias syriaca*. Oecologia 109:286–290. (Part 3)

Hunter, M.D., S.B. Malcolm, and S.E. Hartley. 1996. Population-level variation in plant secondary chemistry and the population biology of herbivores. Chemoecology 7:45–56. (Ch. 6)

Ibarra G., M.V. 2011. Conformación del espacio social de los bosques del ejido del Rosario, Michoacán, 1938–2010. Bol. Inst. Geogr. Univ. Nac. Auton. Mex. 75:75–87. (Ch. 13, 15)

[IBBA] International Butterfly Breeders Association. 2012. http://butterflybreeders.org/. Accessed November 2012. (Ch. 1)

[IELP] International Environmental Law Project and Xerces Society for Invertebrate Conservation. 2012. The legal status of monarch butterflies in California. http://www.xerces.org/wp-content/uploads/2008/09/legal-status-of-california-monarchs.pdf. (Ch. 2, 3, 12)

INEGI. 2012. Datos del territorio de Michoacán y Estado de México. In http://cuentame.inegi.org.mx/ (Ch. 15)

IPCC. 2000. Emission scenarios: Summary for policymakers. A special Report of IPCC Working Group III. United Nations Environmental Program. Intergovernmental Panel on Climate Change. Nairobi, Kenya. (Ch. 13)

IPCC. 2007. Climate Change 2007: The physical science basis. Contribution of working group I to the fourth assessment report of the Intergovernmental Panel on Climate Change. Solomon, S., D. Qin, M. Manning, Z. Chen, M. Marquis, K.B. Averyt, M. Tignor, and H.L. Miller, eds. Cambridge: Cambridge University Press. (Ch. 8)

Izquierdo, I., J.L. Martin, N. Zurita, and M. Arechavaleta, eds. 2004. Lista de especies silvestres de Canarias (hongos, plantas y animales terrestres). Consejería de Medio Ambiente y Ordenación Territorial. Gobierno de Canarias. La Laguna. (Ch. 22)

Jablonski, D. 2008. Biotic interactions and macroevolution: Extensions and mismatches across scales and levels. Evolution 62:715–739. (Ch. 4)

Jactel, H., Birgersson, G., Andersson, S., and F. Schlyter. 2011. Non-host volatiles mediate associational resistance to the pine processionary moth. Oecologia 166:703–711. (Ch. 14)

James, D.G. 1993. Migration biology of the monarch butterfly in Australia. In S.B. Malcolm and M.P. Zalucki, eds., Biology and conservation of the monarch butterfly, pp. 189–200. Los Angeles: Natural History Museum of Los Angeles County. (Ch. 9, 23)

James, D.G. 2012. Attraction of beneficial insects to flowering endemic perennial plants in the Yakima Valley. Annual Report for Western Sustainable Agriculture Research and Education. Logan: Utah State University. 12 pp. (Ch. 21)

James, D.G. 2013. https://www.facebook.com/#!/MonarchButterfliesInThePacificNorthwest. Accessed September 2013. (Ch. 21)

Jardel P., E.J. 2010. Planificación del manejo del fuego. Universidad de Guadalajara, Fundación Manantlán para la Biodiversidad de Occidente A.C., Consejo Civil Méxicano para la Silvicultura Sostenible A.C., Fondo Méxicano para la Conservación de la Naturaleza A.C. 59 pp. (Ch. 15)

Jardel P., E.J., F. Castillo, R. Ramírez, J.C. Chacón, and O.E. Balcázar. 2004. Los incendios forestales en la Reserva de la Biosfera Sierra de Manantlán, Jalisco y Colima. In Villers R., M.L., and J. Lopéz Blanco, eds., Incendios forestales en México, pp. 147–164. Mexico City: UNAM. (Ch. 15)

Jardel P., E.J., E. Alvarado C., J.E. Morfín R., F. Castillo N., and J.G. Flores G. 2009. Regímenes de fuego en ecosistemas forestales de México. In Flores G., J.G., ed., Impacto ambiental de incendios forestales, pp. 73–100. Mexico City: Mundi Presa México. (Ch. 15)

Jesse, L.C.H., and J.J. Obrycki. 2000. Field deposition of Bt transgenic corn pollen: Lethal effects on the monarch butterfly. Oecologia 125:241–248. (Ch. 11)

Jesse, L.C.H., and J.J. Obrycki. 2004. Survival of experimental cohorts of monarch larvae following exposure to transgenic *Bt* corn pollen and anthers. In K.S. Oberhauser and M.J. Solensky, eds., The monarch butterfly: Biology and conservation, pp. 69–75. Ithaca, NY: Cornell University Press. (Ch. 11)

Johnson, C.G. 1963. Physiological factors in insect migration by flight. Nature 198:423–427. (Ch. 22)

Johnson, C.G. 1969. Migration and dispersal of insects by flight. London: Methuen. (Ch. 20)

Journey North. 2012a. http://www.learner.org/jnorth/tm/monarch/PopulationMexicoAnalyzeGraph.html. Accessed December 2012. (Ch. 14)

Journey North. 2012b. Monarch butterfly. www.learner.org/jnorth/monarch/index.html. Accessed December 2012. (Ch. 2)

Journey North. 2013. Journey North website. http://www.learner.org/jnorth/. Accessed October 2013. (Ch. 8, 18, 19, 24)

Kaiser Family Foundation. 2010. http://www.kff.org/entmedia/entmedia012010nr.cfm. Accessed December 2012. (Ch. 1)

Kaltz, O., and J.A. Shykoff. 1998. Local adaptation in host-parasite systems. Heredity. 81:361–370. (Ch. 7)

Kammer, A.E. 1970. Thoracic temperature, shivering, and flight in the monarch butterfly, *Danaus plexippus* (L.). Z. vergl. Physiol. 68:334–344. (Ch. 9, Part 3)

Kawashima, S., K. Matsuo, M. Du, Y. Takahashi, S. Inoue, and S. Yonemua. 2004. An algorithm for estimating potential deposition of corn pollen for environmental assessment. Environ. Biosafety Res. 3:197–207. (Ch. 14)

Kawecki, T.J., and D. Ebert. 2004. Conceptual issues in local adaptation. Ecol. Lett. 7:1225–1241. (Ch. 7)

Kearney, M., and W.P. Porter. 2009. Mechanistic niche modelling: Combining physiological and spatial data to predict species' ranges. Ecol. Lett. 12:334–350. (Ch. 11)

Keiman, A.F., and M. Franco. 2004. Can't see the forest for the butterflies: The need for understanding forest dynamics at monarch overwintering sites. *In* K.S. Oberhauser and M.J. Solensky, eds., The monarch butterfly: Biology and conservation, pp. 135–140. Ithaca, NY: Cornell University Press. (Ch. 9)

Kelling, S., C. Lagoze, W.K. Wong, J. Yu, T. Damoulas, J. Gerbracht, D. Fink, and C. Gomes. 2013. eBird: A human/computer learning network to improve biodiversity conservation and research. Artif. Intell. 34:10–20. (Ch. 2)

Kingsolver, B. 2012. Flight behavior. New York: Harper. (Ch. 21)

Knight, A.L. 1998. A population study of Monarch butterflies in North-Central and South Florida. M.S., University of Florida, Gainesville. (Ch. 7)

Knight, A., and L.P. Brower. 2009. The influence of eastern North American autumnal migrant monarch butterflies (*Danaus plexippus* L.) on continuously breeding resident monarch populations in southern Florida. J. Chem. Ecol. 35:816–823. (Ch. 7, 19, Part 5)

Knutson, A.E., and L. Tedders. 2002. Augmentation of green lacewing, *Chrysoperla rufilabris*, in cotton in Texas. Southwest. Entomol. 27:231–239. (Ch. 6)

Koch, H., and P. Schmid-Hempel. 2011. Socially transmitted gut microbiota protect bumble bees against an intestinal parasite. Proc. Natl. Acad. Sci. 108:19288–19292. (Ch. 7)

Koch, R.L., W.D. Hutchison, R.C. Venette, and G.E. Heimpel. 2003. Susceptibility of immature monarch butterfly, *Danaus plexippus* (Lepidoptera: Nymphalidae: Danainae), to predation by *Harmonia axyridis* (Coleoptera: Coccinellidae). Biol. Control. 8:265–270. (Ch. 6)

Koch, R.L., R.C. Venette, and W.D. Hutchison. 2005. Influence of alternate prey on predation of monarch butterfly (Lepidoptera: Nymphalidae) larvae by the multicolored Asian lady beetle (Coleoptera: Coccinellidae). Environ. Entomol. 34:410–416. (Ch. 6, 11)

Koch, R.L., R.C. Venette, and W.D. Hutchison. 2006. Predicted impact of an exotic generalist predator on monarch butterfly (Lepidoptera: Nymphalidae) populations: A quantitative risk assessment. Biol. Invasions. 8:1179–1193. (Ch. 6)

Konno, K., C. Hirayama, M. Nakamura, K. Tateishi, Y. Tamura, M. Hattori, and K. Kohno. 2004. Papain protects papaya trees from herbivorous insects: Role of cysteine proteases in latex. Plant J. 37:370–378. (Ch. 4)

Kountoupes, D., and K.S. Oberhauser. 2008. Citizen science and youth audiences: Educational outcomes of the Monarch Larva Monitoring Project. J. Community Engagem. Scholarsh. 1:10–20. (Ch. 2)

Kraaijeveld, A.R., and H.C.J. Godfray. 1997. Trade-off between parasitoid resistance and larval competitive ability in Drosophila melanogaster. Nature 389:278–280. (Ch. 7)

Kriticos, D.J., and R.P. Randall. 2001. A comparison of systems to analyse potential weed distributions. *In* R.H. Groves, F.D. Panetta, and J.G. Virtue, eds., Weed risk assessment, pp. 61–79. Melbourne: CSIRO. (Ch. 11)

Kriticos, D.J., R.W. Sutherst, J.R. Brown, S.A. Adkins, and G.F. Maywald. 2003. Climate change and the potential distribution of an invasive alien plant: *Acacia nilotica* ssp. *indica* in Australia. J. Appl. Ecol. 40:111–124. (Ch. 11)

Kursar, T.A., K.G. Dexter, J. Lokvam, R.T. Pennington, J.E. Richardson, M.G. Weber, E.T. Murakami, C. Drake, R. McGregor, and P.D. Coley. 2009. The evolution of antiherbivore defenses and their contribution to species coexistence in the tropical tree genus *Inga*. Proc. Natl. Acad. Sci. 106:18,073–18,078. (Ch. 4)

Lack, D. 1947. The significance of clutch-size. Ibis 89:302–352. (Ch. 6)

Ladner, D.T., and S. Altizer. 2005. Oviposition preference and larval performance of North American monarch butterflies on four *Asclepias* species. Entomol. Exp. Appl. 116:9–20. (Ch. 7, 14)

Lady Bird Johnson Wildflower Center Native Plant Database. http://www.wildflower.org/plants/. Accessed October 2012. (Ch. 10)

Lafferty, K.D., A.P. Dobson, and A.M. Kuris. 2006. Parasites dominate food web links. Proc. Natl. Acad. Sci. 103:11211–11216. (Ch. 7)

Laine, A.-L. 2007. Detecting local adaptation in a natural plant-pathogen metapopulation: A laboratory vs. field transplant approach. J. Evol. Biol. 20:1665–1673. (Ch. 7)

Laine, A.-L. 2008. Temperature-mediated patterns of local adaptation in a natural plant-pathogen metapopulation. Ecol. Lett. 11:327–337. (Ch. 7)

Lambrechts, L., S. Fellous, and J.C. Koella. 2006. Coevolutionary interactions between host and parasite genotypes. Trends Parasitol. 22:12–16. (Ch. 7)

Lane, J. 1984. The status of monarch butterfly overwintering sites in Alta California. Atala (Portland) 9:17–20. (Ch. 12)

Lane, J. 1985. California's monarch butterfly trees. J. Pac. Discov. 38:13–15. (Ch. 12)

Lane, J. 1993. Overwintering monarch butterflies in California: Past and present. *In* S.B. Malcolm and M.P. Zalucki, eds., Biology and conservation of the monarch butterfly, pp. 335–344. Los Angeles: Natural History Museum of Los Angeles County. (Ch. 12)

Larsen, K.J., and R.E. Lee. 1994. Cold tolerance including rapid cold-hardening and inoculative freezing of fall migrant monarch butterflies in Ohio. J. Insect Physiol. 40:859–864. (Ch. 8, 9, Part 3)

Lasota, J.A., and L.T. Kok. 1986. Parasitism and utilization of imported cabbageworm pupae by *Pteromalus puparum* (Hymenoptera: Pteromalidae). Environ. Entomol. 15:994–998. (Ch. 6)

Lavine, M.D., and M.R. Strand. 2002. Insect hemocytes and their role in immunity. Insect Biochem. Mol. Biol. 32:1295–1309. (Ch. 7)

Lavoie, B., and K.S. Oberhauser. 2004. Compensatory feeding in *Danaus plexippus* (Lepidoptera: Nymphalidae) in response to variation in host plant quality. Environ. Entomol. 33(4):1062–1069. (Ch. 14)

Lawler, J.J., D. White, R.P. Neilson, and A.R. Blaustein. 2006. Predicting climate-induced range shifts: Model differences and model reliability. Global Change Biol. 12:1568–1584. (Part 4)

Lawson, B.E., M.D. Day, M. Bowen, R.D. Van Klinken, and M.P. Zalucki. 2010. The effect of data sources and quality on the predictive capacity of CLIMEX models: An assessment of *Teleonemia scrupulosa* and *Octotoma scabripennis* for the biocontrol of *Lantana camara* in Australia. Biol. Control 52:68–76. (Ch. 11)

Lazzaro, B.P., and T.J. Little. 2009. Immunity in a variable world. Phil. Trans. R. Soc. B 364:15–26. (Ch. 7)

Lazzaro, B.P., H.A. Flores, J.G. Lorigan, and C.P. Yourth. 2008. Genotype-by-environment interactions and adaptation to local temperature affect immunity and fecundity in *Drosophila melanogaster*. PLoS Path. 4:E1000025. (Ch. 7)

Lee, J.C., and G.E. Heimpel, G.E. 2005. Impact of flowering buckwheat on lepidopteran cabbage pests and their parasitoids at two spatial scales. Biol. Control. 34:290–301. (Ch. 6)

Lee, K.P., J.S. Cory, K. Wilson, D. Raubenheimer, and S.J. Simpson. 2006. Flexible diet choice offsets protein costs of pathogen resistance in a caterpillar. Proc. R. Soc. Lond. B 273:823–829. (Ch. 7)

Lee, R.E. Jr. 2010. A primer on insect cold-tolerance. In D.L. Denlinger and R.E. Lee Jr., eds., Low temperature biology of insects, pp. 3–34. Cambridge: Cambridge University Press. (Ch. 8)

Leestmans, R. 1975. Etude biogéographique et écologique des Lépidoptères des îles Canaries (Insecta; Lepidoptera). Vieraea 4(1–2):9–106. (Ch. 22)

Lefèvre, T., L. Oliver, M.D. Hunter, and J.C. de Roode. 2010. Evidence for trans-generational medication in nature. Ecol. Lett. 13:1485–1493. doi: 10.1111/j.1461-0248.2010.01537.x. (Ch. 6, 7, Part 2)

Lefèvre, T., J.C. de Roode, B.Z. Kacsoh, and T.A. Schlenke. 2011a. Defence strategies against a parasitoid wasp in *Drosophila*: Fight or flight? Biol. Lett. 8:230–233. (Ch. 7)

Lefèvre, T., A.J. Williams, and J.C. de Roode. 2011b. Genetic variation for resistance, but not tolerance, to a protozoan parasite in the monarch butterfly. Proc. R. Soc. Lond. B 278:751–759. (Ch. 7)

Lefèvre, T., A. Chiang, M. Kelavkar, H. Li, J. Li, C. Lopez Fernandez de Castillejo, L. Oliver, Y. Potini, M.D. Hunter, and J.C. de Roode. 2012. Behavioural resistance against a protozoan parasite in the monarch butterfly. J. Anim. Ecol. 81:70–79. (Ch. 7, Part 2)

Leong, K.L.H. 1999. Restoration of an overwintering grove in Los Osos, San Luis Obispo County, California. In J. Hoth, L. Merino, K. Oberhauser, I. Pisanty, S. Price, and T. Wilkinson, eds., 1997 North American conference on the monarch butterfly: Paper presentations, pp. 221–218. Montreal: Commission for Environmental Cooperation. (Ch. 12)

Leong, K., D. Frey, and C. Nagano. 1990. Wasp predation on overwintering monarch butterflies (Lepidoptera: Danaidae) in central California. Pan-Pac. Entomol. 66:326–328. (Ch. 6)

Leong, K.L.H., H.K. Kaya, M.A. Yoshimura, and D.Frey. 1992. The occurrence and effect of a protozoan parasite, *Ophryocystis elektroscirrha* (Neogregarinida: Ophryocystidae) on overwintering monarch butterflies, *Danaus plexippus* (Lepidoptera: Danaidae) from two California wintering sites. Ecol. Entomol. 17:338–342. (Ch. 7)

Leong, K.L.H., M.A. Yoshimura, and H.K. Kaya. 1997a. Occurrence of a neogregarine protozoan, *Ophryocystis elektroscirrha* McLaughlin and Myers, in populations of monarch and queen butterflies. Pan-Pac. Entomol. 73:49–51. (Ch. 7)

Leong, K.L.H., M.A. Yoshimura, H.K. Kaya, and H. Williams. 1997b. Instar susceptibility of the monarch butterfly (*Danaus plexippus*) to the neogregarine parasite, *Ophryocystis elektroscirrha*. J. Invertebr. Pathol. 69:79–83. (Ch. 7)

Leong, K.L.H., W.H. Sakai, W. Bremer, D. Feuerstein, and G. Yoshimura. 2004. Analysis of the pattern of distribution and abundance of monarch overwintering sites along the California coastline. In K.S. Oberhauser and M.J. Solensky, eds., The monarch butterfly: Biology and conservation, pp. 177–185. Ithaca, NY: Cornell University Press. (Ch. 12)

Lepage, D., and C.M. Francis. 2002. Do feeder counts reliably indicate bird population changes? 21 years of winter bird counts in Ontario, Canada. Condor. 104:255–270. (Ch. 24)

Letourneau, D.K., and L.R. Fox. 1989. Effects of experimental design and nitrogen on cabbage butterfly oviposition. Oecologia 80:211–214. (Ch. 14)

Levin, S., and D. Pimentel. 1981. Selection of intermediate rates of increase in parasite-host systems. Am. Nat. 117:308–315. (Ch. 7)

LGDFS. 2003. Ley General de Desarrollo Forestal Sustentable. Mexico City: Diario Oficial de la Federación, 25 February. (Ch. 13)

LGEEPA. 2007. Ley General de Equilibrio Ecológico y Protección al Ambiente. Mexico City: Diario Oficial de la Federación, 5 July: 25–27. (Ch. 13)

Li, Z., M.P. Zalucki, H. Bao, H. Chen, Z. Hu, D. Zhang, Q. Lin, F. Yin, M. Wang, and X. Feng. 2012. Population dynamics and 'outbreaks' of diamondback moth, *Plutella xylostella*, in Guangdong province, China: Climate or the failure of management? J. Econ. Entomol. 105:739–752. (Ch. 11)

Lieberman, G., and L.L. Hoody. 1998. Closing the achievement gap: Using the environment as an integrating context for learning. A nationwide study. San Diego: State Education and Environment Roundtable. (Ch. 1)

Liedvogel, M., S. Akesson, and S. Bensch. 2011. The genetics of migration on the move. Trends Ecol. Evol. 26:561–569. (Ch. 23)

Liggieri, C., M.C. Arribere, S.A. Trejo, F. Canals, F.X. Aviles, and N.S. Priolo. 2004. Purification and biochemical characterization of asclepain c I from the latex of *Asclepias curassavica* L. Protein J. 23:403–411. (Ch. 4)

Lindenmayer, D.B., and R.F. Noss. 2006. Salvage logging, ecosystem processes, and biodiversity conservation. Conserv. Biol. 20:949–958. (Part 4)

Lindsey, E.A. 2008. Ecological determinants of host resistance to parasite infection in monarch butterflies. Ph.D. Thesis. Emory University, Atlanta, GA. (Ch. 7)

Lindsey, E., M. Mehta, V. Dhulipala, K. Oberhauser, and S. Altizer. 2009. Crowding and disease: Effects of host density on parasite infection in monarch butterflies. Ecol. Entomol. 34:551–561. (Ch. 2, 7)

Link, W.A., J.R. Sauer, and D.K. Niven. 2006. A hierarchical model for regional analysis of population change using Christmas Bird Count data, with application to the American Black Duck. Condor 108:13–24 (Ch. 24)

Lively, C.M. 1989. Adaptation by a parasitic trematode to local populations of its snail host. Evolution 43:1663–1671. (Ch. 7)

Lively, C.M. 1999. Migration, virulence, and the geographic mosaic of adaptation by parasites. Am. Nat. 153:S34-S47. (Ch. 7)

Lively, C.M., M.F. Dybdahl, J. Jokela, E. Osnas, and L.F. Delph. 2004. Host sex and local adaptation by parasites in a snail-trematode interaction. Am. Nat. 164:S6-S18. (Ch. 7)

Llewelyn, J.T.D. 1876. A foreign visitor (*Danais archippus*). Entomol. Mon. Mag. 13:107–108. (Ch. 22)

Loock, E.E.M. 1950. The pines of Mexico and British Honduras. L.S. Gray, Government Publication, pp. 244. Department of Forestry, Pretoria, Union of South Africa. As cited in Brower 1995. (Ch. 9)

Losey, J.E., L.S. Rayor, and M.E. Carter. 1999. Transgenic pollen harms monarch larvae. Nature 399:214. (Ch. 11, 14)

Lotka, A.J. 1942. The progeny of an entire population. Ann. Math. Stat. 13:115–126. (Ch. 5)

Louv, R. 2008. Last child in the woods: Saving our children from nature-deficit disorder. Chapel Hill, NC: Algonquin Books. (Ch. 1)

Lozano, G.A. 1991. Optimal foraging theory—a possible role for parasites. Oikos 60:391–395. (Ch. 7)

Lozano, G.A. 1998. Parasitic stress and self-medication in wild animals. *In* A.P. Møller, M. Milinksi, and P.J.B. Slater, eds., Stress and behavior, pp. 291-317. San Diego: Academic. (Ch. 7)

Lynch, S.P., and R.A. Martin. 1987. Cardenolide content and thin-layer chromatography profiles of monarch butterflies, *Danaus plexippus* L., and their larval host-plant milkweed, *Asclepias viridis* WALT., in northwestern Louisiana. J. Chem. Ecol. 13:47–70. (Ch. 16)

Lynch, S.P., and R.A. Martin. 1993. Milkweed host plant utilization and cardenolide sequestration by monarch butterflies in Louisiana and Texas. *In* S.B. Malcolm and M.P. Zalucki, eds., Biology and conservation of the monarch butterfly, pp. 107–124. Los Angeles: Natural History Museum of Los Angeles County. (Ch. 6, 16)

Lyons, J.I., A.A. Pierce, S.M. Barribeau, E.D. Sternberg, A.J. Mongue, and J.C. de Roode. 2012. Lack of genetic differentiation between monarch butterflies with divergent migration destinations. Mol. Ecol. 21:3433–3444. (Ch. 7, 23, Part 5)

Maddison, W.P., and L.L. Knowles. 2006. Inferring phylogeny despite incomplete lineage sorting. Syst. Biol. 55:21–30. (Ch. 4)

Maddison, W.P., P.E. Midford, and S.P. Otto. 2007. Estimating a binary character's effect on speciation and extinction. Syst. Biol. 56:701–710. (Ch. 4)

Maelzer, D.A., and M.P. Zalucki. 1999. Analysis and interpretation of long term light trap data for *Helicoverpa* spp. (Lepidoptera: Noctuidae) in Australia: The effect of climate and crop host plants. Bull. Entomol. Res. 89:455–464. (Ch. 11)

Malcolm, S.B. 1987. Monarch butterfly migration in North America: Controversy and conservation. Trends Ecol. Evol. 2:135–138. (Ch. 7)

Malcolm, S.B. 1991. Cardenolide-mediated interactions between plants and herbivores. *In* G.A. Rosenthal and M.R. Berenbaum, eds., Herbivores: Their interactions with secondary plant metabolites, 2nd ed. Vol. I: The chemical participants, pp. 251–296. San Diego: Academic Press. (Ch. 4, 5, 6, 7, 16, Part 2)

Malcolm, S.B. 1995. Milkweeds, monarch butterflies and the ecological significance of cardenolides. Chemoecology 5/6:101–117. (Ch. 4)

Malcolm, S.B., and L.P. Brower. 1989. Evolutionary and ecological implications of cardenolide sequestration in the monarch butterfly. Experientia 45:284–295. (Ch. 6, 7, 17, Part 2)

Malcolm, S.B., and M.P. Zalucki, eds. 1993. Biology and conservation of the monarch butterfly. Los Angeles: Natural History Museum of Los Angeles County. (Preface).

Malcolm, S.B., B.J. Cockrell, and L.P. Brower. 1987. Monarch butterfly voltinism: Effects of temperature constraints at different latitudes. Oikos 49:77–82. (Ch. 11, 16, Part 3)

Malcolm, S.B., B.J. Cockrell, and L.P. Brower. 1989. The cardenolide fingerprint of monarch butterflies reared on the common milkweed, *Asclepias syriaca*. J. Chem. Ecol. 15:819–853. (Ch. 6, 22)

Malcolm, S.B., B.J. Cockrell, and L.P. Brower. 1993. Spring recolonization of eastern North America by the monarch butterfly: Successive brood or single sweep migration? *In* S.B. Malcolm and M.P. Zalucki, eds., Biology and conservation of the monarch butterfly, pp. 253–267. Los Angeles: Natural History Museum of Los Angeles County. (Ch. 4, 6, 10, 11, 16, 17, 20, 24, Part 4)

Manson, J.S., S. Rasmann, R. Halitschke, J.D. Thomson, and A.A. Agrawal. 2012. Cardenolides in nectar are not a mere consequence of allocation to other plant parts: A phylogenetic study of milkweeds (*Asclepias*). Funct. Ecol. 26:1100–1110. (Ch. 4)

Manzanilla, H. 1974. Investigaciones epidométricas y sílvicolas en bosques Mexicanos de abies religiosa. Mexico City: Dirección General de Información y Relaciones Públicas, Secretaría de Agricultura y Ganadería. x + 165 pp. (Ch. 9)

Maravalhas, E. 2003. As borboletas de Portugal. Stenstrup, Denmark : Apollo Books. 455p. (Ch. 22)

Martin, A.J. 2002. El Manejo Forestal Contrastante en Dos Núcleos Agrarios de la Reserva de la Biosfera Mariposa Monarca. Relaciones 23(89):54–82. (Ch. 15)

Martin, A.R., and O.C. Burnside. 1980. Common milkweed: Weed on the increase. Weeds Today 11:19–20. (Ch. 14)

Martin, R.A., and S.P. Lynch. 1988. Cardenolide content and thin-layer chromatography profiles of monarch butterflies, *Danaus plexippus* L., and their larval host-plant milkweed, *Asclepias asperula* subsp. *capricornu* (Woods.) in north-central Texas. J. Chem. Ecol. 14:295–318. (Ch. 16)

Martin, R.A., S.P. Lynch, L.P. Brower, S.B. Malcolm, and T. Van Hook. 1992. Cardenolide content, emetic potency, and thin-layer chromatography profiles of monarch butterflies, *Danaus plexippus*, and their larval host-plant milkweed, *Asclepias humistrata*, in Central Florida. Chemoecology 3:1–13. (Ch. 4, 6)

Martínez, E. 2009. Invaden plagas reserva de la mariposa monarca. La Jornada. 4 August 2009. (Ch. 13)

Masters, A.R. 1993. Temperature and thermoregulation in the monarch butterfly. *In* S.B. Malcolm and M.P. Zalucki, eds., Biology and conservation of the monarch butterfly, pp. 147–156. Los Angeles: Natural History Museum of Los Angeles County. (Ch. 8)

Masters, A.R., S.B. Malcolm, and L.P. Brower. 1988. Monarch butterfly (*Danaus plexippus*) thermoregulatory behavior and adaptations for overwintering in Mexico. Ecology 69:458–467. (Ch. 8, 9, 11, 12, 20, Part 3, 4)

Mathews, A.S. 2003. Suppressing fire and memory: Environmental degradation and political restoration in the Sierra Juárez of Oaxaca 1887–2001. Environ. Hist. 8(1):77–108. (Ch. 15)

Matsuki, M., M. Kay, J. Serin, R. Floyd, and J.K. Scott. 2001. Potential risk of accidental introduction of Asian gypsy moth (*Lymantria dispar*) to Australasia: Effects of climatic conditions and suitability of native plants. Agric. For. Entomol. 3:305–320. (Ch. 11)

Matteson, K.C., D.J. Taron, and E.S. Minor. 2012. Assessing citizen contributions to butterfly monitoring in two large cities. Conserv. Biol. 26:557–564. (Ch. 24)

Maywald, G.F., and R.W. Sutherst. 1991. Users guide to CLIMEX a computer program for comparing climates in ecology, 2nd ed. Report no. 48. CSIRO Australia, Division of Entomology. (Ch. 11)

[MBF] Monarch Butterfly Fund. 2013. Monarch Butterfly Fund projects. monarchbutterflyfund.org/. Accessed January 2013. (Ch. 1, 2)

McCord, J.W., and A.K. Davis. 2010. Biological observations of monarch butterfly behavior at a migratory stopover site: Results from a long-term tagging study in coastal South Carolina. J. Insect Behav. 23:405–418. (Ch. 2)

McCord, J.W., and A.K. Davis. 2012. Characteristics of monarch butterflies (*Danaus plexippus*) that stopover at a site in coastal South Carolina during fall migration. J. Res. Lepid. 45:1–8. (Ch. 18)

McKenna, D.D., K.M. McKenna, S.B. Malcolm, and M.R. Berenbaum. 2001. Mortality of lepidoptera along roadways in central Illinois. J. Lepid. Soc. 55:63–68. (Ch. 18)

McKinnon, L., P.A. Smith, E. Nol, J.L. Martin, F.I. Doyle, K.F. Abraham, H.G. Gilchrist, R.I.G. Morrison, and J. Bêty. 2010. Lower predation risk for migratory birds at high latitudes. Science 327:326–327. (Ch. 7)

McLaughlin, R.E., and J. Myers. 1970. *Ophryocystis elektroscirrha* sp. n., a neogregarine pathogen of the monarch butterfly *Danaus plexippus* (L.) and the Florida queen butterfly *Danaus gilippus berenice* Cramer. J. Protozool. 17:300–305. (Ch. 7, 17)

Meade, D.E. 1999. Monarch butterfly overwintering sites in Santa Barbara County California. Althouse and Meade Biological and Environmental Services, November 1999. 114 pp. (Ch. 12)

Meitner, C.J., 1995. Winged voyagers: A review of literature pertaining to the monarch butterfly's life history, migration, overwintering behavior and conservation including recommendations for habitat management and conservation. Draft Report to the USDA/Forest Service of Hiawatha National Forest and Wildlife Unlimited of Delta County, MI, pp. 35–54. (Ch. 12)

Meitner, C.J., L.P. Brower, and A.K. Davis. 2004. Migration patterns and environmental effects on stopover of monarch butterflies (Lepidoptera: Nymphalidae) at Peninsula Point, Michigan. Environ. Entomol. 33:249–256. (Ch. 2, 24)

Merino, L. 1997. Reserva Especial de la Biosfera Mariposa Monarca: Problemática general de la región. Ponencia presentada en la Reunión de América del Norte sobre la Mariposa Monarca 1997. (Ch. 15)

Merino, L., and M. Hernández. 2004. Destrucción de instituciones comunitarias y deterioro de los bosques en la Reserva de la Biosfera Mariposa Monarca, Michoacán, México. Rev. Mex. Soc. 66(2):261–309. (Ch. 13, 15)

Merlin, C., R.J. Gegear, and S.M. Reppert. 2009. Antennal circadian clocks coordinate sun compass orientation in migratory monarch butterflies. Science 325:1700–1704. (Ch. 23, Part 5)

Meyer, M. 1993. Die Lepidoptera der makaronesischen Region. III. Die Tagfalter des nördlichen Macaronesiens (Madeira, Azoren) aus biogeographischer Sicht (Papilionoidea). Atalanta 24(1/2):121–162. (Ch. 22)

Miller, E., and R. Narayanan, eds. 2008. Great Basin wildfire forum: The search for solutions. Reno: Nevada Agricultural Experiment Station. Available at http://dcnr.nv.gov/wp-content/wildfireforum.pdf. Accessed 8 October 2013. (Ch. 16)

Miller, M.R., A. White, and M. Boots. 2005. The evolution of host resistance: Tolerance and control as distinct strategies. J. Theor. Biol. 236:198–207. (Ch. 7)

Miller, N.G., L.I. Wassenaar, K.A. Hobson, and D.R. Norris. 2011. Monarch butterflies cross the Appalachians from the west to recolonize the east coast of North America. Biol. Lett. 7:43–46. (Ch. 11)

Miller, N.G., L.I. Wassenaar, K.A. Hobson, and D.R. Norris. 2012. Migratory connectivity of the monarch butterfly (*Danaus plexippus*): Patterns of spring re-colonization in eastern North America. PLoS ONE 7. (Ch. 23)

Miller-Rushing, A., R. Primack, and R. Bonney. 2012. The history of public participation in ecological research. Front. Ecol. Environ. 10:285–290. (Ch. 2)

Minnich, R.A., M.G. Barbour, J.H. Burk, and J. Sosa-Ramírez. 2000. Californian mixed-conifer forest under unmanagement fire regimes in the Sierra San Pedro Martir, Baja California, Mexico. J. Biogeogr. 27(1):105–129. (Ch. 15)

Missrie, M. 2004. Design and implementation of a new protected area for overwintering monarch butterflies in Mexico. *In* K.S. Oberhauser and M.J. Solensky, eds., The monarch butterfly: Biology and conservation, pp. 141–150. Ithaca, NY: Cornell University Press. (Ch. 3, Part 4)

Missrie, M., and K. Nelson. 2007. Direct payments for conservation: Lessons from the Monarch Butterfly Conservation Fund. *In* A. Usha, ed., Bio-diversity and conservation: International perspectives, pp. 189–212. Hyderabad, India: The Icfai University Press. (Ch. 13)

Mitchell, S.E., E.S. Rogers, T.J. Little, and A.F. Read. 2005. Host-parasite and genotype-by-environment interactions: Temperature modifies potential for selection by a sterilizing pathogen. Evolution 59:70–80. (Ch. 7)

[MLMP] Monarch Larva Monitoring Project. 2013. http://www.mlmp.org/. Accessed October 2013. (Ch. 2, 8, 14, 19, 24)

Monarch ESA Petition. 2014. Petition to protect the monarch butterfly (*Danaus plexippus plexippus*) under the Endangered Species Act. Submitted to Secretary of the US Department of the Interior, by the Center for Biological Diversity and Center for Food Safety, joined by the Xerces Society and Dr. Lincoln Brower, August 26, 2014. http://www.centerforfoodsafety.org/files/monarch-esa-petition-final_77427.pdf. (Ch. 10, Part 4)

Monarch Joint Venture. 2013. www.monarchjointventure.org. Accessed November 2013. (Ch. 3, 21, Preface)

Monarch Lab. 2012. Monarch meeting website. www.monarchlab.org/mn2012/Presentations.aspx. Accessed January 2013. (Ch. 2)

Monarch Net. 2012. The North American network of monarch butterfly monitoring programs. monarchnet.org/. Accessed December 2012. (Ch. 2)

Monarch Parasites. 2012. Monarch Health and Monarch Parasites web page. monarchparasites.org/. Accessed December 2012. (Ch. 2)

Monarch Watch. 2001. Season summaries 1993–2001. monarchwatch.org/season-summaries. Accessed December 2012. (Ch. 2)

Monarch Watch. 2011. http://www.monarchwatch.org/biology/westpop.html. (Ch. 23)

Monarch Watch. 2012. Peak migration dates: When will the migration peak in my area? http://www.monarchwatch.org/tagmig/peak.html. Accessed September 2012. (Ch. 2, 17)

Monarch Watch. 2013a. http://monarchwatch.org/. Accessed January 2013. (Ch. 2, 21)

Monarch Watch. 2013b. http://www.monarchwatch.org/waystations/registry. Accessed March 2013. (Ch. 3)

Monroe, M., D. Frey, and S. Stevens. 2014. Western Monarch Thanksgiving Count data from 1997–2012. http://www.xerces.org/western-monarch-thanksgiving-count/. Accessed October 2014. (Ch. 12)

Montllor, C.B., and E.A. Bernays. 1994. Invertebrate predators and caterpillar foraging. *In* N.E. Stamp and T.M. Casey, eds., Caterpillars: Ecological and evolutionary constraints on foraging, pp. 170–202. New York: Chapman & Hall. (Ch. 5)

Mooney, K.A., P. Jones, and A.A. Agrawal. 2008. Coexisting congeners: Demography, competition, and interactions with cardenolides for two milkweed-feeding aphids. Oikos 117:450–458. (Ch. 4)

Morey, A.C., W.D. Hutchinson, R.C. Venette, and E.C. Burkness. 2012. Cold hardiness of *Helicoverpa zea* (Lepidoptera: Noctuidae) Pupae. Environ. Entomol. 41(1):172–179. (Ch. 8)

Morris, G. 2012. Southwest monarch study. www.monarchlab.org/mn2012/Presentations.aspx. Accessed December 2012. (Ch. 2)

Moskowitz, D., J. Moskowitz, S. Moskowitz, and H. Moskowitz. 2001. Notes on a large dragonfly and butterfly migration in New Jersey. Northeast. Nat. 8:483–490. (Ch. 18)

Mouritsen, H., and B.J. Frost. 2002. Virtual migration in tethered flying monarch butterflies reveals their orientation mechanisms. Proc. Natl. Acad. Sci. 99(15):10162–10166. (Ch. 20)

Muggeridge, J. 1933. The white butterfly (*Pieris rapae*) and its parasites. N. Z. J. Agric. 47:135–142. (Ch. 6)

Murphy, S.M., and P.P. Feeny. 2006. Chemical facilitation of a naturally occurring host shift by *Papilio machaon* butterflies (Papilionidae). Ecol. Monogr. 76:399–414. (Ch. 4)

Myers, J.H. 1987. Population outbreaks of introduced insects: Lessons from the biological control of weeds. *In* P. Barbosa and J. Schultz, eds., Insect outbreaks, pp. 173–193. New York: Academic. (Ch. 22)

Nagano, C.D., and C. Freese. 1987. A world safe for monarchs. J. New Sci. 1554:43–47. (Ch. 12)

Nagano, C.D., and J. Lane. 1985. A survey of the location of monarch butterfly (*Danaus plexippus* [L.]) overwintering roosts in the state of California, U.S.A.: First year 1984/1985. Report to the World Wildlife Fund-US. 71 pp. (Ch. 12)

Nagano, C.D., W.H. Sakai, S.B. Malcolm, B.J. Cockrell, J.P. Donahue, and L.P. Brower. 1993. Spring migration of monarch butterflies in California. *In* S.B. Malcolm and M.P. Zalucki, eds., Biology and conservation of the monarch butterfly, pp. 219–232. Los Angeles: Natural History Museum of Los Angeles County. (Ch. 7, 16)

Nation, J.L. 2002. Insect physiology and biochemistry. New York: CRC Press. (Ch. 10)

National Plant Board. 2013. http://nationalplantboard.org/laws/index.html. Accessed March 2013. (Ch. 3)

National Weather Service. 2011. Advanced hydrological prediction service: Medina River at Bandera. NOAA National Weather Service. http://water.weather.gov/ahps2/hydrograph.php?wfo=ewx&gage=bdat2. Accessed 8 November 2011. (Ch. 10)

Navarrete, J.L., M.I. Ramírez, and D.R. Pérez-Salicrup. 2011. Logging within protected areas: Spatial evaluation of the monarch butterfly biosphere reserve, Mexico. For. Ecol. Manage. 262:646–654. (Ch. 9, 13, 15)

Navarro, F.R., F. Cuezzo, P.A. Goloboff, C. Szumik, M. Lizarralde de Grosso, and M.G. Quintana. 2009. Can insect data be used to infer areas of endemism? An example from the Yungas of Argentina. Rev. Chil. Hist. Nat. 82:507–522. (Ch. 20)

Nelson, C.J. 1993. Sequestration and storage of cardenolides and cardenolide glycosides by *Danaus plexippus* (L.) and *D. chrysippus* petilia (Stoll) when reared on *Asclepias fruticosa* (L.) with a review of some factors that influence sequestration. *In* S.B. Malcolm and M.P. Zalucki, eds., Biology and conservation of the monarch butterfly, pp. 91–105. Los Angeles: Natural History Museum of Los Angeles County. (Ch. 4)

Nelson, C.J., J.N. Seiber, and L.P. Brower. 1981. Seasonal and intraplant variation of cardenolide content in the California milkweed *Asclepias eriocarpa*, and implications for plant defense. J. Chem. Ecol. 7:981–1010. (Ch. 4)

Neven, L.G. 2000. Physiological responses of insects to heat. Postharvest Biol. Tec. 21:103–111. (Ch. 8, Part 3)

Neves, V.C., J.C. Fraga, H. Schafer, V. Vieira, A. Bivar de Sousa, and P.V. Borges. 2001. The occurence of the monarch butterfly *Danaus plexippus* L. in the Azores, with a brief review of its biology. Arquipélago (Life Mar. Sci.) 18A:17–24. (Ch. 22, 23)

Nielsen-Gammon, J.W. 2011. The 2011 Texas drought. A briefing packet for the Texas Legislature, October 31, 2011. College Station: Texas A&M University, Office of the State Climatologist. http://climatexas.tamu.edu/. Accessed 10 October 2012. (Ch. 10)

Niitepõld, K., A.D. Smith, J.L. Osborne, D.R. Reynolds, N.L. Carreck, A.P. Martin, J.H. Marden, O. Ovaskainen, and I. Hanski. 2009. Flight metabolic rate and *Pgi* genotype influence butterfly dispersal rate in the field. Ecology 90:2223–2232. (Ch. 23)

NOAA National Climatic Data Center. 2012. US National Oceanic and Atmospheric Administration, US Department of Agriculture, National Drought Mitigation Center, Servicio Meteorológico Nacional, Mexico, and Agriculture and Agri-Food Canada. North American Drought Monitor [map for 31 October 2011]. North American Drought Monitor, National Climatic Data Center. http://www.ncdc.noaa.gov/temp-and-precip/drought/nadm/. Accessed 21 August 2012. (Ch. 10)

NOAA National Climatic Data Center. 2013. Temperature and precipitation time series, Texas, 1985–2012. http://www.ncdc.noaa.gov/temp-and-precip/time-series/. Accessed 21 January 2013. (Ch. 10)

NOM-015-SEMARNAT/SAGARPA-2007. Que establece las especificaciones técnicas de los métodos del uso del fuego en los terrenos forestales y en los terrenos de uso agropecuario. (Ch. 15)

Norris, K., and M.R. Evans. 2000. Ecological immunology: Life history trade-offs and immune defense in birds. Behav. Ecol. 11:19–26. (Ch. 7)

North American Butterfly Association. 2012. www.naba.org. Accessed November 2012. (Ch. 2)

North American Pollinator Protection Campaign. 2013. http://pollinator.org/nappc/. Accessed September 2013. (Ch. 21)

Núñez, J.C.S., and L.V. García. 1993. Vegetation types of monarch butterfly overwintering habitat in Mexico. *In* S.B. Malcolm and M.P. Zalucki, eds., Biology and conservation of the monarch butterfly, pp. 287–293. Los Angeles: Natural History Museum of Los Angeles County. (Ch. 9)

Oatman, E.R. 1966. An ecological study of cabbage looper and imported cabbageworm populations on cruciferous crops in southern California. J. Econ. Entomol. 59:1134–1139. (Ch. 6)

Oberhauser, K.S. 1989. Effects of spermatophores on male and female monarch butterfly reproductive success. Behav. Ecol. Sociobiol. 25:237–246. (Part 3)

Oberhauser, K.S. 1997. Fecundity, lifespan and egg mass in butterflies: Effects of male-derived nutrients and females size. Func. Ecol. 11:166–175. (Ch. 8)

Oberhauser, K.S. 2007. Programma norteamericano de monitero de la mariposa Monarca. *In* I. Pisanty and M. Caso, eds. Especies, espacios y riegos: Monitoreo para la conservación de la biodiversidad, pp. 33–58. Instituto Nacional de Ecoligia (INESEMARNAT). Mexico City. (Ch. 24)

Oberhauser, K.S. 2012. Tachinid flies and monarch butterflies: Citizen scientists document parasitism patterns over broad spatial and temporal scales. Am. Entomol. 58:19–22. (Ch. 2, 5, 6, 17, Part 2)

Oberhauser, K., and D. Frey. 1999. Coerced mating in monarch butterflies. *In* J. Hoth, L. Merino, K. Oberhauser, I. Pisenty, S. Price, and T. Wilkinson, eds., 1997 North American conference on the monarch butterfly: Paper presentations, pp. 67–78. Montreal: Commission for Environmental Cooperation. (Ch. 19)

Oberhauser, K.S., and G. Lebuhn. 2012. Insects and plants: Engaging undergraduates in authentic research via citizen science. Front. Ecol. Environ. 10:318–320. (Ch. 2)

Oberhauser, K.S., and A.T. Peterson. 2003. Modeling current and future potential wintering distributions of eastern North American monarch butterflies. Proc. Natl. Acad. Sci. 100:14,063–14,068. (Ch. 3, 7, 8, 9, 11, 13, 19, Parts 3, 4)

Oberhauser, K.S., and M.D. Prysby. 2008. Citizen science: creating a research army for conservation. Amer. Entomol. 54:97–99. (Ch. 2)

Oberhauser, K.S., and M. Solensky. 2004. The monarch butterfly: Biology and conservation. Ithaca, NY: Cornell University Press. (Preface)

Oberhauser, K.S., M.D. Prysby, H.R. Mattila, D.E. Stanley-Horn, M.K. Sears, G. Dively, E. Olson, J.M. Pleasants, W.F. Lam, and R.L. Hellmich. 2001. Temporal and spatial overlap between monarch larvae and corn pollen. Proc. Natl. Acad. Sci. 98:11913–11918. (Ch. 2, 5, 6, 8, 14, 17, Part 4)

Oberhauser, K.S., I. Gebhard, C. Cameron, and S. Oberhauser. 2007. Parasitism of monarch butterflies (*Danaus plexippus*) by *Lespesia archippivora* (Diptera: Tachinidae). Am. Midl. Nat. 157:312–328. (Ch. 2, 5, 6, 17)

O'Corry-Crowe, G.M., R.S. Suydam, A. Rosenberg, K.J. Frost, and A.E. Dizon. 1997. Phylogeography, population structure and dispersal patterns of the beluga whale *Delphinapterus leucas* in the western Nearctic revealed by mitochondrial DNA. Mol. Ecol. 6:955–970. (Ch. 23)

Ode, P.J. 2006. Plant chemistry and natural enemy fitness: Effects on herbivore and natural enemy interactions. Annu. Rev. Entomol. 51:163–185. (Ch. 6)

Office of Texas State Climatologist. 2011. Texas A&M University. http://atmo.tamu.edu/osc/. Accessed November 2011. (Ch. 19)

ORBIMAGE Inc. 2010. GeoEye Imagery 03/02/2010, po_403156, Monarch 2010. Thornton, CO. (Ch. 13)

Orozco, M.E., A. Guerrero P., E. Cadena V., D. Velázquez T., and J. Colín J. 2008. Supervivencia campesina y conservación de la naturaleza: Santuario del Cerro Pelón (Reserva de la Biosfera Mariposa Monarca), El Capulín, México. Cuad. Desarr. Rural 5(61):131–168. (Ch. 15)

Oschwald, W.R., F.F. Riecken, R.I Dideriksen, W.H. Scholtes, and F.W. Schaller. 1965. Principal soils of Iowa. Report no. 42. Iowa State Univ. Coop. Extension Service. (Ch. 14)

Owen, D.F., and D.A.S. Smith. 1993. The origin and history of the butterfly fauna of the Atlantic Islands. Bol. Mus. Munic. Funchal. Supplement 2:211–241. (Ch. 22)

Oyeyele, S.O., and M.P. Zalucki. 1990. Cardiac glycosides and oviposition by Danaus plexippus on Asclepias fruticosa in south-east Queensland (Australia), with notes on the effect of plant nitrogen content. Ecol. Entomol. 15:177–185. (Ch. 14)

Pagel, M. 1999. Inferring the historical patterns of biological evolution. Nature 401:877–884. (Ch. 4)

Palma, L., and A. Bivar de Sousa. 2003. Colonias reproductoras de *Danaus plexippus* (L.) (Lepidoptera: Nymphalidae, Danainae) em Portugal continental. Bol. Soc. Port. Entomol. 209 (VII-27):329–340. (Ch. 22)

Pamperis, L.N. 1997. The butterflies of Greece. Athens: Bastas-Plessas Graphic Arts. 559 pp. (Ch. 22)

Paradis, E. 2005. Statistical analysis of diversification with species traits. Evolution 59:1–12. (Ch. 4)

Park, A. 2003. Spatial segregation of pines and oaks under different fire regimes in the Sierra Madre Occidental. 169(1):1–20. (Ch. 15)

Parker, B.J., S.M. Barribeau, A.M. Laughton, J.C. de Roode, and N.M. Gerardo. 2011. Non-immunological defense in an evolutionary framework. Trends Ecol. Evol. 26:242–248. (Ch. 7)

Parmesan, C., and G. Yohe. 2003. A globally coherent fingerprint of climate change impacts across natural systems. Nature 421:37–42. (Ch. 8)

Parmesan, C., N. Ryrholm, C. Stefanescu, J.K. Hill, C.D. Thomas, H. Descimon, B. Huntley, L. Kaila, J. Kullberg, T. Tammaru, W.J. Tennent, J.A. Thomas, and M. Warren. 1999. Poleward shifts in geographical ranges of butterfly species associated with regional warming. Nature 399:579–583. (Ch. 8, 11)

Pateman, R.M., J.K. Hill, D.B. Roy, R. Fox, and C.D. Thomas. 2012. Temperature-dependent alterations in host use drive rapid range expansion in a butterfly. Science 336:1028–1030. (Ch. 18)

Pausas, J.G., and J.E. Keeley. 2009. A burning story: The role of fire in the history of life. BioScience 59(7):593–601. (Ch. 15)

Paz, D., O. Ceballos, N. Gallego, and G. Calvo. 2010. Nueva población de *Danaus plexippus* (Linnaeus, 1758) (Lepidoptera: Nymphalidae) en Doñana (Andalucía, España). Bol. Soc. Entomol. Argent. 47:372. (Ch. 22)

Pechan, T., A. Cohen, W.P. Williams, and D.S. Luthe. 2002. Insect feeding mobilizes a unique plant defense protease that disrupts the peritrophic matrix of caterpillars. Proc. Natl. Acad. Sci. 99:13319–13323. (Ch. 4)

Pérez De-Gregorio, J.J., and M. Rondós Casas. 2005. La *Danaus plexippus* (Linnaeus, 1758) en el delta del Ebro, Cataluña (península Ibérica) (Lepidoptera, Danaidae). Bol. Soc. Entomol. Argent. 36:308. (Ch. 22)

Pérez, S.M., and O.R. Taylor. 2004. Monarch butterflies' migratory behavior persists despite changes in environmental conditions. In K.S. Oberhauser and M.J. Solensky, eds., The monarch butterfly: Biology and conservation, pp. 85–88. Ithaca, NY: Cornell University Press. (Ch. 19)

Perez, S.M., O.R. Taylor, and R. Jander. 1997. A sun compass in monarch butterflies. Nature 387:29. (Ch. 20)

Pergams, O., and P. Zaradic. 2007. Evidence for a fundamental and pervasive shift away from nature-based recreation. Proc. Natl. Acad. Sci. 105:2295–2300. (Ch. 1)

Perkovic, D. 2006. *Danaus chrysippus* (Linnaeus, 1758) (Lepidoptera, Danainae) a new species in the fauna of Croatia. Nat. Croat. 15(1–2):61–64. (Ch. 22)

Peterson, A.T. 2003. Predicting the geography of species invasion via ecological niche modeling. Q. Rev. Biol. 78:419–433. (Ch. 11)

Petit, E., and F. Mayer. 2000. A population genetic analysis of migration: The case of the noctule bat (*Nyctalus noctula*). Mol. Ecol. 9:683–690. (Ch. 23)

Phillips, S.J., R.P. Anderson, and R.E. Schapire. 2006. Maximum entropy modeling of species geographic distributions. Ecol. Model. 190:231–259. (Ch. 11)

Pierce, A., M. Zalucki, M. Banguara, M. Udawatta, M. Kronforst, S. Altizer, J. Fernandez-Haeger, and J. de Roode. 2014. Serial founder effects and genetic differentiation during worldwide range expansion of monarch butterflies. Proceedings of the Royal Society Series B. 281: 20142230. (Part 5)

Pinto-Tomás, A.A., A. Sittenfeld, L. Uribe-Lorío, F. Chavarría, M. Mora, D.H. Janzen, R.M. Goodman, and H. Simon. 2011. Comparison of midgut bacterial diversity in tropical caterpillars (Lepidoptera: Saturniidae) fed on different diets. Environ. Entomol. 40:1111–1122. (Ch. 7)

Platt, W.J. 1975. The colonization and formation of equilibrium plant species associations on badger disturbances in a tall-grass prairie. Ecol. Monogr. 45:285–305. (Ch. 14)

Pleasants, J.M., R.L. Hellmich, G.P. Dively, M.K. Sears, D.E. Stanley-Horn, H.R. Mattila, J.E. Foster, P. Clark, and G.D. Jones. 2001. Corn pollen deposition on milkweeds in and near cornfields. Proc. Natl. Acad. Sci. 98:11,919–11,924. (Ch. 14)

Pleasants, J.M., and K.S. Oberhauser. 2012. Milkweed loss in agricultural fields because of herbicide use: Effect on the monarch butterfly population. Insect Conserv. Divers. doi:10.111/j.1752–4598.2012.00196.x. (Ch. 2, 3, 7, 9, 11, 12, 14, 16, 17, 18, 19, 24, Parts 4, 5)

Pline, W.A., Hatzios K.K, and E.S. Hagood. 2000. Weed and herbicide-resistant soybean (*Glycine max*) response to glufosinate and glyphosate plus ammonium sulfate and pelargonic acid. Weed Tech. 14:667–674. (Ch. 14)

Pollard E. 1977. Method for assessing changes in abundance of butterflies. Biolog. Conserv. 12:115-134. (Ch. 2)

Pollinator Partnership. 2012. Annual Report to the Monarch Joint Venture.

Polunin, O., and A. Huxley. 1972. Flowers of the Mediterranean. Chatto and Windus Ltd, London. 260 pp. (Ch. 22)

Portland Tribune, 2006. Planting milkweed may aid dwindling monarch population. 17 July 2006. http://portlandtribune.com/component/content/article?id=99901 (Ch. 21)

Povey, S., S.C. Cotter, S.J. Simpson, K.P. Lee, and K. Wilson. 2009. Can the protein costs of bacterial resistance be offset by altered feeding behaviour? J. Anim. Ecol. 78:437–446. (Ch. 7)

Price, P.W., C.E. Bouton, P. Gross, B.A. McPheron, J.N. Thompson, and A.E. Weis. 1980. Interactions among three trophic levels: Influence of plants on interactions between insect herbivores and natural enemies. Annu. Rev. Ecol. Syst. 11:41–65. (Ch. 5)

Pritchard, J.K., M. Stephens, and P. Donnelly. 2000. Inference of population structure using multilocus genotype data. Genetics. 155:945–959. (Ch. 23)

Prudic, K.L., J.C. Oliver, and M.D. Bowers. 2005. Soil nutrient effects on oviposition preference, larval performance, and chemical defense of a specialist insect herbivore. Oecologia 143:578–587. (Ch. 14)

Prysby, M.D. 2004. Natural enemies and survival of monarch eggs and larvae. *In* K.S. Oberhauser and M.J. Solensky, eds., The monarch butterfly: Biology and conservation, pp. 27–37. Ithaca, NY: Cornell University Press. (Ch. 2, 5, 6, 17, Part 2)

Prysby, M.D., and K.S. Oberhauser. 1999. Large-scale monitoring of larval monarch populations and milkweed habitat in North America. *In* J. Hoth, L. Merino, K. Oberhauser, I. Pisantry, S. Price, and T. Wilkinson, eds., 1997 North American conference on the monarch butterfly: Paper presentations, pp. 379–384. Montreal: Commission for Environmental Cooperation. (Ch. 2)

Prysby, M.D., and K.S. Oberhauser. 2004. Temporal and geographic variation in monarch densities: Citizen scientists document monarch population patterns. *In* K.S. Oberhauser and M.J. Solensky, eds., The monarch butterfly: Biology and conservation, pp. 9–20. Ithaca, NY: Cornell University Press. (Ch. 2, 17, 19, 24)

Puerto, F.J. 2002. Jardines de aclimatación en la España de la ilustración. Ciencias. 68:30–68. (Ch. 22)

Pyle, R.M. 1974. Watching Washington butterflies. Seattle: Seattle Audubon Society. (Ch. 21)

Pyle, R.M. 1997. The historic flight of monarch # 09727. Monarch News 8(3):1, 3–4. (Ch. 21)

Pyle, R.M. 1998. The biogeography of hope: Why transporting butterflies is a bad idea. Monarch News 8(6):6–7. (Ch. 21)

Pyle, R.M. 1999. Chasing monarchs: Migrating with the butterflies of passage. Houghton Mifflin Company, New York. (Ch. 2, 12, 21, Part 5)

Pyle, R.M. 2000. The butterflies of Cascadia. Seattle: Seattle Audubon Society. (Ch. 21)

Pyle, R.M. 2001. A link between western and Mexican monarchs. Sunset (January):8. (Ch. 2, 21)

Pyle, R.M. 2010. Under their own steam: The biogeographical case against butterfly releases. News Lepid. Soc. 52(2):54–57. (Ch. 21)

Pyle, R.M., and M. Monroe. 2004. Conservation of western monarchs. Wings: Essays Invertebr. Conserv. 27:13–17. (Ch. 12)

Pyle, R.M., S.J. Jepsen, S. Hoffman Black, and M. Monroe. 2010. Xerces society policy on butterfly releases. http://www.xerces.org/wp-content/uploads/2010/08/xerces-butterfly-release-policy.pdf. Accessed November 2012. (Ch. 1)

Pyne, S.J. 1996. World fire. The culture of fire on Earth. Seattle: University of Washington Press. 384 pp. (Ch. 15)

Pyne, S.J. 2001. Fire: A brief history. Seattle: University of Washington Press. (Ch. 17)

Pyne, S.J. 2010. America's fire: A historical context for policy and practice. The Forest History Society. 93 pp. (Ch. 15)

Quer, J. 1762. Flora Española, 3:74–77. Madrid: Imprenta Joachim Ibarra. (Ch. 22)

Quinn, M. 2011. Fall monarch migration through Texas (Fall 2000–2004). Texas Monarch Watch, 14 September 2011. http://www.texasento.net/fall_peak.htm. Accessed 9 August 2012. (Ch. 10)

R Core Team. 2012. R: A language and environment for statistical computing. R Foundation for Statistical Computing, Vienna, Austria. ISBN 3-900051-07-0, URL http://www.R-project.org/. (Ch. 24)

R Development Core Team. 2004. R: A language and environment for statistical computing. R Foundation for Statistical Computing. Vienna. (Ch. 13)

Råberg, L., D. Sim, and A.F. Read. 2007. Disentangling genetic variation for resistance and tolerance to infectious disease in animals. Science 318:318–320. (Ch. 7)

Rafter, J., A. Agrawal, and E. Preisser. 2013. Chinese mantids gut toxic monarch caterpillars: Avoidance of prey defense? Ecol. Entomol. 38:76–82. (Ch. 6)

Ramírez, M.I. 2001. Los espacios forestales de la Sierra de Angangueo (estados de Michoacán y México), México. Madrid: Una visión geográfica, Servicio de Publicaciones, Universidad Complutense de Madrid. 329 pp. http://biblioteca.ucm.es/tesis/ghi/ucm-t25398.pdf (Ch. 13)

Ramírez, M.I., and R. Zubieta. 2005. Análisis regional y comparación metodológica del cambio en la cubierta forestal en la Región Mariposa Monarca, Informe Técnico, WWF-Programa México, Convenio KE31, México. http://www.wwf.org.mx/wwfmex/descargas/mmonarca_analisis_cambio_forestal.pdf (Ch. 13)

Ramírez, M.I., J.G. Azcárate, and L. Luna. 2003. Effects of human activities on monarch butterfly habitat in protected mountain forests, Mexico. Forest. Chron. 79:242–246. (Ch. 13, 15)

Ramírez, M.I., R. Miranda, R. Zubieta, and M. Jiménez. 2007. Land cover and road network for the Monarch Butterfly Biosphere Reserve in Mexico, 2003. J. Maps 2007:181–190. (Ch. 13)

Ramsay, G.W. 1964. Overwintering swarms of the monarch butterfly (*Danaus plexippus*) (L.) in New Zealand. N. Z. Entomol. 3:10–16. (Ch. 6)

Rasmann, S., and A.A. Agrawal. 2011. Latitudinal patterns in plant defense: Evolution of cardenolides, their toxicity and induction following herbivory. Ecol. Lett. 14:476–483. (Ch. 4)

Rasmann, S., A.A. Agrawal, S.C. Cook, and A.C. Erwin. 2009. Cardenolides, induced responses, and interactions between above- and belowground herbivores of milkweed (*Asclepias* spp.). Ecology 90:2393–2404. (Ch. 4)

Ravert Richter, M. 2000. Social wasp (Hymenoptera: Vespidae) foraging behavior. Annu. Rev. Entomol. 45:121–150. (Ch. 6)

Rayor, L.S. 2004. Effects of monarch larval host plant chemistry and body size on *Polistes* wasp predation. *In* K.S. Oberhauser and M.J. Solensky, eds., The monarch butterfly: Biology and conservation, pp. 39–46. Ithaca, NY: Cornell University Press. (Ch. 6)

Rea, B., K. Oberhauser, and M. Quinn. 2011. Milkweed, monarchs and more. A field guide to the invertebrate community in the milkweed patch. Union, WV: Bas Relief Publishing Group. 112 pp. (Ch. 2)

Rebel, H., and A.F. Rogenhofer. 1894. Zur Lepidopterenfauna der Canaren. Ann. Naturhist. Mus. Wien 9:1–96. (Ch. 22)

Rehfeldt, G.E. 2006. A spline climate model for western United States. A spline climate model for western United States. General Technical Report 165. Fort Collins, CO: U.S. Department of Agriculture, Forest Service, Rocky Mountain Research Station. (Ch. 13)

Rehfeldt, G.E., N.L. Crookston, M.V. Warwell, and J.S. Evans. 2006. Empirical analyses of plant-climate relationships for the western United States. Int. J. Plant Sci. 167:1123–1150. (Ch. 13)

Reichert, S.E., and T. Lockley. 1984. Spiders as biological control agents. Annu. Rev. Entomol. 29:299–320. (Ch. 5)

Reichstein, T., J. von Euw, J.A. Parsons, and M. Rothschild. 1968. Heart poisons in the monarch butterfly. Science 161:861–866. (Ch. 5, 6)

Reilly, J.R., and A.E. Hajek. 2008. Density-dependent resistance of the gypsy moth *Lymantria dispar* to its nucleopolyhedrovirus, and the consequences for population dynamics. Oecologia 154:691–701. (Ch. 7)

Rendón-Salinas, E., and G. Tavera-Alonso. 2013. Monitoreo de la superficie forestal ocupada por las colonias de hibernación de la mariposa monarca en Diciembre de 2012. Alianza WWF-Telcel/CONANP. 6pp. www.wwf.org.mx/wwfmex/descargas/rep-monitoreo-colonias-Mariposa-Monarca-2012-2013.pdf. Accessed March 2013. (Ch. 1, 2, 11, 14, 24, Parts 3, 4)

Rendón-Salinas, E., G. Ramírez-Galindo, J. Pérez-Ojeda, and C. Galindo-Leal, eds. 2007. Cuarto Foro Regional Mariposa Monarca Memorias. (Ch. 15)

Rendón-Salinas, E., C.A. Valera-Bermejo, M. Cruz-Pina, and F. Martínez-Meza. 2011. Monitoreo de las Colonias de Hibernación de Mariposa Monarca: Superficie Forestal de Ocupación en Diciembre de 2011. Unpublished report. 8 pp. (Ch. 14)

Reppert, S.M., H.S. Zhu, and R.H. White. 2004. Polarized light helps monarch butterflies navigate. Curr. Biol. 14:155–158. (Ch. 23)

Reppert, S.M., R.J. Gegear, and C. Merlin. 2010. Navigational mechanisms of migrating monarch butterflies. Trends Neurosci. 33:399–406. (Ch. 20, Part 5)

Restif, O., and J.C. Koella. 2003. Shared control of epidemiological traits in a coevolutionary model of host-parasite interactions. Am. Nat. 161:827–836. (Ch. 7)

Riddell, C., S. Adams, P. Schmid-Hempel, and E.B. Mallon. 2009. Differential expression of immune defences is associated with specific host-parasite interactions in insects. PLoS ONE 4:e7621. (Ch. 7)

Rincevich, B., and W.E.G. Muller, eds. 1996. Invertebrate immunology. Berlin: Springer. (Ch. 7)

Robb, T., and M. Forbes. 2005. On understanding seasonal increases in damselfly defence and resistance against ectoparasitic mites. Ecol. Entomol. 30:334–341. (Ch. 7)

Robertson, C. 1928. Flowers and insects: Lists of visitors of four hundred and fifty-three flowers. Lancaster, PA: Science Press Printing Co. (Ch. 10)

Robertson, M.P., C.I. Peter, M.H. Villet, and B.S. Ripley. 2003. Comparing models for predicting species' potential distributions: A case study using correlative and mechanistic predictive modelling techniques. Ecol. Model. 164:153–167. (Ch. 11)

Rodriguez, R.R. 2009. Talan 10 mil árboles en la Reserva de la Monarca, para combatir plaga. Cambio de Michoacan, Morelia. 21 October 2009. www.cambiodemichoacan.com.mx/vernota.php?id = 111258. Accessed 30 March 2013. (Part 4)

Rodríguez-Trejo, D.A., and P. Fulé. 2003. Fire ecology of Mexican pines and a fire management proposal. Int. J. Wildland Fire 12(1):23–37. (Ch. 15)

Rodríguez-Trejo, D.A., and R.L. Myers. 2010. Using Oak characteristics to guide fire regime restoration in Mexican pine-oak and oak forests. Ecol. Restor. 28:304–323. (Ch. 15)

Rodríguez-Trejo, D.A., P.A. Martínez-Hernández, H. Ortiz-Contla, M.R. Chavarría-Sánchez and F. Hernández-Santiago. 2011. The Present status of fire ecology, traditional use of fire, and fire management in Mexico and Central America. Fire Ecol. 7(1):40–56. (Ch. 15)

Rogg, K.A., O.R. Taylor, and D.L.Gibo. 1999. Mark and recapture during the monarch migration: A preliminary analysis. In J. Hoth, L. Merino, K. Oberhauser, I. Pisantry, S. Price, and T. Wilkinson, eds., 1997 North American conference on the monarch butterfly: Paper presentations, pp. 133–138. Montreal: Commission for Environmental Cooperation. (Ch. 2)

Rolff, J., and M.T. Siva-Jothy. 2003. Invertebrate ecological immunology. Science 301:472–475.

Rolff, J., and S. Reynolds. 2009. Insect infection and immunity: Evolution, ecology, and mechanisms. Oxford: Oxford University Press. (Ch. 7)

Romo, J.L. 1999. Valuación económica de la migración de las mariposas monarca. In: Economía de la Biodiversidad Memoria del Seminario Internacional de La Paz, BCS. (Ch. 15)

Rousset, F. 2008. GENEPOP '007: A complete reimplementation of the GENEPOP software for Windows and Linux. Mol. Ecol. Resour. 8:103–106. (Ch. 23)

Rowell, A., and P.F. Moore. 1999. Global review of forest fires. Gland, Switzerland: WWF/UICN. 64 pp. (Ch. 15)

Roy, B.A., and J.W. Kirchner. 2000. Evolutionary dynamics of pathogen resistance and tolerance. Evolution 54:51–63. (Ch. 7)

Rudgers, J.A., S.Y. Strauss, and J.F. Wendel. 2004. Trade-offs among anti-herbivore resistance traits: Insights from Gossypieae (Malvaceae). Am. J. Bot. 91:871–880. (Ch. 4)

Russell, A.L., R.A. Medellín, and G.F. McCracken. 2005. Genetic variation and migration in the Mexican free-tailed bat (*Tadarida brasiliensis mexicana*). Mol. Ecol. 14:2207–2222. (Ch. 23)

Sadd, B.M., and P. Schmid-Hempel. 2008. Principles of ecological immunology. Evol. Appl. 2:113–121. (Ch. 7)

Sáenz, J.T., F.J. Villaseñor, J. Jiménez, and M. Gallardo. 2005. Agroforestería para reconversión de suelos con vocación forestal en el oriente de Michoacán. Instituto Nacional de Investigaciones Forestales, Agrícolas y Pecuarias (INIFAP) Centro Regional del Pacífico Centro Campo Experimental Uruapan. Folleto Técnico No. 1. (Ch. 15)

Sáenz-Romero, C., G.E. Rehfeldt, N.L. Crookston, D. Pierre, R. St-Amant, J. Beaulieu, and B. Richardson. 2010. Spline models of contemporary, 2030, 2060 and 2090 climates for México and their use in understanding climate-change impacts on the vegetation. Clim. Chang. 102:595–623. (Ch. 13)

Sáenz-Romero, C., G.E. Rehfeldt, P. Duval, and R.A. Lindig-Cisneros. 2012. *Abies religiosa* habitat prediction in climatic change scenarios and implications for monarch butterfly conservation in Mexico. Forest Ecol. Manage. 275:98–106. (Ch. 9, 11, 13, Parts 3, 4)

Sakai W.H. 1994. Avian predation on the monarch butterfly, *Danaus plexippus* (Nymphalidae: Danainae), at a California overwintering site. J. Lepid. Soc. 48:148–156. (Ch. 6)

Sakai, W.H., and W.C. Calvert. 1991. Statewide monarch butterfly management plan for the state of California

Department of Parks and Recreation. Final Report, June 1991. Interagency Agreement No. 88-11-050 between California Department of Parks and Recreation and Santa Monica College. 160 pp. (Ch. 12)

Sakano, D., B. Lin, Q. Xia, K. Yamamoto, H. Fujii, and Y. Aso. 2006. Genes encoding small heat shock proteins of the silkworm, *Bombyx mori*. Biosci. Biotech. Bioch. 79:2443–2450. (Ch. 8)

Salazar, F., and O.R. Rodriguez. 2011. Mexico drought is worst in 70 years. San Francisco Chronicle. 3 December 2011. http://www.sfgate.com/world/article/Mexico-drought-is-worst-in-70-years-2345008.php. Accessed 22 Sept 2012. (Ch. 10)

Sapolsky, R.M. 1994. Fallible instinct—a dose of skepticism about the medicinal knowledge of animals. The Sciences 34:13–15. (Ch. 7)

Satterfield, D.A., J.C. Maerz, and S. Altizer. 2014. Loss of migratory behaviour increases infection risk for a butterfly host. Proc. R. Soc. Lond. B. In press. (Ch. 7)

Schappert, P. 1996. Status, distribution and conservation of the monarch butterfly, *Danaus plexippus*, in Canada. Montreal: Commission for Environmental Cooperation. 19 pp. (monarch.philschappert.com/report.html.) (Ch. 3)

Schappert, P. 2004. The last monarch butterfly: Conserving the monarch butterfly in a brave New World. Buffalo, NY: Firefly Books. (Ch. 3)

Schmid-Hempel, P. 2005. Evolutionary ecology of insect immune defenses. Annu. Rev. Entomol. 50:529–551. (Ch. 7)

Schmidt-Koenig, K. 1993. Orientation of autumn migration in the monarch butterfly. *In* S.B. Malcolm and M.P. Zalucki, eds., Biology and conservation of the monarch butterfly, pp. 275–283. Los Angeles: Natural History Museum of Los Angeles County. (Ch. 20)

Schulenburg, H., J. Kurtz, Y. Moret, and M.T. Siva-Jothy. 2009. Introduction. Ecological immunology. Phil. Trans. R. Soc. Biol. 364:3–14. (Ch. 7)

Sears, M.K., R.L. Hellmich, D.E. Stanley-Horn, K.S. Oberhauser, J.M. Pleasants, H.R. Mattila, B.D. Siegfried, and G.P. Dively. 2001. Impact of *Bt* corn pollen on Monarch butterfly populations: A risk assessment. Proc. Natl. Acad. Sci. 98:11937–11942. (Ch. 11, 14)

Seattle Times. 2012. Wash. state inmates raise rare frogs, butterflies. http://seattletimes.com/html/localnews/2019231300_apwaprisonconservationists.html. Accessed January 2013. (Ch. 21)

Seiber, J.N., P.M. Tuskes, L.P. Brower, and C.J. Nelson. 1980. Pharmacodynamics of some individual milkweed cardenolides fed to larvae of the monarch butterfly (*Danaus plexippus* L). J. Chem. Ecol. 6:321–339. (Ch. 4)

Seiber, J.N., S.M. Lee, and J.M. Benson. 1983. Cardiac glycosides (cardenolides) in species of *Asclepias* (Asclepiadaceae). *In* R.F. Keeler and A.T. Tu, eds., Handbook of natural toxins, Vol. I: Plant and fungal toxins, pp. 43–83. Amsterdam: Marcel Dekker. (Ch. 4)

Seiber, J.N., L.P. Brower, S.M. Lee, M.M. McChesney, H.T.A. Cheung, C.J. Nelson, and T.R. Watson. 1986. Cardenolide connection between overwintering monarch butterflies from Mexico and their larval food plant, *Asclepias syriaca*. J. Chem. Ecol. 12:1157–1170. (Ch. 14)

SEMARNAP. 1998. Memoria de trabajo del taller para el programa regional de desarrollo sustentable Zona de la Monarca, Michoacán-Estado de México con instituciones. (Ch. 15)

SEMARNAT 2001. Programa de Manejo Reserva de la Biosfera Mariposa Monarca. Mexico City: Secretaríade Medio Ambiente y Recursos Naturales. (Ch. 3)

Serratore V.R., M.P. Zalucki, and P.A. Carter. 2012. Thermoregulation in moulting and feeding *Danaus plexippus* L. (Lepidoptera: Nymphalidae) caterpillars. Aust. J. Entomol. 52:8–13. (Ch. 8)

Settele, J., T. Shreeve, M. Konvicka, and H. Van Dyck. 2009. Ecology of butterflies in Europe. Cambridge: Cambridge University Press. 513 pp. (Ch. 24)

Seubert, M. 1844. Flora Azorica. Bonnae. 50 pp. (Ch. 22)

Shapiro, A.M. 1981. A recondite breeding site for the monarch (*Danaus plexippus*, Danaidae) in the montane Sierra Nevada. J. Res. Lepid. 20:50–57. (Ch. 22)

Sheley, R., J. Mangold, K. Goodwin, and J. Marks. 2008. Revegetation guidelines for the Great Basin: Considering invasive weeds. Agricultural Research Service Publication. ARS-168. (Ch. 16)

Shelford, V.E. 1963. The ecology of North America. Urbana: University of Illinois Press. (Ch. 11)

Shephard, J.M., J.M. Hughes, and M.P. Zalucki. 2002. Genetic differentiation between Australian and North American populations of the monarch butterfly *Danaus plexippus* (L.) (Lepidoptera: Nymphalidae): An exploration using allozyme electrophoresis. Biol. J. Linn. Soc. 75:437–452. (Ch. 22, 23)

Sigala P., and R. Campos. 2001. El entorno socioambiental de la mariposa monarca. Memoria de Experiencia Profesional. Mexico City: Universidad Autónoma Chapingo, Texcoco. (Ch. 15)

Silvers, C.S., J.G. Morse, and E.E. Grafton Cardwell. 2002. Quality assessment of *Chrysoperla rufilabris* (Neuroptera: Chrysopidae) producers in California. Fla. Entomol. 85:594–598. (Ch. 6)

Simmon, R., L.P. Brower, D.A. Slayback, and M.I. Ramírez. 2008. Deforestation in Monarch Butterfly Reserve, March 7, 2008. http://earthobservatory.nasa.gov/IOTD/view.php?id=8506. Last accessed 3 July 2012. (Ch. 9, Part 4)

Simmons, A.M., and C.E. Rogers. 1991. Dispersal and seasonal occurence of *Noctuidonema guyanense*, an ectoparasitic nematode of adult fall armyworm (Lepidoptera, Noctuidae), in the United States. J. Entomol. Sci. 26:136–148. (Ch. 7)

Simms, E.L. 2000. Defining tolerance as a norm of reaction. Evol. Ecol. 14:563–570. (Ch. 7)

Simpson, J.A., and E.S.C. Weiner, eds. 2008. The compact Oxford English dictionary, 2nd ed. Clarendon Press. (Ch. 21)

Simpson, W., and A. TenWolde. 1999. Physical properties and moisture relations of wood. In Wood handbook: Wood as an engineering material. General Technical Report 113, Forest Products Laboratory. Madison, WI: USDA Forest Service. (Ch. 9)

Sims-Chilton, N.M., M.P. Zalucki, and Y.M. Buckley. 2010. Long term climate effects are confounded with the biological control programme against the invasive weed *Baccharis halimifolia* in Australia. Biol. Invasions 12:3145–3155. (Ch. 11)

Slatkin, M. 1995. A measure of population subdivision based on microsatellite allele frequencies. Genetics 139:457–462. (Ch. 23)

Slayback, D.A., and L.P. Brower. 2007. Further aerial surveys confirm the extreme localization of overwintering monarch butterfly colonies in Mexico. Am. Entomol. 53:146–149. (Ch. 9, Part 4)

Slayback, D.A., L.P. Brower, M.I. Ramírez, and L.S. Fink. 2007. Establishing the presence and absence of overwintering colonies of the monarch butterfly in Mexico by the use of small aircraft. Am. Entomol. 53:28–40. (Ch. 9, 11, 13)

Slaymaker, O. 2001. Why so much concern about climate change and so little attention to land use change? Can. Geogr. 45(1):71–78. (Ch. 13)

Smilanich, A.M., L.A. Dyer, M.D. Bowers, and J.Q. Chambers. 2009a. Immunological costs to specialization and the evolution of insect diet breadth. Ecol. Lett. 12:612–621. (Ch. 6)

Smilanich, A.M., L.A. Dyer, and G.L. Gentry. 2009b. The insect immune response and other putative defenses as effective predictors of parasitism. Ecology 90:1434–1440. (Ch. 6)

Smithers, C.N. 1973. A note on natural enemies of *Danaus plexippus* (L.) (Lepidoptera: Nymphalidae) in Australia. Aust. Entomol. Mag. 1:37–40. (Ch. 6)

Smithers, C.N. 1977. Seasonal distribution and breeding status of *Danaus plexippus* (L.) (Lepidoptera: Nymphalidae) in Australia. J. Aust. Entomol. Soc. 16:175–184. (Ch. 2)

Snook, L.C. 1993. Conservation of the monarch butterfly reserves in Mexico: Focus on the forest. *In* S.B. Malcolm and M.P. Zalucki, eds., Biology and conservation of the monarch butterfly, pp. 363–375. Los Angeles: Natural History Museum of Los Angeles County. (Ch. 9)

Solensky, M.J., and E. Larkin. 2003. Temperature-induced variation in larval coloration in *Danaus plexippus* (Lepidoptera: Nymphalidae). Ann. Entomol. Soc. Am. 96:211–216.

Solensky, M.J., and K.S. Oberhauser. 2009a. Male monarch butterflies, *Danaus plexippus*, adjust ejaculates in response to intensity of sperm competition. Anim. Beh. 77:465–472. (Ch. 8)

Solensky, M.J., and K.S. Oberhauser. 2009b. Sperm precedence in monarch butterflies (*Danaus plexippus*). Behav. Ecol. 20:328–334. (Ch. 23)

Soltis, D.E., and R.K. Kuzoff. 1995. Discordance between nuclear and chloroplast phylogenies in the *Heuchera* group (Saxifragaceae). Evolution 49:727–742. (Ch. 4)

Southwest Monarch Study. 2013a. http://www.swmonarchs.org/. Accessed 8 October 2013. (Ch. 16, 23)

Southwest Monarch Study. 2013b. http://www.swmonarchs.org/az-recoveries_wild.php. Accessed January 2013. (Ch. 21)

Stamp, N.E., and M.D. Bowers. 2000. Foraging behaviour of caterpillars given a choice of plant genotypes in the presence of insect predators. Ecol. Entomol. 25:486–492. (Ch. 5)

Stanley-Horn, D.E., G.P. Dively, R.L. Hellmich, H.R. Mattila, M.K. Sears, R. Rose, L.C.H. Jesse, J.F. Losey, J.J. Obrycki, and L. Lewis. 2001. Assessing the impact of Cry1Ab-expressing corn pollen on monarch butterfly larvae in field studies. Proc. Natl. Acad. Sci. 98:11931–11936. (Ch. 14)

Stapel, J.O., J.R. Ruberson, H.R. Gross Jr., and W.J. Lewis. 1997. Progeny allocation by the parasitoid *Lespesia archippivora* (Diptera: Tachinidae) in larvae of *Spodoptera exigua* (Lepidoptera: Noctuidae). Environ. Entomol. 26:265–271. (Ch. 6)

Steiniger, H., and U. Eitschberger. 1989. Nymphalidae, Danaidae, Lybitheidae, Satyridae and Lycaenidae. Atalanta 20:27–37. (Ch. 22)

Stengle, J. 2011. Dead trees will mar Texas landscape for years. The Associated Press, Dallas. 14 October 2011. http://www.newsvine.com/_news/2011/10/14/8323797-dead-trees-will-mar-texas-landscape-for-years. Accessed 22 Sept 2012. (Ch. 10)

Stepek, G., D.J. Buttle, I.R. Duce, A. Lowe, and J.M. Behnke. 2005. Assessment of the anthelmintic effect of natural plant cysteine proteinases against the gastrointestinal nematode, *Heligmosomoides polygyrus*, in vitro. Parasitology 130:203–211. (Ch. 4)

Sternberg, E.D., T. Lefèvre, A.H. Rawstern, and J.C. de Roode. 2011. A virulaent parasite can provide protection against a lethal parasitoid. Infect. Genet. Evol. 11:399–406. (Ch. 6)

Sternberg, E.D., T. Lefèvre, J. Li, C.L. Fernandez de Castillejo, H. Li, M.D. Hunter, and J.C. de Roode. 2012. Food plant derived disease tolerance and resistance in a natural butterfly-plant-parasite interactions. Evolution 66:3367–76 (Ch. 4, 7, 14, Part 2)

Stevens, S., and D. Frey. 2004. How the other half lives: Monarch population trends west of the Great Divide. Biological Sciences Department, California Polytechnic State University. Unpublished report. 7 pp. (Ch. 2, 12)

Stevens, S.R, and D.F. Frey. 2010. Host plant pattern and variation in climate predict the location of natal grounds for migratory monarch butterflies in western North America. J. Insect Conserv. doi 10.1007/s10841–010–9303–5. (Ch. 2, 3, 12, 16, Part 5)

Stinson J., and M. Berman. 1990. Predator induced colour polymorphism in *Danaus plexippus* L. (Lepidoptera: Nymphalidae) in Hawaii. Heredity 65:401–406. (Ch. 6)

Stireman, J.O. III, and M.S. Singer. 2003a. What determines host range in parasitoids? An analysis of a tachinid parasitoid community. Oecologia 135:629–638. (Ch. 6)

Stireman, J.O. III, and M.S. Singer. 2003b. Determinants of parasitoid-host associations: Insights from a natural tachinid-lepidopteran community. Ecology 84:296–310. (Ch. 6)

Stowe, K.A., R.J. Marquis, C.G. Hochwender, and E.L. Simms. 2000. The evolutionary ecology of tolerance to consumer damage. Annu. Rev. Ecol. Syst. 31:565–595. (Ch. 7)

Strand, M.R. 2008. The insect cellular immune response. Insect Sci. 15:1–14. (Ch. 6)

Straub, S.C.K., M. Fishbein, T. Livshultz, Z. Foster, M. Parks, K. Weitemier, R.C. Cronn, and A. Liston. 2011. Building a model: Developing genomic resources for common milkweed (*Asclepias syriaca*) with low coverage genome sequencing. BMC Genomics 12:211. (Ch. 4)

Straub, S.C.K., M. Parks, K. Weitemier, M. Fishbein, R.C. Cronn, and A. Liston. 2012. Navigating the tip of the genomic iceberg: Next-generation sequencing for plant systematics. Am. J. Bot. 99:349–364. (Ch. 4)

Strauss, S.Y., and A.A. Agrawal. 1999. The ecology and evolution of plant tolerance to herbivory. Trends Ecol. Evol. 14:179–185. (Ch. 4)

Strecker, U., and H. Wilkens. 2000. *Danaus plexippus* (Linnaeus, 1758) new for Lanzarote (Canary Islands) (Lepidoptera, Nymphalidae). Atalanta 31:61–62. (Ch. 22)

Sturrock R.N., S.J. Frankel, A.V. Brown, P.E. Hennon, J.T. Kliejunas, K.J. Lewis, J.J.Worrall, and A.J. Woods. 2011. Climate change and forest diseases. Plant Pathol. 60:133–149. (Ch. 13)

Sugihara, N.G., J. Van Wagtendonk, and J. Fites-Kaufman. 2006. Fire as an ecological process. In Fire in California's ecosystems, pp. 58–74. London: University of California Press. (Ch. 15)

Sutherst, R.W. 2003. Prediction of species geographical ranges. J. Biogeogr. 30:805–816. (Ch. 11)

Sutherst, R.W., and G.F. Maywald. 2005. A climate model of the red imported fire ant, *Solenopsis invicta* Buren (Hymenoptera: Formicidae): Implications for invasion of new regions, particularly Oceania. Environ. Entomol. 34:317–335. (Ch. 11)

Sutherst, R.W., F. Constable, K.J. Finlay, R. Harrington, J. Luck, and M.P. Zalucki. 2011. Adapting to crop pest and pathogen risks under a changing climate. WIREs Clim. Change 2:220–237. (Ch. 11)

Sutherst, R.W., G.F. Maywald, and D.J. Kriticos. 2007. CLIMEX version 3: User's guide. Hearne Scientific Software Pty Ltd, www.Hearne.com.au. (Ch. 11)

Sutherst, R.W., J.P. Spradbery, and G.F. Maywald. 1989. The potential geographic distribution of the old world screw worm fly, *Chysomya bezziana*. Med. Vet. Entomol. 3:273–280. (Ch. 11)

Swengel, A.B. 1990. Monitoring butterfly populations using the Fourth of July Butterfly Count. Am. Midl. Nat. 124:395–406. (Ch. 2)

Swengel, A.B. 1994. Population fluctuations of the monarch (*Danaus plexippus*) in the 4th of July butterfly count 1977–1994. Am. Midl. Nat. 134:205–14. (Ch. 21)

Swengel, A.B. 1995. Population fluctuations of the monarch butterfly (*Danaus plexippus*) in the 4th of July Butterfly Count 1977–1994. Am. Midl. Nat. 134:205–214. (Ch. 2)

Swengel, A.B. 2006. NABA Butterfly Count, Column 1: Subregions of eastern monarchs. Am. Butterflies, Fall/Winter 2006:54. (Ch. 2)

Swezey, O.H. 1923. Records of introduction of beneficial insects into the Hawaiian Islands. Proc. Hawaii. Entomol. Soc. 5:299–304. (Ch. 6)

Swezey, O.H. 1927. Notes on the Mexican tachinid, *Archytas cirphis* Curran, introduced into Hawaii as an armyworm parasite (Diptera). Proc. Hawaii. Entomol. Soc. 6:499–503. (Ch. 6)

Takabayashi, J., and M. Dicke. 1996. Plant-carnivore mutualism through herbivore-induced carnivore attractants. Trends. Plant. Sci. 1:109–113. (Ch. 5)

Takagi, M. 1986. The reproductive strategy of the gregarious parasitoid, *Pteromalus puparum* (Hymenoptera: Pteromalidae) 2. Host size discrimination and regulation of the number and sex ratio of progeny in a single host. Oecologia 70:321–325. (Ch. 6)

Tarrier, M. 2000. *Danaus plexippus* L. au Maroc (Lepidoptera, Nymphalidae, Danainae). Alexanor. 21 (1), 1999:61–62. (Ch. 22)

Tarrier, M., and J. Delacre. 2008. Les papillons de jour du Maroc. Guide d'identification et de bio-indication. Biotope, Mèze (Collection Parthénope); Museum national d´Histoire naturelle, Paris, 480p. (Ch. 22)

Tauber, M.J., C.A. Tauber, & S. Masaki. 1986. Seasonal Adaptations of Insects. New York, NY: Oxford University Press. (Ch. 9)

Taylor, A.F., F. Kuo, and W. Sullivan. 2001. Views of nature and self-discipline: Evidence from inner city children. J. Environ. Psychol. 22:49–63. (Ch. 1)

Texas Parks and Wildlife Department. 2012. An analysis of Texas waterways. A report on the physical characteristics of rivers, streams, and bayous in Texas. http://www.tpwd.state.tx.us/publications/pwdpubs/pwd_rp_t3200_1047/21_c_tx_medina_navidad_nueces.phtml. Accessed 26 October 2012. (Ch. 10)

Thaler, J.S. 1999. Jasmonate-inducible plant defenses cause increased parasitism of herbivores. Nature 399:686–689. (Ch. 5)

Thompson, J.N. 1994. The Coevolutionary Process. University of Chicago Press, Chicago. (Ch. 7)

Thrall, P.H., J.J. Burdon, and J.D. Bever. 2002. Local adaptation in the *Linum marginale-Melampsora lini* host-pathogen interaction. Evolution 56:1340–1351. (Ch. 7)

Thuiller, W. 2003. BIOMOD—optimizing predictions of species distributions and projecting potential future shifts under global change. Global Change Biol. 9:1353–1362. (Ch. 11)

Tian H., L.C. Stige, B. Cazelles, K.L. Kausrud, S. Svarverud, N.C. Stenseth, and Z. Zhang. 2011. Reconstruction of a 1,910-y-long locust series reveals consistent associations with climate fluctuations in China. Proc. Natl. Acad. Sci. 108:14521–14526. (Ch. 2)

Tooker, J.F., P.F. Reagel, and L.M. Hanks. 2002. Nectar sources of day-flying Lepidoptera of Central Illinois. Ann. Entomol. Soc. Am. 95:84–96. (Ch. 10)

Trejo, S.A., L.M.I. Lopez, C.V. Cimino, N.O. Caffini, and C.L. Natalucci. 2001. Purification and characterization of a new plant endopeptidase isolated from latex of *Asclepias fruticosa* L. (*Asclepiadaceae*). J. Protein Chem. 20:469–477. (Ch. 4)

Tucker, C.M. 2004. Community institutions and forest management in Mexico's Monarch Butterfly Reserve. Soc. Nat. Resour. (17):569–587. (Ch. 15)

Tumlinson, J.H., W.J. Lewis, and L.E.M.Vet. 1993. How parasitic wasps find their hosts. Sci. Am. 268:100–106. (Ch. 5, 6)

Turchin, P. 2003. Complex population dynamics: A theoretical/empirical synthesis. Princeton, NJ: Princeton University Press. (Ch. 5)

Turchin, P., S.N. Wood, S.P. Ellner, B.E. Kendall, W.W. Murdoch, A. Fischlin, J. Casas, E. McCauley, and C.J. Briggs. 2003. Dynamical effects of plant quality and parasitism on population cycles of larch budmoth. Ecology 84:1207–1214. (Ch. 5)

Turlings, T.C.J., J.H. Loughrin, P.J. McCall, U.S.R. Rose, W.J. Lewis, and J.H. Tumlinson. 1995. How caterpillar-damaged plants protect themselves by attracting parasitic wasps. Proc. Natl. Acad. Sci. 92:4169–4174. (Ch. 5)

Tuskes, P.M., and L.P. Brower. 1978. Overwintering ecology of the monarch butterfly, *Danaus plexippus* L., in California. J. Ecol. Entomol. 3:141–153. (Ch. 12, 23)

Udayagiri, S., and R.L. Jones. 1992. Role of plant odor in parasitism of European corn borer by braconid specialist *Mac-*

rocentrus grandii Goidanich: Isolation and characterization of plant syndromes eliciting parasitoid flight response. J. Chem. Ecol. 18:1841–1855. (Ch. 6)

UNESCO. 2010. World Network of Biosphere Reserves 2010: Sites for Sustainable Development. Man and Biosphere Program. United Nations Education, Scientific and Cultural Organization. (Ch. 13)

UNESCO. 2013. Monarch Butterfly Biosphere Reserve. http://whc.unesco.org/en/list/1290 (Ch. 15)

Urquhart, F.A. 1960. The monarch butterfly. Univ. Toronto Press, Toronto. (Ch. 2, 6, 18, 20, 21)

Urquhart, F.A. 1976. Found at last: The monarch's winter home. Natl. Geogr. 150:160–173. (Ch. 2, 3, 13, 15)

Urquhart, F.A. 1987. The monarch butterfly: International traveler. Chicago: Nelson-Hall. (Ch. 2, 21)

Urquhart, F.A., and N.R. Urquhart. 1977. Overwintering areas and migratory routes of the monarch butterfly (*Danaus p. plexippus*, Lepidoptera: Danaidae) in North America, with special reference to the western population. Can. Entomol. 109:1583–1589. (Ch. 2, 21, 23)

Urquhart, F.A., and N.R. Urquhart. 1978. Autumnal migration routes of the eastern population of the monarch butterfly (*Danaus plexippus*) (L.) (Danaidae:Lepidoptera) in North America to the overwintering site in the neovolcanic plateau of Mexico. Can. J. Zool. 56:1754–1764. (Ch. 7, 23, Part 2)

Urquhart, F.A., and N.R. Urquhart. 1979. Vernal migration of the monarch butterfly (*Danaus plexippus* Lepidoptera, Daniadae) in North America from the overwintering site in the neo-volcanic plateau of Mexico. Can. Entomol. 111:15–18. (Ch. 23)

[USDA] United States Department of Agriculture, National Agricultural Statistics Service. 2011. usda01.library.cornell.edu/usda/current/Acre/Acre-06-30-2011.pdf. Accessed June 2011. (Ch. 14)

USDA, Economic Research Service. 2012. Adoption of genetically engineered crops in the U.S. http://www.ers.usda.gov/data-products/adoption-of-genetically-engineered-crops-in-the-us.aspx. Accessed July 2012. (Ch. 14)

USDA, Conservation Programs. 2013. http://www.fsa.usda.gov/FSA/webapp?area=home&subject=copr&topic=rns-css. Accessed January 2013. (Ch. 14)

[USDA-NRCS] USDA Natural Resources Conservation Service. 2013. http://www.nrcs.usda.gov/. Accessed 27 September 2013. (Ch. 16)

USDA Plants Data Base. http://plants.usda.gov. Accessed October 2012. (Ch. 10)

[USFS] U.S. Forest Service. 2011. Reports to the Monarch Joint Venture on 2010 monarch conservation activities of the U.S. Forest Service. http://www.fs.fed.us/wildflowers/pollinators/monarchbutterfly/conservation/mjvaccomplishments.shtml. (Ch. 3)

USFS. 2012. Reports to the Monarch Joint Venture on 2011 monarch conservation activities of the U.S. Forest Service. http://www.fs.fed.us/wildflowers/pollinators/monarchbutterfly/conservation/mjvaccomplishments.shtml. (Ch. 3)

[USFWS] U.S. Fish and Wildlife Service. 2012a. Report on USFWS FY09 pollinator conservation accomplishments. (Ch. 3)

USFWS. 2012b. Report on USFWS FY10 pollinator conservation accomplishments. (Ch. 3)

[USGS] United States Geological Survey. 2012a. USGS surface-water monthly statistics for the nation. USGS 08178880 Medina River at Bandera, TX. waterdata.usgs.gov/nwis/monthly?. Accessed 17 October 2012. (Ch. 10)

USGS. 2012b. USGS Current conditions for Texas. USGS 08178880 Medina River at Bandera, TX. URL: http://nwis.waterdata.usgs.gov/tx/nwis/uv?. Accessed 17 October 2012. (Ch. 10)

USGS. 2012c. USGS 08195000 Frio River at Concan, TX. URL: http://nwis.waterdata.usgs.gov/nwis/uv?. Accessed 26 October 2012. (Ch. 10)

USGS. 2012d. USGS global visualization viewer. http://glovis.usgs.gov. Accessed 28 March 2012. (Ch. 13)

Van Baalen, M., and M.W. Sabelis. 1995. The dynamics of multiple infection and the evolution of virulence. Am. Nat. 146:881–910. (Ch. 7)

Van de Merendonk, S., and J. van Lenteren. 1978. Determinination of mortality of greenhouse whitefly *Tiraleurodes vaporarium* (Westwood) (Homoptera: Aleyrodidae) eggs, larvae and pupae on four host-plant species: Eggplant (*Salonum melongena* L.), cucumber (*Cucumus sativus* L.), tomato (*Lycopersicum esculentum* L.) and paprika (*Capsicum annuum* L.). Med. Fac. Landbouww. Rijksunviv. Gent. 434210429. (Ch. 5)

Vandenbosch, R. 2007. What do monarch population time series tell us about eastern and western population mixing? J. Lepid. Soc. 61:28–31. (Ch. 23)

Van der Heyden, T. 2009. Bemerkungen zur Biologie, zur Ökologie und zur Verbreitung von *Danaus chrysippus* Linnaeus, 1758 im Mittelmeerraum, insbesondere in der Türkey (Lepidoptera: Nymphalidae, Danainae). Nachr. Entomol. Ver. Apollo, N.F. 30(3):173–176. (Ch. 22)

van der Meijden, E., M. Wijn, and H.J. Verkaar. 1988. Defense and regrowth, alternative plant strategies in the struggle against herbivores. Oikos 51:355–363. (Ch. 4)

van der Ploeg, R.R., W. Böhm, and M.B. Kirkham. 1999. On the origin of the theory of mineral nutrition of plants and the law of the minimum. Soil Sci. Soc. Am. J. 63:1055–1062. (Ch. 11)

Vane-Wright, R.I. 1993. The Columbus hypothesis: An explanation for the dramatic 19th century range expansion of the monarch butterfly. *In* S.B. Malcolm and M.P. Zalucki, eds., Biology and conservation of the monarch butterfly, pp. 179–187. Los Angeles: Natural History Museum of Los Angeles County. (Ch. 19, 23)

Van Gils, J.A., V.J. Munster, R. Radersma, D. Liefhebber, R.A.M. Fouchier, and M. Klaassen. 2007. Hampered foraging and migratory performance in swans infected with low-pathogenic avian influenza A virus. PLoS ONE 2:e184. (Ch. 7)

Van Hook, T. 1990. A phenological study of monarch butterflies at a Gulf-coast Florida temporary clustering site during the annual fall migration: 1988–1989. Report to St. Marks National Wildlife Refuge. U.S. Fish and Wildlife Service. (Ch. 2)

Van Hook, T. 1996. Monarch butterfly mating ecology at a Mexican overwintering site: Proximate causes of non-random mating. Ph.D. thesis, University of Florida, Gainesville. (Ch. 8) van Klinken, R.D., G. Fichera, and H. Cordo. 2003. Targeting biological control across diverse landscapes: The release, establishment and early success of two insects on mesquite (*Prosopis*) in rangeland Australia. Biol. Control 2:8–20. (Ch. 11)

Vannette, R.L., and M.D. Hunter. 2011a. Genetic variation in expression of defense phenotype may mediate evolutionary adaptation of *Asclepias syriaca* to elevated CO_2. Global Change Biol. 17:1277–1288. (Part 3)

Vannette, R.L., and M.D. Hunter. 2011b. Plant defence theory re-examined: Nonlinear expectations based on the costs and benefits of resource mutualisms. J. Ecol. 99:66–76. (Part 2)

Van Strien, A.J., R. van de Pavert, D. Moss, T.J. Yates, C.A.M. van Swaay, and P. Vos. 1997. The statistical power of two butterfly monitoring schemes to detect trends. J. Appl. Ecol. 34:817–828. (Ch. 24)

Van Zandt, P.A., and A.A. Agrawal. 2004. Community-wide impacts of herbivore-induced plant responses in milkweed (*Asclepias syriaca*). Ecology 85:2616–2629. (Ch. 7)

Vázquez, P.M. 2009. Legisladores piden cuentas a autoridades por plagas en zona de la Monarca. Cambio de Michoacán, Morelia, Michoacán, 30 August 2009. Accessed 24 July 2012. doi: http://www.cambiodemichoacan.com.mx/vernota.php?id=107979 (Ch. 9)

Vega, F.E., and H.K. Kaya. 2012. Insect pathology. New York: Academic. 490pp. (Ch. 7)

Velázquez, A., E. Durán, I. Ramírez, J.F. Mas, G. Bocco, G. Ramírez, and J.L. Palacio. 2003. Land use-cover change processes in highly biodiverse areas: The case of Oaxaca, Mexico. Global Environ. Change 13:175–184. (Ch. 13)

Venegas, Y. 2010. La implementación del programa de conservación para el desarrollo sostenible. Estudio de su focalización en la Reserva de la Biosfera Mariposa Monarca. Tesis de Maestría. Facultad Latinoamericana de Ciencias Sociales. (Ch. 15)

Venegas-Pérez, Y., S. Rodríguez, and D.T. López. 2011. Análisis Base para el diseño de la Estrategia de Reforestación de la Reserva de la Biosfera Mariposa Monarca. Michoacán, México. Monarch Butterfly Fund—Dirección de la Reserva de la Biosfera Mariposa Monarca. 88p. http://monarchbutterflyfund.org/sites/default/files/Analisis_Reforestacion_RBMM_MBF_2011(1).pdf. Accessed 13 August 2012. (Ch. 13)

Vermeij, G.J. 1994. The evolutionary interaction among species: Selection, escalation, and coevolution. Annu. Rev. Ecol. Syst. 25:219–236. (Ch. 4)

Vetter, J. 2011. Introduction: Lay participation in the history of scientific observation. Sci. Context 24:127–41. (Ch. 2)

Vidal y López, M. 1921. Material para la flora marroquí. Bol. R. Soc. Esp. Hist. Nat. 21:274–281. (Ch. 22)

Vidal, O., and E. Rendón-Salinas. 2014. Dynamics and trends of overwintering colonies of the monarch butterfly in Mexico. Biological Conservation, 180: 165–175. (Part 5)

Vidal, O., J. Lopez-Garcia, and E. Rendón-Salinas. 2014. Trends in deforestation and forest degradation after a decade of monitoring in the Monarch Butterfly Biosphere Reserve in Mexico. Conserv. Biol. 28:177–186. doi: 10.1111/cobi.12138. (Part 4)

Vieira, V. 1997. Lepidoptera of the Azores islands. Bol. Mus. Munic. Funchal 49(273):5–76. (Ch. 22)

Vieira, V. 1999. New records of Lepidoptera from Porto Santo island (Insecta: Lepidoptera). SHILAP-Rev. Lepid. 27(107):319–326. (Ch. 22)

Villablanca, F. 2010. Monarch Alert annual report: Overwintering population 2009–2010. http://monarchalert.calpoly.edu/pdf/Monarch-Alert-Report-2010.pdf. Accessed October 2012. (Ch. 12)

Villablanca, F. 2012a. Monarch movement between coastal California overwintering sites: The super-population. http://www.monarchlab.org/mn2012/Presentations.aspx. Accessed December 2012. (Ch. 2)

Villablanca, F. 2012b. Monarch butterfly occupancy and habitat utilization at Sweet Springs East, Los Osos, CA. Biological Resources Assessment # DRC2011–00013 for the Morro Coast Audubon Society. Unpublished report. (Ch. 2)

Voltera, V. 1926. Fluctuations in the abundance of a species considered mathematically. Nature 118:558–560. (Ch. 5)

Walford, P. 1980. Lipids in the life cycle of the monarch butterfly, *Danaus plexippus*. Undergraduate honors thesis, Amherst College, Amherst, MA. (Ch. 10)

Walker, J.J. 1886. *Anosia plexippus* L. (*Danaus plexippus*, F.) at Gibraltar. Entomol. Mon. Mag. 23:162. (Ch. 22)

Walker, P.A., and K.D. Cocks. 1991. HABITAT: A procedure for modelling the disjoint environmental envelope for a plant or animal species. Global Ecol. Biogeogr. 1:108–118. (Ch. 11)

Walton, R.K., and L.P. Brower. 1996. Monitoring the fall migration of the monarch butterfly *Danaus plexippus* L. (Nymphalidae: Danaidae) in eastern North America: 1991–1994. J. Lepid. Soc. 50:1–10. (Ch. 2)

Walton, R., L.P. Brower, and A.K. Davis. 2005. Long-term monitoring and fall migration patterns of the monarch butterfly (Nymphalidae: Danainae) in Cape May, NJ. Ann. Entomol. Soc. Am. 98:682–689. (Ch. 2)

Wason, E.L., and M.D. Hunter. 2013. Genetic variation in plant volatile emission does not result in differential attraction of natural enemies in the field. Oecologia. doi 10.1007/s00442-013-2787-4. (Ch. 4, 6)

Wason, E.L., A.A. Agrawal, and M.D. Hunter. 2013. A genetically-based latitudinal cline in the emission of herbivore-induced plant volatile organic compounds. J. Chem. Ecol. 39:1101–1111. (Ch. 6)

Wassenaar, L.I., and K.A. Hobson. 1998. Natal origins of migratory monarch butterflies at wintering colonies in Mexico: New isotopic evidence. Proc. Natl. Acad. Sci. 95:15436–15439. (Ch. 3, 11, 14, 17, 18, 24, Part 4)

Watson, H. 1844. Notes on the botany of the Azores. Hooker's Lond. J. Bot. 3:582–617. (Ch. 22)

Watson, H. 1870. Botany of the Azores. In F.D.-C. Godman, ed., Natural history of the Azores or Western Islands, pp. 113–288. London: John Van Voorst & Paternoster Row. (Ch. 22)

Watt, M.S., D.J. Kriticos, S. Alcaraz, A. Brown, and A. Leriche. 2009. The hosts and potential geographic range of *Dothistroma* needle blight. Forest Ecol. Manage. 257:1505–1519. (Ch. 11)

Webber, B.L., C.J. Yates, D.C. Le Maitre, J.K. Scott, D.J. Kriticos, N. Ota, A. McNeill, J.J. Le Roux, and G.F. Midgley. 2011. Modelling horses for novel climate courses: Insights from projecting potential distributions of native and alien Australian acacias with correlative and mechanistic models. Divers. Distrib. 17:978–1000. (Ch. 11)

Weir, J. 1894. Proc. S. Lond. Entomol. Nat. Hist. Soc. 1892–1893:34–35. (Ch. 20)

Weiss, S.B. 1998. Habitat suitability, restoration, and vegetation management at Monach Grove Sanctuary. Pacific Grove, California. Report prepared for Thomas Reid Associates (for City of Pacific Grove). 34 pp. + 32 appendices. (Part 4)

Weiss, S.B. 2005. Topoclimate and microclimate in the monarch butterfly biosphere reserve. Report to World Wildlife Fund. 37 pp. (Ch. 9)

Wells, H., P.H. Wells, and P. Cook. 1990. The importance of overwinter aggregation for reproductive success of monarch butterflies (*Danaus plexippus* L.). J. Theor. Biol. 147:115–131. (Part 3)

Wells, S.M., R.M. Pyle, and N.M. Collins. 1983. The IUCN invertebrate red data book. International Union for Conservation of Nature and Natural Resources. Gland, Switzerland: IUCN. (Ch. 3, 12, Part 1)

Wenner, A.M., and A.M. Harris. 1993. Do California monarchs undergo long-distance directed migration? *In* S.B. Malcolm and M.P. Zalucki, eds., Biology and conservation of the monarch butterfly, pp. 209–218. Los Angeles: Natural History Museum of Los Angeles County. (Ch. 16, 23)

Werner, E.E., and S.D. Peacor. 2003. A review of trait-mediated indirect interactions in ecological communities. Ecology 84:1083–1100. (Ch. 7, Part 2)

Wharton, T.N., and D.J. Kriticos. 2004. The fundamental and realized niche of the Monterey pine aphid, *Essigella californica* (Essig) (Hemiptera: Aphididae): Implications for managing softwood plantations in Australia. Divers. Distrib. 10:253–262. (Ch. 11)

White, D.S., and O.J. Sexton. 1989. The monarch butterfly (Lepidoptera: Danaidae) as prey for the dragonfly *Hagenisu brevistylus* (Odanata: Gomphidae). Entomol. News. 100:129–132. (Ch. 6)

White, J. 1983. Regional and local variation in composition and structure of the tallgrass prairie vegetation of Iowa and eastern Nebraska. MS thesis. Iowa State University. (Ch. 14)

Wikler, K., T.R. Gordon, A.J. Storer, and D.L. Wood. 2003. Pitch canker. Integrated pest management for gardeners and landscape professionals. Pest Notes, University of California Davis Publication 74107. (Ch. 12)

Wilbur, H.M. 1976. Life history evolution in seven species in the genus *Asclepias*. J. Ecol. 64:223–240. (Ch. 14)

Williams, C.B. 1958. Insect migration. London: Collins, The New Naturalist, xiv + 235 pages. (Ch. 20)

Williams, C.B., G.F. Cockbill, M.E. Gibbs, and J.A. Downes. 1942. Studies on the migrations of Lepidoptera. Trans. R. Entomol. Soc. Lond. 92(1):101–283. (Ch. 22)

Williams, J.J., D.A. Stow, and L.P. Brower. 2007. The influence of forest fragmentation on the location of overwintering monarch butterflies in central Mexico. J. Lepid. Soc. 61:90–104. (Ch. 9)

Willkomm, M. 1893. Suplementum Prodromi Florae Hispanicae. Stuttgart. (Ch. 22)

Wilson, L.T., J.M. Smilanick, M.P. Hoffmann, D.L. Flaherty, and S.M. Ruiz. 1988. Leaf nitrogen and position in relation to population parameters of Pacific spider mite, *Tetranychus pacificus* (Acari: Tetranychidae) on grapes. Environ. Entomol. 17:964 968. (Ch. 5)

Wiemers, M. 1995. The butterflies of the Canary Islands. A survey on their distribution, biology and ecology (Lepidoptera: Papilionoidea and Hesperioidea). Linn. Belg. 15(2):63–118. (Ch. 22)

Wink, M. 2003. Evolution of secondary metabolites from an ecological and molecular phylogenetic perspective. Phytochemistry 64:3–19. (Ch. 4)

Wisconsin Butterflies. 2012. Wisconsin butterflies. http://wisconsinbutterflies.org/. Accessed January 2013. (Ch. 2)

Wojcik, V.A., and S. Buchmann. 2012. A review of pollinator conservation and management on infrastructure supporting rights-of-way. J. Pollinat. Ecol. 7:16–26. (Ch. 3)

Wolinska, J., and K.C. King. 2009. Environment can alter selection in host-parasite interactions. Trends Parasitol. 25:236–244. (Ch. 7)

Woodson, R.E. Jr. 1954. The North American species of *Asclepias* L. Ann. Mo. Bot. Gard. 41:1–211. (Ch. 4, 19, 20, Part 2)

Wootton, J.T. 1994. The nature and consequences of indirect effects in ecological communities. Annu. Rev. Ecol. Syst. 25:443–466. (Ch. 7)

Worner, S.P., and M. Gevrey. 2006. Modelling global insect pest species assemblages to determine risk of invasion. J. Appl. Ecol. 43:858–867. (Ch. 11)

Worrall, J.J., L. Egeland, T. Eager, R.A. Mask, E.W. Johnson, P.A. Kemp, and W.D. Shepperd. 2008. Rapid mortality of *Populus tremuloides* in southwestern Colorado, USA. For. Ecol. Manage. 255:686–696. (Ch. 13)

Wrangham, R.W., and T. Nishida. 1983. *Aspilia* spp. Leaves: A puzzle in the feeding behavior of wild chimpanzees. Primates 24:276–282. (Ch. 7)

Wrede, J. 2010. Trees, shrubs and vines of the Texas Hill Country: A field guide, 2nd ed. College Station: Texas A&M University Press. (Ch. 10)

[WWF] World Wildlife Fund Mexico. 2004. La Tala Ilegal y su Impacto en la Reserva de la Biosfera Mariposa Monarca. 37 pp. (Ch. 15)

WWF. 2012a. Degradación y pérdida forestal en la zona núcleo de la reserva de la biosfera mariposa monarca 2011–2012. Informe presentado en el Comité Técnico del Fideicomiso del Fondo Monarca. 2 pp. (Ch. 3)

WWF. 2012b. En 2011–2012, y por primera vez desde que se decretó la Reserva Mariposa Monarca en el 2000, no se detectó tala ilegal en su zona núcleo. Press release, 15 August 2012. Fondmonarca, CONANP, SEMARNAT, WWF, Telcel, Governor of the State of México. www.wwf.org.mx/wwfmex/archivos/bm/120815-baja-tala-ilegal-en-zona-nucleo-reserva-monarca.php. Accessed March 2013. (Part 4)

Wyatt, R., and S.B. Broyles. 1994. Ecology and evolution of reproduction in milkweeds. Annu. Rev. Ecol. Syst. 25:423–441. (Ch. 22)

[Xerces] The Xerces Society for Invertebrate Conservation. 2012a. http://www.xerces.org/. Accessed January 13, 2013. (Ch. 21)

Xerces. 2012b. Western monarch Thanksgiving count. http://www.xerces.org/butterfly-conservation/western-monarch-thanksgiving-count/. Accessed December 2012. (Ch. 2)

Xerces. 2012c. Western monarchs in peril. http://www.xerces.org/wp-content/uploads/2010/11/western-monarchs1.pdf. (Ch. 3)

Xerces. 2012d. Database of western monarch overwintering locations. Unpublished data. (Ch. 12)

Xerces. 2013. Project Milkweed. http://www.xerces.org/milkweed. Accessed 13 January 2013. (Ch. 16)

Yakubu, A.A., R. Sáenz, J. Stein, and L.E. Jones. 2004. Monarch butterfly spatially discrete advection model. Math. Biosci. 190:193–202. (Ch. 11)

Yenish, J.P., T.A. Fry, B.R. Durgan, and D.L. Wyse. 1997. Establishment of common milkweed (*Asclepias syriaca*) in corn, soybean and wheat. Weed Sci. 45:44–53. (Ch. 14)

Yonow, T., and R.W. Sutherst. 1998. The geographical distribution of Queensland fruit fly, *Bactrocera (Dacus) tryoni*, in relation to climate. Aust. J. Agric. Res. 49:935–953. (Ch. 11)

Yonow, T., D.J. Kriticos, and R.W. Medd. 2004. The potential geographic range of *Pyrenophora semeniperda*. Phytopathology 94:805–812. (Ch. 11)

York, H.A., and K.S. Oberhauser. 2002. Effects of duration and timing of heat stress on monarch butterfly (*Danaus plexippus*) (Lepidoptera: Nymphalidae) development. J. Kans. Entomol. Soc. 75:290–298. (Ch. 8, 11, Part 3)

Young, A.M. 1982. An evolutionary-ecological model of the evolution of migratory behavior in the monarch butterfly, and its absence in the queen butterfly. Acta Biotheor. 31:219–237. (Ch. 20)

Zahl, P.A. 1963. Mystery of the monarch butterfly. Natl. Geogr. 123:588–98. (Ch. 21)

Zalucki, M.P. 1981a. Temporal and spatial variation of parasitism in *Danaus plexippus* (L.) (Lepidoptera: Nymphalidae: Danainae). Aust. Entomol. Mag. 8:3–8. (Ch. 6)

Zalucki, M.P. 1981b. The effects of age and weather on egg laying in *Danaus plexippus* L. (Lepidoptera: Danaidae). Res. Popul. Ecol. 23:318–327. (Ch. 8, 11, Part 3)

Zalucki, M.P. 1982. Temperature and rate of development in *Danaus plexippus* L. and *D. chrysippus* L. (Lepidoptera: Nymphalidae). J. Aust. Entomol. Soc. 21:241–46. (Ch. 8, 11, 17, Part 3)

Zalucki, M.P., and L.P. Brower. 1992. Survival of first instar larvae of *Danaus plexippus* (Lepidoptera: Danainae) in relation to cardiac glycoside and latex content of *Asclepias humistrata* (Asclepiadaceae). Chemoecology 3:81–93. (Ch. 5, 6)

Zalucki, M.P., and A.R. Clarke. 2004. Monarchs across the Pacific: The Columbus hypothesis revisted. Biol. J. Linn. Soc. 82:111–121. (Ch. 19, 22, 23, Part 5)

Zalucki, M.P., and M.J. Furlong. 2005. Forecasting *Helicoverpa* populations in Australia: A comparison of regression based models and a bio-climatic based modelling approach. Insect Sci. 12:45–56. (Ch. 11)

Zalucki, M.P., and M.J. Furlong. 2008. Predicting outbreaks of a migratory pest: An analysis of DBM distribution and abundance. *In* A.M. Shelton, H.L. Collins, Y. Zhang, and Q. Wu., eds., The management of the diamondback moth and other crucifer insect pests: Proceedings of the fifth international workshop, pp. 122–131. Beijing: China Agricultural Science and Technology Press. (Ch. 11)

Zalucki, M.P., and M.J. Furlong. 2011. Predicting outbreaks of a migratory pest: An analysis of DBM distribution and abundance revisited. *In* R. Srinivasan, A.M. Shelton, and H.L. Collins, eds., Proceedings of the sixth International workshop on management of the diamondback moth and other crucifer insect pests, pp. 8–14. Taiwan: AVRDC. (Ch. 11)

Zalucki, M.P., and R.L. Kitching. 1982a. Dynamics of oviposition in Danaus plexippus (Insecta: Lepidoptera) on milkweed, Asclepias spp. J. Zool. 198:103–116. (Ch. 14, 17)

Zalucki, M.P., and R.L. Kitching. 1982b. Temporal and spatial variation of mortality in field populations of *Danaus plexippus* L. and *D. chrysippus* L. larvae (Lepidoptera: Nymphalidae) Oecologia 53:201–207. (Ch. 5, 6, 8)

Zalucki, M.P., and J.H. Lammers. 2010. Dispersal and egg shortfall in monarch butterflies: What happens when the matrix is cleaned up? Ecol. Entomol. 35:84–91. (Ch. 11, 16, 22)

Zalucki, M.P., and S.B. Malcolm. 1999. Plant latex and first instar monarch larval growth and survival on three North American milkweed species. J. Chem. Ecol. 25:1827–1842. (Ch. 5, 6, 14, Part 2)

Zalucki, M.P., and W.A. Rochester. 1999. Estimating the effect of climate on the distribution and abundance of the monarch butterfly, *Danaus plexippus* (L.): A tale of two continents. *In* J. Hoth, L. Merino, K. Oberhauser, I. Pisanty, S. Price, and T. Wilkinson, eds., 1997 North American conference on the monarch butterfly: Paper presentations, pp. 151–163. Montreal: Commission for Environmental Cooperation. (Ch. 11, 19, 22, Part 3)

Zalucki, M.P., and W.A. Rochester. 2004. Spatial and temporal population dynamics of monarchs down-under: Lessons for North America. *In* K.S. Oberhauser and M.J. Solensky, eds., The monarch butterfly: Biology and conservation, pp. 219–228. Ithaca, NY: Cornell University Press. (Ch. 8, 11, 22, Part 3)

Zalucki, M.P., and Y. Suzuki. 1987. Milkweed patch quality, adult population structure and egg laying in *Danaus plexippus* (Lepidoptea: Nymphalidae). J. Lepid. Soc. 41:13–22. (Ch. 14)

Zalucki, M.P., and R. van Klinken. 2006. Predicting population dynamics and abundance of introduced biological agents: Science or gazing into crystal balls? Aust. J. Entomol. 45:331–344. (Ch. 11)

Zalucki, M.P., S. Oyeyele, and P. Vowles. 1989. Selective oviposition by *Danaus plexippus* L. (Lepidoptera: Nymphalidae) in a mixed stand of *Asclepias fruticosa* and *A.curasavica* in southeast Queensland. J. Aust. Entomol. Soc. 28:141–146. (Ch. 22)

Zalucki, M.P., L.P. Brower, and S.B. Malcolm. 1990. Oviposition by Danaus plexippus in relation to cardenolide content of three Asclepias species in the southeastern U.S.A. Ecol. Entomol. 15:231–240. (Ch. 7, 14)

Zalucki, M.P., J.M. Hughes, J.M. Arthur, and P.A. Carter. 1993. Seasonal variation at four loci in a continuously breeding population of *Danaus plexippus* L. Heredity 70:205–213. (Ch. 23)

Zalucki, M.P., L.P. Brower, and A. Alonso. 2001a. Detrimental effects of latex and cardiac glycosides on survival and growth of first-instar monarch butterfly larvae *Danaus plexippus* feeding on the sandhill milkweed *Asclepias humistrata*. Ecol. Entomol. 26:212–224. (Ch. 4, 5, 6, 7)

Zalucki, M.P., S.B. Malcolm, T.D. Paine, C.C. Hanlon, L.P. Brower, and A.R. Clarke. 2001b. It's the first bites that count: Survival of first-instar monarchs on milkweeds. Aust. Ecol. 26:547–555. (Ch. 5, 6, 7, 11)

Zalucki, M.P., A.R. Clarke, and S.B. Malcolm. 2002. Ecology and behavior of first-instar larval Lepidoptera. Annu. Rev. Entomol. 47:361–393. (Ch. 5)

Zalucki, M.P., S.B. Malcolm, C.C. Hanlon, and T.D. Paine. 2012. First-instar monarch larval growth and survival on milkweeds in southern California: Effects of latex, leaf hairs and cardenolides. Chemoecology 22:75–88. (Ch. 4, 5, 6, 11)

Zehnder, C.B., and M.D. Hunter. 2007. Interspecific variation within the genus *Asclepias* in response to herbivory by a phloem-feeding insect herbivore. J. Chem. Ecol. 33:2044–2053. (Ch. 7)

Zhan, S., and S.M. Reppert. 2013. MonarchBase: The monarch butterfly genome database. Nucleic Acids Res. 41:D758–D763. (Part 5)

Zhan, S., C. Merlin, J.L. Boore, and S.M. Reppert. 2011. The monarch butterfly genome yields insights into long-distance migration. Cell 147:1171–1185. (Ch. 13, 23, Part 5)

Zhan, S., W. Zhang, K. Niitepõld, J. Hsu, J. Fernández Haeger, M. Zalucki, S. Altizer, J. de Roode, S. Reppert, and M. Kronforst. 2014. The genetics of monarch butterfly migration and warning coloration. Nature. 514: 317–321. (Part 5)

Zhen, Y., M.L. Aardema, E.M. Medina, M. Schumer, and P. Andolfatto. 2012. Parallel molecular evolution in an herbivore community. Science 337:1634–1637. (Ch. 7, 23)

Zhu, H., A. Casselman, and S.M. Reppert. 2008a. Chasing migration genes: A brain expressed sequence tag resource for summer and migratory monarch butterflies (*Danaus plexippus*). PLoS ONE 1:e1345. (Ch. 23)

Zhu, H., I. Sauman, Q. Yuan, A. Casselman, M. Emery-Le, P. Emery, and S.M. Reppert. 2008b. Cryptochromes define a novel circadian clock mechanism in monarch butterflies that may underlie sun compass navigation. PLoS Biol. 6:e4. (Ch. 7, 23)

Zhu, H., R.J. Gegear, A. Casselman, S. Kanginakudru, and S.M. Reppert. 2009. Defining behavioral and molecular differences between summer and migratory monarch butterflies. BMC Biol. 7:14. (Ch. 19, 23)

Zipkin, E.F., L. Ries, R. Reeves, J. Regetz, and K.S. Oberhauser. 2012. Tracking climate impacts on the migratory monarch butterfly. Global Change Biol. 18:3039–3049. (Ch. 2, 11, 16, 24, Part 3)

Zonneveld, M. van, A. Jarvis, W. Dvorak, G. Lema, and C. Leibing. 2009. Climate change impact predictions on *Pinus patula* and *Pinus tecunumanii* populations in Mexico and Central America. For. Ecol. Manage. 257:1566–1576. (Ch. 13)

Zuk, M., J.T. Rotenberry, and R.M. Tinghitella. 2006. Silent night: Adaptive disappearance of a sexual signal in a parasitized population of field crickets. Biol. Lett. 2:521–524. (Ch. 7)

Contributors

Anurag A. Agrawal
Department of Ecology and Evolutionary Biology
 and Department of Entomology
Cornell University
Ithaca, New York, USA

Jared G. Ali
Department of Entomology
Michigan State University
East Lansing, Michigan, USA

Sonia Altizer
Odum School of Ecology
University of Georgia
Athens, Georgia, USA

Michael C. Anderson
Eden Prairie, Minnesota, USA

Sophia M. Anderson
Eden Prairie, Minnesota, USA

Kim Bailey
Environmental Protection Division
Georgia Department of Natural Resources
Atlanta, Georgia, USA

Rebecca Batalden
Department of Ecology, Evolution, and Behavior
University of Minnesota
St. Paul, Minnesota, USA

Kristen A. Baum
Department of Zoology
Oklahoma State University
Stillwater, Oklahoma, USA

Scott Hoffman Black
The Xerces Society for Invertebrate Conservation
Portland, Oregon, USA

Brianna Borders
The Xerces Society for Invertebrate Conservation
Portland, Oregon, USA

Lincoln P. Brower
Department of Biology
Sweet Briar College
Sweet Briar, Virginia, USA

Wendy Caldwell
University of Minnesota
St. Paul, Minnesota, USA

Mariana Cantú-Fernández
Centro de Investigaciones en Ecosistemas
Universidad Nacional Autónoma de México
Morelia, Michoacán, México

Nicola Chamberlain
Faculty of Arts and Sciences Center for Systems
 Biology
Harvard University
Cambridge, Massachusetts, USA

Sonya Charest
Educational programs
Montreal Insectarium
Montreal, Québec, Canada

Andrew K. Davis
Odum School of Ecology
University of Georgia
Athens, Georgia, USA

Alma De Anda
Covina, California, USA

Guadalupe del Rio Pesado
Alternare, A.C.
México, D.F, México

Juan Fernández-Haeger
Department of Botany, Ecology, and Plant Physiology
University of Córdoba
Córdoba, Spain

Linda S. Fink
Department of Biology
Sweet Briar College
Sweet Briar, Virginia, USA

Mark Fishbein
Department of Botany
Oklahoma State University
Stillwater, Oklahoma, USA

Eligio García Serrano
Coordinador del Fondo Monarca
Zitácuaro, Michoacán, México

Mark Garland
Cape May Monarch Monitoring Project
Cape May Point, New Jersey, USA

Brian Hayes
Monarch Teacher Network
Education Information Resource Center
Mullica Hill, New Jersey, USA

Elizabeth Howard
Journey North
Norwich, Vermont, USA

Mark D. Hunter
Department of Ecology & Evolutionary Biology
University of Michigan
Ann Arbor, Michigan, USA

Sarina Jepsen
The Xerces Society for Invertebrate Conservation
Portland, Oregon USA

Diego Jordano
Department of Botany, Ecology, and Plant Physiology
University of Córdoba
Córdoba, Spain

Matthew C. Kaiser
Graduate Program in Ecology, Evolution, and Behavior
University of Minnesota
St. Paul, Minnesota, USA

Ridlon J. Kiphart
Texas Master Naturalists
Boerne, Texas, USA

Marcus R. Kronforst
Department of Ecology & Evolution
University of Chicago
Chicago, Illinois, USA

Janet Kudell-Ekstrum
USDA Forest Service
Hiawatha National Forest
Rapid River, Michigan, USA

Eric Lee-Mäder
The Xerces Society for Invertebrate Conservation
University of Minnesota Department of Entomology
Portland, Oregon, USA

Jim Lovett
Monarch Watch
University of Kansas
Lawrence, Kansas, USA

Stephen B. Malcolm
Department of Biological Sciences,
Western Michigan University,
Kalamazoo, Michigan, USA

Héctor Martínez-Torres
Centro de Investigaciones en Ecosistemas
Universidad Nacional Autónoma de México
Morelia, Michoacán, México

Susan Meyers
Education Department
Stone Mountain Memorial Association
Stone Mountain, Georgia, USA

Erik A. Mollenhauer
Monarch Teacher Network
Education Information Resource Center
Mullica Hill, New Jersey, USA

Mía Monroe
California Monarch Campaign
The Xerces Society
Portland, Oregon, USA

Eneida B. Montesinos-Patino
Monarch Butterfly Fund
Tustin, California, USA

Gail M. Morris
Southwest Monarch Study
Chandler, Arizona, USA

Elisha K. Mueller
Department of Zoology
Oklahoma State University
Stillwater, Oklahoma, USA

Kelly R. Nail
Conservation Biology Graduate Program
University of Minnesota
St. Paul, Minnesota, USA

Karen S. Oberhauser
Department of Fisheries, Wildlife and Conservation Biology
University of Minnesota
St. Paul, Minnesota, USA

Diego R. Pérez-Salicrup
Centro de Investigaciones en Ecosistemas
Universidad Nacional Autónoma de México
Morelia, Michoacán, México

Amanda A. Pierce
Department of Biology
Emory University
Atlanta, Georgia, USA

John Pleasants
Ecology, Evolution and Organismal Biology Department
Iowa State University
Ames, Iowa, USA

Victoria Pocius
Department of Ecology and Evolutionary Biology
University of Kansas
Lawrence, Kansas, USA

Robert Michael Pyle
Northwest Lepidoptera Survey
Gray's River, Washington, USA

M. Isabel Ramírez
Centro de Investigaciones en Geografía Ambiental
Universidad Nacional Autónoma de México
Morelia, Michoacán, México

Sergio Rasmann
Department of Ecology and Evolutionary Biology
University of California Irvine
Irvine, California, USA

Gerald Rehfeldt
Rocky Mountain Research Station
USDA Forest Service
Moscow, Idaho, USA

Eduardo Rendón-Salinas
World Wildlife Fund–Mexico
Zitácuaro, Michoacán, México

Leslie Ries
National Socio-Environmental Synthesis Center
Annapolis, Maryland, USA

Jacobus C. de Roode
Department of Biology
Emory University
Atlanta, Georgia, USA

Richard G. RuBino
Department of Urban and Regional Planning
Florida State University
Tallahassee, Florida, USA

Ann Ryan
Monarch Watch
University of Kansas
Lawrence, Kansas, USA

Cuauhtémoc Sáenz-Romero
Instituto de Investigaciones Agropecuarias y Forestales
Universidad Michoacana de San Nicolás de Hidalgo
Tarímbaro, Michoacán, México

Lidia Salas-Canela
Centro de Investigaciones en Geografía Ambiental
Universidad Nacional Autónoma de México
Morelia, Michoacán, México

Phil Schappert
Biophilia Consulting
Halifax, Nova Scotia, Canada

Priya C. Shahani
Institute for Natural Resources
Oregon State University
Portland, Oregon, USA

Benjamin H. Slager
Department of Biological Sciences,
Western Michigan University,
Kalamazoo, Michigan, USA

Michelle J. Solensky
Department of Biology
University of Jamestown
Jamestown, North Dakota, USA

Douglas J. Taron
Chicago Academy of Sciences/Peggy Notebaert
 Nature Museum
Chicago, Illinois, USA

Orley R. Taylor
Monarch Watch
University of Kansas
Lawrence, Kansas, USA

Rocío Treviño
Programa Correo Real
Protección de la Fauna Mexicana A.C.
Saltillo, Coahuila, México

Francis X. Villablanca
Biological Sciences Department
California Polytechnic State University
San Luis Obispo, California, USA

Dick Walton
Monarch Monitoring Project
New Jersey Audubon / Cape May Bird Observatory
Cape May Point, New Jersey, USA

Ernest H. Williams
Department of Biology
Hamilton College
Clinton, New York, USA

Elisabeth Young-Isebrand
Department of Fisheries, Wildlife and Conservation
 Biology
University of Minnesota
St. Paul, Minnesota, USA

Myron P. Zalucki
School of Biological Sciences
The University of Queensland
Brisbane, Australia

Raúl R. Zubieta
Posgrado en Ciencias Biológicas
Centrode Investigaciones en Ecosistemas
Universidad Nacional Autónoma de México
Morelia, Michoacán, México

Index

Abies religiosa. See oyamel fir
agricultural fields
　egg density, 176–177
　milkweed in, 169, 171, 174–177, 198, 208
allozyme markers, 258–259, 265, 267
Alternare A.C., 34–35
annual cycle, 268–271
aphids, 44–45, 46, 60, 63, 65, 69, 192
Asclepias
　angustifolia, 20
　asperula, 54, 56, 193–194, 218, 221
　barjoniifolia, 228, 234
　boliviensis, 228, 234
　bracteolata, 228
　californica, 54
　candida, 228
　cordifolia, 194
　curassavica, 28, 44, 52, 56, 71, 76–77, 81, 89, 93, 100, 196, 216–224, 228, 234, 248–252, 254–256
　eriocarpa, 194
　fascicularis, 194, 238
　fruticosa, 196
　geographic diversity, 49–50
　hallii, 55
　humistrata, 58, 193
　incarnata, 44, 58, 71, 77, 89, 191
　latifolia, 194
　lemmonii, 54
　linaria, 20, 56, 58
　mellodora, 228
　as monarch hosts, 52
　nivea, 55
　oenotheroides, 218, 221, 223
　perennis, 56, 193
　phylogeny, 47–49, 53
　physocarpa, 89, 196
　pilgeriana (flava), 228
　pseudofimbriata, 49
　pumila, 58
　purpurascens, 55–56
　solanoana, 55–56
　speciosa, 58, 77, 191, 194, 238, 244
　subulata, 20, 56, 194
　subverticillata, 56, 58
　sullivantii, 191
　syriaca, 55–56, 58, 62, 69, 71, 74, 76–77, 81, 100, 131, 169, 176, 190–191, 197
　tuberosa, 57–58, 176, 191, 193
　verticillata, 77, 176, 191
　vestita, 54
　viridiflora, 176
　viridis, 191, 193, 197–201
　See also milkweed

Badgett, Gina, 21
biocenosis, 34
biocontrol
　non-target species, effects on, 73–74
　See also parasitoids; *names of biocontrol agents*
Brower, Lincoln, 20
Bt crops, 139–140, 169–170
Butterflies and moths of North America (BAMONA), 15, 27, 28, 245
butterflies as symbols, 1–2
butterfly monitoring programs, 15, 24–27, 271–277, 280
　absence data, importance of, 24–25
　counts, 25–26
　opportunistic, 25–27
　transects, 25–26

California Natural Diversity Database (CNDDB), 150
California population. *See* western population
Cape May Monarch Monitoring Project, 15, 20–21, 273, 276, 277
cardenolides. *See* cardiac glycosides
cardiac glycosides, 43–44, 46, 50–52, 54, 55, 57, 58, 88
　effects on monarchs, 88–89
　levels in milkweed, 77–78
　monarch sequestration, 44–45, 52, 55–59, 71, 73, 88
Chincoteague National Wildlife Refuge, 16, 29
citizen science, 1, 271–276.
　censuses, 20, 21
　conservation impacts of, 28–29, 140
　contributions of, 2, 13, 124, 204, 206, 214, 273, 281
　data gaps, 27–28
　definition of, 13
　educational value of, 8
　history of, 13–14
　maps of data collection sites, 17–20, 24
　super volunteers, 23
　supporting volunteers, 29
　and technology, 14
　variability of data, 271, 273
　See also individual program names
climate change
　milkweed, effects on, 99
　monarch, effects on, 95–98, 99, 107–108, 130–131, 214, 224, 245–246
　oyamel fir range, effects on, 97, 108, 115, 145, 158–159, 164, 166, 167, 224
　species responses to, 107, 145, 273
　wintering monarchs, effects on, 108
　See also extreme weather events; precipitation; temperature
CLIMEX, 97, 133–140, 248, 250
coastal zone, 152

cold hardening, 96, 110
cold tolerance, 100, 107
Commission for Environmental Cooperation (CEC), 33
CONANP (Comisión Nacional de Áreas Naturales Protegidas), 22, 33, 34
conferences on monarchs, ix
Conservation Reserve Program (CRP), 176–178
Cook, David, 21
Correo Real, 2, 7, 15, 16–18, 27, 28, 29
Cynanchum
  acutum, 249
  louiseae, 32
  rossicum, 32

degree days, 101, 136
diapause, 97, 111, 215–216, 218, 221–223, 253
drought, 118, 123, 124–126, 128–129, 145, 245

eButterfly, 15, 26, 40, 245
ecological niche models, 98, 100, 107, 134, 160
ejidos, 33, 34, 185
environmental education, value of, 8, 9, 11, 12
extreme weather events, 96

fall breeding, 197, 200–201, 215–216, 219, 222–224, 270
Farm Bill, U.S., 38
fire in winter sites, 144, 182
  causes, 189
  characteristics, 187
  education, 186, 188
  management, 145, 180–185, 187–188
  natural fire regime, 180
  restoration after, 186–187
  social perception of, 187
Fondo Mexicano para la Conservación de la Naturaleza, A.C. (FMCN)
  Monarch Fund, 33
  payments to landowners, 34
Frey, Dennis, 21, 22
frostweed. See *Verbesina virginica*

garden grant programs, 8
genetic bottleneck, 260
genetic clustering analysis, 260, 263
genetic drift, 258, 267
genetically modified crops. See herbicide-tolerant crops; Bt crops
genome, monarch, 258, 266–267
Gibbs, Denise, 29
Glassberg, Jeffrey, 25
glyphosate-resistant crops. See herbicide-tolerant crops
Gomphocarpus, 253–254
  fruticosus, 248–252, 254–256
  physocarpus, 249–251, 254
  See also milkweed

habitat certification programs, 6, 9
habitat loss
  breeding, 31, 37, 97, 143–144, 280
  See also winter sites
habitat restoration, 191
Harmonia axyridis, 74

herbicide-tolerant crops, 31, 139, 169–170, 190, 198, 269–270
  direct effects on monarchs, 170
herbicide use
  effects on milkweed, 171, 174
  effects on monarchs, 173–175
Howard, Elizabeth, 9, 16

Illinois Butterfly Monitoring Network. See butterfly monitoring programs
immune defenses, 86–88
  temperature, effects of, 88
invasive species. See *Cynanchum*

James, David, 244–245
Journey North, 7, 9, 15, 16, 27, 29, 30, 207–208, 216–217
  symbolic migration, 7, 9

La Cruz Habitat Protection Project, 34
lacewings (*Chrysoperla rufilabris*)
  monarch egg consumption rates, 79, 81
  as monarch predators, 65, 74, 78–79
Landry, Pat, 21
Lespesia archippivora. See tachinid flies
lipids
  accumulation of, 117–118, 129
  burning of, 115
  conversion of sugars into, 117, 127
  as energy for winter survival, 110–111, 128
  levels in monarchs, 118–127
  measuring, in monarchs, 124
  precipitation, effects on, 97
  storage, 117
  utilization, 117

marginal parasitism rate, 75
Meitner, C.J., 21
metabolic rate, 111
metapopulation, 252, 255
microclimate
  definition, 109
  features of, 110
  under tree canopy, 111–114
microsatellite markers, 258, 260, 265, 267
migration, 90, 109, 131–132, 203–204, 222, 239, 241, 257
  costs of, 92
  as defense against parasites, 90–92
  as driver of monarch numbers, 280
  as endangered phenomenon, 32, 93
  energy/lipid gain during, 110, 123, 129, 138
  evolution of, 92, 216
  flight orientation, 242–243
  genes, 265
  loss of, 93
  nectar sources during, 118, 280
  pace, 207–209, 211–212
  and parasitism, 24
  timing, 123, 137, 223, 270
  See also annual cycle
migratory culling, 91–92, 223
migratory escape, 91, 223
milkweed
  abundance, 220–221
  burning, 199, 201–202, 238

clonality in, 58
control of, 170–171, 238
decline, 190
defenses against herbivores, 43–44, 47–49, 50–52, 55, 58
historical distribution, 176
latex, 43, 50–51, 54–55, 57, 59, 69, 97
management, 198, 256
monarch defenses against, 44
monarch performance on, 52–55
mowing, 199, 201–202, 238
native, 191, 195, 218, 220–223, 245
as noxious weeds, 32, 40, 256
patch occupancy, 252, 255
patch size, 252
production, 192
range, 198, 236–237, 254
regrowth ability, 52
restoration, 191, 194–196
in roadsides, 178, 238
root-to-shoot ratio, 58
seed availability, 38, 190–191
seed collection, 192
in South America, 225, 228
specialist herbivores of, 52, 192
species diversity, 43
temperature tolerance, 223–224
tolerance of herbivory, 52, 57–59
trichomes, 51, 54, 55, 62
in Upper Midwest, 169, 174–175
See also *Asclepias*; *Gomphocarpus*
mitochondrial DNA, 258, 260, 265, 267
modeling, bioclimatic, 132, 247
annual growth index, 136
climate envelope, 132
ecoclimatic index, 133, 250
growth index, 250
molecular genetics, 205, 258
monarch
curricula, 7–11
declining numbers of, 130, 139, 143, 175, 190, 205, 269, 280
dispersal
panmictic, 263
undirected, 264
wind assisted, 254
flight speed, 212–213
gene flow, 86, 205, 244, 257–258, 260, 262–263, 265
interstate transfer, 205, 244–245
legal protection for, 33, 34, 36, 39
legal status, 154
as model system, 92
subspecies, 228, 263
threats, 32, 130
value to people, 12
Monarch Alert, 15, 22
Monarch Butterfly Biosphere Reserve, 33, 35, 157–158, 179–184, 186–189, 271
monitoring in, 22, 271
workshops in, 7, 10–11
Monarch Butterfly Fund, 22, 36
monarch conservation
1980 decree, 33
1986 decree, 33
2000 decree, 33, 34
in Canada, 39–40
international efforts, 32–33, 146
in Mexico, 33–36
sister protected area network, 32–33, 37, 40
in U.S., 36–38, 153–154, 196, 245
monarch distribution
global, 204, 248, 259
historical, 248
Monarch Health, 15, 23–24, 29
Monarch Joint Venture (MJV), ix, x, 18, 36–38, 128, 190
Monarch Larva Monitoring Project (MLMP), 8, 21, 23, 25, 27, 28, 29, 30, 99, 175, 216–217, 220, 273–277
Monarch Monitoring Project. *See* Cape May Monarch Monitoring Project
Monarch Net, 14
Monarch Teacher Network, 7
Monarch Watch, 2, 6–7, 15, 18–19, 21, 28, 38, 190, 245
See also tagging programs
Monarch Waystation Program, 6, 28, 38
monarchs, sales and mass-rearing, 7, 8, 11
benefits of, 12
concerns about impacts, 11–12, 243–244
Monarchs Across Georgia, 7, 9–10
monarchs by location
Arizona, 19, 194, 242
Australia, 14, 255, 259
Azores, 247–249
Balearic Islands, 249
British Columbia, 236–237, 241–242
British Isles, 248–249, 254
Canary Islands, 249
Gibraltar, 248, 254
the Great Basin, 194, 239
Hawaii, 259–262
Idaho, 236, 239, 241–242
Madeira, 248–249
Morocco, 247–250, 255
New Zealand, 260–262
Oregon, 236–237, 241
Portugal, 247–249, 254–255
Southern Great Plains, 197
Spain, 247–251, 255–256
Upper Midwest, 169, 174–175
Utah, 241–242
Washington, 236–237, 239, 241
West Indies, 260
Monarchs in the Classroom, 2, 7, 8, 28
Monarchs Without Borders, 2, 7, 8–9
Monroe, Mia, 21
Montesiños, Eneida, 10
Morris, Gail, 242

NOM-015, 188–189
North American Butterfly Association (NABA), 15, 25–27, 28, 271, 273–277, 279, 280
North American Monarch Conservation Plan (NAMCP), ix, 33, 36

Oberhauser, Karen, 8, 23, 33
Ohio Butterfly Monitoring Network. *See* butterfly monitoring programs

Ophryocystis elektroscirrha (OE), 23, 24, 44, 45, 80, 83–89, 196, 201–202, 205, 223
    assessing infection levels, 85
    effects of milkweed toxicity on, 81
    effects on monarchs, 24, 44, 84–85
    host immunity to, 88–89
    host tolerance of, 86
    life cycle, 84
    monarch resistance to, 85–86, 87
    self-medication behavior by monarchs, 44, 88–89, 92–93
    transmission of, 83–84
    worldwide prevalence of, 84, 91, 93
    See also virulence
outdoor learning, value of, 5–6
overcollection, 155
overwintering
    densities, 131
    forest canopy, protective effects of, 96, 110–111
    forest quality, importance of, 115–116
    microclimatic requirements, 96
    mortality, 96, 280
    storms, effects of, 115
    See also winter sites
oyamel fir, 110–111, 157, 181
    assisted migration, 145
    boughs, clusters of monarchs on, 111–112
    decreasing density and size, 110, 113, 115
    location of monarchs on, 113–115
    reforestation priority sites, 165, 167
    specific heat of, 112–113
    thermal buffering by trunks, 112–113
    trunks, clusters of monarchs on, 111–113

parasitoids, 44
    biocontrol, use in, 73–74
    host defenses against, 80
    host plants, effects on numbers, 77–78
    tachinid flies, 44, 77–78
    wasps, 44
    See also names of individual species
Payments for Ecosystem Services (PES), 188
Peninsula Point Monarch Research Project, 15, 21, 273, 276, 277
Pgi gene, 265–266
phenology, 173
plain tiger (African monarch, *Danaus chrysippus*), 250, 255
Pollinator Conservation Program, 191, 195
Pollinator Partnership, 38
population size, 205
precipitation
    flooding, 253
    lipid levels, effects on, 97
    milkweed quality, effects on, 97, 219–220
    monarch abundance, effects on, 97
    mortality during winter, effects on, 111
predation, 44, 68–70
    ants, 44, 60, 66, 69
    black-backed orioles, 71
    black-headed grosbeaks, 71
    effects on monarchs, 68–70
    by generalist natural enemies, 81
    lacewing larvae, 65, 74, 78–79
    list of known predators, 72–73
    paper wasps, 44, 71, 75, 79, 81
    research questions, 45, 60
    spiders, 44, 66, 69
    stink bugs, 66
premigrants. *See* fall breeding
Project Milkweed, 146, 191, 195
Prysby, Michelle, 23
Pteromalus spp., 75–76, 79–81
    cassotis, 74–76, 79
    puparum, 74, 79
pupae
    natural enemies of, 74–76, 79–80, 82
    pupation success, 101–102, 106
pyrrolizidine alkaloids (PAs), 232–233

queens (*Danaus gilippus*), 83, 192

rearing programs, 241, 244
Red Monarca (Monarch Network), 35, 36
Rendón, Eduardo, 22
reproductive status. *See* diapause
research, suggested, 68, 80, 82, 92, 97, 107, 224, 262, 266, 281
roosting, 208
    locations, 209–210
    roost duration, 209, 211, 213
    threats to, 238
    trees, 207
Roundup Ready crops. *See* herbicide-tolerant crops
RuBino, Richard, 21, 23

sequestration of toxins. *See* cardiac glycosides; pyrrolizidine alkaloids
southern monarch (*Danaus erippus*), 204, 225–226
    distribution, 227–228, 233–234
    fat content, 229, 231, 233
    flight orientation, 228–229, 232
    flight speed, 229
    migration, 225, 233–234
    nectaring, 231–234
    and *Notiochelidon cyanoleuca*, 229
    roosting, 231–234
    wing length, 229–231, 233
southern U.S. breeding. *See* fall breeding; winter breeding
Southwest Monarch Study (SWMB), 15, 19–20, 28, 242
spring breeding, 192–193, 200, 215
St. Marks National Wildlife Refuge (SMNWR), 15, 21
stopover sites, 129
supercooling points, 96, 100, 104–105
survival and mortality rates of monarchs, 44, 60–64, 71
    daily rates, 63–65
    factors affecting, 63–64, 66–70, 71

tachinid flies
    Compsilura concinnata, 73
    egg numbers, effects on survival, 80
    Lespesia archippivora, 73–74, 77–78, 80, 81, 202
    parasitism rates in monarchs, 73
    toxic hosts, effects of, 73, 81
tagging programs
    Insect Migration Association, 16, 18
    Monarch Watch, 6, 8, 18

recoveries, 239–242, 262
recovery rate, 18, 21
St. Marks National Wildlife Refuge, 21
Taron, Doug, 26
Taylor, Orley "Chip," 6, 18
temperature, effects on monarchs
    acclimation, 96, 102, 107
    colonization, 255
    as cue for physiological changes, 97
    development rates, effects on, 95–96, 100–108, 223
    development thresholds, 100
    egg production, effects on, 95
    heat shock proteins, 96, 102
    monarch abundance effects on, 97
    mortality, effects on, 95–96, 100–108, 110–111
    pupation success, effects on, 101–102, 106
    size, effects on, 95, 102, 106
    sublethal effects, 100, 106
    thermal limits, 99, 100, 102, 105–106
    thermoregulation, 96, 115
    See also microclimate
Texas Hill Country, 117, 123–124
Texas Monarch Watch, 123
Treviño, Rocio, 7, 16–17

Urquhart, Fred, 14, 16, 18, 157, 241–242
    See also tagging programs: Insect Migration Association
U.S. Department of Agriculture
    Natural Resources Conservation Service (NRCS), 38, 191, 193–195
U.S. Fish and Wildlife Service, 36
    National Wildlife Refuge, 37
    Wildlife Without Borders, 36
U.S. Forest Service, 37, 153
    International Programs, 36

Van Hook, Tonya, 21
Verbesina virginica, 118, 124, 126, 127
virulence, 88, 92
    evolution of, 93
    geographic variation in, 92
    indirect effects on, 88
volatile organic compounds, 74
    frass, 76, 79

Walton, Dick, 20
Western Monarch Thanksgiving Count, 15, 21–22, 28, 29, 147, 148, 150
western population
    breeding habitat, 154, 194
    database, 151–152
    decline, 150
    legal status, 154–155
    management, 156
    monitoring, 152
    See also winter sites
Wildlife Habitat Council, 38
winter breeding, 100, 215–216, 219, 223–224, 253, 268
    risks of, 100, 223
winter sites
    area of influence, 161
    blanket-like insulation in, 111, 116, 138
    buffer zone, 158, 179
    in California, 21, 38–39, 148, 149, 154–155
    core zone, 158, 179
    decrease in tree density and size in Mexico, 110, 113, 115
    discovery of in Mexico, 32
    forest loss, 161–163
    history of, 158
    illegal logging in, 115, 143–144, 158, 165, 188
    locations in Mexico, 109–110, 131
    plant species in Mexico, 110–111
    radiational cooling in, 113, 115
    temperature in Mexico, 109–110, 165, 167
    threats to in California, 152–153
    threats to in Mexico, 31, 97, 108, 115–116
    tree species in California, 21, 153, 259
    tree species in Mexico, 110–111, 181
    umbrella-like cover in, 111, 116, 138
    vertical temperature profiles, 113–115
    windbreak effects of, 111
World Wildlife Fund
    Canada, 40
    Mexico, 22, 28, 33, 271, 276–277
    Telcel Alliance, 34

Xerces Society, 21, 22, 25, 28, 38, 39, 147, 190–191, 195, 245